Argumentation Library

Volume 29

Since 1986 Springer, formerly Kluwer Academic Publishers, publishes the international interdisciplinary journal *Argumentation*. This journal is a medium for distributing contributions to the study of argumentation from all schools of thought. From a journal that published guest-edited issues devoted to specific themes, *Argumentation* has developed into a regular journal providing a platform for discussing all theoretical aspects of argumentative discourse. Since 1999 the journal has an accompanying book series consisting of volumes containing substantial contributions to the study of argumentation. The *Argumentation Library* aims to be a high quality book series consisting of monographs and edited volumes. It publishes texts offering important theoretical insights in certain major characteristics of argumentative discourse in order to inform the international community of argumentation theorists of recent developments in the field. The insights concerned may pertain to the process of argumentation but also to aspects of argumentative texts resulting from this process. This means that books will be published not only on various types of argumentative procedures, but also on the features of enthymematic argumentation, argumentation structures, argumentation schemes and fallacies. Contributions to the series can be made by scholars from a broad variety of disciplines, ranging from law to history, from linguistics to theology, and from science to sociology. In particular, contributions are invited from argumentation theorists with a background in informal or formal logic, modern or classical rhetoric, and discourse analysis or speech communication. A prerequisite in all cases is that the contribution involved is original and provides the forum of argumentation theorists with an exemplary specimen of advanced scholarship. The *Argumentation Library* should enrich the study of argumentation with insights that enhance its quality and constitute a fruitful starting point for further research and application. All proposals will be carefully taken into consideration by the editors. They are to be submitted in fourfold. If the prospects for including a certain project in the series are realistic, the author(s) will be invited to send at least three representative chapters of their manuscript for review to the editors. In case the manuscript is then judged eligible for publication, the complete manuscript will be reviewed by outside expert referees. Only then a final decision can be taken concerning publication.

More information about this series at http://www.springer.com/series/5642

Andrea Rocci

Modality in Argumentation

A Semantic Investigation of the Role of Modalities in the Structure of Arguments with an Application to Italian Modal Expressions

Springer

Andrea Rocci
Istituto di Argomentazione,
Linguistica e Semiotica
Università della Svizzera italiana University
of Lugano
Lugano, Ticino, Switzerland

ISSN 1566-7650 ISSN 2215-1907 (electronic)
Argumentation Library
ISBN 978-94-024-1061-7 ISBN 978-94-024-1063-1 (eBook)
DOI 10.1007/978-94-024-1063-1

Library of Congress Control Number: 2016963798

Printed on acid-free paper

This Springer imprint is published by Springer Nature
The registered company is Springer Science+Business Media B.V.
The registered company address is: Van Godewijckstraat 30, 3311 GX Dordrecht, The Netherlands

Acknowledgments

This book would not have been started without the instigation of Eddo Rigotti, my teacher and then colleague and friend. He introduced me to linguistics, to semantics, and to the treasures of ancient and medieval thinking about language, logic, argument and rhetoric. And he bluntly told me that this book had to be written. He instigated so many good things in my life that it will be hard to thank him enough in this life. I'm happy to be able to do that for this book.

This book would never have been finished without the encouragement, wisdom, and friendship of Frans van Eemeren. Without Frans I would not be doing argumentation theory in the first place – and the field itself would be very different without his monumental contribution. It took a long time to write this book, and there were times when I was close to giving up on the project. Frans' belief in the project was a key factor in bringing it to completion.

My Lugano colleague Johanna Miecznikowski, a true linguist and one of the nicest persons I have ever met, is also to thank. She was a driving force in the research projects from which this book is born and coauthor of important papers. But I thank her even more for the many pleasant conversations on abstruse issues of modality through these years. Our former PhD student Elena Musi came to share these conversations and, after moving to the USA, she is bringing forward the research program at the heart of this book in ways I had not foreseen. Besides being an energetic researcher and coauthor, Elena took the time to help me with the task of bringing this ponderous manuscript, evolved through many drafts, to publishable form. When Elena moved to the USA, Chiara Pollaroli, another former PhD student and research collaborator, took up this task and was at my side in the latest editing stages. I am enormously grateful for that.

This research would not have been possible without the generous support of the Swiss National Science Foundation (SNF) with the grants 100012-120740/1 and 141350.

The University of Lugano – also known by its official name of *Università della Svizzera italiana* (USI) – is a lively interdisciplinary working environment where, thanks to the initiative of Eddo Rigotti, it was possible to develop argumentation as a field of research within the Faculty of Communication Sciences. I am grateful to several Lugano colleagues and former colleagues, but particularly to Sara Greco and Rudi Palmieri, with whom I have shared the development of the argumentative adventure started by Eddo.

To my wife Alessandra and to my children Giovanni, Stefano, and Maria goes my deepest gratitude and love, not because they endured the writing of this long book with its troublesome impact on family life (which they did), but rather because they are the warp and weft of a life where even such an endeavor can be meaningful.

***

Being the final report of about a decade of research, this book also incorporates and updates results that have seen publication elsewhere. In particular, the initial sections of Chap. 3 develop ideas initially presented in:

– Rocci, Andrea. 2008. Modality and its conversational backgrounds in the reconstruction of argumentation. *Argumentation* 22: 165–189.

Similarly, the analyses of the Italian modal verbs proposed in Chap. 6 are expanded and revised versions of those presented in previous publications. In particular, the chapter reproduces passages from:

– Rocci, Andrea. 2012. Modality and argumentative discourse relations: a study of the Italian necessity modal dovere. Journal of Pragmatics 44(15): 2129–2149.
– Rocci, Andrea. 2013. Modal conversational backgrounds and evidential bases in predictions: The view from the Italian modals. In Time: language, cognition & Reality, eds. Kasia M. Jaszczolt, and Louis de Saussure, 128–153. Oxford: Oxford University Press.

Contents

Chapter 1
Introduction

Abstract This short chapter introduces the notions of *modality* and *argumentation* and lays down the central research questions of the book, which revolve around modality in relation to the essential structure of arguments. The *Introduction* also serves as a disclaimer explaining how the book is neither a contribution to modal logic nor to the theory of non-monotonic, presumptive reasoning. The final section provides a reading guide previewing the following chapters.

> τά γὰρ πολλὰ περὶ ὧν αἱ κρίσεις καὶ αἱ σκέπσεις, ἐνδέχεται καὶ ἄλλως ἔχειν
> "for most of the things which we judge and examine can be other than they are" (Aristotle, *Rhetoric*, I, 1357a, transl. by J.H. Freese, LOEB[1])

> These terms – 'possible', 'necessary' and the like – are best understood, I shall argue, by examining the functions they have when we come to set out our arguments. (Toulmin 2003 [1958]: 17)

> In fact, one of the main virtues of a premise semantics for modality is that it links the semantics of modals to general principles of rational inquiry that apply whenever we reason from a set of premises. (Kratzer 2012 [1977]: 10)

This book examines *modality* – the semantic category encompassing notions of possibility and necessity – looking at its role in *argumentation*, that is in discourse where arguments (or reasons) are given in support of a standpoint (or conclusion).[2]

One of the founders of modern argumentation theory, Stephen E. Toulmin, quoted above in epigraph, chose to open his inquiry into the "working logic"[3] of

---

[1] See Aristotle (1926) in the references. Throughout the book, for the translation of passages from the works of Aristotle, I will normally refer to the Revised Oxford Translation edited by Jonathan Barnes (see Aristotle 2014). On a few occasions, like the quote in epigraph to this *Introduction*, I will refer to another translation that appears more transparent for the purposes at hand.

[2] This book represents the (belated) "final report" of the research project *Modality in Argumentation. A semantic-argumentative study of predictions in Italian economic-financial newspapers*. The project was generously supported by the Swiss National Science Foundation (Grant: 100012-120740/1) from September 2008 to December 2011. Andrea Rocci was the principal investigator, Eddo Rigotti co-applicant. Johanna Miecznikowski was senior researcher and Gergana Zlatkova participated as Ph.D. student.

[3] The expression "working logic" is used in the title of the IV essay of the book, where it is opposed to the "idealised logic" informed by the standard of logical necessity/ deductive validity (cf. Toulmin 2003 [1958]: 135).

A. Rocci, *Modality in Argumentation*, Argumentation Library 29,
DOI 10.1007/978-94-024-1063-1_1

real-life "fields of argument" with a discussion of the meaning of modal expressions. If he chose to open *The Uses of Argument* with such a discussion and then continued to refer to modality throughout the book, it is certainly because he believed that modal notions are particularly revealing of the nature of ordinary arguments. In particular, in the first essay the discussion of the meaning of the modals was instrumental to introduce the notion of different "fields" from which real world arguments draw the criteria that justify their force. It is according to these field-dependent criteria, Toulmin insists, that real arguments should be evaluated and not according to the logician's ideal of deductive validity.

Yet, it is interesting to observe that Toulmin, in the quote above, does not say that one can better *understand arguments* by examining the modals occurring in them. He takes the converse perspective, saying that the *modal expressions "are best understood* [...] by examining the functions they have when we come to set out our arguments". What follows this quote is an insightful, yet problematic, semantic analysis of the English modal expression *cannot*, which I will dissect in Chap. 3, § 3.2.3. Similarly, the second essay of Toulmin's book reads equally as an analysis of the semantics and pragmatics of the adverb *probably* and as a discussion of a notion of probability relevant for ordinary arguments.

The present investigation also takes both perspectives. On the one hand, it examines what modals tell us about arguments, and reads as a contribution to argumentation theory and to the analysis of argumentative discourse. On the other hand, it shows that looking at the roles modal expressions can have in arguments reveals crucial aspects of the meaning and use of the modals themselves. Thus, this book can be read also as a contribution to the semantics and pragmatics of modality. One could swap *Modality in Argumentation* for *Argumentation in Modality* as the title of this book.

Let us first have a brief look at argumentation and modality, before saying what they have to do with each other and how this book addresses their relationship.

1.1 Argumentation and Modality: Working Definitions

In this book argumentation, or argumentative discourse, means a discursive activity in which an *arguer* puts forward a series of propositions – the *arguments* or reasons – as an attempt to effect the *reasonable acceptance* of a *standpoint* by a *critic.* The standpoint is a proposition to which the arguer is committed and which is not yet accepted, videlicet is *doubted*, by the critic. The reasonable acceptance is secured by making the standpoint *follow inferentially* from arguments that are either already accepted by the critic (i.e. are part of the *common ground* of the arguer and of the critic) or themselves supported as (sub-)standpoints by arguments ultimately grounded in the common ground of the participants.

This is not the kind of definition one finds in a logic textbook. It is couched in communicative, dialogical terms and recovers ideas that harken back to the rhetorical and dialectical traditions of Antiquity, bringing them together in a way that is typical of a contemporary perspective on argument that is usually characterized as *normative pragmatic* or *pragma-dialectic*.[4]

The domain of modality is vast. It encompasses the concepts expressed by words such as *possibility, necessity, probability*, but it also covers notions such as *ability, need, obligation* or *risk*, and, as we will see, quite a few others. Its linguistic manifestation is manifold, including both lexical means – ranging from modal verbs to adverbs, adjectives and nouns – and grammatical means, such as grammatical mood and tense. As one could expect with such a broad notional category and so intertwined with the linguistic systems of human languages, modality has been apprehended and defined in different ways, with some views considering it a fuzzy category or, even, simply a bundle of related notions chained by family resemblances and partially sharing their means of linguistic expression (cf. Bybee et al. 1994). Needless to say, entering such a debate would require a book on its own and would lead us astray.

Instead, I adopt a concept of modality inherited from a broad philosophical tradition along with a few paradigmatic examples of its linguistic expression. In due course this working notion will be made more precise as I seek to elucidate its relevance for understanding how arguments work. Along the way a broader range of – sometimes less obvious – linguistic exponents of the category will be discussed. According to this view, modality is the semantic category associated with the basic human cognitive ability of thinking that *things might be otherwise*, that is thinking of *alternatives*: situations other than what is the case.[5] Modality refers generally to the linguistic means that allow "one to say things *about*, or *on the basis of*, situations which need not be real" (Portner 2009: 1, emphasis is ours). The *on the basis of* part of Portner's definition is important because it allows us to recognize the modal nature of examples such as (1):

(1) John had to resign.

Typically, this kind of English sentence is uttered in a situation where John did, in fact, resign. Most native speakers of English would find a continuation of (1) with *but he didn't* unacceptable or at least strange. Yet, even if the event of John's resignation is a fact in the speaker's world, (1) obliges us to think of the alternatives. It does not mean simply that he resigned, it means that, for some contextually relevant class of alternatives, there was no other possible alternative. There was no viable *legal / moral / practical / expedient / stylish,* etc. alternative to resignation. Modal

[4] The definition offered here is close, in several respect, to the one adopted by one specific normative pragmatic theory, namely Pragma-Dialectics (cf. van Eemeren and Grootendorst 2004: 1). The definition also notably draws on ideas found in Rigotti and Greco-Morasso (2009), Jacobs (2000) and Pinto (1996), especially in what concerns the relationship between argumentation and inference.

[5] One prominent logico-philosophical tradition developed the theoretical construct of *possible world* to deal with reasoning about alternative situations. David Lewis' (1973) work on counterfactuals is one of the most influential in-depth discussions of possible worlds in modern philosophical logic. The book is also noteworthy for the impact it had on the linguistic semantic analysis of modality and conditional constructions.

expressions have another important characteristic that (1) showcases: they operate some form of *quantification* over the relevant alternatives. In (1) we have to do with a *universal* quantification: *all the viable alternatives* involve John's resignation, or equivalently *there is no viable alternative that does not* involve John's resignation.

It is worth to start noting that as a standpoint (1) requires the support of different arguments than the simple *descriptive* proposition that John resigned. Being able to show a letter of resignation signed by John would be enough to support the descriptive proposition, but not the modal one. Modalities contribute to define argumentatively relevant basic semantic classes of propositions, which Freeman (2000, 2005: 93–113), following a rhetorical tradition (cf. Kruger 1975, Fahnestock and Secor 1982), calls *types of statement*. Rhetoricians have long since observed that there are different broad classes of arguable statements, such as, for instance, evaluative vs. descriptive statements, which require different means of proof when put forth as a standpoint. By contributing to the determine the type of statement, modalities indirectly determine *what counts as evidence* for a given standpoint. This is an important and often overlooked aspect of the interaction between modalities and arguments. But modalities can be relevant to arguments also in other, more direct, ways, of which something will be said in the following sections of this introduction.

1.2 Modality and the Essential Structure of Arguments: A First Reconnaissance

How *relevant* is modality for argumentation theory? Obviously, it will be the task of the entire book to provide answer to this question, but we need a provisional one to justify that the journey is worth taking. Consider the following dialogue:

(2) A: I heard that Suzy got caught shoplifting yesterday afternoon.
B: She **can't** have done it. She and I were at the movies all afternoon.

The turn of speaker B contains an argument: the standpoint that Suzy is innocent of shoplifting yesterday afternoon is supported by the alibi that Suzy was at the movies with speaker B all afternoon. The turn of speaker B also contains a modal (*can't*), which refers to some kind of impossibility. Intuitively we recognize that this notion of impossibility plays a role in the argumentation. But what role is it? The modal appears in the sentence that manifests the standpoint defended by B, but it is not entirely clear whether the modal should be considered part of the content of that standpoint: should we consider that B was defending the standpoint 'It is impossible that Suzy got caught shoplifting yesterday afternoon' or the standpoint 'Suzy wasn't caught shoplifting yesterday afternoon'? If the modal is not part of the *content* of the standpoint we might consider the hypothesis that it signals something like the force of conviction with which B upholds the standpoint, or the force that the

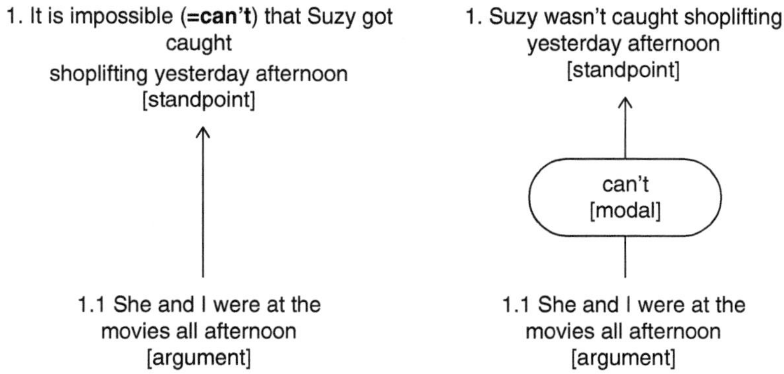

Fig. 1.1 Structural graph representations of the alternative analyses of example (2)

argument lends to the standpoint. The alternative analyses can be visualized as in Fig. 1.1, employing a style of representation commonly used in argumentation theory (cf. Snoeck Henkemans 1997; Freeman 1991, 2011; Vorobej 2006).[6]

For instance, we might compare the argument in (2) and the one in (3), below, and observe that the former is presented as making a *stronger case* for Suzy's innocence.

(3) A: I heard that Suzy got caught shoplifting yesterday afternoon.
B: That's ***unlikely***. She's not that kind of girl.

We might as well consider that the impossibility manifested by *can't* has to do not only with the force but also with the *kind* of argument used, with the precise way in which the argument presented works in supporting a conclusion: the alibi works as an argument because it's *physically impossible* for people to be in two places at the same time. Then the analysis would look quite different from either option in Fig. 1.1. Figure 1.2, provides a *rough* sketch of what an analysis of the argument along these lines would look like.

In this case we would say that the modality expressed by *can't* has to do with what Ancient and Medieval theorists called the *topos* or *locus* from which the argument is drawn (Rigotti and Greco-Morasso 2010). Some modern argumentation theorists here would prefer to speak of *argument schemes* (van Eemeren 2010) or *argumentation schemes* (Walton et al. 2008). Now consider the substitution with another modal expression, *certainly*, as shown in (4).

(4) A: I heard that Suzy got caught shoplifting yesterday afternoon.
B: **Certainly** she did not do it. She and I were at the movies all afternoon.

---

[6]The specific type of graph representations used in the book derives from the view of argumentation structure espoused by the Pragma-Dialectical theory of argumentation (cf. Snoeck Henkemans 1997; van Eemeren et al. 2007, Chapter 7).

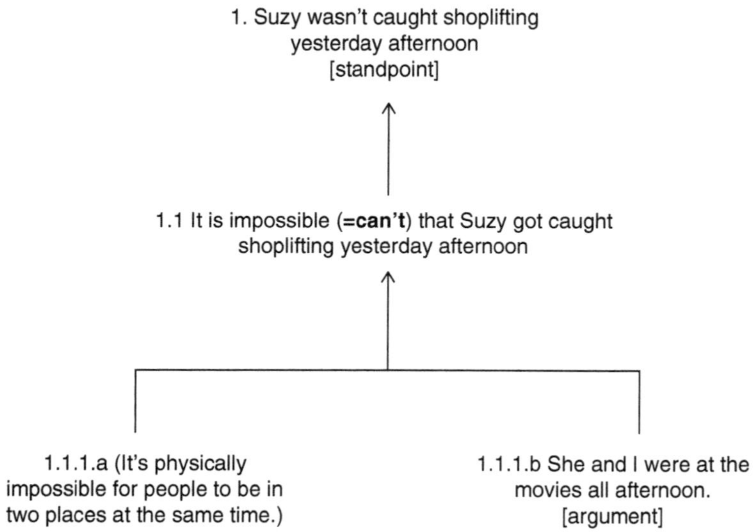

Fig. 1.2 A different graph structure for the reasoning in example (2)

Leaving aside, for the moment, finer semantic and pragmatic considerations that can explain why *certainly* feels somewhat awkward in B's answer, let us focus on one feature of the modal *certainly*: it seems to tell us about *how strong* a case the arguer is making (a rather strong case), but, contrary to what happens with *can't*, we do not feel that the modality is also pointing to the *kind* of argument used.

Words like *certainly* are often said to express a *doxastic* modality, a modality concerned with the alternatives that are admissible within the belief system of a given agent or, more simply, the alternatives consistent with the beliefs of that agent. Another term frequently used is *epistemic*. The term *epistemic* is however ambiguous between a broad reading where it is coextensive with doxastic and a strict reading where it refers only to what is actually *known*, rather than simply believed. The adverb *certainly* directly refers to the arguer's mental state of 'certainty' and conveys the intended strength of the argument via the reference to the arguer's subjective strength of conviction. What *certainly* does not show is that the arguer's belief in Suzy's innocence derives from a special kind of impossibility. Consider now another variant of our example, presented in (5), where the semantic contribution of the modal also diverges from (2), but in a direction that seems quite opposite to what we have in (4).

(5) A: I heard that Suzy got caught shoplifting yesterday afternoon.
B: She **could not** do it. She and I were at the movies all afternoon.

The difference between *could not do* and *can't have done* is easily perceivable but difficult to pinpoint. In order properly draw this distinction one needs theoretical machinery that will be introduced in the following chapters (see, in particular, Chap. 5, § 5.2.4.4). But there are certain informal remarks that can enrich this preliminary illustration of how the meaning of modal expression can be finely intertwined with the structure of arguments.

We can start by observing that in (5) the temporal operator takes scope over the modal, while in (1) it is the other way round. In (5) the impossibility of shoplifting is placed in time: *there was a time when it was impossible for her to shoplift*. In (2), on the contrary, the impossibility that *she shoplifted in the past* is seen as present and, perhaps, eternal. The speaker abstractly considers that the proposition 'Suzy shoplifted yesterday afternoon' is *incompatible* with another proposition which is a known fact ('She and I were at the movies all afternoon'), given a certain accepted ontology of the physical world. The temporal version in (5) brings in an idea of *causality*, which is absent from (2): being at the movies is not just something that is *incompatible* with shoplifting but something that *prevents* it from happening. In real discussion contexts this might trigger additional conversational implicatures. It might suggest that Suzy is somewhat inclined to that sort of vice, and she could have given in to that kleptomaniac inclination, weren't it for the particular situation that *prevented* her from doing so. The *doxastic* modality in (4) is about the relationship between the proposition and speaker's internal belief states, while the modality in (2) concerns the logical incompatibility between the proposition and known facts in a certain ontology of the physical world – some scholars would characterize it as simultaneously *alethic* and *epistemic*. It concerns the relationship of the proposition with certain facts of the world, namely the world's physical make-up. It is therefore an *ontological* modality. Another, more widespread, term for these modalities is *alethic*,[7] as facts are propositions that are the case in the world. Since these facts are *known* facts, the modality can be characterized also as *epistemic*, both in the broad and in the strict sense. Example (5) is about the relationship between the proposition – or perhaps the event it denotes – and the circumstances – some students of modality would call it *circumstantial*, but, again, some would also insist it is *alethic/ontological* too, as circumstances are facts.[8] If we look at it from the viewpoint of argumentation, this last kind of modality seems to be primarily about the real-world relation that underlies the argument (a circumstance preventing an event), secondarily about the corresponding logical relations (the incompatibility between the two propositions), and only in a very indirect way about the beliefs of the speaker (his/her dismissal of the proposition). This book adopts the hypothesis that *all* these levels of consideration plus a further one – speech acts – are relevant for understand-

[7] The term *alethic modality* (from Gr. *alétheia* 'truth') is widely used in the philosophical and logical literature (but less so in linguistics) to refer to what is possible or necessary in view of true facts. Different kinds of alethic modality can be envisaged according to the range of facts that are taken into consideration: physical, metaphysical, logical, etc. Alethic modalities will be examined in detail in Chap. 5 as part of a semantic typology of modalities geared at capturing their role in arguments.

[8] On the vast range of everyday alethic modalities see the brilliant remarks in Lycan (1994: 171–178). Note that among certain students of linguistic semantics (cf., for instance, Lyons 1977: 791; Palmer 1986: 11; Papafragou 2000; Nuyts 2006: 8) *alethic* modality is understood in a much narrower sense as synonymous with *logical modality*. For a more thorough discussion of varieties of alethic modality among the linguistically and argumentatively relevant "flavors" of modality the reader is directed to Chap. 5.

ing how arguments work and modal expressions, of the different kinds, have an interesting story to tell about them.

For the moment it is not necessary to push the analysis of these examples further. The main point it was meant to make is simple: the argument in (2) does not simply happen to refer to *impossibilities* just like it happens to refer to *shoplifting* and *going to the movies*. Rather, *can't* and the other modals examined appear closely connected to the *essential* properties of the argument. That is to what makes it an argument and to what makes it an argument of a certain kind. This book thus examines *modality in relation to the essential structure of arguments.*

1.3 Aims of the Study

The exploration of modality's relation to the essential structure of arguments will be guided by two related questions. These questions were first raised by Stephen Toulmin and have thence lingered under various guises in argumentation theory without finding a satisfying answer.

(Q1) **How are modals similar to arguments?**

In the first essay of the *Uses of Argument* not only Toulmin argues that modals are best understood by looking at their role in arguments, he goes on to establish if not a complete identity, at least an *isomorphism* between the structure of modal meanings and the structure of arguments, distinguishing in both modal meanings and argumentative moves an invariant *force* and a variable, contextual, set of criteria depending on the *field* to which the propositions concerned belong. I take up from Toulmin the question of the relationship between the semantic analysis of modality and the *structure of arguments*. At first sight, it may seem a most abstruse question, at least from the viewpoint of someone who is mostly interested in the analysis and criticism of actual arguments, rather than in fine points of semantics. In fact, it turns out to be a very productive question to look into.

(Q2) **Is there a special place for modality in the structure of arguments?**

In the third essay of the *Uses of Argument* – the most widely cited in the argumentation literature – Toulmin finds a special slot for modalities, the *qualifier* in his analysis of the fundamental "layout" of arguments – the eponymous *Toulmin model*. This slot is distinct from both the standpoint and the argument.[9] The graph representations of arguments used by argumentation theorists, which we introduced in the previous section, derive, in part, from Toulmin's own view of argument layout. Toulmin sees the qualifier as an indication of the *strength* of the inferential step

[9] In the eponymous *Toulmin model* the (modal) *qualifier*, indicating the strength of the inferential step, is a slot of its own and is neither part of the claim (that is the standpoint) nor of the data that support it.

from the argument to the conclusion.[10] The special place of the qualifier in the Toulmin model brings us to the question of the distinctive place, or role, of modality in the fundamental structure of arguments.

This is perhaps the question that argumentation scholars have more explicitly considered. Students of argumentation have typically dealt with the role of modality in the analysis of arguments in two ways. Either modality is seen as an indication of the *strength of support* that the premises offer to the conclusion (cf. Freeman 1991, 2011) or alternatively, in pragmatic terms, as an indication of the *degree of commitment* of the arguer towards the propositional content of the standpoint being advanced (cf. Snoeck Henkemans 1992; Tseronis 2009). In this book this second question will be examined and reframed in the light of the answer that will be given to the first question.

To these two Toulminian questions I add a third one, which is best answered not by considering modality generally, as a semantic category, but by looking closely into the linguistic semantics of specific modal expressions in a given language.

(Q3) **What does the use of modal expressions in an ordinary discourse tell us about the arguments being put forth in the discourse?**

This question bridges research on *argumentative indicators* (van Eemeren et al. 2007) in argumentation theory with current research on the discourse semantics of linguistic constructions. An argumentative indicator is a word or expression that can tell us something important about the argumentation manifested by a text or dialogue. Argumentative connectives marking an utterance either as an argument (e.g. *because, since*) or as a conclusion (e.g. *therefore* or *so*) are perhaps the simplest example of argumentative indicators.

Yet, the notion of indicator is much broader than that, and the indication could concern any kind of information relevant for *reconstructing* how an argument works. The notion of indicator is intimately tied to the idea that the superficial verbal form of arguments is not immediately transparent and that their fundamental organization needs to be reconstructed in order to be able to evaluate them. Research on indicators takes the perspective of the analyst for whom reconstruction is a necessary step of argument criticism. As a consequence, the research on indicators takes a rather practical stance considering linguistic structures: whatever can help the analyst to establish with reasonable probability the commitments of the arguers can be an indicator. As observed in Rocci (2008: 166), the notion of indicator is not restricted to expressions that *refer* to arguments. Often an indicator does not *refer* to the argumentative move itself but carries some other semantic, pragmatic or even stylistic feature, which, for a variety of reasons, can happen to be often— not necessarily always —concomitant with that argumentative move. So, a verbal expression can be an indicator *per accidens* because of its significant correlation with the performance of an argumentative move. I call these indicators

[10] More precisely the qualifier indicates the *strength* conferred by the warrant to the inferential step from the data to the claim (Toulmin [1958] 2003: 94). The place of modality in the Toulmin model will be discussed at length in Chap. 3.

indirect (cf. Rocci 2008: 167). As long as we deal with indirect indicators the point of view of the analyst and that of the communicators may partially diverge.

On the one hand, the indicators will be, in every case, based on the meanings to which the arguers are committed. In interpreting these meanings the analyst will rely on "linguistic knowledge" and on "pragmatic knowledge of how what is said should be interpreted in the context" (Snoeck Henkemans 1992: 103). This is the linguistic and pragmatic knowledge that is involved in the normal process of meaning construction and interpretation, as stressed by Snoeck Henkemans (1992: 103):

> Arguers who are engaged in a critical discussion aimed at resolving a dispute must observe the pragma-linguistic conventions for appropriate use of the words and expressions they use in defending theirs standpoints. Since the analyst may assume that the arguer has observed these conventions, he will attempt to give an interpretation of the arguer's standpoint[11] that is plausible in the light of the conventions.

Yet, in the case of *indirect indicators*, once the meanings of the arguers have been thus clarified, the analysts use these meanings as indirect evidence for reconstructing the argument, following practical rules that are not necessarily part of the normal process of discourse understanding.

For instance, van Eemeren et al. (2007: 172–173) observe that the use of future tense markers (*will*, *shall*, *be going to*, etc.) is an indicator that some type of *causal argumentation scheme* is being used to support the standpoint. In reality, the link between the use of the future and the argument scheme in question is indirect and connected with different forms of causal reasoning in different ways. On the one hand, statements in the future tense are often predictions, and predictions are often justified causally by arguments from cause to effect—such as in (6.a) below. On the other hand, deliberation on future actions is often justified by the positive/negative nature of the (later) future consequences of these actions. In this second case both the standpoint and the argument refer to the future, and the inferential link goes from the evaluation of the effect to the evaluation of the cause—as shown in example (6.b).

(6.a) Arg: Storms have devastated the crops throughout the country. Stp: The price of vegetables *will* increase.

(6.b) Stp: Travelling with Mark is definitely a bad idea. Arg: He *will* annoy you to no end with his strange fixations about hygiene on buses and airplanes.

The correlation between the use of the future and causal reasoning is a "rule of thumb" that analysts can profitably follow, but it does not correspond to *one* meaning construction and meaning interpretation path by the arguers.

There are, however, words and grammatical constructions whose business consists, at least in part, in telling something *about* the arguments being put forth in discourse. I call these linguistic structures *direct argumentative indicators.*

In this book I investigate the hypothesis that at least *some* modal constructions function semantically as fine grained direct indicators of argumentation, guiding the

[11] In the context of Snoeck Henkeman's original discussion only the reconstruction of standpoints was at issue, but the same could be said of the reconstruction of arguments.

addressee towards the intended interpretation of argumentative connections between utterances in discourse. It is clear, though, that by saying that modal expressions *tell something about arguments*, I do not mean that they are like meta-argumentative expressions, such as *premise, conclusion, valid, deductive, concession*, and so on. One could say they are epi-argumentative, somewhat akin to performative expressions and even more to illocutionary modifiers, in that they flag the very utterance in which they appear as having a given argumentative function or guide the inferences of the interpreter by restricting the range of possible interpretations.

One key difference between these direct indicators and the most classic examples of performative and illocutionary modifying expressions is that they do not concern the pragmatic function of the utterance taken in isolation – if ever there is such a function – but contribute to clarifying how the utterance contributes to a coherent discourse conveying an argument. A second difference concerns the way in which the conventional linguistic meaning of the direct indicators interacts deeply with finely specified types of contextual information that the addressee is invited to recover from the ongoing discourse or situation.

Contrary to what happens with the indirect indicators, here the perspective of the *analyst*, the perspective of the *addressee* and the perspective of the *communicator* trying to constrain the interpretation process of the addressee are *tightly* coupled. Understanding the indicator is part and parcel of normal discourse understanding rather than a specialized activity of the analyst. This makes direct argumentative indicators a particularly precious resource in view of a pragmatically realistic reconstruction. In this book I will typically adopt the perspective of the addressee, assuming that what is clearly relevant *for the natural understanding of an argument as an argument* is, *a fortiori*, relevant for its reconstruction and subsequent criticism.

Direct indicators are also the terrain where the research on argumentative indicators can meet a large body of recent research on the semantics of discourse understanding centered on the contribution of the conventionally encoded meanings of linguistic constructions to the understanding of broad stretches of discourse. Bringing together these two perspectives is an intended methodological contribution of this book.

Not surprisingly, in the argumentation literature the role of modals as indicators has been mostly tied to the speaker's indication either of the *strength of support* that a standpoint enjoys or of the *degree of commitment* towards its content. Some authors have gone further taking the indication of the strength of support as an indirect clue for understanding *argument (macro-)structure*, that is how the different reasons that make up an extended argumentation fit together to lend support to the conclusion (Cf. Freeman 1985, 2011; Snoeck Henkemans 1992). As for the modal expressions taken into consideration, the modal adverb *probably* is the only one that received several fully fledged semantic analyses by argumentation scholars (cf. Toulmin [1958] 2003; Ennis 2003; Pinto 2007; Tseronis 2009).

Through the semantic analysis of different modal expressions in English and in Italian this book will show that this focus on force and degrees of commitment presents a *far too restrictive* view of modals as argumentative indicators, and that there is definitely more that they can tell us about the workings of arguments. Again, the strategy for the analysis of modality developed in answering our first question will serve us well in exploiting modals as indicators.

The present investigation on modal constructions as *direct argumentative indicators* will both benefit from and contribute to a relatively recent surge of research on the semantics of modality in relation to contextual and discursive information. The book draws extensively on the theories and techniques developed in linguistic semantics – in particular in the tradition of formal semantics – for the analysis of modal meanings as well as on the author's own work on the analysis of modal meanings in discourse context.

In addressing the three questions outlined above, I will keep a broader purpose in mind. I want to provide an illustration of the proper place of semantic considerations in argumentative analyses, and particularly in those analyses which seek to capture the dynamics of argumentative discussions in the concrete contexts of social life in which they arise. As it discusses matters that are of interest for semanticists and discourse analysts, the book remains anchored to the perspective and concerns typical of argumentation theory: the analysis of ordinary arguments oriented towards their critical evaluation. The task of Chap. 2 will be precisely to focus on the specific disciplinary viewpoint of argumentation theory on the analysis and evaluation of arguments and make explicit the role that semantic and pragmatic considerations can play from that specific viewpoint.

1.4 A Twofold Research Strategy

An analogy could help clarifying further the perspective adopted by this book: modality is investigated here in relation to argumentation very much like the notional semantic category of *time* could be investigated in relation to *narrative*. That is by asking how temporal relations between states and events – as expressed by tenses, temporal adverbials and other means – contribute to make a narrative discourse what it is.[12] With modality and arguments things are however more complex than the analogy would suggest. The discussion of our initial set of examples already showed that modality comes in different "flavors" – as semanticist von Fintel (2006) would put it. We have already distinguished *alethic*, *epistemic* and *doxastic* modalities and observed that there are different sorts of alethic modalities. This book will develop the hypothesis that these different kinds or "flavors" make a distinct contribution with what I have called the essential structure of arguments.

[12] The comparison with the role of time in narrative can be more telling if we consider that *narrativity* requires temporality – a text that does not order events in time cannot be a story – but temporality does not represent a sufficient condition: it is not enough to order events in time to tell a story. For instance, narrative relations have a pragmatic facet, they are an action of a narrator telling a story to the narratee, and the picture can be further complicated by the presence of a point of view in the narrative and by the evocation of multiple voices, which aren't identified with the narrator nor with the narratee.

Providing a clear definition of these "flavors" and, even more crucially, explaining how they are obtained in the interpretation of an utterance is a crucial task for a linguistic semantics of modality. In order to account for the fact that many modal expressions, such as modal verbs or adjectives, can manifest according to their discourse context a seemingly boundless variety of modal "flavors", the formal semantics tradition has developed a treatment of modal semantics as *relational* and *context dependent*. This theory, which goes under the name of Relative Modality was first proposed by the German linguist Angelika Kratzer (1977, 1981, 1991), in a close dialogue with the contemporary logical and philosophical work on possible worlds semantics that was being developed by David Lewis ([1973] 2007). This theory forms the starting point of virtually all the work currently being done on modality in formal semantics (See von Fintel 2006; Kaufmann et al. 2006 and, in particular, Portner 2009 for recent syntheses of this body of work) and has inspired more informal linguistic work on modality, especially in relation to the "interface" between linguistic meanings and pragmatics (cf. Papafragou 2000).

The central idea of this theory is that the modals are relational predicates of the form M (B, p). They take (at least) two arguments: the propositional content in their scope – the *prejacent*[13] proposition p – and a set of propositions, called the *conversational background* (B). Modal expressions of necessity like the modal auxiliary *must* can be understood in terms of the *logical consequence* of the prejacent from the conversational background, while possibility expressions such as the modal auxiliaries *can* or *may* are to be conceived in terms of the *logical compatibility* between the prejacent and the background:

'necessary'/ 'must' (B, p): p is a logical consequence of B
'possible'/ 'can'/ 'may'(B, p) : p is logically compatible with B

Assuming a possible world reading of logical consequence, saying that p is a logical consequence of B is equivalent to saying that p holds true in all the alternative worlds or situations where the propositions of the set B are true. While, in the case of logical compatibility, p will be true in at least one of the worlds or situations where the propositions of B are true. For instance, *can't* in example (2) discussed above could receive an interpretation along the following lines:

> 'Suzy was shoplifting yesterday afternoon' is incompatible with (what is known of the whereabouts of Suzy yesterday afternoon & what is known about the functioning of the physical world)

One advantage of this kind of approach – the one that immediately caught the attention of linguists – is that the variety of modal "flavors" results from an invariant modal force and a variety of conversational backgrounds reconstructed in the context of utterance. There are however other interesting advantages of this general approach that began to be noticed. One of them is that it offers a way of describing how modal meanings relate to other information in a developing discourse (cf. Roberts 1989; Geurts 1999).

[13] I follow von Fintel (2006) in recovering this term introduced by medieval logicians.

This book will develop the idea that this style of semantic treatment of modality is not only successful in its original endeavor of accounting for the different modal flavors but provides us with the basic tools for explaining the interaction of modality with discourse context. In particular, the relational and contextual approach will allow to investigate how the relational semantic structure of modal meanings contributes to define *argumentatively relevant relations* in discourse at different levels including "real world" relations, logical relations, inferential relations and speech-act level relations.

The investigation presented in the book also adopts a second perspective on the relationship between modality and argumentation, which complements the one outlined above. To use again the *time-narrative* analogy: once it has been established that in a language – say, French – grammatical tenses are one of the main means for expressing time relations, one could examine more closely the use of *tenses* in narratives, perhaps to discover that their business there is not limited to expressing time relations and that they contribute also other information – perhaps somewhat related to time, perhaps arbitrarily packaged together with it – which is important for making a story what it is. In other words, with reference to a specific language, the meaning of the grammatical morphemes of *tense* can be seen, in part, as contributing directly to the discursive construction of *narrative*, without the mediation of a referential category such as time. An extreme, not necessarily advisable, example is represented by Harald Weinrich's (2004 [1964]) well known attempt to characterize the contribution of French tenses to narrative and to other modes of discourse completely dispensing of time reference.[14] Similarly, in connection with the third research question of the book constructions based on *modal verbs* and *mood morphemes* will be examined here for their full potential as direct argumentative indicators. Any way in which they shape or constrain the interpretation of the arguments in which they appear is relevant here, without limiting the observations to those aspects of an argument – such as strength of support – that seem more intuitively or more directly connected with the concept of modality. In doing so, we can end up learning more about modality and related notions and refine our methodological tools for argument criticism.

Drawing from the terminology of linguistic semantics, we may call the perspective moving from the notional category of modality *onomasiological* and the second perspective, which investigates the argumentative functions of linguistic forms, *semasiological.*

One of the main findings of this second research strategy in the book is that epistemic and doxastic modal expressions are found to be particularly revealing of the

[14] In the strongly anti-referentialist structuralist climate of the early 1960s, Harald Weinrich (2004 [1964]), a student of French linguistics, famously set out to create a structuralist text linguistics that sought to connect the structure of narrative directly with tenses, dispensing of the troublingly referential notion of time. In the process Weinrich discovered interesting non temporal properties of French (and Italian) tenses, which he (ironically) conceptualized in terms of different attitudes of the emitter (narrating vs commenting) towards the (reference!) world. Later on, Weinrich had to soften considerably his once belligerent stance on the irrelevance of time, while keeping the other interesting aspects of his analysis.

functioning of the arguments in which they appear because they typically also convey specific *evidential* meanings. *Evidentiality* is the semantic category corresponding to the indication of the speaker's source of knowledge (evidence) of the propositional content of the utterance (cf. Chafe and Nichols 1986; Dendale 1994; Squartini 2004, 2008). In titling an early and influential collection of studies Chafe and Nichols (1986) evocatively characterized evidentiality as "the linguistic coding of epistemology".

In sharp contrast with the bi-millennial scholarly history of modality, the notion of *evidentiality* originated fairly recently from a field of inquiry far removed from the logical and philosophical analysis of language. The idea first emerged in the grammatical description of certain native American languages where the grammatical marking of the information source – typically distinguishing *direct perception, hearsay* and *inference* – is obligatory to produce a grammatically acceptable sentence. The notion and term were then used in typological linguistics to refer to the (many) languages of the world where such a grammatical system exists (Guentcheva 1996; Aikhenvald 2004). However, English and the other Western-European languages are not among those where evidentiality is fully grammaticalized.

Possibly because of its disciplinary origin and of the scarcity of studies on the evidential strategies of English and of other major European languages, the analysis evidentiality was absent for a long time from theoretical discussions in semantics and in the philosophy of language. Its impact on argumentation studies was nonexistent, despite the obvious connection between the linguistic marking of types of evidence in the utterance and the discursive presentation of the evidence itself in the form of arguments.

It is only recently that indicators of the source of evidence have become the object of focused attention in semantics, also in European languages (such as English, Spanish, French or Italian) were their presence is not obligatory (cf. Dendale and Tasmowski 1994; Aijmer 2009; Cornillie 2007; Pietrandrea 2007; Squartini 2004, 2008) and their close connection with the semantics and pragmatics of epistemic and doxastic modality has begun to be explored in theoretical studies (cf. Faller 2002; McCready and Ogata 2007; Sbisà 2014).

Remaining focused on the central notions of the category of modality such as necessity and possibility, this book offers an in-depth discussion of the conceptual relationship between evidentiality and argumentation and a detailed analysis of how the evidential component in the semantics of several markers of modality interacts with argumentation.[15]

[15] In fact, the line of research developed in the present book was the inspiration of a closely related but autonomous research focusing *specifically* on evidential constructions as direct argumentative indicators carried out by Johanna Miecznikowski and Elena Musi within the research project *From perception to inference. Evidential, argumentative and textual aspects of perception predicates in Italian* (Swiss National Science Foundation Grant n. 141350). See, in particular, Miecznikowski and Musi (2015), Musi (2014, 2015).

1.5 A Focus on Italian Linguistic Structures and on Contextualized Discourse Data

It is clear from the above remarks that this kind of fine grained investigation of direct argumentative indicators needs to be carried out with reference to the semantics of given linguistic units in a target language, a language well-known to the analyst, ideally his/her native tongue. It also needs to be carried out using authentic data considered in their rich context, and this context needs to be reasonably known to the analyst so that it can be unraveled when reconstructing underlying reasoning requires so.

In this book the languages investigated are Italian, and, to a lesser extent, English. While the theoretical discussion of modality in relation to the essential structure of arguments will mostly exploit English examples, both invented and selected ad hoc from various sources, the study of modal markers will be largely devoted to the Italian modal system, bringing together the results of several previous semantic studies, leveraging on my native intuitions and exploiting systematically rich corpus evidence.

So, as we move from English to Italian data through the book, we also move from out of context illustrative examples to examples extracted from a large carefully designed corpus, representative of a family of closely related discourse genres and of a given context of social interaction. The corpus, which represents the main source of Italian language data for this book, was collected for the research project *Modality in Argumentation. A semantic-argumentative study of predictions in Italian economic-financial newspapers.*[16] The corpus amounts to roughly 4 million token words and consists of journalistic articles collected from three specialized Italian economic-financial dailies (*Il Sole 24 Ore*, *Italia Oggi* and *MF/Milano Finanza*).

On the one hand, the extensive discussion of Italian expressions of modality and the analysis of samples of argumentative discourse in Italian are important features of this work, and I sincerely hope they will be of interest also for linguists working on the semantics and pragmatics of Italian modal expressions.

On the other hand, this decision may seem to restrict the appeal of the final part of the book to scholars working with Italian or other Romance languages. In fact, I believe it is not necessarily so: the book is accessible to a broad range of readers and remains, first of all, a semantic contribution to the core of argumentation theory. Readers unfamiliar with Italian will be perfectly able to follow the analyses through the English translation of the examples and, where necessary, the glosses. They will be lead to compare and contrast them with parallel English expressions. More explicit comparisons with English and other languages will be drawn when relevant, with pointers to the literature.

[16] The project *Modality in Argumentation. A semantic-argumentative study of predictions in Italian economic-financial newspapers* was supported by the Swiss National Science Foundation (Grant: 100012-120740/1) from September 2008 to December 2011. See also footnote 1.

This will *not* take completely away a feeling of *strangeness* that readers unfamiliar with Italian will get from the final parts of the book and the impression that some of the underlying theoretical distinctions may well be language specific and that they may turn out not to be extensible or useful in approaching argumentative discourse in English or in other languages. This is, however, a *healthy* feeling. This is an often ignored caveat that *all* argumentation scholars should bear in mind when they go about developing conceptual distinctions on the sole basis of analyses of English argumentative discourse. At the same time, if the fine analyses of the English adverb *probably* which have been offered by Toulmin and by other authors reviewed here have been rightly considered not just as accounts of peculiarities of a curious English idiom,[17] but as insightful discussions of the role of modality in argument, there is no reason of principle for considering accounts of Italian modal verbs or adverbs more parochial. If anything the parallel consideration of English and Italian found in several passages of this book (with the occasional nod to French) should sharpen our ability to capture general underlying dynamics, even if an explicit and systematic comparative investigation of modality in the two languages lies well beyond the scope of the book.

The journalistic corpus employed as the main source of data allows to focus on the role of modality in arguments within a given context of social interaction, namely the media coverage of the economy and, more specifically, of the financial markets.

The study of *argumentation in context*, a burgeoning area of argumentation studies, appears *prima facie* far removed from issues of semantics. In fact, it is not so. In Chap. 2 I will argue that semantic considerations, in general, can play an important role in understanding how arguments relate to their social contexts, and modality, in particular, is systematically tied to the dynamics of contextualized activity types, to the point that it can function as a highly sensitive *contextualization cue* (cf. Gaik 1992) for argumentation. More specifically, Chap. 6 will provide a short overview of the broader dynamics of financial communication to which the business-financial journalistic coverage participates (cf. Rocci 2014; Rocci and Luciani 2016; Westbrook 2014) highlighting the correlation between the modal semantic structures deployed in that discourse genre and a number of social, physical and epistemic constraints of the underlying field of social interaction.

1.6 What This Book Is Not About

Modality and argumentation are two hugely vast domains: investigations into each of them span over two millennia, across a range of disciplines encompassing philosophy, logic, rhetoric and linguistics, with modern ramification into artificial

[17] It is worth noting how, at the time of publication of *The Uses of Argument*, the allegation of playing otiosely with English idioms as hobbyist lexicographers was a common (and for the most part unfair) accusation leveled against ordinary language philosophers by more logically oriented analytic philosophers.

intelligence and the mathematical theory of probability. The aims and limitations of the present study will emerge more distinctly once we have cleared the way of a few big issues that lay at the intersection of the two vast domains, which are not addressed directly in the book. These issues are not attacked up-front, yet they will continue to lurk in the background to surface again at various points in the thread of the main argument of the present work.

Thus, I take the liberty of evoking briefly some of the issues the present book *is not* directly about.

1.6.1 This Book Is Not About Modal Logic

It should be said from the outset that this book is not primarily about the *logic* of modalities. The specificities of arguments containing *modal propositions* have been addressed since Aristotle and *modal logic*, understood as the normative study of the validity of arguments containing modal propositions, flourished particularly in the Middle Ages, to be later almost forgotten and emerge again in the Twentieth Century in the context of formal logic.

For the purposes of the present research, modal propositions can be defined as complex propositions formed by a modal operator taking a prejacent proposition in its scope. This is a modern view of modality. Antiquity and, more explicitly, the Middle Ages viewed modality under many respects equivalently, but through a different set of categories and defined modal propositions accordingly. The anonymous treatise *De propositionibus modalibus* (cf. Anonymous 2006), traditionally attributed to St. Thomas Aquinas, defines a modal proposition as a proposition where a "mode" (i.e. a "determining attribute") is attributed neither to the subject of the proposition (as in *Socrates is white*) nor to the predicate (as in *Socrates runs well*), but to "the composition itself of the predicate with the subject, as when it is said 'that Socrates runs is impossible'" (cf. Uckelman 2009b: 2 for the English translation).

In chapters 12–13 of the book *On Interpretation* (cf. Hintikka 1973: 41–61) Aristotle captures the *immediate* inferences (conversion and opposition) that can be established with regards to propositions constructed with modalities and negation.[18] As nicely put by Hacquard (2011: 1486), "under all types of interpretations, possibility and necessity modals enter into patterns of entailments and logical equivalences similar to those involving universal and existential quantifiers". As it is well known, since late Antiquity, these patterns have been beautifully mapped onto the traditional *square of opposition* (cf. Horn 2001 [1989], 2014) (Fig. 1.3).

Contradictory propositions, connected by the diagonals of the square, entail each other's falsity, while they are, in turn, entailed by the negation of their opposite.

[18] According to the respective scope of the negation and the modal, the negation can be either *internal* (when the negated proposition forms the prejacent of the modal) or *external* (when the whole modal proposition is negated).

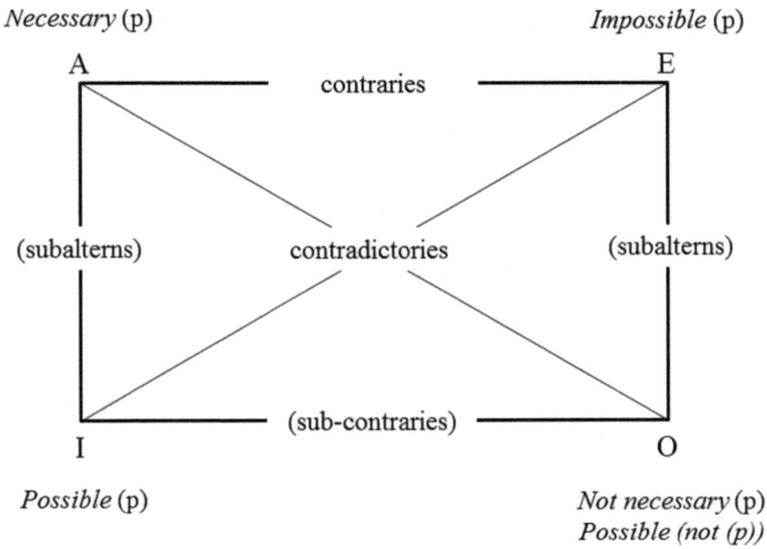

Fig. 1.3 The modal square of opposition

Thus, contradictories are logically equivalent (=) to the negation of their opposite on the diagonal of the square:

(7) Possible p = Not impossible p
(8) Necessary p = Not possible not p

As noted by Hacquard (2011) cited above, these inferences remain valid across all flavors of terms such as *possible* and *necessary* and across a wide range of isomorph modal concepts. This includes not only the alethic and epistemic modalities that have been introduced in the previous sections, but also *deontic* concepts, corresponding to those moral, legal or practical alternatives that were evoked in commenting the necessity of John's resignation in example (1). Let us take *obligatoriness* (a deontic necessity) and *permissibility* (a deontic possibility): from the obligatoriness of wearing a bike helmet I can deduce that it is not permissible not to wear one, and vice versa. From the permissibility of smoking in the dedicated smokers' areas of the airport, I can deduce that smoking is not forbidden there.

Contrary propositions on the upper corners of the square do entail each other's falsity. Yet, there are not equivalent to the negation of their opposite. Traditionally put, contraries cannot be true at the same time, but can be false at the same time.

(9) Necessary p ⇒ Not impossible p
(10) Impossible p ⇒ Not necessary p

To exemplify: from the obligatoriness of helmets I can infer that they are not forbidden (9), but I cannot infer obligatoriness from the lack of a prohibition. From the prohibition of smoking I can infer that it is false that smoking is mandatory (10), yet from knowing that it is false that smoking is mandatory, I cannot infer that it is indeed prohibited. Additionally, the propositions in the upper corner entail their subaltern propositions in the corresponding lower corner:

(11) Necessary p ⇒ Possible p
(12) Impossible p ⇒ Possible not p

If I know that helmets are compulsory, I can immediately infer that they are also allowed (11), and if I know that smoking is forbidden on the restaurant's premises I can infer that it is not mandatory (12). All these inferences are deductively valid and are immediate, requiring only one premise rather than the composition of two or more statements.

As it is well known, the immediate inferences enshrined by the square assume that modal expressions such as *Possible p* can refer also to states of affairs that are necessarily the case and not only to those that are merely contingent. In fact, Aristotle saw possibility expressions such as Greek *dunatón* 'possible' and *endechómenon* 'contingent' as "homonymous" (cf. *Prior Analytics* I, 3, 25a37–40), i.e. as being ambiguous between two readings, so that he recognized two distinct notions of possibility, albeit without providing a distinct term for each (Hintikka 1973: 29). We can speak of *possibility proper* or *unilateral possibility* for the notion captured by the square, and of *bilateral possibility* (see also Horn 2001 [1989]) for the notion that is incompatible with necessity. The term *contingent* (Lat. *contingens*) started to be specialized to indicate bilateral possibility in the works of Late Medieval logicians (cf. Knuuttila 1982).

Modern linguists have offered pragmatic explanations of the two readings of possibility expressions (Horn 2001 [1989]; van der Auwera 1996), according to which the unilateral meaning is basic and the bilateral meaning arises as the result of a Gricean pragmatic inference (a *scalar Quantity implicature*). The utterance of the weaker, less informative, proposition *Possible p* invites the hearer to conclude that the stronger, more informative, proposition *Necessary p* is not the case, unless there is evidence of the contrary. Thus, a layer of defeasible *pragmatic inferences* is superposed to the logical immediate inferences captured by the square. Neither the immediate inferences captured by the square, nor the scalar pragmatic inferences superposed to them occupy the central stage in the present work.

In contrast, much will be said about the systematic ambiguity or, better, underspecification of many modal expressions between the different flavors (e.g. alethic, epistemic, deontic), about the pragmatic and discursive processes invoked to resolve it. In this context, a little will be also said about the logical and pragmatic inferences that "bridge" different flavors. We have already seen, in Sect. 1.2, that alethic impossibility pragmatically implicates epistemic impossibility. It also pragmatically implicates doxastic impossibility.[19] For instance, the deontic obligation to perform an action presupposes that the action is alethically possible in a very specific sense, i.e. that it is doable by the intended subject of the obligation. On the other hand, in

[19] This is a quality implicature, as one cannot coherently *say* that a state of affairs is alethically impossible and then admit it as a belief. Doing so would amount to presenting oneself as either *insincere* (I say what I don't believe) or *irrational* (I admit beliefs that I know to be impossible).

many contexts the assertion of an alethic possibility pragmatically implicates also an epistemic uncertainty (two-sided possibility), via a bridging quantity implicature, as in (13).

(13) They can win the match

The assertion of (13), in most contexts implies that the speaker does not know that the team *will* certainly, or most probably, win the match – for instance because the match has been fixed.

In the *Prior Analytics* (I, 9–12) Aristotle also addressed *mediated* inferences involving modalities and tries to develop a modal syllogistics, that is a "theory of syllogisms from premisses some or all of which are apodeictic (necessary) or problematic (possible) as distinguished from plain (assertoric) premisses" (Hintikka 1973: 135). An example of modal syllogism, derived from *Prior Analytics* (I, 15), is the following, which Aristotle considered valid (cf. Hintikka 1973: 137):

(14)

(14.a) It is *possible* that all female spectators in the theater like the movie. (A possibly applies to all B)

(14.b) All spectators in the first row are female. (B applies to all C)

(14.c) It is *possible* that all spectators in the first row like the movie (A possibly applies to all C)

Contrary to the immediate inferences of the square, Aristotle's modal syllogistic was not to become a stable acquisition for logic. The exact interpretation of Aristotle's modal syllogisms remains to this day obscure and the claim to validity of several syllogistic forms presented, including the one illustrated in (1),[20] is regarded as questionable by modern logicians (cf. also Rescher 2006b).

In the Middle Ages important developments in modal logic were fueled by the needs of metaphysical and properly theological speculation. The very idea that we should recognize different sorts of modalities rather than just one kind of possibility and necessity emerges with the medieval thinkers.

On the one hand, Medieval philosophers include in their treatment of modalities a variety of predicates expressing *knowledge*, *doubt*, *opinion*, *appearance*, *volition* and *agency* (cf. Uckelman 2012; Rocci 2005a: 79; Colmegna 1984: 174–178) somewhat anticipating the modern conception on non-alethic modalities (such as the deontic concepts of obligatoriness and permissibility used in the examples above).

[20] The problem with (14) lies in premise (14.b): the conclusion (14.c) is validly deduced only if (14.b) holds true in *all relevant alternatives* rather than simply happening to be the case contingently. In other words, (13.b) needs to be covertly understood as *necessary* in some relevant sense. "Otherwise – as observed by Hintikka (1973: 137)— it might happen that then the Bs to which A possibly applies are not Cs any more when this possibility is actualized". In term of the movie theater example: suppose that the possibility that all female spectators like the movie is indeed realized only *tomorrow*: the fact that *today* all first row spectators happen to be female does not tell us anything about the occupation of the first row tomorrow. To infer (14.c) we would instead need to know that the seat of the first row are *always* reserved to female spectators.

On the other hand, when it comes to alethic possibility and necessity, they are brought by their philosophical and theological needs to broaden the range of possibility. While for Aristotle, a possibility is, has been or will be actualized at some moment in time (Hintikka 1973: 95), the philosophers of the High Middle Ages come to conceive of possibilities that never become the case but are still envisaged by God's omniscience and *could* become the case God willing. John Duns Scotus (1266–1308) comes to conceive the contingent (bi-lateral possible) not as something that is not always or necessarily the case, but as something whose opposite could be actual *at the very moment when it occurs*, thus introducing an idea of synchronic alternatives that anticipates the modern notion of possible worlds (cf. Knuuttila 2013, 1982). Scotus calls this broader, more abstract, notion of possibility *possibilitas logica* or *potentia logica*. For Scotus (cf. Vos 2006: 175), logical possibility is a quality (*modus*) of the composition of terms (subject and predicate) which are not incompatible with each other (*non sibi invicem repugnant*) and can be meaningfully joined to form a proposition. Such a possibility, Scotus explains, does not require any reality apart from the semantic compatibility of the terms and should be contrasted with real possibility (*potentia realis*), which takes into account the structure of the natural world.

In recent years, the sophisticated contributions of the medieval scholars to the field have been increasingly the object of investigation by logicians re-examining them under the lens of contemporary formal logic (cf. Uckelman 2009a). Also in this book, on occasion, it will be useful to revisit Medieval authors' insights regarding certain key distinctions in the domain of modality.

Indeed, as for several areas of philosophical logic, modal logic was almost reborn well into the Twentieth Century, after long period of stagnation and disregard, as part of the new – mathematically informed – formal logic. During the first part of the century, many formal logicians followed Russell's example in regarding modal notions as leftovers of a traditional metaphysics which had no place in mathematical philosophy (cf. Rescher 2006c [1979]). Starting with the late 1940s, however, modal logic witnessed a rapid growth thanks to the huge impulse of the *possible world semantics* developed by logicians such as Saul Kripke (1963) and Jaakko Hintikka (1962). Modal logic is now a huge field, with wide ranging applications, including the formalization of different aspects of argumentation in ordinary discourse.

While in this book I draw from possible world semantics in Chap. 4 in order to provide a semantic analysis of modal expressions in natural language, the book does not directly contribute to modal logic. Most importantly, the present study will not deal, except marginally, with modal logic's concern for determining which arguments containing modal propositions are deductively valid.

As we have seen in the previous sections, the present investigation is concerned not with modality in the *content* of certain arguments, but rather about what modals can say about the essential structure of arguments and what role or part modality might have in such a structure. Yet, as it will emerge, over the course of the investigation, such a clear-cut distinction between modality in the *content* and modality as part of the structure can become at times quite difficult to draw.

1.6.2 *This Book Is Not About Non-demonstrative and Non-deductive Reasoning*

There is a kind of modality that clearly contributes to the form of the argument rather than to its content. Irrespectively of the modal or non-modal nature of the propositions composing an argument, however, one could say that deductive validity itself should be seen as a modal notion, namely as a form of *necessity*. This necessity of the inference emerges clearly in Aristotle's characterization of the opposition between demonstrative and dialectical reasoning:

> Now a deduction is an argument in which, certain things being laid down something other than these necessarily comes about through them. It is a demonstration, when the premisses from which the deduction starts are true and primitive, or are such that our knowledge of them has originally come through premisses which are primitive and true; and it is a dialectical deduction, if it reasons from reputable opinions (*Topics*, I 1, 100a25-26, transl. by W.A. Pickard-Cambridge, Revised Oxford Translation).

It interesting to note how Aristotle in this passage clearly separates the necessary nature of the inference from the nature of the premises: the necessity of the inference characterizes both demonstrative arguments based on primary evidence and dialectical arguments, based on "reputable" or "generally accepted" (endoxical) premises.

This is the kind of necessity found in textbook definitions of deduction: "a deductively valid argument is an argument such that it is not possible both for its premises to be true and its conclusion to be false argum" (Lepore 2000: 10, cf. also Vorobej 2006: 53). Thus, in a deductively valid argument the conclusion is logically necessary *relative* to the premises. In the following chapters it will appear clearly how *relative* modal notions such as the "logical" necessity that obtains in valid arguments and the idea of separation between the relative modality of the inference and the status of the premises plays a key role in relying modality and arguments in an insightful way.

Unfortunately, confusion regarding this basic distinction creeps into many discussions of modality. When linguists working on the semantics of modality talk of *logical modality* (cf., for instance, Lyons 1977: 791; Palmer 1986: 11; Papafragou 2000; Nuyts 2006: 8), they often remain ambiguous between two very different notions: sometimes the *logical necessity* of a proposition p is taken so-to-say 'absolutely' to mean its *analyticity* (either purely logical or semantic analyticity), sometimes the same term *logical necessity* is taken 'relatively' to mean that a proposition p is deduced from a set of propositions which aren't necessarily logical axioms, or primary in Aristotle's sense. In Chap. 3 we will see that even Toulmin, in his critique of logical necessity, might be guilty of such a confusion.

In the first book of Aristotle's *Rhetoric* we find a passage that adopts a quite different take on *necessity* and that could be even construed as suggesting a view at odds with the notion of *necessity* of the inference in *Topics* (I, 1, 100a25–26). It is a quite powerful, suggestive, passage but much less perspicuous than the text of the *Topics*:

> There are few facts of the necessary type that can form the basis of rhetorical deductions. Most of the things about which we make decisions, and into which we inquire, present us with alternative possibilities. For it is about our actions that we deliberate and inquire, and all our actions have a contingent character; hardly any of them are determined by necessity. Again, conclusions that state what holds for the most part and is possible must be drawn from premisses that do the same, just as necessary conclusions must be drawn from necessary premisses; this too is clear to us from the Analytics. It is evident, therefore, that the propositions forming the basis of enthymemes, though some of them may be necessary, will in the main hold for the most part. (*Rhetoric* 1357a, transl. by W.R. Roberts, Revised Oxford Translation)

Here Aristotle seems to weave together key aspects of reasoning in public life: (1) the ontology of the subject matter, which concerns contingent human decisions and actions rather than the basic ontology of the world, (2) the ontological and epistemic status of the premises which describe possibilities or regularities rather than necessary propositions, and (3) the consequent ontological and epistemic status of the conclusions. In fact, the modal notions in this passage seem to apply equally to the *ontological* level, to the *logical* level and to the *epistemic* one.

Suggestive as it may be, the passage is not of straightforward interpretation. It speaks of the necessity of conclusions or lack thereof, but it is not entirely clear whether it says anything about the force of the inferential connection between premises and conclusion, which emerged from. If we read the passage as suggesting a complete parallelism of the three levels (ontological, logical and epistemic) we get the highly problematic idea that inferences that take ontologically contingent propositions as premises cannot have deductive force. Chapter 3, in the context of the Toulminian discussion of modality's place in the essential structure of arguments, will examine a bit more in-depth this issue, with the help of other Aristotelian passages, showing how Aristotle did, in fact, admit, in the dialectical domain, necessary (i.e. deductive) inferences from contingent premises.

The suggestion of *Rhetoric* (I, 1357a) is certainly operating in the very first section of Perelman's and Olbrechts-Tyteca's *Traité de l'argumentation*, where the authors contrast the use of reason in deliberation and argument with the prevailing Cartesian view of reason of modern Western philosophy:

> Although it would scarcely enough occur to anyone to deny that the power of deliberation and argumentation is a distinctive sign of a reasonable being, the study of the methods of proof used to secure *adherence* has been completely neglected by logicians and epistemologists for the last three centuries. This state of affairs is due to the noncompulsive element of the arguments adduced in support of a thesis. The very nature of deliberation and argumentation is opposed to necessity and self-evidence. The domain of argumentation is that of the credible, the plausible, the probable, to the degree that the latter eludes the certainty of calculations (Perelman and Olbrechts-Tyteca 1969: 1).

Of the different levels underlying *Rhetoric* I, 1357a, the authors of the *New Rhetoric* seem to focus on the epistemic one, opposing the *credible, plausible* and non-calculable *probable* to the *necessary* and self-evident. Once argumentation theory considers reasoning outside of formal logic and pure mathematics to address deliberation in social life, it cannot be content of the necessity and self-evidence of demonstration, and needs to open up, so-to-say, the paradigm of epistemic modalities.

Taken together with the two previous Aristotelian quotes, the introductory passage of the *New Rhetoric* highlights quite aptly a set of three important issues, lying at the intersection of modality and argumentation, that are *not* addressed directly in this book.

The first two issues concern how we should characterize the epistemic status, respectively, of *premises* and *conclusions* in the arguments that are exchanged in the various arenas of social life, what range or paradigm of modal terms is more appropriate to capture it, in view of the reconstruction and critical evaluation of ordinary arguments and what kind of semantics we should attribute to these meta-argumentative terms. What does it mean, exactly, that a premise is *plausible* or *presumed*? For what kind of conclusions should we reserve the qualifier of *probable*? Should statistics play any role in ascribing probability to a conclusion?

This idea of the need of a broader paradigm of modalities to characterize the starting points of arguments as well as the results that they aim to effect in their addressees is taken to the extreme consequences in an insightfully provocative essay by Robert Pinto. Pinto (2001: 10) advocates a progressively "generalized view of argument" as an attempt "to modify conscious attitudes through rational means", where attitudes are not limited to firm unqualified belief, but run the full gamut of *doxastic* attitudes (*considering possible p, suspecting p, having really no idea whether p*), including attitudes that are not limited to a doxastic component (*fearing that p, hoping that p*).[21] In view of the range of attitude associated with premises and conclusions, Pinto (2001: 14) sees the task of argumentation theory as investigating "the ways in which qualified belief in premisses affects the attitudes which it is rational to take toward the conclusion". Pinto gives one simple example of the kind of rules that could guide the critical evaluation arguments with a broadened range of doxastic attitudes:

> For example, a rule that suggests itself is that the doxastic attitude toward a conclusion should be no stronger than the doxastic attitude toward the weakest of the premisses (Pinto 2001: 14).

The relationship between the qualified premises and the qualified conclusions brings us to the third issue, which concerns the possibility of a modal paradigm characterizing not only in the premises and the conclusions themselves, but also the strength of the inferential connection between the two. As shown above, Aristotle in *Topics* (I, 1, 100a25–26) feels no need of weakening the truth-preserving necessary link between premises and conclusion in dialectical arguments: dialectical arguments are *non-demonstrative* because of the endoxical rather than primary nature of their premises, but they can still be deductive thanks to the necessary nature of the connection. Today, the idea of *deductive non-demonstrative inferences* features prominently in Relevance Theory (Sperber and Wilson 1995 [1986]), an influential cognitive theory of pragmatics which views discourse comprehension as based on

[21] Truth be told, Pinto includes in his discussion also evaluative attitudes that are clearly non-doxastic and seeks to even non-propositional attitudes (Pinto 2001: 15–17). I do not consider this further extension for my present purposes.

fast and unconscious non-demonstrative deductive inferences, taking as their premises assumptions entertained with different degrees of strength of belief by the communicators. This is a minority view in modern pragmatics, with other major theories advocating a *non-deductive* approach to pragmatic inferences (cf. Levinson 2000, Asher and Lascarides 2003).

Similarly, in contemporary argumentation theory too, the idea that *deductive non-demonstrative* inferences could be sufficient to account for ordinary arguments seems to be a minority view. The prevailing view is that also the nature of the inferential link should be weakened accordingly. The need of addressing non-deductive forms of reasoning is quite apparent in Toulmin (2003 [1958]) and, in fact, his interest towards modal notions such as (im-)possibility, probability and necessity is partly motivated by this exigency.

Terms like *ampliative*, *defeasible*, *non-monotonic*, *plausible*, *presumptive* and *probabilistic* reasoning have been used, in a partially overlapping fashion, by philosophers, formal and informal logicians and students of artificial intelligence and have wide currency in the argumentation theory literature (cf. Woods 2010). Only a small fraction of this sea is reducible to "inductive or Bayesian logic" (Woods 2010: 239), which is traditionally seen as the only alternative to deductive reasoning. To capture this vast ocean, formal systems aimed of defeasible reasoning have been developed (cf. Pollock 1987, 2010; van Benthem1996, 2009), while other logicians and argumentation scholars have offered conceptual accounts of the forms and epistemological underpinnings of presumptive reasoning (cf. Walton 1996; Rescher 2006a, d [1979]).

While these matters certainly overlap the scope of the present investigation and will surface again in the following chapters, the vast literature on non-deductive inference in argumentation is not addressed systematically in the present work, which takes another route. The investigation remains anchored to the analysis of the semantics of modal constructions in natural languages. It takes a broader view of modality by considering the full range of modal flavors – or, at least, a very wide one – and not limiting the inquiry to the epistemic, doxastic and logical modalities that have been seen used meta-argumentatively by theorists to characterize non-demonstrative and/or non-deductive reasoning. It considers a broader range of argumentatively relevant functions of the modals and not just their indication of the strength of support or of the epistemic status of premises and conclusions. It dwells primarily on matters of reconstruction without addressing explicitly the problem of the *evaluation* of the merits of non-demonstrative and non-deductive arguments. That said, I hope that the semantic investigations contained herein will be of use for those that work in these key areas of argumentation theory.

1.7 Structure of the Volume and Reading Guide

I close this introduction by previewing how the questions and research agenda sketched here will be pursued in remainder of the book. The book is quite long and I don't presume all readers will endure it from cover to cover. So, since I was not able to provide a shorter book, I will at least provide a summary guide to orient the reader.

Chapter 2 outlines the profile of argumentation theory as scientific enterprise and discusses how the study of meaning in its semantic and pragmatic aspects can contribute to the descriptive and normative study of arguments expressed in ordinary discourse. In doing so, the main features of normative-pragmatic theories of argumentation are introduced, with particular reference to the theory Pragma-Dialectics. Two other general theoretical tools used in the book are also introduced, namely Congruity Theory, a broad approach to representing semantic and pragmatic meanings in discourse, and the Argumentum Model of Topics (AMT), a semantically informed approach to the inferential schemes underlying ordinary arguments. The last section of the chapter considers how the semantic considerations that occupy the center stage in this book also contribute to an understanding of how argumentation is deployed in concrete contexts of social interaction.

Linguists who approach this book from the angle of modality should read this chapter – even if there is not much about modality in it – to be introduced to argumentation theory and consequently better understand the angle taken by the book. Having a normative facet, theorizing in the argumentation tradition can differ sharply from theoretical linguistics. Readers that are very familiar with argumentation theory should also profit from reading the chapter, especially as it discusses the role of semantic considerations in argumentation theory with unusual depth. Those who are in a hurry and decide to skip the chapter might still want to go back to it later and check the sections on Congruity Theory, on the Argumentum Model of Topics and Freeman's semantic taxonomy of statements, all of which are necessary to understand the analyses deployed later in the book.

The key research questions of this book were inspired, to a large extent, by Stephen Toulmin's ideas on modality and argumentation. Chapter 3 contains a detailed examination of Toulmin's extensive and multi-faceted treatment of modality in the different chapters of *The Uses of Argument*. The chapter also considers the reception of Toulmin's ideas in later studies, as well as their relation to current contributions to argumentation theory and linguistics.

Of all the chapters, Chap. 3 is the one that can be better read also in isolation. Readers interested in Toulmin's contribution to argumentation theory are likely to find a new "semantic" angle on some key Toulminian ideas (from field dependency to the elements of the Toulmin model) as well as critical remarks on the internal consistency of Toulmin's views of modality.

The chapter can be also read in lieu of a state of the art of the treatment of modality in argumentation theory. While its coverage of the, rather dispersed, subsequent discussions of modality by argumentation theorists is not fully exhaustive, it uses Toulmin's ideas on modality and the issues they raise to fame in an insightful manner both the subsequent treatments in argumentation theory and the research gap addressed by this book.

Chapter 4 draws on studies of the semantics of modality in the formal semantics tradition. Moving from the need of restricting the domain of quantification in a possible worlds semantics of modality, it presents a relational context-dependent semantics of modality inspired by the theory of Relative Modality. This approach is immediately applied to the analysis of conditionals and to the modeling of anaphora and rhetorical relations in discourse.

With this basic theoretical machinery in place, Chap. 5 can move to reconstructing the most relevant "flavors" of modality (alethic, deontic, deontic-practical, anankastic and epistemic) in terms of a *relational*, *discourse sensitive*, *context dependent*, *procedural* approach to the interpretation of modal constructions. The approach, while inspired by Relative Modality and the formal semantics tradition, integrates pragmatic insights, is sketched in a rather informal metalanguage and is clearly cast as a tool for the actual analysis of discourse and argumentation.

Throughout Chaps. 4 and 5, argumentative discourse relations between standpoints and premises are discussed in the light of the relational semantic analyses of the modal expressions introduced therein. Argumentative discourse relations are defined at the level of speech acts; at the level of the inferential processes involved, and at the level of the semantico-ontological relations of "conveyance" – or *topics* – supporting the inferential processes. A framework is sketched for investigating how the relational meanings of the modals may contribute to each of the three levels.

In Chap. 5, in particular, the typology of modal "flavors" is used to investigate the interaction of modality and argumentation with a special focus on the *topics* underlying argumentative inferences. The semantic category of *evidentiality*, which is closely related to epistemic modality, is also briefly examined in relation to argumentation, with a survey of the most basic categories of evidence sources encoded by languages (*perception*, *testimony*, *inference*).

Chapter 6 examines how the linguistic expressions of epistemic and doxastic modality in a specific language can function as rich and finely grained indicators for the reconstruction and subsequent evaluation of arguments put forth in ordinary discourse. In order to do so the chapter examines a segment of the Italian modal system, with a particular focus on different constructions with the verb and *dovere* 'must' – and, to a lesser extent with *potere* 'may, can', which have epistemic interpretations. In these constructions the basic context-dependent relational semantics is enriched by subtle constraints on the evidential source. As a result, different modal constructions end up guiding the establishing of argumentative relations of very specific kinds, differing not only in the expressed strength of support or degree of commitment towards the conclusion, but also along an impressive series of parameters including the nature and epistemic status of the premises, the presence of rebuttals, the semantic type of the standpoint and the argumentative *locus* that is invoked.

The case studies in Chap. 6, however, are not confined to the examination of the potential of linguistic expressions as argumentative indicators. It is also aimed at understanding how the social and material ontology of the contexts in which argumentative discussions take place is directly reflected in the semantic structure of argumentative discourse in the area of modality. These semantic reflections of context, it is argued, are essential to properly reconstruct issues and standpoints of the discussion as well as to properly evaluate the constraints and affordances with which arguers are confronted when constructing an argumentative strategy. In order to do so, it examines how argumentation and modality intertwine in the genre of financial newspaper articles on the backdrop of the broader activity system of financial communication. The large corpus of Italian financial newspaper articles used as data for the semantic investigation also allows to study how context-dependent modal expressions are employed in arguments supporting economic-financial forecasts and other business predictions.

The *Conclusion* of the book summarizes the main results of the investigation and mentions some spin-offs and ongoing developments of the research by myself and other colleagues.

References

Aijmer K (2009) Seem and evidentiality. Funct Lang 16(1):63–88. doi:10.1075/fol.16.1.05aij
Aikhenvald AY (2004) Evidentiality. Oxford University Press, Oxford
Anonymous (2006) De propositionibus modalibus. In: Alarcón E (ed) Corpus Thomisticum. St. Thomae de Aquino Opera omnia. Fundación Tomás de Aquino 2000–2013. http://www.corpusthomisticum.org/. Accessed 5 Oct 2016
Aristotle (1926) The "art" of rhetoric (LOEB Classical Library vol. 193). Freese JH (ed) Harvard University Press, Cambridge (MA)
Aristotle (2014) The complete works of Aristotle: the Revised Oxford Translation, One-volume digital edition. Barnes J (ed) Princeton, Princeton University Press (originally published in 1984 as 2 volumes)
Asher N, Lascarides A (2003) Logics of conversation. Cambridge University Press, Cambridge
Bybee J, Perkins R, Pagliuca W (1994) The evolution of grammar: tense, aspect and modality in the languages of the world. The University of Chicago Press, Chicago
Chafe W, Nichols J (1986) Evidentiality. The linguistic coding of epistemology. Ablex, Norwood
Colmegna M (1984) Enunciazioni circa il modo. Per una grammatica della logica delle modalità. Versus Quaderni di Studi Semiotici 38(39):149–189
Cornillie B (2007) Evidentiality and epistemic modality in Spanish (Semi)Auxiliaries. Mouton de Gruyter, Berlin/New York
Dendale P (1994) Devoir épistémique, marqueur modal ou évidentiel? Lang Fr 102(1):24–40
Dendale P, Tasmowski L (1994) Présentation. L'évidentialité ou le marquage des sources du savoir. Lang Fr 102:3–7
Ennis RH (2006) Probably. In: Hitchcock D, Verheij B (eds) Arguing on the Toulmin model. Springer, Dordrecht, pp 145–164
Fahnestock J, Secor M (1982) A Rhetoric of argument. Random House, New York
Faller MT (2002) Semantics and pragmatics of evidentials in Cuzco Quechua. Ph.D dissertation, Stanford University, Stanford
Freeman JB (1985) Dialectical situations and argument analysis. Informal Logic 7(2–3):151–162
Freeman JB (1991) Dialectics and the macrostructure of arguments. A theory of argument structure. Foris Publications, Berlin
Freeman JB (2000) What type of statements are there? Argumentation 14(2):135–157
Freeman JB (2005) Acceptable premises: an epistemic approach to an informal logic problem. Cambridge University Press, Cambridge
Freeman JB (2011) Argument structure: representation and theory. Springer, Amsterdam
Gaik F (1992) Radio talk-show therapy and the pragmatics of possible worlds. In: Duranti A, Goodwin C (eds) Rethinking context: language as an interactive phenomenon. Cambridge University Press, Cambridge, pp 271–289
Geurts B (1999) Presuppositions and pronouns. Elsevier, Amsterdam

Guentchéva Z (1996) L'énonciation médiatisée. Peeters, Leuven
Hacquard V (2011) Modality. In: Maienborn C, von Heusinger K, Portner P (eds) Semantics: an international handbook of natural language meaning. Mouton de Gruyter, Berlin/New York, pp 1484–1515
Hintikka J (1962) Knowledge and belief. an introduction to the logic of the two notions. Cornell University Press, London
Hintikka J (1973) Time and necessity. Oxford University Press, Oxford
Horn L (2001 [1989]) A natural history of negation. Stanford, CSLI Publications
Horn L (2014) The cloud of unknowing. In: Hoeksema J, Gilbers D (eds) Black book: a festschrift in honor of Frans Zwarts. University of Groningen, Groningen, pp 178–186
Jacobs S (2000) Rhetoric and dialectic from the standpoint of normative pragmatics. Argumentation 14:261–286
Kaufmann S, Condoravdi C, Harizanov V (2006) Formal approaches to modality. In: Frawley W (ed) The expression of modality. Mouton de Gruyter, Berlin/New York, pp 71–105
Knuuttila S (1982) Modal logic. In: Kretzmann N, Kenny A, Pinborg J (eds) The cambridge history of later medieval philosophy. Cambridge University Press, Cambridge, pp 342–357
Knuuttila S (2013) Medieval theories of modality. In: The stanford encyclopedia of philosophy (eds) Edward N. Zalta et al. Stanford. Standford University Press. http://plato.stanford.edu/archives/fall2015/entries/modality-medieval. Accessed 5 October 2016
Kratzer A (1977) What "must" and "can" must and can mean. Linguist Philos Int J 1:337–355
Kratzer A (1991) Modality. In: von Stechow A, Wunderlich D (eds) Semantik/Semantics: an international handbook of contemporary research. Mouton de Gruyter, Berlin/New York, pp 639–650
Kratzer A (1981) The notional category of modality. In: Eikmeyer H-J, Rieser H (eds) Words, worlds, and contexts: new approaches in word semantics. Mouton de Gruyter, Berlin/New York, pp 38–75
Kripke SA (1963) Semantical analysis of modal logic I. Normal modal propositional calculi. Math Logic Q (Formerly: Zeitschrift für mathematische Logik und Grundlagen der Mathematik) 9(5–6):67–96
Kruger AN (1975) The nature of controversial statements. Philos Rhetor 8(3):137–158
Lepore E (2000) Meaning and argument. Blackwell, Oxford
Levinson SC (2000) Presumptive meanings: the theory of generalized conversational implicature. MIT Press, Cambridge, MA
Lewis D (2007) Counterfactuals. Blackwell, Oxford
Lycan WG (1994) Relative modality. In: Lycan W (ed) Modality and meaning. Kluwer, Dordrecht, pp 171–200
Lyons J (1977) Semantics. Cambridge University Press, Cambridge
McCready E, Ogata N (2007) Evidentiality, modality and probability. Linguistics and Philosophy 30(2):147–206
Miecznikowski J, Musi E (2015) Verbs of appearance and argument schemes: Italian *sembrare* as an argumentative indicator. In: Garssen B, van Eemeren FH (eds) Reflections on theoretical issues in argumentation theory. Springer, Amsterdam, pp 259–278
Musi E (2014) Evidential modals at the semantic- argumentative interface: appearance verbs as indicators of defeasible argumentation. Informal Logic 34(4):417–442
Musi E (2015) Dalle apparenze alle inferenze: i verbi sembrare ed apparire come indicatori argomentativi. Ph.D. dissertation, Università della Svizzera italiana
Musi E, Ghosh, D, Muresan, S (2016) Towards feasible guidelines for the annotation of argument schemes. In: Proceedings of the third workshop on argument mining (ArgMining2016): 82–93. http://aclweb.org/anthology/W/W16/W16-2810.pdf. Accessed 5 Oct 2016
Nuyts J (2006) Modality: overview and linguistic issues. In: Frawley W (ed) The expression of modality. Mouton De Gruyter, Berlin/New York, pp 1–26
Palmer F (1986) Mood and modality. Cambridge University Press, Cambridge
Papafragou A (2000) Modality: issues in the semantics-pragmatics interface. Elsevier, Amsterdam

Perelman C, Olbrechts-Tyteca L (1969) The new rhetoric. In: Wilkinson J, Weaver P (eds) A treatise on argumentation. University of Notre Dame Press, Notre Dame
Pietrandrea P (2007) The grammatical nature of some epistemic-evidential adverbs in spoken Italian. Ital J Linguistics 19(1):39–64
Pinto RC (1996) The relation of argument to inference. In: van Benthem J et al (eds) Logic and argumentation. North-Holland, Amsterdam
Pinto RC (2001) Generalizing the notion of argument. In: Pinto R (ed) Argument, inference and dialectic. Collected papers on informal logic. Kluwer, Dordrecht, pp 10–20
Pinto RC (2007) On understanding 'probably' and other modal qualifiers. In: Hansen HV et al (eds) Dissensus and the search for common Ground, CD ROM. OSSA, Windsor, pp 1–10
Pollock JL (1987) Defeasible reasoning. Cogn Sci 11:481–518
Pollock JL (2010) Defeasible reasoning and degrees of justification. Argum Computation 1:7–22
Portner P (2009) Modality. Oxford University Press, Oxford
Rescher N (2006a) Presumption and the practices of tentative cognition. Cambridge University Press, Cambridge
Rescher, N (2006b [1971]) On Aristotle's apodeictic syllogism. In: Studies in the history of logic, Ontos Verlag,Frankfurt, 1–13
Rescher, N (2006c [1979]) Russell and modal logic. In: Studies in the history of logic, Ontos Verlag, Frankfurt, 159–171
Rescher, N (2006d [1979]) Default reasoning. In: Studies in the history of logic, Ontos Verlag, Frankfurt, 173–183
Rigotti E, Greco Morasso S (2010) Comparing the argumentum model of topics to other contemporary approaches to argument schemes: the procedural and material components. Argumentation 24(4):489–512
Roberts C (1989) Modal subordination and pronominal anaphora in discourse. Linguistics and philosophy 6:683–721
Rocci A (2005) La Modalità Epistemica tra Semantica e Argomentazione. I.S.U. Università cattolica, Milano
Rocci A (2008) Modality and its conversational backgrounds in the reconstruction of agumentation. *Argumentation* 22(2):165–189
Rocci A (2014) The discourse system of financial communication. Cahiers de l'ILSL34: 201–221
Rocci A, Luciani M (2016) Economic-financial journalists as argumentative intermediaries. J Argumentation in Context 5(1):88–111
Sbisà M (2014) Evidentiality and illocution. Intercult Pragmat 11(3):463–483
Snoeck Henkemans F (1992) Analysing complex argumentation. SicSat, Amsterdam
Snoeck Henkemans F (1997) Analysing complex argumentation: the reconstruction of multiple and coordinatively compound argumentation in a critical discussion. SicSat, Amsterdam
Sperber, Dan, and Deirdre Wilson. 1995. Relevance. Communication and cognition. Blackwell, Oxford
Squartini M (2004) Disentangling evidentiality and epistemic modality in romance. Lingua: Int Rev Gen Linguistics 114(7):873–895
Squartini M (2008) Lexical vs. grammatical evidentiality in French and Italian. Lingua 46(5):917–947
Toulmin SE (2003 [1958]) The uses of argument. Updated ed. Cambridge University Press, Cambridge
Tseronis A (2009) Qualifying standpoints. Stance adverbs as a presentational device for managing the burden of proof. LOT Dissertation Series 233. Netherlands Graduate School of Linguistics, Utrecht
Uckelman SL (2009a) Modalities in medieval logic. Institute for Logic, Language and Computation, Amsterdam. https://www.illc.uva.nl/Research/Publications/Dissertations/DS-2009-04.text.pdf. Accessed 5 Oct 2016
Uckelman SL (2009b) On modal propositions. http://lyrawww.uvt.nl/~sluckelman/latex/modalibus/modalibus.pdf. Accessed 13 Feb 2012 (no longer available)

Uckelman SL, (2012) Modal logic in the metalogicon. In: New perspectives on John of Salisbury (Durham University 19 April 2012), 1–9. http://lyrawww.uvt.nl/~sluckelman/latex/jos-modal/jos-modal.pdf. Accessed 4 June 2012 (no longer available)
van Benthem J (1996) Logic and argumentation. In: van Benthem J, van Eemeren FH, Grootendorst R, Veltman F (eds) Logic and argumentation. North-Holland, Amsterdam, pp 27–42
van Benthem J (2009) One logician's perspective on argumentation. Cogency J Reasoning and Argumentation 1(2):13–25
van der Auwera J (1996) Modality: the three-layered scalar square. J Semant 13(3):181–195
van Eemeren FH (2010) Strategic maneuvering in argumentative discourse. John Benjamins, Amsterdam/Philadelphia
van Eemeren FH, Grootendorst R (2004) A systematic theory of argumentation: the pragma-dialectical approach. Cambridge University Press, Cambridge
van Eemeren FH, Houtlosser P, Henkemans FS (2007) Argumentative indicators in discourse a pragma-dialectical study. Springer, Amsterdam
von Fintel K (2006) Modality and language. In: Borchert DM (ed) Encyclopedia of philosophy, 2nd edn. Detroit, MacMillan Reference USA. http://web.mit.edu/fintel/fintel-2006-modality.pdf. Accessed 5 Oct 2016
Vorobej M (2006) A theory of argument. Cambridge University Press, Cambridge
Vos A (2006) Philosophy of John Duns Scotus. Edinburgh University Press, Edinburgh
Walton D (1996) Argumentation schemes for presumptive reasoning. Studies in argumentation series. Lawrence Erlbaum, Mahwah
Walton D, Reed C, Macagno F (2008) Argumentation schemes. Cambridge University Press, Cambridge
Weinrich H (2004) Tempus. Le Funzioni dei Tempi nel Testo, Bologna/Il Mulino
Westbrook I (2014) Strategic financial and investor communication: the stock price story. Routledge, London
Woods J (2010) Defeasible reasoning. In: Reed C, Tindale CW (eds) Dialectics, dialogue and argumentation. An examination of Douglas Walton's theories of reasoning and argument. College Publications, London, pp 239–261

Chapter 2
Meaning and Argumentation

Abstract The chapter outlines the profile of argumentation theory as a scientific enterprise and discusses how the study of meaning in its semantic and pragmatic aspects can contribute to the descriptive and normative study of arguments expressed in ordinary discourse. In doing so, the main features of normative-pragmatic theories of argumentation are introduced, with particular reference to the theory Pragma-Dialectics. Two other general theoretical tools used in the book are also introduced, namely Congruity Theory, a broad approach to representing semantic and pragmatic meanings in discourse, and the Argumentum Model of Topics (AMT), a semantically informed approach to the inferential schemes underlying ordinary arguments. The last section of the chapter considers how the semantic considerations that occupy the center stage in this book also contribute to an understanding of how argumentation is deployed in concrete contexts of social interaction.

2.1 The Viewpoint of Argumentation Theory

Having painted in broad strokes the aims, research strategies and limitations of this study, it becomes important now to consider more in detail how the semantic research proposed fits with the disciplinary goals of argumentation theory. This book can, hopefully, be of interest to different types of readers, including linguists unfamiliar with argumentation theory and argumentation theorists who may or may not be familiar with specific areas of linguistics, and with semantics in particular. This chapter is therefore devoted to create the necessary common ground.

In this first section, I provide a sketch of the argumentation theory enterprise focusing on its primarily *critical* goals and consequently *normative* orientation.

The subsequent section will be devoted to examining what it means to do *analysis* in argumentation theory, a task that is shaped both by the critical goal of argumentation theory and by the object to which it is applied, which is ordinary discourse naturally occurring in context.

The linguistic and discursive nature of argumentation will occupy the center stage of the third section of the chapter, which will further clarify the role that specifically *semantic* considerations can play in the analysis of arguments, alongside the *pragmatic* considerations that have played a much greater and more visible role in the argumentation literature of the last thirty years.

A. Rocci, *Modality in Argumentation*, Argumentation Library 29,
DOI 10.1007/978-94-024-1063-1_2

In the final section of the chapter, I will argue that semantic considerations can play an important role also in understanding how arguments relate to their social contexts.

As the chapter develops this broad perspective on the role of the analysis of meaning in the argumentation enterprise, it introduces specific theories and analytical tools that will be employed in the following chapters. These specific tools include the *Argumentum Model of Topics* as a means to reconstruct the inferential configuration of enthymematic arguments (§ 2.2.3), an argumentatively relevant typology of proposition types developed by James Freeman (§ 2.3.3) and *Congruity Theory* (§ 2.3.4) as a broad strategy for representing predicate-argument structures at a propositional and discursive level.

2.1.1 Argumentation Theory Is Normative

Argumentation theory is, as Walton (2013:1) puts it, "a rich interdisciplinary area of research straddling philosophy, communication studies, linguistics, psychology and artificial intelligence that has developed context-sensitive practical methods" for the identification, analysis and evaluation of arguments.

In fact, irrespectively of the disciplinary affiliation of its practitioners, argumentation theory is best defined as a distinctive scholarly enterprise by its concern for the critical evaluation of reasoning expressed in ordinary discourse in a variety of spheres of human activity. That is to say, argumentation theory looks for principled ways for establishing whether or at what conditions an argument is a good (i.e. reasonable) one. In short, argumentation theory involves a *normative* component.

Perhaps the notion that best represents the normative orientation of argumentation theory is the notion of *fallacy*. The definition of *fallacy* and the analysis of specific fallacies are profusely discussed in the argumentation literature.[1] Let us take, as an example, recent papers by Lewiński (2011) and Walton and Macagno (2010) discussing respectively the *straw man* fallacy and the related fallacy of the *wrenching from context*. Both fallacies concern a situation in which arguers misrepresent the position of their opponent in order to be able to more easily refute the misrepresented position instead of the one actually expressed by the opponent. An argumentation-theoretic treatment of a fallacy typically includes a principled – non *ad hoc* – account of why the fallacy should be regarded as unreasonable and criteria for establishing when exactly the fallacy can be said to occur that can be applied to the task of evaluating naturally occurring instances of argumentative discourse. For instance, a treatment of the above mentioned fallacies might include criteria for

[1] As an indication of the prominence of this topic it can be observed that a simple search for the term *fallacy* in the content of the journal *Argumentation* reveals that the word appears in 533 articles of the 1,095 published so far by the journal (that is in 48.6% of the articles). The data refer to the situation in August 2016.

deciding what counts as *misrepresentation* and a discussion of whether a reformulation like (1.a) is an unfair representation of the original position taken in (1.b). The well known case in example (1) is extensively discussed in Walton and Macagno (2010).

(1)

(a) Al Gore claimed to have *invented* the Internet.

(b) "During my service in the United States Congress, I took the initiative in creating the Internet. I took the initiative in moving forward a whole range of initiatives that have proven to be important to our country's economic growth and environmental protection, improvements in our educational system." (Al Gore on CNN's "Late Edition" program on 9 March 1999, quoted from Walton and Macagno 2010: 289).

The *normative* orientation makes the argumentation enterprise akin to logic[2], on the one side, and to branches of ethics such as the ethics of discourse and communication on the other side. To some extent argumentation theory also exhibits a kinship with disciplines, such as artificial intelligence, which also have practical and design goals and may be concerned with the question whether a certain kind of reasoning is *good enough for a given task*, more or less broadly defined. This same evaluative and normative orientation sets apart argumentation from disciplines such as cognitive psychology and linguistics, which tend to approach reasoning and discourse with a purely descriptive and explanatory purview. In discourse analysis – a field that derives most of its analytical tools from linguistics – works that pursue a critical and evaluative aim alongside a descriptive and explanatory one are not infrequent and a broad *Critical Discourse Analysis* (CDA) movement has developed.

Traditionally, discourse analysts, even those who see themselves as *critical*, have been fairly inexplicit about the meaning of *criticism* and the norms that guide their evaluations. When CDA scholars do address normativity they adopt a variety of stances towards it[3], with some qualifying if not outright rejecting the characterization of their work as normative. Recently explicit discussions of the normative component of CDA have become more common, as witnessed by Chilton's (2005, 2009) attempt at providing cognitive foundations for criticism and even more by Fairclough

---

[2] There are several competing views of what logic is about. The traditional view of logic as the "art of reasoning" is obviously a normative one – it is also a methodological one, as logic is seen as an instrument, a tool (*organon*) for knowing. As for the contemporary views of the disciplines, it is true that (formal) logic can be conceived merely as the study of the formal properties of formal systems, which does not entail any normativity, but the issue of normativity remains central for most philosophically significant conceptions of logic, in particular for those centered on the notion of logical consequence and the study of formally valid inferences. See Hofweber (2014) for a discussion.

[3] For instance, while T. van Dijk, one of the founding figures of the Critical Discourse Analysis movement, squarely declares that CDA is certainly "unabashedly normative" and "presupposes an applied ethics" (van Dijk 1993: 253), R. Wodak, another major figure in the CDA arena, in a recent interview (Kendall 2007), warns against the risk of dogmatism associated with normativity and assumes a culturally relativist view of normativity. See O'Regan (2006) for a discussion of different views of normativity in CDA.

and Fairclough's (2012) recent merger of the CDA and argumentation theory perspectives for the analysis of political discourse.

Also in view of these disciplinary differences – which in the experience of the author have been more than once the occasion of misunderstandings – it is useful to consider in detail how the normative perspective of argumentation theory articulates with semantic and pragmatic considerations in the analysis and evaluation of arguments. It is only in this context that the present investigation of the role of modality in the functioning of argumentative discourse can be fully appreciated.

2.1.2 Dialectic and Normative Pragmatics

The development of a normative perspective in argumentation theory can be traced back to the motives that animated two foundational books published in 1958, one by Stephen E. Toulmin ([1958] 2003) and one by Chaïm Perlelman and Lucie Olberchts-Tyteca ([1958] 2008). As is well known, they independently pleaded for a new approach to the study of reasoning and argument out of a dissatisfaction with logic, identified with the formal study of deductively valid demonstrative argument.[4] In the view of both Toulmin and Perelman a logic exclusively preoccupied with formal criteria of validity risked abandoning outside the pale of rationality most of the arguments that are exchanged in ordinary life – indeed everything outside mathematics and (perhaps) the natural sciences. Both took law and judicial practice as a paradigm of a rationality that could not be reduced to formal logic. Perelman explicitly linked the new enterprise to Aristotle's views of *dialectic* and *rhetoric* as expressed in the *Topics* and in the *Rhetoric*. In choosing a name for his endeavor he settled for the "New Rhetoric" discarding *dialectic* as a term misappropriated by unrelated enterprises in contemporary philosophy. In later writings (Perelman 1977), however, he continued to insist on the importance of recovering the original notion of dialectic of the Ancients, seen as reasoning from commonly accepted premises.

While the attempts of Toulmin and Perelman are certainly informed by the desire of extending the scope of the critical scrutiny of arguments to all arenas of intellectual and social life, they do not offer a normative theory (Van Eemeren and Grootendorst 2004: 127–131). Toulmin connects the different components of his "layout" of arguments – the Toulmin model – with matching critical questions, sug-

[4] In North America the fledging discipline of argumentation theory joined forces and partially merged with the "informal logic movement". In the 1970s "informal logicians" began to observe that contemporary formal logic was ill suited for the teaching of introductory logic courses aimed at sharpening the students' critical skills and equipping them for the rational scrutiny of arguments in everyday life and pleaded for an alternative approach. As a matter of fact, many of the major contributors to the current argumentation theory enterprise, including James Freeman, Ralph Johnson and Douglas Walton define themselves also as informal logicians or have adopted that label at a certain point in their career.

gesting a dialectical view of argument evaluation, but ultimately the evaluation of an argument is deferred to criteria that belong to the different *fields* of argument. Perelman substitutes validity with the persuasiveness of the argument for a given audience, and escapes the equation of reasonableness with mere effectiveness by introducing the notion of a *universal audience*. While philosophically interesting, the universal audience does not provide much in the way of actual guidance to the task of evaluating the reasonableness of arguments.

About a decade later, Australian logician Charles L. Hamblin, with his influential book on *Fallacies* (Hamblin 1970), suggested a different perspective on the relationship between dialectic and logic and, indirectly, on the way in which the criticism of ordinary arguments might be carried out. In this view the dialectical study of arguments is more an extension of their logical treatment than an alternative to it. Hamblin moves from a criticism of what he considers the traditional definition of fallacies: arguments that seem to be valid but are not.[5] He observes that fallacies cannot be consistently seen as logically invalid arguments. Of some, like, for instance, the use of threats (*argumentum ad baculum*), one could be inclined to say that they are not arguments at all. Others, like begging the question (*petitio principi*), are formally valid arguments. The *straw-man* fallacy discussed earlier is also interesting in this respect: it is not in itself an invalid argument, but one which "refutes" a proposition other than the actual standpoint of the other party. Hamblin argues that a unified treatment of fallacies can be developed by examining their role as moves in a dialogue and introduces the idea of formal dialectical systems to model the rules of certain types of discussion.

The idea of *commitment* to the truth of a proposition and the modeling of its dynamics plays a central role in formal dialectical systems. In these systems the proponent of a thesis is bound to prove it *ex concessis* – that is deriving it from the propositions to which the other participant in the dialogue (the opponent) is committed. Both the proponent and the opponent are bound to keep the consistency of their commitments. If the thesis is entailed by the commitments of the opponent, the opponent must either accept the thesis or retract some of the commitments. If the thesis is found to be inconsistent with one of the commitments of the proponent, either the thesis or that commitment must be retracted. Hamblin models these dynamics via the notion of a *commitment store*: a set of statements containing the commitments of each participant, which is updated at each move in the dialogue – a notion closely related to the concept of *common ground* developed in pragmatics and dynamic semantics.[6] One can see how the concepts of *commitment* and *commit-*

[5] Cf. Hamblin (1970: 12): "A fallacious argument, as almost every account from Aristotle onwards tells you, is one that seems to be valid but is not so". Here I am interested in highlighting the developments that Hamblin's work sparked in argumentation theory rather than in the accuracy and fairness of his assessment of previous views of fallacies. There are in fact reasons to think, with Hansen (2002), that Hamblin's summary of the traditional conception quoted above is far from being accurate.

ment store can be used to evaluate "informal fallacies" like the straw man: by attacking a straw man the critic attributes to the arguer commitments that are not really part of the commitment store.

Hamblin's work on the dynamics of commitment in idealized argumentative dialogues and other works in the same *formal dialectic* vein were important sources of inspiration of *normative pragmatic* theories of argumentation that currently dominate the scene of argumentation theory. This orientation was inaugurated by Frans van Eemeren and Rob Grootendost (1984, 2004) who proposed a theory (Pragma-Dialectics) that seeks to bridge the gap between formal dialectical systems and real world discussions by integrating Searlean speech-act theory and Gricean pragmatics with a normative dialectical model. Van Eemeren and Grootendorst (2004: 21) regard argumentation as "part of an explicit or implicit discussion between parties who try to resolve a difference of opinion (which may be implicit) by testing the acceptability of the standpoints concerned".

The working definition of argumentative discourse given earlier in the *Introduction* (§ 2.1) corresponds to a discourse that is intended as a relevant, cooperative contribution to such a discussion. To the extent to which the participants have committed themselves to the goals of this *critical discussion* they incur in a number of entailed *dialectical obligations* depending on the role they take in the discussion and the current stage of the discussion. These obligations are spelled out informally in a set of rules that define a code of conduct for reasonable discussants.

It is not possible here to go through the full set of rules proposed by Pragma-Dialectics. It is useful however to select four dimensions of evaluation which are particularly revealing of the normative pragmatic stance in general: *freedom*, *responsibility*, *acceptability* and *relevance*. These dimensions of evaluation are not discussed as such in the pragma-dialectic literature, but each corresponds to one or more rules in the code of conduct for reasonable discussants.

The first dimension is the *freedom* for arguers to put forth standpoints and the corresponding freedom for critics to cast doubt on these standpoints. Tactics such as the *argumentum ad baculum* are fallacious in that they hinder a reasonable discussion by preventing the parties to present standpoints or raise doubts about them. With the freedom of advancing standpoints comes the dialectical *responsibility* that is associated with it: arguers by advancing a standpoint commit themselves to supporting it with arguments if asked to do so, that is to carry the *burden of proof*. This

[6] The notion of a commitment store and its update dynamics in Hamblin's dialectical system should ring familiar to readers acquainted with the work on dynamic semantics in logic and linguistics as well as with formal treatments of pragmatics. Certainly the commitment store bears a certain similarity with the notion of *common ground* as devised by R. Stalnaker in his works on presupposition and assertion, as well as with the notion of *conversational scorekeeping* as presented in David Lewis. As we will see in Chap. 4, the treatment of modality in the relative modality theory includes a comparable notion of "store" – the *conversational background* – which is amenable to a dynamic interpretation.

obligation can be seen as stemming directly from the felicity conditions of the assertive speech act of advancing a standpoint.[7] Conversational moves such as (2) are fallacious in that they violate this commitment:

(2) A: Journalists get more subjective every day.
B: How come?
A: Well, maybe you can explain that they are as objective as they used to be! (from van Eemeren et al. 2003: 282)

Arguments that are presented in support of a standpoint are evaluated along two dimensions: their *acceptability* and their *relevance*.[8] Acceptability pertains to the *material* aspect of the argument, while relevance pertains, so to say, to the *formal* one. Acceptability and relevance can be considered both from an *epistemic* and from a *dialectical* point of view. I borrow the distinction between the two perspectives from Freeman (2005).

From a dialectical viewpoint, considering the acceptability of an argument (or premise) means asking whether its propositional content is already part of the common ground, that is it is accepted as true – or at least as likely, expected, probable or plausible– by the discussants. From an epistemic viewpoint, a premise's acceptability corresponds to its being based on an adequate "belief generation mechanism" (Freeman 2005).

Apart from being plausible, arguments need to be *relevant* to the standpoint they support, so that the standpoint can be inferred from the arguments as a conclusion. This is possible if there is an argument scheme, corresponding to a deductive rule or to some other kind of rule of inference (e.g. inductive or abductive) that can be applied to the premises to obtain the standpoint as a conclusion. From a dialectical viewpoint, the relevance of an argument is guaranteed when there is an *intersubjectively accepted* argument scheme or inference rule that can be applied to the premises to obtain the standpoint as a conclusion. From an epistemic viewpoint, the relevance is ensured by the applicability of an argument scheme or inference rule capable of *reliably generating correct conclusions from correct premises.*

It is important to stress that both epistemic and dialectical considerations are necessary in the evaluation of the acceptability and relevance of arguments. In Pragma-Dialectics this idea is expressed by considering *both* "problem-solving validity" and "conventional validity" as necessary for an argument to be considered reasonable. Van Eemeren (2010: 1) characterizes argumentation as "basically aimed at resolving a difference of opinion about the acceptability of a standpoint by making an appeal to the other party's reasonableness". It is interesting to observe here,

---

[7] See also Houtlosser (2002: 171) who sees the obligation to defend as the "essential condition" of the speech act of "advancing a point of view": "Advancing a point of view counts as taking responsibility for a positive in respect of O [= the "formulated opinion" making up the propositional content], i.e. assuming an obligation to defend a positive position in respect of O if requested to do so."

[8] Freeman (2011: 2) speaks here of "premise acceptability" and "connection adequacy".

following Rigotti's (2011: 268) remarks on conventional validity in Pragma-Dialectics, that such an appeal to reasonableness cannot be successful if "the acceptance takes place without the scrutiny of the other's reason even if the argument is in itself valid". In other words: the critic cannot reach a truly reasonable assent without actually inferring the standpoint from premises that she accepts following an inference procedure that she accepts. Thus, conventional validity is a necessary condition of the reasonableness of arguments, even if it is not a sufficient one.[9]

The brief discussion above should be sufficient to appreciate how a normative pragmatic perspective, rather than representing an alternative to the purely logical analysis of arguments, should be seen as an extension: a more encompassing view of the reasonableness of argumentation as a communicative activity aimed at resolving a difference of opinion on the merits.[10] Firstly, a normative pragmatic perspective enables us to consider issues that go beyond the relationship between premise and conclusion by addressing the so-called *informal fallacies* on essentially Gricean grounds: fallacies are uncooperative moves, which hinder the accomplishment of the goal to which both participants in an argumentative dialogue are committed (i.e. the resolution of a difference of opinion on the merits). Secondly, a normative pragmatic perspective provides a framework for addressing issues of intersubjective, conventional validity as regards the acceptability of premises and the inferential procedures that are intended to guarantee the relevance of the arguments.

Let us consider more closely the first level, by examining the grounding of dialectical rules in pragmatics. It is indeed possible to see fallacies as a breach of the Cooperative Principle (CP), demanding that participants make their contribution "such as is required, at the stage at which it occurs, by the accepted purpose of the talk exchange" (Grice 1975: 45). In order to do so, however, one needs to take very seriously Grice's reference to "the accepted purpose of the talk exchange" in his formulation of the principle and pursue it "in new ways", as suggested by Herbert Clark (1996: 140) in an influential book on language use as joint action. As poignantly observed by Clark (1996: 141–142, 146), Grice does not really develop a theory of where the common purpose and the idea of contributing to the common purpose are central, instead he translates the CP in a series of maxims that are "rules of thumb" largely independent from the purpose of a conversation. Apart from Clark's own approach to conversation as guided by "joint projects", the idea of purpose-specific cooperation is pursued by a family of pragmatic theories, notably including Mann (1988, 2002), that see conversational cooperation in terms of differ-

[9] On the insufficiency of conventional validity see van Eemeren (2010: 137): "Granting that 'conventional validity' based on intersubjective agreement is indeed a prerequisite for reaching a conclusive judgment concerning the acceptability of argumentative moves, I would like to emphasize that, because of its overriding importance determining their 'problem-solving validity' should come first".

[10] According to van Eemeren (2010: 53) in Pragma-Dialectics, as well as in other dialectical approaches "the logical perspective is incorporated in the dialectical perspective".

ent frameworks, or *dialogue games*, defined as sets of *joint goals*, *joint commitments* and of the "dialogical obligations" that derive from them.

I believe that the Pragma-Dialectical notion of a *critical discussion* can be assimilated to one such dialogue game, defined *solely* by the joint goal of resolving a difference of opinion on the merits, by the sub-tasks that are functional to this overarching goal (e.g. making the difference of opinion mutually clear, establishing the relevant common ground, etc.) and by the dialogue obligations entailed by the goal and subtasks.

Regarding this view of the critical discussion as a dialogue game there is, however, an important caveat to be made. Treating the critical discussion as a dialogue game is, I believe, fully consistent with the way in which this notion is defined and employed by the proponents of Pragma-Dialectics as long as we maintain that it represents an *idealization* and not a *description* of a culturally shared and institutionally sanctioned "activity type" (Levinson 1979). Indeed, as amply illustrated by van Eemeren (2010: 130–162), there are many socially prominent activity types (e.g. a criminal trial) that inherently involve sets of commitments that *closely match* those of a critical discussion. But these concrete activity types always involve *additional rules* that further constrain the allowable contributions of the participants and that may occasionally deviate from what is optimal in view of a critical discussion. Other normative pragmatic approaches to argumentation (Walton and Krabbe 1995, Walton 1998) have taken a different route, refusing to identify the normative dimension of argumentation with *one* idealized framework of dialogical cooperation. Instead, they define several different, but still quite abstract, "dialogue types" (e.g. *persuasion, negotiation, deliberation, inquiry*, etc.) each providing its own normative framework.

This alternative direction is not pursued in this study and a unity of the defining features of argumentation at a dialectical and pragmatic level is maintained. While the conceptual analysis of the defining features of dialogue games such as *negotiation* or *deliberation* at a middle level of abstraction is certainly worth pursuing alongside the ethnographic description of concrete activity types, this is not necessarily an alternative to defining argumentation as its own, highly abstract, dialogue game. In fact, one can maintain that wherever argumentation is functional to the joint goals of negotiation or deliberation dialogues it means that a game of critical discussion is being embedded within these games. This option allows for the possibility that not everything in a negotiation or deliberation is necessarily argumentative and frees argumentation theory proper from the burden fully accounting for these games. In Sect. 2.4, I will briefly discuss the importance of considering the actual contexts of interaction in which argumentation takes place in terms of concrete *activity types*.

2.1.3 The Inferential Dimension

Not only logic is compatible with normative pragmatics, it is also necessary to it. Since the correct application of rules of inference that are not only intersubjectively accepted but also epistemically reliable is essential in view of the problem-solving validity of an argument, logical considerations remain a necessary component of a normative pragmatic view of argumentation.

In this respect it must be said that the landscape of logic has changed since Toulmin and Perelman raised their allegations of irrelevance to deductive logic. As observed in the introduction there are several lines of research within formal logic proper and in the neighboring fields of artificial intelligence and formal (linguistic) semantics that have blurred the clear cut line that the ideal founders of modern argumentation studies had drawn between formal, deductive, necessary arguments on the one side and substantial, non-deductive, non-necessary arguments on the other side (cf. van Benthem 1996, 2009, Woods 2010 for a survey). One of these developments, as van Benthem (1996, 2009) also reminds us, is represented by a series of related approaches to non-demonstrative reasoning, identified by labels such as *non-monotonic logic, defeasible logic* and *default logic* (cf. Woods 2010), which are intertwined with the development of commonsense reasoning models in artificial intelligence (Pollock 1987, 2010). As explained in the *Introduction* (§ 1.6.2), these logics remain mostly outside the scope of the present investigation.

Apart from non-monotonic logics, van Benthem (1996) mentions two other developments contributing to reduce the gulf between logic and argumentation. One is the – then fledging – interest of formal semantics for the structure of discourse, which later would give its mature fruits in the form of linguistically realistic semantic treatments of "rhetorical relations" between sentences in discourse and dialogue (Asher and Lascarides 2003) – and the "dynamic turn" in logic and semantics corresponding to a shift from a "declarative" approach focused on "static truth-conditions" (van Benthem 1996: 37) to a "procedural" one that sees utterances and discourses as denoting changes in human information states or in the common ground of the communicative interaction. On the one hand, the semantic work on discourse structure is important because it begins to bridge the gap between the sentence level with which logic an semantics have traditionally been concerned with the complex texts making up real arguments. This work can be seen as convergent, at least in principle, with the work done in argumentation theory on the *macrostructure* of arguments (Freeman 1991, 2011; Soneck Henkemans 1992, see Sect. 2.4 below). On the other hand, the dynamic view of discourse is interwoven with the development of formal dialectic, with its notion of an updated commitment store, and, indirectly, to normative pragmatics. Both these lines of development have had an influence on the present book.

In fact, the path taken by this book curiously reflects this story of estrangement and successive rapprochement between logic and argumentation theory. As we will see in greater detail in Chap. 3 when Toulmin (2003 [1958]) discusses the intimate relationship between modals and moves in an argument, he sketches an analysis of the context dependence of modal meanings and establishes a link between the different kinds of modality and what he calls the different *fields* of argument. Then the analysis of modality features prominently in Toulmin's criticism of formal logic. For Toulmin, formal logicians, with their exclusive focus on analytic criteria, have disregarded the field dependence of the criteria for possibility and necessity considering instead only one kind of possibility – consistency of the claim with the data – and one kind of necessity – logical consequence of the claim from the data – which are, in themselves useless "to show us that any conclusion we shall be concerned in practice is genuinely possible, impossible or necessary", at least as long as "substantial" analytic arguments are concerned (Toulmin 2003 [1958]: 157). In this book I show – ironically perhaps– that semantic tools largely developed in the formal semantic tradition, building on those very notions of consistency and logical consequence, can go a long way in explaining the context dependency of modals and the different flavors they can acquire, providing at the same time the tools for understanding their relationship with the developing common ground of a discussion and their contribution to the signaling of the discourse relations that make up the macrostructure of argumentative texts.

This rapprochement of logic to concerns that are closer to those of argumentation theory combines with the work that has been done within argumentation theory on the rules of inference supporting ordinary arguments. This area of research often goes under the label of *argumentation schemes* (Walton 1996, Walton et al. 2008), where the term *scheme* refers to the rules of inference, while other authors approach these concerns drawing on the rich Ancient, Medieval and Renaissance tradition of the *topics* and use the notion of a *topic* (alternatively, τόπος, 'topos' or *locus*) as their main theoretical construct (Rigotti 2009; Rigotti and Greco-Morasso 2010).

This strand of inquiry overlaps only in part with the concerns of logical research for non-deductive reasoning. It is true that often argument schemes are presented as non-deductive rules (be they inductive, abductive or presumptive), what is even more characteristic of them, however, is that they are not formulated in terms of abstract logical forms, but at a more concrete level, involving a rich inventory of semantic-ontological relations on which the inference is based. One speaks, for instance, of arguments *from cause to effect* or *from parts to wholes* and so on. Katzav and Reed (2004) call "relations of conveyance" those relations between facts such as *cause*, *part*, *class membership* on which an argument scheme can be based. Rigotti and Greco-Morasso (2010) simply call these relations *loci*, recovering one of the readings of the *topos/locus* in the Medieval tradition.

In the wholly informal account of Perelman and Olbrechts-Tyteca ([1958] 2008) these more concrete ontological relations on which ordinary arguments are based seem to stand *in lieu of* schemes of inference based on logical form. This appears clearly when the authors contrast "quasi-logical arguments" with the "arguments based on the structure of reality", which include, for instance, arguments based on causal relations: while the former are based on their relationship with "logical and mathematical formulas" the latter exploit the structure of reality to establish a "solidarity" between the standpoint and what is accepted by the audience (cf. Olbrechts-Tyteca [1958] 2008: 351). In contrast, in recent approaches to argument schemes (cf. for instance Walton et al. 2008, Rigotti 2007b, 2009, Rigotti and Greco-Morasso

2010) it often appears that schemes, while described at the level of more concrete semantic-ontological relations, *do* have an underlying logical form, even though not necessarily a deductive one.

In this book I will refer to the *Argumentum Model of Topics* (AMT) and its notion of *locus* (Rigotti and Greco-Morasso 2010), which takes into account both the semantic-ontological content of argument schemes and their logical form. Considering both levels will be essential for a clear and satisfying account of the argumentative role of the vast range of modalities which do not correspond to logical modalities nor to epistemic and doxastic modalities. In short, of those modalities that Toulmin considered guiltily neglected by formal logicians. As I have anticipated above, the nature of the account will be in many ways quite unlike what Toulmin had in mind.

2.2 Argumentative Analysis

2.2.1 Analysis as a Precondition of Critical Evaluation

The critical evaluation of an argumentative discourse presupposes its interpretation. As Walton (2007: 247) puts it, the possibility of criticism rests on the assumption that "a natural language text or discourse can be interpreted by the would-be critic as expressing some clearly attributable argument or set of assertions". In fact, the kind of normative model outlined in the previous sections – that is a normative pragmatic framework integrated with an account of the inferential dimension based on the topics – needs to be fed with rich and subtle information about the meaning of the discourse. It presupposes that the meaning of the discourse has been made explicit and transparent to an extent that is far from trivial.

Ideally, the result of the analysis should represent *explicitly* all those aspects of a discourse that are relevant for its evaluation as an attempt at effecting the reasonable acceptance of a standpoint, *filtering out* irrelevant features and *integrating* relevant information implicit in the discourse. Using a metaphor from early transformational grammar, van Eemeren and Grootendorst (2004: 95), see the results of this analysis as the "argumentative 'deep structure' of the discourse or text".[11] The term they use to refer to this "deep structure" is *reconstruction* (cf. also van Eemeren, et al. 1993) The term reconstruction evokes the idea that the analysis, in order to enable evalua-

[11] It is tempting to compare the representation resulting from an argumentative reconstruction with the traditional idea of a *logical form* in formal logic and its cognate notions in linguistic semantics and pragmatics (e.g. LF, semantic representation, propositional form). In fact, this comparison would draw us into one of the debates that marked the emergence of argumentation theory as the nature of logical form and the extent to which it is relevant for assessing the reasonableness of arguments is one of the central issues discussed in Toulmin (1958). As it is well known, Toulmin rejects formal validity as a "working" criterion of reasonableness and recasts "logical form" as proper "formalities" along a jurisprudential analogy.

tion, must go beyond the description of what is available at the surface of discourse or those units and categories that are immediately recognizable by the participants. In Pragma-Dialectics the term *reconstruction* also relates to the degree of idealization inherent in this kind of argumentative analysis, which aims to reconstructs a discourse, to construe it, "*as if it were* a critical discussion" (van Eemeren et al. 1993: 38).

2.2.2 What an Argumentative Analysis Needs to Capture

In discussing argumentative reconstruction I will focus on the inferential dimension of argumentation, which is the one primarily considered in the present investigation of modality in arguments. With respect to this dimension, one has thus to consider what kinds of information about the meaning of a discourse we need to have specified in order to evaluate the *relevance* and *acceptability* of arguments that are presented in support of a standpoint.

This information will include, first of all, (I) the fact that there is a standpoint that is argued for in the discourse, which represents a precondition to reconstructing the text or discourse as argumentative in the first place. Then, one needs to know (II/a) what is the propositional content of this standpoint. Most theories of argument acknowledge that there are different semantic types of standpoints (II/b). For instance, Pragma-Dialectics (cf. van Eemeren 2010: 2, 15, 235) distinguishes between *descriptive*, *evaluative* and *prescriptive* standpoints. As it will clearly emerge later in this chapter, such semantic distinctions are relevant to argumentation as the different types of standpoint require different kinds of evidence to support them.

Once the standpoint has been established, one needs to ascertain (III) what utterances in the text are presented as arguments, and (IV) specify the propositional content of the argument. Most of the arguments that we encounter in real discourse are complex, involving several utterances that are ultimately intended to lend support to the standpoint. It is not possible to assess the acceptability and relevance of the complex argument as a whole without working out (V) how its components fit together in supporting the standpoint.

This means that we need to know whether (V/a) each premise presented in the text is taken as independently grounded in the existing common ground or represents and intermediate standpoint whose acceptability depends, in turn, from other premises supporting it *(subordinative argumentation)*. Furthermore, we need to know (V/b) whether the premises supporting the standpoint or an intermediate standpoint are taken to be independently relevant for supporting it (*multiple argumentation*) or are intended to be relevant only if taken jointly (*compound argumentation*). This is what argumentation scholars call *argumentation structure* (Snoeck Henkemans 1992) or, to contrast it with the propositional logical forms that are the traditional province of logicians, *macrostructure* (Freeman 1991, 2011). Until the basic questions of macrostructure are resolved, it is not possible to address finer points of inferential *relevance* or to meaningfully compare the argument with what is *accepted* in the common ground of the discussion.

Consider the following basic argumentative text taken from a media release issued by American financier Carl Icahn in 2007:

(3) Carl C. Icahn today announced that Institutional Shareholder Services (ISS) has recommended that stockholders of Motorola, Inc. vote their shares FOR the election of Mr. Icahn to the Motorola board of directors on the GOLD proxy card at Motorola's May 7, 2007 Annual Meeting of Stockholders. ISS, one of the nation's leading independent proxy voting advisors, provides voting recommendations to more than 1,500 major institutional funds, mutual funds and fiduciaries. (Business Wire, April 25, 2007)

At the time Carl Icahn, who was a Motorola shareholder had decided to apply for a seat in the Board of Directors of the company and was campaigning among shareholders to get voted to the post at the 2007 Annual Meeting of Shareholders of Motorola. Given this contextual information it is clear that the content of the press release is being offered in support of an implicit standpoint, which will count as advanced in the reconstructed "deep structure" of the argument even if the act of advancing the standpoint is not performed explicitly in the text (I). The propositional content of this standpoint (II) needs to be unembedded from the advice of ISS. We can reconstruct it as (4):

(4) Motorola shareholders should vote for the election of Carl Icahn to the board of directors at the 2007 Annual Shareholder Meeting of Motorola.

Two explicit arguments are presented in the text in support of this standpoint (III, IV). The arguments, reproduced in (5) and (6) simply correspond to the two sentences of our excerpt, minus the metapragmatic formula "Carl C. Icahn today announced that" (cf. Jacobs 2000).

(5) Institutional Shareholder Services (ISS) has recommended that stockholders of Motorola, Inc. vote their shares FOR the election of Mr. Icahn to the Motorola board of directors on the GOLD proxy card at Motorola's May 7, 2007 Annual Meeting of Stockholders.

(6) ISS, one of the nation's leading independent proxy voting advisors, provides voting recommendations to more than 1,500 major institutional funds, mutual funds and fiduciaries.

How do these two premises fit together (V) in the argument? Neither seems apt to support the standpoint on its own. So, they are not *multiple* independent arguments. At the same time, they do not seem to stand in an inferential premise-conclusion relation among themselves: the fact that 'ISS provides voting recommendations to more than 1,500 institutional investors' in not a reason for concluding that *then* 'ISS has recommended to vote for Icahn'! Clearly, the macrostructure is not *subordinative*. Rather, the argumentation is *compound* as the fact that ISS did advise to vote Icahn (5) and the fact that ISS is an authoritative advisor (6) need to be taken together to produce a compelling reason in support of the standpoint. This compound macrostructure can be represented as in Fig. 2.1.

Here I follow Palmieri (2014: 38–40) in adopting a style of diagramming the macrostructure of arguments which takes inspiration from Pragma-Dialectics (van

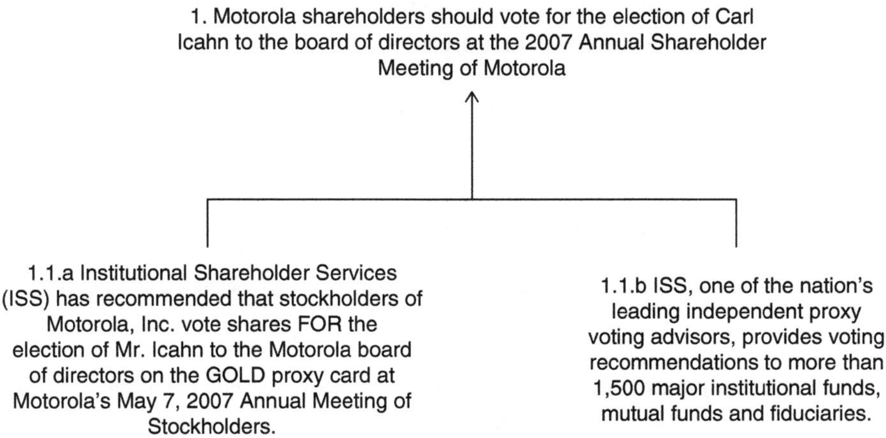

Fig. 2.1 Argumentation structure as a first step towards the reconstruction of the inferential dimension

Eemeren et al. 2002: 63–73).[12] The notation is straightforward. Standpoints are placed on the top of the diagram and receive a number. When only one argument is being analyzed this number is 1. Arguments supporting the standpoint are signaled by appending a number to the one indicating the standpoint, separated by a "decimal" sign (e.g. 1.1). When *multiple* arguments are offered in support of a standpoint they will be indicated as 1.1, 1.2, 1.3, etc. In *compound argumentation*, the linked premises will be indicated as 1.1a, 1.1b, etc. Upward pointing arrows represent and inferential move from premises to conclusion. Since in compound structures premises are taken jointly to infer the conclusion only one arrow joins them to the standpoint. Multiple structures, in contrast, will feature multiple arrows connecting each argument to the standpoint.

Once macrostructure is in place we need to know (VI) how exactly premises are taken to be – jointly or independently – inferentially relevant for the conclusion. In other words, we need to work out, for each arrow, what kinds of loci, argument schemes, inference rules are invoked. At this level, it will be also necessary (VI/b) to make explicit any contextually relevant tacit premise the inference relies upon.

[12] A partially different style of diagramming is offered is proposed by Freeman (1991, 2011) and adopted by Vorobej (2006). There are underlying subtle theoretical differences between Freeman's approach to macrostructure and the one advocated by Pragma-Dialectics, which I cannot discuss here. They are elaborated at length in Snoeck Henkemans (1992, 2001) and Freeman (2011). The view of macrostructure adopted in this book is the one presupposed by the Argumentum Model of Topics. As Palmieri (2014: 38–40) explains, the AMT view is inspired by Pragma-Dialectics but differs from the standard pragma-dialectical treatment in one area. In AMT compound structures, such as the one in Figure 2.1, are considered a special case of single argumentation where more than one premise is made explicit at the surface of the discourse. Freeman (2011: 89–128) would call these premises *linked.* In contrast, the standard pragma-dialectic analysis sees *coordinative* arguments as separate arguments pooled together for reasons of sufficiency. Freeman (2011) would call the latter *convergent.* In fact, AMT analyses *do not* postulate convergent structures.

Table 2.1 Required reconstruction in view of the evaluation of relevance and acceptability

I	Recognize the advancement of a standpoint
II	Determine the propositional content of the standpoint (II/a) and its semantic type (II/b)
III	Recover the utterances that function as premises in the discourse
IV	Determine the propositional content of the discursive premises
V	Determine argumentation structure, including subordinative (V/a), multiple and compound structures (V/b).
VI	Recognize the argumentative *locus* licensing the inference (VI/a) and reconstruct the full inferential configuration, including any implicit premise (VI/b)
VII	Determine how strong a case the argument is making for the standpoint in the arguer's view.

Last, but certainly not least in the context of this book, we need (VII) to establish *how strong a case* the premises make for the standpoint according to the arguer. Roughly speaking, the actual strength of support of a standpoint is a function both of the strength of the inferential link and of the epistemic status enjoyed by the premises. Thus, in evaluation, the reasonableness of the epistemic/ doxastic attitude expressed or implied by the speaker towards the standpoint will have to be measured against the strength that is warranted by these two factors. Table 2.1, below, summarizes the requirements that the critical evaluation of relevance and acceptability imposes on argumentative reconstruction.

2.2.3 Using AMT to Reconstruct the Inferential Configuration of Enthymematic Arguments

Developing the treatment of level VI, let us briefly examine how the *Argumentum Model of Topics* approach (AMT) can be used to represent the inferential configuration of an argument expressed in ordinary discourse with a degree of explicitness sufficient to evaluate it in terms of *acceptability* and *relevance*.

The aim of this presentation is chiefly expository. For a full discussion of the merits of AMT see Rigotti and Greco Morasso (2010), which also includes a comparison with Ancient and Medieval views, as well as with competing modern theories of argument schemes. As remarked above, in order to fully appreciate the contribution of modalities to the essential structure of arguments we need to have in place a theory of argument schemes taking into account the semantico-ontological makeup of arguments at a level that is less abstract than the mere logical form, and I believe that AMT fits the bill. I do not claim that there aren't other theories that could be used. My presentation broadly follows the lines of Palmieri (2014), which contains a clear, concise and up-to-date presentation of the model. The differences reflect a still ongoing conversation about AMT with the proponents of the theory and with other researchers.[13]

[13] In particular, my understanding of AMT benefited from extended exchanges with Eddo Rigotti, Sara Greco Morasso, Rudi Palmieri and Elena Musi. None of these people, however, should be held responsible for the views expressed here.

2.2.3.1 Enthymemes and *endoxa*

An AMT analysis moves from the assumption that arguments expressed in ordinary discourse are *enthymemes*, both in the so-called "standard definition" that we inherit from the tradition and in the more authentically Aristotelian sense. The standard definition of an enthymeme, in fact, dates backs to ancient authors such as rhetorician Quintilian:

> [...] the incomplete syllogism known as an *enthymeme* [...] Some call the *enthymeme* a *rhetorical syllogism,* while others regard it as a part of the *syllogism,* because whereas the latter always has its premises and conclusion and effects its proof by the employment of all its parts, the *enthymeme* is content to let its proof be understood without explicit statement." (Quintilian *Institutio Oratoria* V, 14, 183 and 194)

In the late Antiquity this definition was adopted by Boethius and was inherited by medieval logic finding its way into modern logic textbooks. Of this definition AMT retains the idea that the manifestation of ordinary arguments on the surface of discourse is incomplete and their interpretation as arguments requires the recovery of *implicitly communicated propositions.*

Implicit premises are a much debated issue in argumentation theory and the very idea of propositions that are not linguistically manifested but should be counted as *premises* appears almost self contradictory to some argumentation scholars (cf. in particular Hitchcock 1985). In contrast, from the viewpoint of (post-) Gricean pragmatics implicit premises become simply a special case of *implicature* that is of propositions that do not receive linguistic manifestation in discourse but count as communicated by relying on the interpretive inferences of the hearer based on assumptions of cooperation between communication participants. This means that tacit premises are inevitably *inferred premises*, as pointed out in Rocci (2006). In every enthymematic argument, in fact, there are two intertwined inferential processes going on: (1) the inference that is communicated by the arguer to the audience (*communicated inference* or *argumentative inference*) and (2) the inference that the audience is invited to draw in order to fully recover the pragmatically implicated part of the argument and understand the argumentative inference. This second inference can be called *communicative* or *pragmatic* inference. Implicit premises are premises of the argumentative inference but they are conclusions of the pragmatic inference (for an in-depth discussion see Rocci 2006: 421–424).[14]

[14] While Rocci (2006) contains a more extensive notes and examples of the distinction between argumentative inference and pragmatic inference, it does not provide a full treatment of pragmatic inference seen in connection to argumentation. Such an endeavor has not, to my knowledge, been attempted yet. Should it be attempted, one would have to choose between several theories of implicature, differing in theoretical assumptions, degree of formal explicitness and breadth of empirical coverage. One family of approaches, however, does appear promising and broadly compatible with the semantico-pragmatic approach to argumentative discourse expounded in this chapter. These approaches explain (most) implicatures as resulting from the abductive accommodation of discursive presuppositions at various levels. See, in particular, the foundational paper of Thomason (1990).

Implicitly communicated propositions in enthymematic arguments are not limited to premises. Implicit conclusions/ standpoints are also common in ordinary discourse (cf. Rocci 2006: 422–424). The press release in (3), analyzed above in Sect. 2.2.2, is a case in point: two premises are made available in the text but the intended conclusion they license is not explicitly stated. In this case, argumentative inference and pragmatic inference seem to run parallel towards the conclusion.

There is a wide consensus among scholars of Aristotelian rhetoric (cf. Piazza 2000: 145–151) that the implicit/incomplete nature of the deductive structure (i.e. the "standard definition") is *not* the defining characteristic of the enthymeme for Aristotle. Rather, for Aristotle this is a corollary deriving from a more fundamental property concerning the epistemic status of the premises.

Contrary to the primary truths making up the premises of analytical syllogisms, the premises of enthymemes are not always necessary but can be just *probable* propositions, εἰκότα, *eikota* (from εἰκός = *likely*). In Aristotle *eikota* refers in particular to general statements that hold true *most of the times* (cf. Piazza 2004: 120). These generalizations take the role of major premises in the syllogistic structure of the enthymeme. Furthermore, the status of these premises is specified with respect to the intersubjectivity of the arguers and of the community to which the arguers belong. Enthymematic premises are *endoxa.* In Aristotle the adjective ἔνδοξος (from εν 'in' and δόξα 'opinion' or 'fame') refers to propositions that are in the common opinion and, as a consequence are generally accepted within a community. In the *Topics*, Aristotle gives an articulated definition of the *endoxa*:

> Those opinions are reputable (*endoxa*) which are accepted by everyone or by the majority or by the wise—i.e. by all, or by the majority, or by the most notable and reputable of them." (*Topics* I 100b 21–23, transl. by W.A. Pickard-Cambridge, Revised Oxford Translation), see Aristotle 2014.

Endoxical premises are not always left implicit. If we go back again to the example of Carl Icahn's press release we can see that the premise 1.1.b in (6) can safely be considered an *endoxon*. After all, investors *should* know that ISS is the leading advisor on voting for corporate board posts, or, at least the majority of investors should know, or expert professional investors, or the majority among the expert professional investors, or the most distinguished, like those "major institutional funds, mutual funds and fiduciaries" mentioned in the passage.

If one or more premises on which the enthymematic inference is based can be left implicit it is precisely because they are *endoxa*. As Aristotle puts it:

> The enthymeme must consist of few propositions, fewer often than those which make up a primary deduction. For if any of these propositions is a familiar fact, there is no need even to mention it; the hearer adds it himself. (*Rhetoric*, I, 1357a transl. by W. Rhys Roberts, Revised Oxford Translation, see Aristotle 2014.)

Bitzer's (1959) influential interpretation of the Aristotelian enthymeme sees it as dialogically constructed through the joint efforts of speaker and audience:

> to say that the enthymeme is an 'incomplete syllogism' - that is, a syllogism having one or more suppressed premises - means that the speaker does not lay down his premises but lets his audience supply them out of its stock of opinion and knowledge (Bitzer 1959: 407)

Bitzer's understanding of the enthymeme anticipates pragmatic views of the implicit where interpretive inferences are drawn in a framework of dialogical cooperation and communication takes place on the backdrop of a *common ground* of shared knowledge, belief and values (cf. in particular Clark 1996). The *endoxa* functioning as enthymematic premises, in particular, appear connected to what Clark (1996: 100) calls the *communal common ground*, which dialogue participant infer to be shared thanks to their common belonging to a variety of cultural communities.[15] However, if one keeps Aristotle's definition of *endoxon*, which I have illustrated above with the (explicit) endoxical premise in (6), endoxical propositions cannot be always straightforwardly identified with proposition that are common cultural knowledge. There is also a normative facet to the notion of *endoxon* connected with the social organization of knowledge in a field or community such that endoxicality extends also to those propositions one *should* be ready to accept, given their widespread or authoritative acceptance within the community.[16]

Later on in this book, when dealing with epistemic modalities, it will be clear that the two dimensions along which enthymematic premises are characterized respectively as *eikota* and as *endoxa* are both important for a fine grained analysis of epistemic modal expressions: both the degree of certainty or probability and its subjective or intersubjective nature are relevant for the functioning of a propositional content in arguments.

2.2.3.2 The Procedural Component of AMT: *loci* and *maxims*

An AMT analysis moves from the macrostructural representation of an argument to zoom into each arrow signaling an inferential step from discursively manifested premises to a conclusion. The goal of the analysis is to make explicit how the inferential step is warranted from a relevance and acceptability viewpoint. According to the AMT perspective the functioning of an argument scheme cannot be captured in terms of a single inference warrant licensing the step from the argument to the standpoint. As observed by Rigotti and Greco Morasso (2010: 493), this view contrasts with those of several "modern authors" who "conceive of argument schemes as the whole bearing structures that connect the premises to the standpoint or conclusion in a piece of real argumentation".

Taking inspiration from the pragma-dialectical distinction between the *material starting points* and *procedural starting points* in the opening stage of an argumenta-

[15] Clark (1996: 100) illustrates the notion of communal common ground as follows: "We often categorize people by nationality, profession, hobbies, language, religion or politics as a basis for inferring what they know, believe or assume. When I meet Ann at a party and discover she's a classical music enthusiast my picture of her suddenly expands. I assume she knows everything any such enthusiast would know – and that is a great deal."

[16] One could speculate that *endoxa* deprived of cognitive salience and doxastic force in the audience but still enjoying their normative force in the community have every chance of appearing as *explicit* premises in the enthymeme.

tive discussion, Rigotti and Greco Morasso (2010: 493) propose "to apply the notions of material and procedural starting points to identify the different nature of premises at work in the argument scheme".

The procedural starting point coincides with the abstract inferential license (*maxim*, from the Latin *maxima propositio* 'proposition of greatest extension') activated in the argument scheme, while the material starting point consists of premises guaranteeing the applicability of the maxim to the actual context. These material premises, epistemically and normatively rooted in the context of the argumentative discussion, are the often-implicit *endoxon* – discussed above – and a minor premise, called *datum*, which often represents the only discursively manifested premise of the argument.

The procedural component of AMT, through which the semantic-ontological "relations of conveyance" (cf. Katzav and Reed 2004) of the *loci* are brought to bear on the material premises in an inferential path, can be broken down in three levels:

(I) *Locus*,
(II) *Maxim*,
(III) *Logical Form*.[17]

The Latin term *locus* and the Greek *topos* are based on 'place' metaphor which highlights their role in rhetorical invention. As Cicero puts it:

> 'Places' is the name Aristotle gave those locations, so to speak, from which we can draw arguments (*Topica*, 7, see Cicero 2003)

More directly relevant to the functioning of the procedural component of AMT is the Medieval characterization of the *loci* as *habitudines* ('relations'). Rigotti (2008: 563) highlights how Peter of Spain (XXIII Century) in his *Summule Logicales* defines the loci *From efficient cause* and *From definition*:

> The locus from efficient cause is the relation (*habitudo*) of the efficient cause to its effect. [...]
>
> A definition is an utterance which is meaning the mode of being of a thing. The locus from definition is the relation (*habitudo*) of the definition to the defined. (*Summule Logicales* 5.10)

[17] I will follow rather closely Rigotti and Greco Morasso (2010: 493–498) in the presentation of the first two levels, diverging in the account of the logical form. The whole AMT framework is a theory in development. Some areas have just been sketched but have not yet achieved the desirable degree of explicitness. The sketchy treatment of logical forms in Rigotti and Greco (2010: 495) is not entirely satisfactory. It suggests that logical forms should be characterized in terms of the valid inference rules of classic propositional logic, such as *modus ponens* and *modus tollens* and of deductively invalid forms such as the "abductive" *false modus ponens*. However, if taken literally, this suggestion raises technical issues with the functioning of the 'Y-structure' model detailed in the next section. In short, we will see that, in order to produce the conclusion of the inference, the maxim needs to interact with non-propositional terms – which is simply not possible if we maintain that the form truly is a *modus ponens* or any other propositional rule. Additionally, the choice of extensional rules based on material implication obscures the intensional, "modal" nature of the ontological relations captured by the *loci*.

According to AMT, *locus* is a high level relation in the world's ontology. One can see loci as the high level skeleton of a folk ontology of the metaphysical, physical and social world. Starting from the macrostructure of an argument, the first step of an AMT analysis consists in associating an underlying ontological relation to the inferential move from the argument to the standpoint. Consider the simple example in (7), drawn from Palmieri (2014: 32).

(7) We cannot expand our business because we do not have enough capital.

Starting from the macrostructure in Fig. 2.2, the AMT analysis will associate the arrow to the *locus from the instrumental cause*, thanks to the relation holding between a semantic term of the standpoint (*the action of expanding a business*) and a term of the argument (*capital*): in relation to expanding the business capital is a *means* or *instrument* – an Aristotelian instrumental cause. Schematically:

INSTRUMENTAL_CAUSE ('capital', 'expanding a business')

If we look back at the argument in Carl Icahn's press release (Fig. 2.1) we find a very different kind of relation underlying the inference, taking scope on terms whose semantic type is very different from the kind of terms we find in the *Locus from the instrumental cause.* In this case the relevant relation is that between an authoritative source (ISS) and the propositional content of a discourse ('Stockholders of Motorola, Inc. should vote for the election of Mr. Icahn to the Motorola board of directors'). We have here a *Locus from authority* in its expert opinion variant. The contrast between these two very disparate relations brings us to the issue of the inventory and classification of loci.

Several classifications of loci have been attempted since Cicero by Ancient, Medieval and Renaissance scholars (see the discussion of select examples in Rigotti 2007a, b, 2008, 2013 and a synopsis in Walton et al. 2008: 275–300). Modern authors, starting from Perelman and Olbrechts-Tyteca (2008 [1958]), have also proposed lists and taxonomies of argument schemes, ranging from the very compact pragma-dialectic tripartition in *symptomatic*, *causal* and *comparative* arguments (cf. Garssen 2001), to the extensive "compendium" of 60 argument schemes offered by Walton et al. (2008: 308–346).

The AMT approach relies on a taxonomy which revisits the Ancient and Medieval tradition both in the list of *loci* and in their groupings, highlighting the semantic-ontological criteria that inform this tradition. The version of the AMT taxonomy of

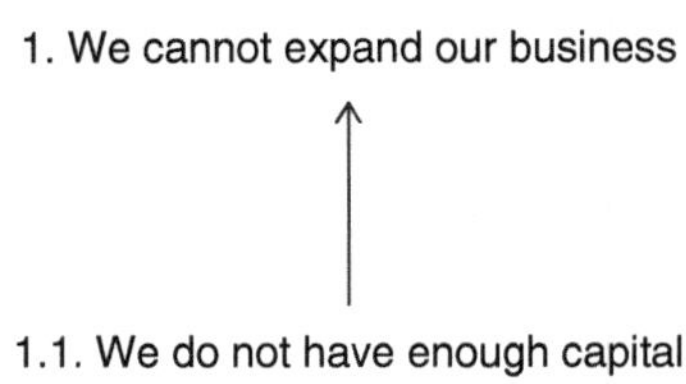

Fig. 2.2 The macrostructure of example (7)

loci displayed below in Fig. 2.3 elaborates on earlier versions in Rigotti (2007a) and Palmieri (2014).

Cicero (*Topica*, 8) was the first to introduce a distinction between *intrinsic* loci which "are attached to the subject under discussion itself" and *extrinsic* ones, "drawn from without", "which stand apart and are clearly dissociated". The first group included "arguments drawn from the whole, from its parts, from a distinctive quality [= nota] , and from those things that are somehow related to the subject at issue".[18] Boethius and Medieval authors maintained the bipartition in extrinsic and extrinsic, sometimes making more explicit the internal subdivision or cline within intrinsic loci from those more closely attached to the substance to those more weakly related to it.

The taxonomy presented here follows this tradition, making its semantic-ontological rationale more explicit. The intrinsic loci include, first of all, a group of *definitional loci*. These cover the relations such as semantic tautology between a predicate and a set of necessary and sufficient traits for its attribution (*From definition*), and a set of semantic entailment relations that modern formal ontologies conceptualize in terms of inheritance of traits between classes (sorts, kinds, categories) and instantiation of classes by individuals (*From genus and species*). To use again a term from formal ontologies, definitional loci do not involve *attributes* relating one object to other things, aspects or parts. This is the province of *circumstantial loci.* On the border between definitional and circumstantial loci stands the *locus From Derivates*, licensing inferences based on the conservation of information across semantic types (e.g. from the desirability of *reason* to the desirability of *reasoned argument*).

From a modern viewpoint, *circumstantial loci* are best understood in terms of Frame Semantics (cf. Fillmore 1984, 2003, Fillmore and Baker 2009, Rocci and Luciani 2016b) as relations between frame elements within a frame, understood as a definition or characterization of a scene, situation or extended state of affairs. These include mereological relations (*From Whole and Parts*), the entailment relations resulting from conversives in a frame (e.g. *teacher – students*) captured by the *locus From Correlates*, the implications of the spatial and temporal coordinates of an event, down to the merely typical or plausible associations (e.g. *wearing mittens – wearing a scarf*) of the *locus From Concomitances.*

A remarkably cohesive grouping of intrinsic loci is represented by those traditionally associated with the Aristotelian causes. These loci are, so-to-say, less intrinsic than the previous two groups. Relations between cause and effect (*Efficient Cause*), between material and artifact (*Material Cause*), between instrument and action (*Instrumental Cause*), between action and goal (*Final Cause*), between design and realization (*Formal Cause*) invoke the ontology of events and thus changes to the world, rather than the mere extension of a situation or scene or its location in space and time. More importantly, in the context of a study of modality,

[18] Here I diverge slightly from Reinhardt's translation of the passage (pp. 119–121). Reinhardt renders *ex nota* as "from etymology".

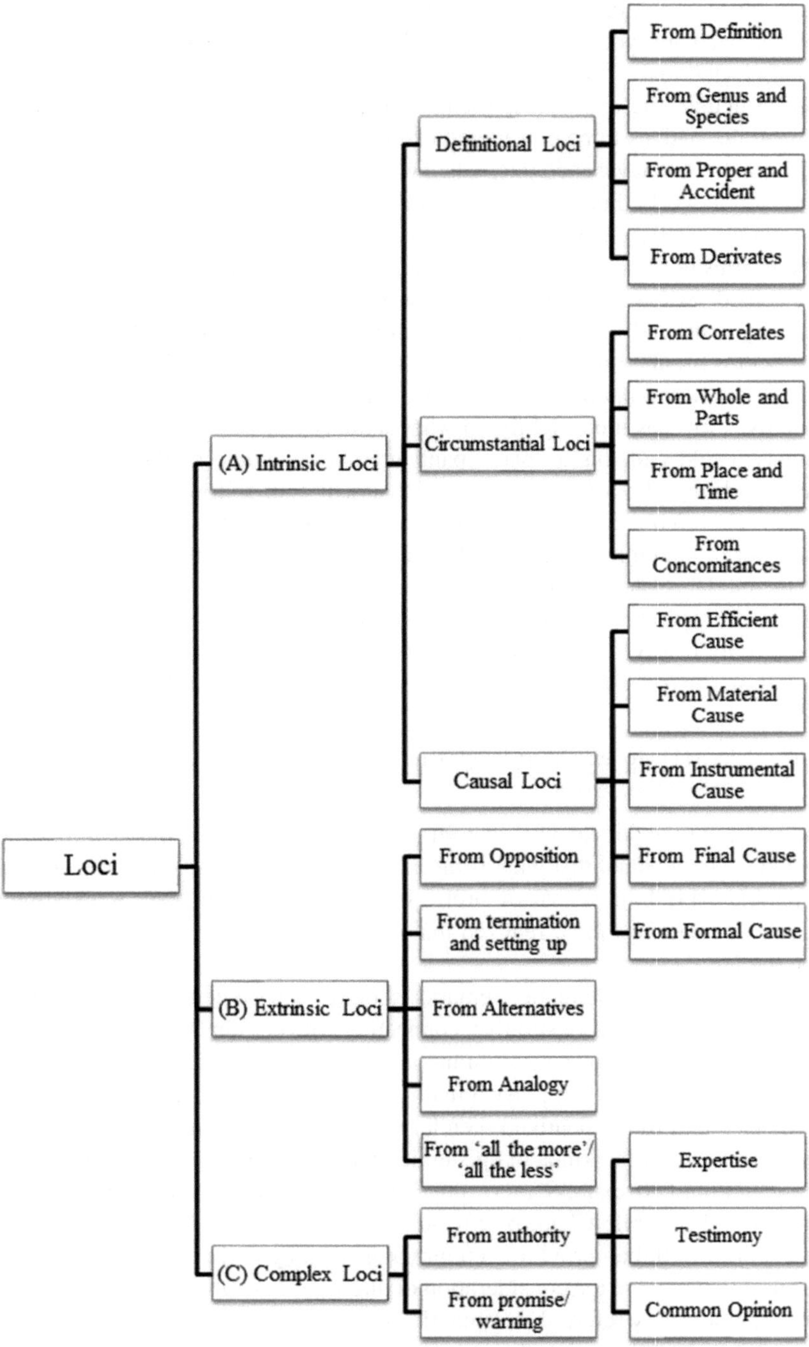

Fig. 2.3 A revised AMT taxonomy of *loci*, from Rigotti (2007a, 2009) and Palmieri (2014), with major adaptations

causal concepts require that human ability of "thinking of alternatives", of situations that need not (yet) to be real, which I have placed at the root of modal concepts in the Introduction.

In particular, the causal loci focus on the world of human actions, where the kinds of alternatives to be considered extend beyond the simpler alethic, ontological, alternatives that we encounter when we conceptualize simple efficient causation in the natural world. For instance, consider that reasoning and talk about action and deliberation involve envisaging a possible, non actual, state of affairs corresponding to the *desires* of the agent, which is usually called *bouletic* modality. This desire is to be considered on the backdrop of what is generally possible in the future, which is a fairly restricted kind of alethic modality– a special kind of *potentia realis* of Duns Scotus – sometimes called by linguists *circumstantial modality* (Kratzer 1981, 2012) or *dynamic modality* (Palmer 2001, Portner 2009). Typically, consideration of alternatives focuses more tightly on what actually *can* be done by the agent (some linguists would call this a "participant internal" *dynamic* modality) and what is *needed* to bring about the desire – which I will call, following Kronning (1996, 2001), a *deontico-practical* modality. If we go back to the modal *cannot* in the standpoint of example (7) above ("We cannot expand our business"), we find that here the modal relates to some kind of *dynamic/circumstantial* impossibility of performing the action. Interestingly, discovering the precise type of circumstantial impossibility involved is one and the same thing as discovering that the arguments draws from the *locus From Instrumental Cause*. Consider another already familiar example:

(8) John had to resign. Otherwise his family would have suffered in the spotlight of the scandal-hungry media.

Here we need to understand to what set of alternatives does the modal *have to* refer. As soon we understand that what is meant is that resigning was the only alternative compatible with John's goal of protecting his family from the "scandal-hungry media", we have *ipso facto* realized that the argument in (8) is based on the *locus From the Final Cause*.

These initial, unsystematic, remarks should be sufficient to suggest that the causal loci have a deep, constitutive, connection with certain flavors of modality that are most prominent in discourse on human actions. In fact, it will turn out that the interaction with causal loci plays a major role in accounting for the argumentative significance and the indicative potential of non-epistemic modal expressions and even of certain specific kinds of epistemic modal constructions. The semantic scaffolding required to express more precisely these intuitions will be provided in Chaps. 4 and 5, while Chap. 6 will offer corpus-based analyses of actual modal constructions in Italian providing empirical grounding to them.

Causal loci involve the consideration of alternative situations or *possible worlds*, as they are normally referred in semantic and philosophical treatments of modality. Given that all causal relations involve the reference to situations that are non-actual from the viewpoint of "the subject under discussion itself", these loci are clearly less intrinsic than the definitional and circumstantial. Yet, the range of possible worlds they take into account is constrained in quite specific ways. Consider, for instance, the relation of cause to effect in the *locus From efficient cause*. Here, the only *accessible* alternatives that need be taken into account are real possibilities to be realized in time given the actual situation, or, in other words, alternative continuations of the history of the world of the actual situation. Counterfactual, synchronic, alternatives to the actual situation are not relevant. To introduce another technical term from possible worlds semantics, there are quite specific *accessibility relations* constraining the alternatives taken into consideration in causal reasoning. While these causal accessibility relations might not be easy to characterize technically[19], we do have an intuitive grasp that they concern, so-to-say, developments of the same world.

These constraints do not hold in the case of extrinsic loci. These cover relations between *alternative situations* which are conceived synchronically rather than as developments one of the other and *comparable situations* in which isomorphism and mapping of features is relevant rather than their belonging to a chain of developments of the same world. The *locus From analogy* is a case in point: the analog may belong to the same world of the situation in the standpoint, to fiction and fable, or to a realm set up solely for the purpose of mental experiment. In every case, what counts is not the transition of a situation into another, but the structural similarities, the "mappings" between the situations.[20] To use a metaphor, in extrinsic loci what "moves" between alternatives is only the mind of the arguers.

Ancient and Medieval classifications often complement the intrinsic and extrinsic loci with a third category of *loci medii* or *mixti*, which should combine aspects of the former two basic categories. While the idea of the combination is attractive, in practice, the loci included in the *mixti* form indeed a mixed bag, for which is hard to envisage a convincing rationale. In AMT the third group, renamed of *complex loci*, has been repurposed to host those loci which have a *pragmatic* rather than a semantic-ontological connection with the standpoint. These include the arguments *From Authority* – with the variants of expertise, testimony and common opinion. As well as the inferences relying on broadly *commissive* speech acts (*From Promise and Warning*). As it will emerge from the empirical investigation in Chap. 6 both these loci have close connection with modality: the former is closely tied to the *reportative evidential* values of modal constructions, while the latter is tied to deon-

[19] Accessibility relations are introduced in Chap. 4 which includes some basic discussion of important formal properties of broad classes of accessibility relations. The issue of a realistic – or at least useful – formal model of world accessibility for causal relations is not even touched in this book, as the discussion of modal semantics in Chaps. 4 and 5 – while using insights from formal semantics – is conducted at what could be described as a level of semi-formal schematization. An attempt at a fully formal model of causality is found in Hobbs (2005).

[20] Pollaroli and Rocci (2015) tries to capture this "mapping" dimension of the *locus* of *From Analogy* through the theories of *Mental Spaces* and *Conceptual Integration* developed by Fauconnier and Turner (2006).

tic modals referring to institutional commitments, the social and legal obligations they generate and the ensuing expectations.

The taxonomy of *loci* presented above has already been used to systematically annotate arguments supporting standpoints introduced by the Italian modal-evidential verb *sembrare* (cf. Eng. *seem*) in Musi (2015) and by the Italian modal-evidential adverbs *evidentemente* and *ovviamente* and their English counterparts *evidently* and *obviously* (Musi and Rocci 2016). I will come back to the results of these corpus based research in the *Conclusions* of this book. A slightly simplified version of the AMT taxonomy has been used in a pilot annotation study of 30 persuasive essays using multiple minimally trained non-expert annotators (Musi *et al.* 2016). More extensive annotation experiments using AMT are underway as I write.

One distinctive feature of the AMT approach is the idea that singling out the locus-relation relevant in an argument is not sufficient, in itself, to reconstruct its inferential configuration. In order to do so it is necessary to move to the level of *maxims*. Maxims are abstract inference warrants in conditional form which hold for the terms of a given locus-relation. For each locus, a bundle of maxims can be conceived. In the case of example (7), the following maxim of the locus *From instrumental cause* is being invoked:

(9) If the necessary means to realize an action are absent, then the action cannot be undertaken.

Each locus gives rise to several maxims. For instance, the same locus *From instrumental cause* also gives rise to the maxims in (10), (11), (12) and (13). Some maxims have the force of necessity, such as (10) and (11) and can be used in deductive reasoning, others, like (12) and (13), could support only a presumptive inference (cf. Rescher 2006: 76–87).

(10) If an action has been performed, then the necessary means have been used.

(11) If an action has to be performed, then the necessary means have to be used.

(12) If an action is *ceteris paribus* desirable, then *ceteris paribus* the use of sufficient means is desirable.

(13) If normally sufficient means have been obtained, then the undertaking of the action is *ceteris paribus* possible.

In (12) the use of the means is desirable only if the means do not produce side effects that offset the desirability of the action (cf. Rigotti 2008 and Palmieri 2014: 34–35 for a discussion). On the other hand, (13) captures the idea that in the real world no instrument is ever *absolutely* sufficient to perform an action. At best, tools can be considered up to the job within a range of normal operating conditions. Knowledge of the maxims can be considered part and parcel of our ordinary grasp of the locus-relation. Knowledge of the maxims (10)-(13) can be considered part of *knowing what an instrument is*. They are, so-to-say, theorems derived from the semantic axioms that define the locus-relation itself. As such, they are independent from cultural knowledge and beliefs concerning specific instantiations of the relation (e.g. knowing what are the proper tools for preparing ravioli). As Palmieri (2014: 34) observes, the maxim in (9) can be instantiated by any of the following concrete arguments, where the roles of *necessary means* and *action* receive quite diverse concrete fillings:

(14) John cannot slice the salami because he does not have a knife.
(15) Mary cannot dry her hair because the blow dryer is broken.
(16) Luke cannot read because he forgot his glasses at home.

An important issue with which theories of topics and argument schemes are faced (cf. Rigotti and Greco Morasso 2010: 503–507) is the gap between the abstractly formulated inference warrant represented by the maxim, which relates means and action, and the concrete terms we find in the explicitly formulated premises. We need to infer about capital and expanding business, knife and slicing salami, but what we have is a warrant concerning instrument and action.

Bridging this gap is the job of the third level of the procedural component of AMT: the *Logical Form.* AMT relies on a complex syllogistic logical form, called *Y-structure*, to bring the maxim to bear on the material component (*endoxon* and *datum*) and derive the standpoint as a conclusion.

2.2.3.3 Mapping the Inferential Configuration of Arguments Using AMT's Y-Structure Diagram

The Y-structure of AMT consists of two syllogistic logical forms. A categorical syllogism combines the two material premises taking the *endoxon* as major premise and the datum as minor. For example (7) Palmieri (2014: 36) reconstruct the following *endoxical syllogism*:

(17) *endoxical syllogism*
Major Premise (*endoxon*): Capital is a necessary means to expand a business
Minor Premise (datum): We do not have enough capital (= capital is absent)
Conclusion: The means to expand our business is absent

The conclusion of the endoxical syllogism provides a proposition which can interact with the maxim in a second syllogistic structure, which can be referred to as the *topical syllogism.* Here the conclusion of the endoxical syllogism takes the role of minor premise:

(18) *topical syllogism*
Major Premise (maxim): (For all actions and respective means) if the necessary means to realize an action are absent, then the action cannot be undertaken.
Minor Premise (from endoxical syllogism): The necessary means for (the action of) expanding our business is absent
Conclusion (standpoint): We cannot expand our business (= The action of expanding our business cannot be undertaken) (From Palmieri 2014:35, with adaptations)

The integration of the two syllogistic structures – two categorical syllogisms – can be represented graphically by a Y-shaped diagram, whence the eponymous Y-structure in Fig. 2.4.

In the diagram the simple lines connecting the premise boxes correspond to a connective of logical conjunction, while the arrows indicate a logical consequence relation from the joined premise boxes to the conclusion box pointed by the arrow. The dotted arrow going from the locus to the maxim indicates that the maxim is a semantic entailment of the locus relation. The Y shape of the diagram highlights how the line of inference issued from the material starting points interacts with the procedural starting point (*maxim*) to license the conclusion/ standpoint of the argument.

To offer a further illustration, in Fig. 2.5 the Y-structure diagram is applied to the authentic example of Carl Icahn's press release presented earlier in example (3). In this case both the datum and the *endoxon* are made explicit in the text, as shown by the compound structure in Fig. 2.1.

Thanks to the Y-structure, AMT offers an explicit way to combine the abstract inferential warrants of the tradition of research on the *topics* with the socially shared *endoxical* premises prominent in the rhetorical reflection on enthymemes. As Rigotti

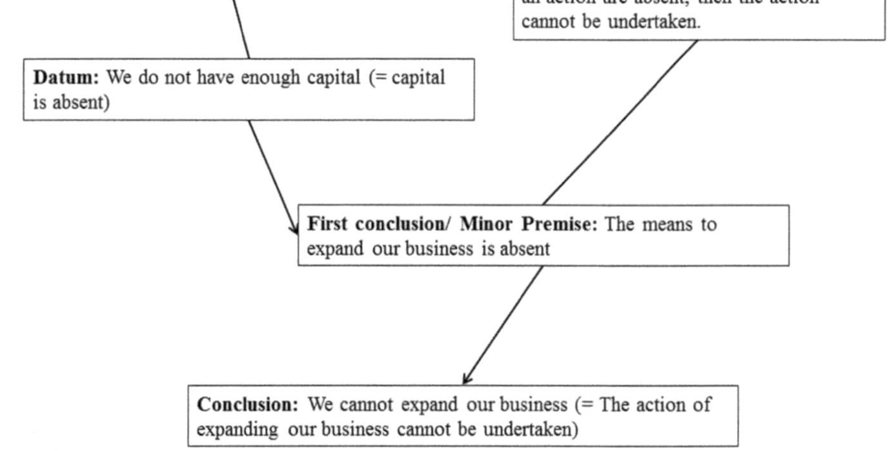

Fig. 2.4 AMT's Y-structure mapping of the inferential configuration of (7), adapted from Palmieri (2014: 37)

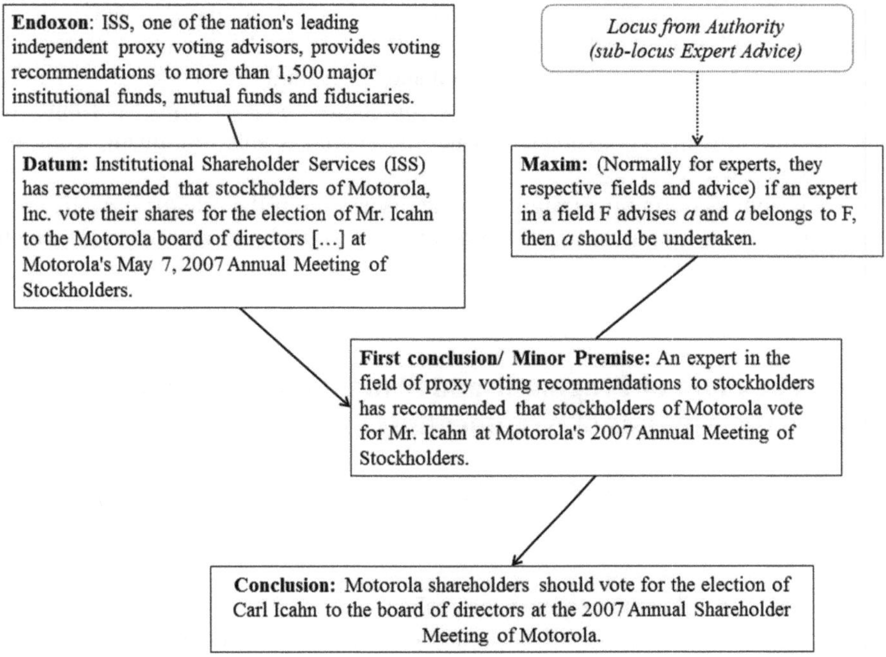

Fig. 2.5 AMT's Y-structure mapping of the inferential configuration of (3)

and Greco Morasso (2010: 507–510) observe, AMT is quite unique in merging these two major lines of thinking about premises in ordinary arguments. In doing so, the AMT approach becomes a particularly useful in the critical appreciation both of the *acceptability* and of the *relevance* of arguments. Rigotti and Greco Morasso (2010), in particular, stress how for each component of the Y-structure specific critical questions can be fashioned to evaluate the argument, according to a method of argument criticism that has been favored by argumentation theorists at least since Toulmin (see also Freeman 1991, Walton et al. 2008: 3 and *passim*). Keeping *endoxon* and *maxim* separated allows for sharper questions. Two kinds of critical questions of relevance concerning the procedural component are worth highlighting.

Firstly, one could question the grounding of the maxim in the locus. Is the maxim a *necessary semantic truth* (cf. Freeman 2005: 115) given the accepted meaning of the locus-relations? Or is the maxim only something expected to be generally, normally, usually, typically or by default the case when the locus-relations holds? Or, finally, are we in the presence of a mere *pseudo-maxim*, lacking real topical rooting? Here, as Rigotti and Greco Morasso (2010) suggest throughout their paper, argument criticism ultimately coincides with what they call *semantic analysis* and ordinary language philosophers of Toulmin's time would have perhaps called *conceptual analysis*.

Secondly and relatedly, once "the foundation of the *force* of the argument towards the standpoint" (Rigotti and Greco Morasso 2010: 509) has been made clear through semantic analysis, and the maxim formulated accordingly in deductive or merely presumptive terms, one could ask whether the modal force the arguer attributes to the standpoint really corresponds to what the maxim is able to transmit from the material premises.

These remarks bring us straight into the discussion of the role of semantic considerations in an evaluation oriented analysis of argumentative discourse, which makes the object of the next section.

2.3 The Place of Pragmatic and Semantic Considerations in the Analysis of Argumentative Discourse

2.3.1 The Interplay of Meaning Analysis and Argument Criticism: An Illustration

To provide a first illustration of how the analysis and criticism of arguments depends on a careful reconstruction of the meaning of utterances in argumentative discourse, let us consider again the case of Al Gore's speech presented earlier in example (1) – which I borrowed from Walton and Macagno (2010). Suppose someone utters (19) as part of an ongoing heated political discussion:

(19) Certainly, Al Gore did not invent the Internet.

Before evaluating the reasonableness of (19) we need to establish which kind of move, which kind of speech-act, the utterer can be taken to perform. For instance, we could conclude that the speaker, by uttering (19) intends to simultaneously perform two acts:

(a) casting doubt on the claim, attributed to Al Gore, that 'he invented the Internet';
(b) putting forward, with certainty, the contradictory standpoint that 'he did not invent the Internet'.

In terms of the summary of the kinds of argumentatively relevant information provided in Table 2.1, this meaning attribution relates to level (I): the recognition that a standpoint has been advanced. However, given the proper context and co-text, the same sentence could be taken as performing an altogether different act. The speaker could be inviting her dialogue partner to clarify the meaning of Al Gore's obscure statement and proposing to rule out one interpretation as absurd:

(20)
(a) He "took the initiative in creating the Internet"? Certainly, Al Gore did NOT invent the Internet.
(b) He "took the initiative in creating the Internet"? What the heck does he mean by that? Certainly, Al Gore did not INVENT the Internet.

The contrast between (20.a) and (20.b) shows that the amount of co-text available can make the attribution of this second intention more or less easy. In speech, prosodic signals such as a *contrastive focus* on the word *invent* may also serve as pointers in the direction of this second interpretation.

The choice of interpretation can have decisive consequences on the result of the evaluation. For instance, if we take the speaker of (19) as casting doubt and advancing a contradictory standpoint we might have a case for considering her guilty of the *straw man* fallacy. On the other hand, if the speaker is requesting clarification, contributing to clarify, or even just venting frustration at the obscurity of Al Gore's claim, the allegation of *straw man fallacy* cannot even be considered. In his study on the straw man fallacy in Internet political forums, Lewiński (2011) finds a number of authentic examples that can be taken either as instances of the straw man or as legitimate requests of clarification of the commitments of the interlocutor.

Examples (19) and (20) show that in order to critically evaluate an argumentative discourse we need a representation of discourse meaning specifying what kinds of dialectically relevant speech acts are being performed. By *dialectically relevant* speech-acts here I mean specifically those speech acts defined as relevant to the resolution of the dispute in the Pragma-Dialectical model of the critical discussion (cf. van Eemeren and Grootendorst 2004: 67). A functional pragmatic analysis of speech acts is a precondition to determining their dialectical import as advocated by Pragma-Dialectics since its inception (cf. van Eemeren and Grootendorst 1984).

Argumentatively relevant speech acts include not only such acts as *advancing a standpoint*, *drawing a conclusion* from available premises or *presenting an argument* in support of a standpoint, but also a variety of other speech acts[21]: like *casting doubt* on the interlocutor's standpoint, *demanding a clarification* of the content of the interlocutor's standpoint, *conceding* a proposition as part of the common ground, presenting a *refutation* of an argument presented by the interlocutor.

As a side note, it can be observed that the full consideration of these dialectically relevant pragmatic aspects of meaning is in fact much broader than what is captured by levels (I) and (III) in Table 2.1 in § 2.2.2. This is because Table 2.1 does not cover the full task of reconstructing an argumentative exchange in view of its evaluation as a critical discussion, but only the meaning reconstruction needed in order to evaluate the *relevance* and *acceptability* of arguments that are presented in support of a standpoint. In short, Table 2.1 is focused on the inferential dimension of argumentation, which occupies the center stage in this book.

Most, if not all, the speech acts that are important for the analysis of argumentation are *relational* in nature: drawing a conclusion *from premises*, presenting an

[21] See van Eemeren and Grootendorst (2004: 67) for a list of speech acts relevant in the context of a critical discussion. It should be observed that some of the examples that I mention here (e.g. *drawing a conclusion* and *presenting a refutation*) do not appear in van Eemeren's and Grootendorst's list. This happens because, in view of their contribution to a critical discussion, they can be seen as variants of more general speech acts. *Drawing a conclusion* can be a way of manifesting one's standpoint in the confrontation stage or to establish the results of a discussion in the concluding stage.

argument *in support of a standpoint*, casting doubt *on the interlocutor's standpoint*, demanding a clarification *of the content of the interlocutor's standpoint*, presenting a refutation *of an argument*. They either refer back to a previous utterance in the discourse or anticipate one. The relation can be dialogical when the resumption or anticipation refers to an utterance of the other dialogue partner, or monological when the connection is internal to the speaker's discourse.

It becomes therefore convenient to approach the representation of the meaning of an argumentative discourse from the angle of those theories (like, for instance, Asher and Lascarides 2003) that unify the treatment of speech acts with that of the so-called *coherence relations* or, as the influential work of Mann and Thompson (1987, 1988) calls them, *rhetorical relations*. In the following sections I will briefly present one such approach: *Congruity Theory* (Rigotti 1993, 2005; Rigotti and Rocci 2001, 2006a). Congruity Theory is an approach to the semantic and pragmatic facets of meaning providing an integrated treatment of *speech acts* and *discourse relations* which is particularly apt to capture descriptively their nuanced pragmatic functioning in connected discourse, while still providing a sound basis for the recognition of those dialectically significant commitments that are relevant in the context of the pragma-dialectic ideal model of the critical discussion.[22]

A semantic analysis of the propositional content of the argumentative discourse is also essential to carry out its critical evaluation (cf. levels II and IV in Table 2.1, § 2.2.2). We can go back once more to our Al Gore example to see how a consideration of semantics provides insights that are complementary to the consideration of speech acts. Since the straw man fallacy has to do with the misattribution of commitments it is interesting to look into what kind of propositional content we can take Gore to be committed to, given his utterance.

In order to do that I will use some basic ideas from *compositional semantics*, i.e. the theory of "how the meanings of individual words combine to form the meaning of larger meaningful components of discourse" (Asher 2013: 16). In our case these larger meaningful components are the propositional contents of the dialectically relevant speech acts discussed above. The basic notions I introduce here play a role in Congruity Theory but they are certainly not exclusive to it.[23]

Predicates are the basic meaning carrying units, they behave like *functors* that require application to one or more *argument*[24] terms in order to yield a larger unit of

---

[22] In terms of the tentative list of factors that an argumentative analysis needs to capture (Table 1, in section 2.1) these speech-act level considerations are relevant for levels I and III.

[23] The notions introduced here are, in fact, found is several major approaches to semantics, such as Seuren (1985), Asher (2011) and Mel'čuk (2012). For an accessible short introduction to semantics as a tool for the analysis of verbal communication see Rocci and Luciani (2016b).

[24] For the sake of clarity: the *arguments* of semantics, sometimes also called *semantic actants* (Mel'čuk 2004) of the predicate, have quite simply *nothing* to do with the *arguments* of argumentation theory. Here I stick to the current semantic nomenclature despite the ambiguity rather than use substitutes such as actant. The context will be usually sufficient to disambiguate, but when confusion is possible I will disambiguate by appending a subscript 's' to argument_S in the semantic sense.

meaning such as the propositional content. Predicates are said to have one or more *argument places* or slots, according to the argument terms that are necessary for them to yield a propositional meaning. So, *to sleep* is a one-place (monadic) predicate, requiring a single argument term, while *brother* is a two-place (diadic) predicate (as one is always the brother *of someone else*) and *to give* is a three-place (triadic) place.

What is interesting here is that the application of predicates to arguments is subject to *selectional restrictions*. As Asher (2013: 16) puts it "not every word can serve as an argument to any other word as a predicate", some selectional restrictions correspond to basic logical and ontological types, while others are subtler and more fine grained. In philosophy, as we will see in Chap. 3 selectional restrictions of predicates have been used for a long time as a tool for conceptual analysis, drawing on Husserl's idea that there is a "pure grammar" that dictates how concepts can be combined to form meaningful wholes (cf. Rigotti and Rocci 2001). In this tradition of linguistic and philosophical analysis the creation of nonsensical, incongruous, mismatched applications of predicates to argument terms reveals the selectional restrictions and, in doing so, elucidates aspects of the predicate concepts themselves.

The selectional restrictions of the predicate *to create* and of the predicate construction *to take the initiative in doing something* will be crucial in the analysis of our case. First of all, we can make the not too surprising observation that *to create* requires that what is created did not exist before. For instance, uttering something like (21) is incongruous:

(21) *Al Gore created the Internet, even though it already existed.

The resulting incongruity tells us that this requirement has the status of a necessary semantic implication of the meaning of the predicate *to create*. More precisely, these inferences originating from the selectional restrictions have the status of *presuppositions* (cf. Seuren 1985, Rigotti and Rocci 2001, Asher 2011).

Yet, Gore did not say he *created* the Internet, he said that he *took the initiative* in creating it. Could that mean, as Walton and Macagno (2010: 289) maintain, that Gore "was responsible for helping to create the environment (in an economic and legislative sense) that fostered the development of the Internet"? Certainly, this reading, where Gore is *helping* or *fostering* the creation of the Internet, is better supported by context than the one where he is taken to claim to have invented the Internet – we all know that Gore is a politician and not an engineer or scientist. For Walton and Macagno (2010) here the straw man is set up by "wrenching from context" the words of Gore. It could be interesting to see, however, whether this more reasonable and historically accurate reading is consistent with semantics. Consider other authentic examples of our construction:

(22)

(a) France proposed the no-fly zone with Britain and took the initiative in launching air strikes at the forces loyal to Gaddafi (WebCorp[25]).

(b) In 1902 President Roosevelt took the initiative in opening the international Court of Arbitration at The Hague, which, though founded in 1899, had not been called upon by any power in its first three years of existence (WebCorp).

In (22.a) the construction is used in a context where the very first air strike to be launched was in fact carried out by French planes. Example (22.b) presents an interesting distinction: the Court of Arbitration had existed for three years but did not actually sit until the United States and Mexico had recourse to it to solve a difference between the two states. In every case the construction *to take the initiative in V-ing* presupposes that the activity denoted by the gerund was not going on before and entails that the subject of the construction is the initiator of the activity – compare (23.a) and (23.b) below.

(23)

(a) *France took the initiative in launching air strikes against Lybia. When France joined, Britain had been bombing loyalist forces for two days.

(b) France helped to launch air strikes against Lybia. When France joined, Britain had been bombing loyalist forces for two days.

Interestingly, the quick reconnaissance of the semantics of the two predicates in question seems to tell us that *taking the initiative in creating something* is neither compatible with a scenario where that something is already there, nor with a scenario where other people are already busy creating that something. In order to interpret Gore's claim in terms of *helping*, as Walton and Macagno (2010) propose, a charitable interpreter has to *override* these semantic constraints. And this is not something we normally do without good reason.

In a broadly Gricean view of verbal communication, which here I take to be basically correct, apparent violations of semantic constraints need to be accounted for in terms of communicative effects: metaphor, hyperbole and other traditional tropes are paradigmatic examples of semantic violations that prompt a search for additional implications. This is not one of those cases. In the absence of any additional *overt* communicative effect that could have been reasonably sought through the violation one must conclude that the speaker was being uncooperative, either by inadvertence or by design.[26]

[25] This example has been collected from Internet sources using the WebCorp tool for the creation of linguistic corpora and the extraction of concordances from the World Wide Web.

[26] A discussion of the *covert* – and thus not fully cooperative – effects that can be obtained by putting forth anomalous utterances whose semantic constraints do not find straightforward satisfaction in the context of utterance would lead us astray from our path. One key concept for understanding these effects is the psycholinguistic notion of *shallow processing* (Sanford and Graesser 2006). On the *overt / covert* distinction in relation to verbal communication and persuasive effects see Crook (2004).

Interestingly, a reading where Gore *invented* the Internet provides an interpretation that *does* satisfy the presuppositional selectional restrictions outlined above (Internet was not there and nobody before Gore was creating it), even if it does so by enriching the vague meaning of the statement in a most implausible and hence uncharitable manner.

If we now go back to the two possible speech acts that the utterance of (19) could realize we can see how looking at the semantic content of the original utterance illuminates them. The use of (19) as a way to prompt clarification of the original statement can be harsh but is ultimately resolution oriented because it brings to light the deceptive potential of the original utterance by taking to the extreme of implausibility the idea that 'Internet was not there and nobody before Gore was creating it' that the semantics of the utterance conveys. On the other hand, if (19) is meant as casting doubt on a standpoint attributed to Gore (and then easily refute it), then it is not defensible as a reasonable move, irrespectively of the deceptive potential of the original utterance, which needs to be clarified or rectified rather than refuted on an implausible interpretation. Performing a fallacy in retaliation of another does not make it a reasonable move.

2.3.2 Pragmatics and Semantics in Argumentative Reconstruction

In the previous section I used pragmatic and semantic insights to reconstruct the illocutionary force and propositional content of a single speech act so that its reasonableness in the context of a critical discussion can be evaluated. The illustration was limited to moves that are ideally situated in the *confrontation stage* of a critical discussion, where "it is made clear that a standpoint is not accepted because it encounters doubts or objections, which imply a difference of opinion" (van Eemeren et al. 2007: 11). Pragmatic and semantic considerations helped to establish that (19) could be reasonable (if somewhat harsh) as a request of clarification ("request of a usage declarative" in pragma-dialectic terms) but would not be reasonable as an expression of doubt, because it selects an implausible interpretation and consequently targets a propositional content which was not clearly advanced as a standpoint.

The idea that, in order to establish a representation of the discourse specifying the types information considered in Sect. 2.2 (cf., in particular, the levels I-VII listed in Table 2.1) the analyst will have to rely on pragmatic insights on discourse interpretation is relatively well established in some areas of argumentation studies, starting from Pragma-Dialectics and fellow normative pragmatic approaches (See, in particular, van Eemeren et al. 1993).

Pragmatic inference intervenes to complement what is linguistically expressed in the discourse at various levels. Standpoints (I, II) are often not made explicit and can only be reconstructed by accounting for the pragmatic inferences through which an ordinary interpreter of the discourse can recover this information. Establishing the utterances that are presented as premises (III), their propositional contents (IV), how they fit together in supporting the standpoint (V) requires a principled reconstruction of discourse structure based on constructs of discourse relations or relational speech acts of the type mentioned in the previous sections.

The establishment of discourse relations, while linguistically constrained in different ways (discourse connectives, tense and aspect, general constraints on attachment and anaphora resolution), crucially relies on the pragmatic inferences (communicative inferences) of the discourse participants (Cf. Asher and Lascarides 2003, Webber et al. 2003). Moreover, as anticipated in Sect. 2.3.1, the recovery of the implicit components of enthymematic arguments require the appeal to pragmatic principles of discourse interpretation. In this area the recourse to pragmatics is far from new in argumentation studies: pragmatic principles have been invoked to provide a clearer understanding of the Aristotelian notion of enthymeme at least since the seminal paper of Jackson and Jacobs (1980) and a quite detailed discussion of the "problem of unexpressed premises" based on Gricean principles can be found in van Eemeren and Grootendorst (1984: 119–149).

Each of the aspects cursorily mentioned above would deserve an extensive discussion, but the general point I want to make here should already be clear: in order to be able to evaluate an argument, argumentation scholars cannot avoid the intricacies of discourse interpretation. All too often the reasonableness or fallaciousness of an argument hinges on a more or less subtle matter of interpretation. Hamblin (1970: 224) observed that is it often difficult to "nail a fallacy" because it depends on the text being understood as an argument and perpetrators often leave themselves open a way for retreat: "In many cases of supposed fallacy it is possible for the alleged perpetrator to protest, with an innocent face, that he cannot be convinced because he was not been arguing at all". As observed by Jacobs (2000: 265) it is often difficult to distinguish between bad arguments and non-arguments, and messages might attempt to persuade by combining more explicit arguments with secondary lines of argumentation (Rocci 2008), which are at the same time less sound and less overtly communicated, requiring a high degree of interpretive cooperation from the hearer. At the same time, however, in order to be successful the manipulative, fallacious arguer *does* need to convey (overtly or covertly) his communicative intention and in doing so leaves traces for the addressee that can be discovered by analysts paying attention to textual cues (cf. Rocci 2005; Jacobs 2000). Uncooperative, fallacious argumentation typically involves a variety discursive attempts at "eating one's cake and having it" of which the traditional *fallacy of equivocation* represents just a special case. Most of the times these attempts cannot be uncovered without reconstructing how interpreters are invited to "fill-in the blanks" to recover the intended meanings.

It would appear natural, in this context, to pay attention also to the semantic aspects of discourse interpretation and to their interaction with pragmatics. Yet, semantics has not received systematic attention in argumentation studies to the point that the very use of the term is rare in the discipline.[27] The present book, by addressing the role of modality and modal expressions in arguments, intends to fill this research gap, showing that explicit linguistic semantic considerations can be precious in view of the argumentative reconstruction. The study of modality provides the occasion for laying down a systematic map of the ways in which semantics can be relevant to argument reconstruction. It is useful here to anticipate a first sketch of this map. I take the role of semantics in argumentative reconstruction to be basically four-fold. Semantics is relevant

1) To clarify the propositional commitments of the arguers;
2) To determine the semantic type and inferential properties of the propositions functioning as standpoints and as premises and, in particular, to determine which *loci*-relations are compatible (congruous) with these propositions and which specific maxims are possible.
3) To account for expressions that either denote *loci*-relations holding between propositional contents (e.g. causal or final constructions) or partially constrain discourse interpretation as far as the recognition of these semantic-ontological relations is concerned;
4) To account for expressions that either denote argumentative speech acts or partially constrain discourse interpretation as far as the recognition of argumentative speech acts and argumentative macrostructure.

Let us work out in greater detail what each level is about:

(1) At a first level linguistic semantics can serve, together with pragmatics, to determine to what propositional contents the arguers can be reasonably held committed to. Here semantic analysis serves to resolve the polysemy and underspecification of the linguistic forms employed in the argument as well as to pinpoint fallacious shifts

[27] There are, as always, exceptions that should be duly noted. The French tradition of research on *argumentation dans la langue* ('argumentation within language' or, more precisely, 'argumentation within the language system') initiated by Anscombre and Ducrot (1983) is the most notable. The theory of argumentation within language is first and foremost a semantic theory seeking to provide, on the one hand, a systematic treatment for that non truth conditional evaluative component conventionally attached to *lexical predicates* which is traditionally vaguely discussed in terms of connotation and, on the other hand, to outline the constraints that *discourse connectives* impose on the argumentative function of the segments they relate. The semantic model proposed to account for these phenomena ends up rejecting denotative semantics altogether and – in particular in its later incarnations – sharply divorcing argumentation from logic and inferential processes. In fact, it treats relations between arguments and conclusions essentially as encoded into the system of the langue, given that the linguistic meaning of a sentence is equated with the virtual conclusions it "argues" for – its "argumentative orientation". The impact of this body of work on argumentation theory, as opposed to linguistics, has been limited. While the overall theory, given its radical semantic anti-realism, seems to be squarely incompatible with the very idea of critically evaluating the quality of arguments (See Rocci 2006.b for a discussion) the early insightful analyses of discourse connectives have made their way into research on *argumentative indicators* (cf. van Eemeren et al. 2007). The present treatment of modality is indebted to earlier works of Ducrot (1980) for the notion of *instructional / procedural* meaning.

in meaning, strategic exploitation of vagueness and more generally to the resolution of fallacies *in dictione*.

In the case of polysemy a semantic analysis can work out separate semantic representations specifying, at the same time, their contextual constraints, starting from the presuppositional constraints associated to argument places (cf. Rigotti et al. 2006). These constraints can then be used to support the attribution of a particular value in the reconstruction of the propositional commitments of the arguers and to pinpoint cases of equivocation, where an expression is used in contexts satisfying mutually incompatible constraints.

Linguistic semantics also developed sophisticated treatments of forms of underspecification other than polysemy, distinguishing, in particular *context dependency* as requiring special consideration. Context dependent expressions (cf. Recanati 1989, Kay 1997) have incomplete linguistic meanings that set up variable slots to be saturated by contextual information of a specific kind to obtain a full-fledged semantic representation. There is an unexpectedly vast range of subtly context dependent expressions in natural languages, and – as Toulmin ([1958] 2003) correctly hypothesized – the most basic modal expressions are among them. These expressions can undergo meaning shifts in discourse that are considerably subtler than those realized through polysemy. Thus, recognizing context dependent expressions and working out how their saturation can be plausibly accounted in a discourse situation can be important for reconstruction.

(2) The second level at which semantic considerations can affect the reconstruction of arguments has to do with what Freeman (2011) calls the *microstructure* of arguments. Semantic representations of linguistic expressions are important because the meanings of the component expressions contribute to determine the inferential properties of the propositions in which they appear. Logicians have traditionally dealt with this aspect by focusing on the semantics of a few "logical words" – such as the existential and universal quantifiers – as the only ones responsible for the determination of the logical form relevant for establishing the deductive validity of the argument. Modal words, or at least *some* of them, have featured in this list together with the aforementioned quantifiers, truth-functional connectives and negation. The parallel developments in linguistic semantics and in argumentation theory contributed in different ways to alter this picture.

On the one hand, semanticists have increasingly rejected the distinction between "logical" and "non-logical" vocabulary (cf. McCawley 1993: xviii) as they discovered non-trivial inferential properties in a wide range of linguistic expressions and recognized that also the so-called logical words – like the modals – required more sophisticated semantic representations.

On the other hand, argumentation theoretic research on topics and argument schemes has shifted the attention from the abstract requirements of logical form to the richer, more specific requirements that argumentation schemes impose on the propositions that instantiate them in discourse, or on specific semantic constituents of those propositions. In other words, theories such as AMT, by invoking "relations of conveyance" such as *cause-effect*, *parts-whole*, *means-ends*, use a richer conceptual system – some semanticists would say a "richer ontology" – and, as a consequence, the evaluation of the correct application of the schemes needs to be done *in terms of this conceptual system*. This move makes a rich semantic analysis of all sorts of "non-logical words" immediately relevant for judging the correct application of argument schemes.

While the correct application of a given argument scheme may be determined by the presence of very specific features in the semantic makeup of the premises and the standpoint, there are certain aspects of the semantics of propositions that have broader consequences on their role in arguments. The idea is that arguable propositions come in different basic *semantic types* and that these types determine what kind of arguments can be used to support them as standpoints.

In the argumentation literature one finds several different versions of this idea. At a certain moment Toulmin ([1958] 2003: 13) connects his *fields of argument* with different "logical types" of propositions, each calling for its own specific criteria of evidence in argumentation.

A different take on the idea of semantic types is found in James Freeman (2000, 2005), who builds on a rather vague modern rhetorical tradition to develop a semantically consistent and argumentatively relevant classification of *types of statements*. The structure and rationale of this typology is presented fully in the next section, but we can immediately illustrate the kind of distinctions drawn by Freeman with the very first example found in this book: example (1) in the *Introduction*. As observed commenting that example, it is intuitively clear that two standpoints like (24.a) and (24.b) require not just a different *amount*, but different *kinds* of evidence to be supported:

(24)
(a) John resigned
(b) John *had* to resign

The difference is connected to the fact that (24.a) is a simple *description* of facts, while the modal proposition in (24.b) implies either a causal *interpretation* of the situation or, under a different reading an *evaluation* of it with respect to a system of norms. The semantics of modality, as the following chapters will show, is deeply involved in the determination of the argumentatively relevant semantic types of propositions. As observed earlier in Sect. 2.3.2, discovering the specific flavor of modality involved in (24.b) is one and the same thing as ascertaining which *loci* can be exploited to infer it as a standpoint.

(3) This brings us to the third level at which semantic considerations are relevant in argumentative reconstruction. Certain linguistic expressions and constructions can express directly and more or less unambiguously the *loci*-relations, while others may partially constrain discourse interpretation as far as the establishment of these relations is concerned. One of the hypotheses that will be examined in this book is that modal expressions such as the modal verb *have to* in (24.b) belong to the latter group.

(4) The fourth level is the analysis of linguistic expressions whose meaning consists wholly or in part in manifesting the performance of argumentatively relevant speech acts or the macrostructure of arguments. This level is close to the kind of semantic considerations that can be found in recent work on *argumentative indicators* (van Eemeren et al. 2007) or in earlier work on the "lexicon of argumentation" (Stati 1998). Most existing argumentative treatments of modality in terms of

strength of support that a standpoint enjoys or of the expression of the arguer's *degree of commitment* towards its content belong essentially here.

There are different modes of linguistically expressing aspects of argumentative speech-acts and of the discursive macrostructure of arguments. A first distinction can be drawn between the performative use of meta-argumentative expressions (e.g. *it is reasonable to conclude that..., we can argue..., we must conclude..., a further independent reason for that is...*) and the signaling of argumentation structure and argumentative speech acts by less transparent discourse markers and grammatical means. The latter can hardly be made the focus of attention and comment by ordinary language users and typically exhibit a *procedural* functioning (cf. Ducrot 1980, Kay 2003, Nølke et al. 2004, Wilson and Sperber 2012).

Procedural expressions (also called *instructional*) are deictic, context-dependent expressions. They have very abstract skeletal linguistic meanings containing variables that have to be saturated in context by the addressee. At the same time, they encode instructions that guide and restrict how context is to be searched to fill in the variable slots. As a consequence, these words influence *how the relevant discourse context of an utterance is to be construed* by the interpreter, leading to the establishment of discourse relations and guiding the recovery of implicatures.

One kind of linguistic units that has been successfully analyzed as procedural is discourse connectives, and in particular "argumentative connectives" such as English *but* and its equivalents in other languages (See for instance Nølke et al. 2004 for instructional analyses of French *mais* 'but' and *donc* 'so / therefore'). In this book it will be argued that several key expressions of modality – like modal verbs in English and in other languages – can be successfully analyzed as context-dependent procedural units that can guide the establishment of discourse relations between argument and standpoint and the recovery of implicit premises. Furthermore, the procedural component of modal expressions may restrict the kind of propositions that can be recovered or reconstructed as premises.

Having completed this map of the role of semantic considerations in reconstruction, it becomes interesting to situate previous work of modality and the present contribution with respect to it.

Previous treatments of modality in argumentation take into consideration mostly level (4): modals as linguistic indicators of relevant aspects of argumentatively relevant speech act and of argumentation macrostructure. As previously observed, with the partial exception of Toulmin ([1958] 2003), they focus exclusively on two closely related "flavors" of modality: the *epistemic* (alternatives consistent with what is known) and the *doxastic* (alternatives consistent with what is believed). Epistemic/ doxastic modals are found to be relevant chiefly to the determination of the strength of commitment towards the proposition expressed by the standpoint (VII). Secondarily, they are sometimes seen as indicators of the advancement of a standpoint (I), or as offering – via their strength – an indirect indication relevant to establish argumentative macrostructure.[28]

[28] Cf. Freeman (2011: 94–103). This issue is discussed in Chap. 3, Sect. 3.2.5.

In the present book the relevance of modal meanings for reconstructing arguments is examined not only at the level of argumentative speech-acts, but at each of the four levels examined above. The examination reveals that modals can play a distinctive role in retrieving or reconstructing *all* the kinds of information (I-VII) listed in Table 2.1. Two decisive factors in achieving this result will be the working assumption that all flavors of modality are, in principle, relevant for understanding arguments and the adoption of the theory of Relative Modality (Kratzer 1981, 2012). This theory, whose significance has been already briefly hinted at in the *Introduction*, postulates a semantics for the modals that is both relational and context dependent, allowing for the discursive saturation of their underspecified relational meanings and for the establishment of relations both at the semantic-ontological level of the *loci*-relations, and at the pragmatic level of the relational speech-acts corresponding to dialectically significant moves.

2.3.3 Semantic Types of the Propositions Functioning as Standpoints and as Premises

In the previous section it was suggested that semantics can help argumentative reconstruction by determining argumentatively relevant semantic types to which propositions belong. Modal operators, by combining with their prejacent proposition create a modal proposition whose semantic type is different from the prejacent. Example (24), in the previous section, illustrates very well how this can affect argumentation: being able to show a letter of resignation signed by John would be enough to support the standpoint *John resigned* (24.a), which is a mere statement of fact. The same evidence would be irrelevant if the goal is to defend as a standpoint the modal proposition *John had to resign* (24.b). But here the tricky thing is that, in order to find out what kind of arguments are relevant to support (24.a), we need to establish more precisely what kind of modal "flavor" is conveyed by the English modal *have to* in the context of utterance. Suppose that we can establish that the modal conveys a meaning of moral obligation, then an argument such as the one in (25) could fit the bill.

(25) John had to resign. He could not remain complicit of the unethical business practices of the company.

But, if the meaning of *have to* pertains instead to legal obligation, we would need an argument referring to legal provisions and statutes mandating the resignation of officials under certain circumstances and factual premises telling us that John finds himself in such circumstances. There could be also situations in which *have to* is neither legal nor moral, but refers to what is necessary in view of certain goals or subjective needs of the agent, as it happens in (26) or (27). This, again, has consequences on the kind of arguments that can reasonably support the modal proposition as a standpoint.

(26) John had to resign. They had threatened his family.
(27) John had to resign. He could not tolerate the pressure anymore.

In order to account for these differences in a systematic, non *ad hoc*, fashion we need a typology of propositions that, on the one hand, is able to capture relevant distinctions from the viewpoint of argumentation and, on the other hand, can interface in a rather precise manner with the different types of modal propositions that are obtained by applying modalities of different flavors to their prejacents.

Several – yet not all – theories of argumentation have recognized that reasoned argument does not always concern matters of fact. In interpersonal arguments, in the life of organizations as well as in public life, people often discuss about the evaluation of past and present states of affairs or about the expediency of a possible course of action. Pragma-Dialectics readily recognizes that there are different types of standpoints and that they all pertain to the purview of a theory of argumentative discussions:

> When theorizing about argumentation it is vital to realize that the standpoints at issue in a difference of opinion can pertain to any kind of subject and that these standpoints can be descriptive as well as evaluative or practical. In argumentative discourse, an evaluative judgment, such as "The film Infamous is brilliant," or a practical prescription to do something, like "You should join me to that meeting," can be just as well at issue as a descriptive claim about a factual state of affairs such as "Amsterdam is much bigger than Rotterdam" (van Eemeren 2009: 70).

For Pragma-Dialectics, accepting that the theory should cover *descriptive*, *evaluative* and *prescriptive* standpoints means avoiding the Scylla of the quasi-positivist position of those "epistemic" argumentation theorist who consider reasonable discussion about practical or evaluative matters impossible, and the Charybdis of those rhetorical theorists who limit the object of their investigation to calls to action (cf. van Eemeren 2009: 71).

While Pragma-Dialectics regularly refers to a tripartite typology of standpoints distinguishing between descriptive, evaluative and descriptive ones (cf. van Eemeren 2010: 2, 15, 235 and van Eemeren 2013) the semantic aspects of this distinction have not been systematically worked out. This is, in a sense, natural, given the pragmatic nature of the theory and its ability to treat all types of standpoint with the same pragmatic conceptual tools. In fact, in Pragma-Dialectics the speech-act of advancing a standpoint as *acceptable* is not affected by the descriptive, prescriptive or evaluative nature of its content.[29]

[29] Van Eemeren and Grootendorst (2004: 63) observe that not all assertives "pronounce the truth of a proposition", as some "express a judgment on its acceptability in a wider sense". The two examples offered are *In my opinion, no exceptions are possible to the freedom to express one's opinion* (a normative statement) and *I think Baudelaire is the best French poet* (an evaluative one). While the tripartite typology is not evoked in the passage, it seems clear to me that the authors intend to ensure that the category of assertive speech-acts cover not only descriptive but also prescriptive and evaluative statements.

While the semantic aspects are not discussed and the pragmatic core of Pragma-Dialectics can be applied indifferently to all types of standpoints, the tripartite typology has been given attention in recent pragma-dialectical work on the characterization of concrete argumentative activity types in different social contexts, as specific activity types typically "revolve around" a specific kind of standpoint. For instance, according to van Eemeren (2013: 280), "in disputation aimed at establishing the truth of a claim in the scientific domain [...] the standpoint at issue is prototypically descriptive", while "in adjudication aimed at reaching a verdict in the legal domain the standpoint is prototypically evaluative". This important connection between standpoint types and the concrete social context of argumentative discussion will be dealt with in § 2.4 later in this chapter.

How does the tripartite typology of standpoints fare as a tool for capturing the semantic types created by modalities and their argumentative consequences? It gives mixed results. It can be argued that a tripartite typology is successful in capturing how *John had to resign* differs from its non-modal counterpart (*John resigned*) under the interpretations highlighted in examples (25), (26) and (27). The modal propositions can be understood as different subtypes of *evaluative* standpoints, and thus contrasted with the *descriptive* modal proposition. Note that in these example it would not make sense to speak of *prescriptive* standpoints, given that the prejacent is a past event and no future course of action is at issue. Note, also, that the evaluative interpretation is possible with (26) and (27) because the goals or subjective needs of the agent are somehow endorsed by the speaker. However, it is not always possible to characterize the argumentatively relevant sematic types created by modal propositions in terms of the tripartite typology of standpoints. Suppose that (28) is put forth as a standpoint:

(28) In order to steal the diamonds, the thief *had to* break into the bank's vault.

Unless the speaker shares the goal of stealing the diamonds, (28) would not be considered an *evaluative* standpoint, but rather a *descriptive* one. Very often both the modal and the non-modal statement will be lumped together into the *descriptive standpoint* category. To take another familiar example discussed in the *Introduction*, both (29.a) and (29.b) would fall under the same descriptive label:

(29)
(a) Suzy *could* not steal from the department store yesterday afternoon.
(b) Suzy did not steal from the department store yesterday afternoon.

Generally speaking, this is not a problem for the tripartite classification, which remains perfectly consistent. Simply, the *descriptive standpoint* category is very broad and includes both purely extensional states of affairs and quite diverse modal propositions such as (28) and (29). However, for the purpose of studying the argumentative consequences of the semantic types created by modalities a finer grained classification of propositional types is needed.

I believe that key elements of such a typology can be found in the semantic "typology of statements" developed by James Freeman (2000, 2005). One advantage of Freeman's typology for the present investigation is that it is constructed using basically the same intensional semantics approach that forms the basis of formal semantic analyses of the modals.

In this section, I present briefly the origin and structure of Freeman's typology, on which the following chapters will build a more detailed nuanced account of the types corresponding to the different modal flavors. The "typology of statement" was originally presented in article form (Freeman 2000) and then integrated in a major work proposing an epistemic approach to the problem of the acceptability of arguments (see § 2.1.2, above).

It has to be said that, contrary to other scholars in argumentation and rhetoric, Freeman does not have the intention to use his typology of statements to classify standpoints or claims. Rather, he wants to see what kind of statements can function as *premises* and look at what kind of basic "belief generation mechanisms" (Freeman 2005: 41) can vouch for them. He regards his epistemic viewpoint on the acceptability of premises as complementary to the dialectical one:

- From a dialectical viewpoint, a proposition is a starting point because it is part of the common ground of the discussants. They are explicitly or implicitly jointly committed to accepting it. Once a proposition is a starting point it can be used in argument.
- From an epistemic point of view a proposition can be a starting point in argument or in intrapersonal reasoning because I can reasonably believe it without the need of supporting argument. Traditionally, this was said to happen when a proposition is *evident in itself*.

However, evidence is a problematic notion and surely the starting points of our reasoning are for the most part not strictly speaking evident. Freeman (2005) tries to develop a notion of basic acceptability that is weaker than evidence by looking at what kind of "basic belief generating mechanism" create a reasonable favorable presumption towards a proposition without the need of an argument. It turns out that according to the semantic type of proposition the relevant *belief generating mechanisms* cannot be but different. For instance, we do not derive our basic moral notions and our basic certainty about what occurs in the physical world from the same source.

Despite Freeman's goal of examining the starting points of argumentation, his classification can be used as well to examine how different semantic types of standpoint constrain the types of loci that can be invoked to license the inference and, indirectly, the types of premises that are congruent with these *loci*. In a sense, it is not so surprising that there is a certain parallelism between the primary belief generation mechanisms and the secondary belief generation processes at work in argumentation. Furthermore, applying Freeman's classification to standpoints brings it back to its rhetorical roots.

In fact, the inspiration of Freeman's typology comes from modern typologies of claims[30] developed in by students of rhetoric, such as Kruger's (1975) and Fahnestock and Secor (1982). In turn, the source of inspiration of these modern rhetorical works is the Ancient rhetorical doctrine known in Greek as *stases* (sing. *stasis*) and in Latin *status causae*. The *stases* defined a fine grained typology of argumentative situations and of the issues that they generate, which was largely oriented towards the judicial practice. Kruger (1975: 137) clearly refers to a segment of *stasis* theory when he mentions "three basic types of disputes" discussed by early Latin rhetoricians:

1) *An sit?* – Kruger's question of fact, conjectural stasis, according to Ancient terminology (e.g. *Did Brutus kill Caesar?*);
2) *Quid sit?* – question of definition or definitional stasis (e.g. *If it is granted that Brutus killed Caesar, what kind of act was it? Was the act murder or self-defense?*);
3) *Quale sit?* – question of quality or qualitative stasis (e.g. *If it was in fact murder, how should we evaluate this act? Was it honorable or dishonorable? Was it expedient or useless? Was it lawful or unlawful?*).

The three Latin categories evoked by Kruger[31], and some of their modern avatars, are the starting point of Freeman's reflections and still leave traces in his more comprehensive typology. I will not go through the arguments that motivate Freeman to rework rhetorical classifications into his typology, moving instead to the presentation of the final product, which is summarized below in Fig. 2.6.

The classification is based on a series of dichotomies. First, Freeman sets apart those statements whose truth can be established exclusively on the basis of their form and meaning, without any recourse of referring to the state of the world at the moment of utterance. Freeman gives examples such as the following:

[30]Toulmin's term *claim* is roughly equivalent to Pragma-Dialectic's *standpoint*. Both *standpoint* and *claim* are dialectical terms, while *conclusion* focuses on the inferential aspects of the argument. See § 2.3.4.3 for a discussion of the speech acts of concluding and of advancing a standpoint from the viewpoint of Congruity Theory. On *claims* and *conclusion* see also Freeman (1991: 49–50).

[31]The three types of disputes evoked by Kruger (1975) are but a fragment of the elaborate systems of *stases* developed by Latin and, especially, Greek rhetoricians. These systems, often based on a series of dichotomies had both a taxonomical and a dynamic facet. Each node in the taxonomy presupposes the resolution of the issues featuring above it, and thus can be read both as a type of argumentative discussion and as a stage in an argumentative discussion, presupposing the resolution of the stages above. For instance, issues of *definition* presuppose the resolution of issues of fact, and issues of *evaluation* presuppose the resolution of issues of definition. One of the most elaborate systems of *stases* that was transmitted to us is Hermogene's treatise (II century A.D.), which discusses no less than thirteen *stases* (See Patillon in Hermogenes 2009: XLIII, for a synoptic scheme of the stases).

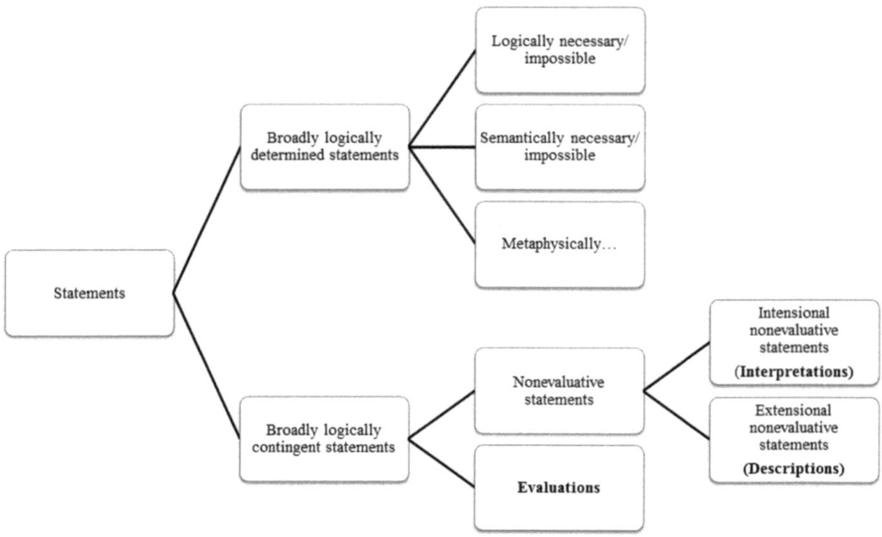

Fig. 2.6 Argumentatively relevant typology of statements according to Freeman (2005: 93–113)

Necessary/ impossible statements (broad logical, semantic and conceptual necessity)

(30.a) Either it is raining or it is not raining

(30.b) All bachelors are unmarried

(30.c) No natural numbers are colored

(30.d) $2 + 3 = 5$

These propositions are always true (or always false) or, more precisely, they are *necessarily* true or *necessarily* false. They are true (or false) no matter how the world might turn out to be, or, equivalently, in all possible worlds; where, by *all possible worlds*, one has to understand all conceivable, consistent worlds according to a broad notion of possibility, akin to the *possibilitas logica* of Duns Scotus.[32] Freeman's idea of logically determinate statements is pretty broad: it is not limited to formally true statements such as (30.a) as it includes also statements that are necessarily true given the semantic system of the language (30.b), or more encompassingly what we might call the conceptual system, the basic ontology or metaphysics of the world (30.c). In fact, as Freeman (2005: 117) recognizes, "the principles of constructibility" are part and parcel of the semantico-syntactic knowledge of a language so that the idea that "color properties cannot be ascribed to natural number" should be part of our ability of constructing semantically *congruous*

[32] See the *Introduction*, § 1.6.1.

linguistic expressions.[33] Mathematical statements, such as (30.d) are also considered necessarily true (or false in this case!).

Note that even logically necessary statements are arguable statements, in the sense that we have to recognize whether candidate necessary statements in fact are such.

The second dichotomy separates among the logically contingent statements the evaluative and non-evaluative ones. Freeman's examples of the former are reproduced below in (31).

Evaluations

(31.a) It was good for me to be afflicted (Psalm 119:71a NIV)

(31.b) It is better to be Socrates unsatisfied than a pig satisfied (J.S. Mill)

(31.c) You should repay that debt

(31.d) Extremism in the defense of liberty is no vice and moderation is no virtue (Sen. Barry Goldwater) (Freeman 2000: 146)

Freeman (2000: 146, 2005: 101) considers evaluative statements easy to single out: "we recognize evaluative predicates straight off". Evaluative statements characterize a state of affairs as good or bad, better or worse, as something to prefer or rather to avoid; they qualify an act as either right or wrong, obligatory, permissible or forbidden, a person as praiseworthy or blameworthy. Also from the examples it is clear that the category of evaluations is vast and encompasses both the *evaluative* and the *prescriptive* standpoints recognized by Pragma-Dialectics.[34] It will be therefore important to recognize *different sub-types of evaluation* as we consider the inferential properties of propositions constructed by modal predicates of different flavors. Freeman's own extensive treatment of the typology of evaluations (Freeman 2005: 218–280) will be one of the sources I will rely on in this task.

The most innovative and insightful aspect of Freeman's typology is represented by the dichotomy he proposes to further subdivide the area of the logically contingent non-evaluative statements. He moves from the ancient distinction between statements of fact (*descriptions*) and statements of definition (*interpretations*), but

---

[33] Again, Duns Scotus' notion of *possibilitas logica* was closely tied to the idea of constructibility of semantically congruous expressions (cf. *Introduction*, § 1.6.1). An interpretation of the same notion of congruous construction is the cornerstone of the semantic approach of Congruity Theory (see below § 2.3.4). For a discussion of the philosophical and linguistic roots of Congruity Theory see Rigotti and Rocci (2001).

[34] One aspect of Hermogene's typology is worth mentioning in view of the modern discussions about the rhetorical classification of claims reported by Kruger (1975) and by Freeman (2000): modern authors sometimes hesitate between including *evaluative claims* (also called *value claims*) and including *practical* (or *policy*) claims in their classifications; some authors include both, others chose one or the other. The typology of *stases* developed by Hermogenes deals elegantly with the relationship between evaluative and practical claims: the qualitative stases are subdivided so that the evaluation of *future* states of affairs pertains to the *pragmatic stasis* and the evaluation of accomplished states of affairs pertains to the stasis of *judicial qualification* (cf. Patillon in Hermogenes 2009: XLIII).

he gives an altogether different rationale for interpretations, which end up to cover much more than definitions in the strict sense. For descriptions, Freeman offers examples like the following:

Descriptions

(32.a) A bus is passing my office window.

(32.b) There was a red apple on the windowsill.

(32.c) An alien spaceship passed over my house last night.

(32.d) All the subjects in the experiment developed cold symptoms.

(32.e) All swans are white (understood as asserting only that there are no swans which fail to be white)

Descriptive statements are contingent and extensional. Their truth depends entirely on what is the case in the world, they do not involve consideration of what *could* have been the case or what *should* have been the case. They have a natural connection to the testimony of our senses. Freeman's class of description interprets the idea of factual claims in a much more stringent way. Statements of fact like (32.c) may be highly controversial and unlikely, but its eventual truth would depend only on the state of the world at a given time. Purely empirical generalizations such as (32.d) and (32.e) also fall in the descriptive category, inasmuch they function as summaries of several reports of individual happenings without claiming a stronger connection between an antecedent and a consequent.

By stronger connection between an antecedent and a consequent Freeman means the kind of connection that can be established by a causal predication or by a counterfactual conditional. As in (33.a, b and c):

(33)

(a) John developed cold symptoms *as a consequence of* his participation in the experiment

(b) *If* John *had* participated in the experiment, he *would have* developed cold symptoms

(c) All the subjects in the experiment *will necessarily* develop cold symptoms.

All the statements in (33) are, in fact, *interpretations.* Freeman defines interpretations as contingent, non-evaluative *intensional* statements. These are propositions whose truth depends not only what is the case in the world (whether John actually participated in the experiment or not, whether he developed symptoms of a cold or not), but also on what could be could have been the case (like what kind of symptom he would have developed in case he participated). In contrast, being non-evaluative, they do not depend on what should be or should have been the case (e.g. that John should have participated in the experiment, or that it is unfortunate that he has cold symptoms).

It appears that the non-evaluative modal examples in (28) and (29.a) fit Freeman's definition of interpretations as *non-evaluative*, *intensional* statements. For instance, the truth or falsity of (29.a) depend not only from Suzy's actual actions, but also on the alternative courses of action were available to her in the given circumstances. At the same time, the truth of (29.a) is not determined by which courses of action were right or wrong. Similarly, the truth of (28) depends on the consideration of the alternative courses of action enabling the burglar to steal the diamonds. Yet, it does not depend on whether stealing the diamonds was a good idea or not, nor on the moral evaluation of the act of breaking into the bank's vault.

Freeman refines further his definition of interpretive statement in a way that suggests even more clearly a connection with the semantics of modality. According to Freeman (2005: 108) interpretations either assert or presuppose *nomic regularities*, that is "subjunctive conditionals" (i.e. counterfactual in the broad sense) of necessity or possibility "making a claim about all possible worlds of some sort". Their truth does not depend just on what is the case in a world of evaluation but also on what happens in a range of relevant alternative worlds for which the nomic generalization holds.

These "modal" conditionals are, of course, not reducible to the extensional *material implication*, merely indicating that it is not the case that the antecedent is true and the consequent false. At the same time, they cannot be understood in terms of an unrestricted necessary implication (*strict implication*) stating that in all the worlds (no matter how remote and implausible) where the antecedent is true the consequent also is. Interpretations correspond to "subjunctive conditionals" where the antecedent necessitates the consequence only on a backdrop of assumptions:

> Consider, for example, "If you were to turn the key in the ignition, the car would start" This presupposes that the car is in proper working order, has some fuel in the tank, and that the laws of physics governing hoe the car's ignition mechanism brings about its starting remain the same. In all worlds where these conditions hold true, then if the ignition is activated the car starts. We are not concerned with worlds where these conditions are somehow altered (Freeman 2005: 150, paraphrasing Daniels and Freeman 1980)

I will not delve here into the possible worlds semantics employed by Freeman to formalize these conditionals. Chapter 4 will introduce the relevant machinery of possible worlds semantics, on which also Relative Modality theory relies. The Relative Modality analysis of modal meanings will also show how the ideas of necessity and possibility of a proposition *on a backdrop of contextual assumptions* are key for understanding the functioning of modality in natural language (cf. Chap. 4, § 4.2.7). Not only that, one of the central ideas of Relative Modality is actually that, once we understand modal operators as relational predicates relying a proposition and a background of assumptions, the semantics of modal operators and that of the context-dependent "subjunctive conditionals" discussed by Freeman can be given a unified semantics (cf. Chap. 4, § 4.2.8).

It can be observed that the modal proposition in (28) cannot be reduced to purely causal connections: it states what is necessary in a certain situation in view of the *goals* of an agent. In fact, propositions stating or presupposing a causal connection are a prominent type of interpretation, but they are not the only one. Statements, such as (34.b) below, ascribing intention, significance or meaning to an act also fall under Freeman's idea of interpretive statements. In order to better appreciate the argumentative implications of Freeman's idea of interpretation it is useful to examine more closely this example in comparison with the descriptive statement in (34.a).

(34.a) There was a red apple on the windowsill

(34.b) Horatio placed that red apple on the windowsill to show his love for Ophelia

The descriptive statement in (34.a) only contains observational predicates, like "x *is an apple*, y *is a windowsill*, x *is on* y". The example in (34.b), on the other hand, invokes non observational concepts such as Horatio's *love* – a disposition – and the apple *being a sign* of that love. These are intensional concepts and make the statement an intensional one, whose truth conditions depend not only from the state of the actual world but also of other possible worlds.

Imagine, as Freeman asks us, that the two statements are put forth as part of an argumentative exchange. Freeman observes that the addressee of (34.a) and (34.b) will proceed rather differently in determining whether these statements are acceptable – i.e. have an "epistemic presumption" (Freeman 2005: 21–37) in their favor. While knowledge of (34.a) can be based on direct perception by the speaker, and count as "personal testimony" for the addressee, (34.b) is more complex because the speaker "is not reporting what he has seen but explaining Horatio's overt act in terms of Horatio's dispositions and intentions" (Freeman 2005: 93). For the addressee this would never count as a simple *testimony* based on direct perception; rather it could be accepted as an *expert opinion* if the addressee is "aware of the proponent's having expertise or at least special credentials concerning Horatio's intimate affairs" (Ibid.). For Freeman the different conditions of presumption of the two statements closely mirror the different logical type of the two propositions.

In the following chapters Freeman's semantically motivated typology of statements will be used as a point of reference in examining the specificities of the argumentative functioning of the different kinds of modal propositions resulting from the different flavors of modality.

2.3.4 Congruity Theory

The semantic and pragmatic analysis of meaning requires an adequate metalanguage. The aim of bringing the analyses of modal expressions to bear on argumentative reconstruction in the respects anticipated in the previous sections (see, in particular, § 2.3.2) demands that this language be flexible, capable of covering different aspects of meaning, but also not overly ambitious in its requirements of explicitness and formality. As it will be shown in Chap. 4 modality is one of those areas of meaning were models deriving from formal logic, such as Relative Modality, have enjoyed a remarkable descriptive and explanatory success. Formal approaches of this kind, issued from the logical tradition, emphasize the requirement of explicitness to a degree that is uncommon in most areas of the humanities and social sciences. For instance, when Portner (2009: 106–107) compares formal and functional approaches to modality he sets the bar of explicitness very high for an explanatory account to be called a theory:

> Whatever the basic nature of semantic facts is [...] a semantic theory is a system of ideas which predicts those facts, makes additional predictions about what other facts we may discover, and does all this in a completely mechanical way, that is, without relying on a prior understanding of the language being analyzed or of the theoretical terms involved. Of course, in order to be a candidate for the right theory, it should not make any predictions which we know to be incorrect.

This is a view that formal semantics directly inherits from Richard Montague's (1970) program of treating English "as a formal language". Portner is ready to admit that currently there are no theories of semantics that conform to this ideal. There are, however, completely explicit "fragments" of such an ideal theory and there are "theories which are not completely mechanical, but which could be made mechanical with a little bit of work" (Portner 2009: 107). Both can be called *theories* in a looser sense, according to Portner.

Most dialectical, pragmatic –let alone rhetorical– approaches to argumentation are far from qualifying even for this "looser" use of *theory*. They are not mechanizable in their present state of development – think, for instance, of the powerful and quite unrestricted "transformations" that both Pragma-Dialectics and AMT allow to bridge natural discourse and their respective canonical reconstructed "deep structures". Perhaps more importantly, I am not entirely sure that most theories of argumentation do strive (or should strive) towards such an ideal of mechanization, except in a very approximate manner and only for practical purposes such as the automatic extraction of arguments from texts (*argument mining*). Contrary to formal semantics, they typically sacrifice explicitness for a broader coverage of the vast and largely uncharted territory of argumentation, remaining content of a more moderate understanding of explicitness as cleanness and consistency of definitions and employing mere shorthand schemes rather than fully fledged formalisms.

In fact, I believe, contra Portner, that full mechanization of a theory or theory fragment is not the only possible goal of formalisms. Precise, highly explicit representations, have a role to play also in approaches that remain far removed from the restrictive ideal outlined above. As Devlin and Rosenberg (1996) suggested, formalization can be a precious tool to "zoom-in" and achieve precision in considering aspects of complex processes of meaning construction and communicative interaction, which, as a whole, are not amenable to formal treatment. This is what can be gained from the explicitness of formal theories of modality in the present discussion of modality in relation to the essential structure of arguments in natural discourse.

As anticipated in Sect. 2.3.1, I will use elements Congruity Theory both to present semantic analyses of lexical and constructional expressions of modality and to represent speech-act level discourse relations in argumentative discourse. Congruity Theory provides an adequate language for bringing together the semantics of modality and argumentative reconstruction; not only because it represents a middle-ground informal approach where quasi-formal, symbolic representations are used in the spirit of "zooming" and clarification of specific nodes of meaning construction; but also for some of its substantial features that make it apt to capture how modal expressions interact with context via their presuppositions and contribute to manifest relations between utterances in discourse.

2.3.4.1 Congruity Theory and Lexical Semantics

The basic considerations on the semantics of *to create* and *take the initiative in V-ing* presented in § 2.3.1 are based on Congruity Theory's approach to the semantics of lexical, morphological and constructional predicates, in particular as regards the attention devoted to the selectional restrictions associated to the argument places of predicates (cf. Rigotti 1993, 2005; Rigotti and Rocci 2001, 2006a; Rigotti et al. 2006; Rocci 2005, Rocci and Luciani 2016b).

An extended notion of *predicate* plays a kernel role in Congruity Theory. Functionally, predicates are functors that require application to one or more arguments in order to obtain a proposition. Ontologically, predicates are seen as *possible modes of being*, a general notion that subsumes more specific distinctions such as those between properties and relations, states and events, actions and non-actions. The selectional restrictions that predicates impose on their argument places are seen as the source of the presuppositional phenomena that demand satisfaction or accommodation in the discourse context. Consider, for instance, the following pairs of sentences:

(35)
(a) Lamine Diack *has resigned* from his position as an honorary member of the International Olympic Committee.
(b) Lamine Diack *has not resigned* from his position as an honorary member of the International Olympic Committee.

Uttering either the affirmative version of the sentence in (35.a) or the negative in (35.b) requires the recognition of a series of implications which survive negation and need to be *satisfied* in the common ground of the exchange or *accommodated* by the audience:

1) Lamine Diack *exists* (existential presuppositions)
2) Lamine Diack *is a person* (categorial presupposition)
3) Honorary membership in the International Olympic Committee *exists* – as a social, institutional, fact (existential presupposition)
4) Honorary membership in the International Olympic Committee *is an office* (categorial presupposition)
5) There was a time prior to the moment of utterance when Lamine Diack held the office of honorary member of the International Olympic Committee (factive presupposition).

Propositions 1–5 are presuppositions of (35.a) and (35.b). Admitting them is a precondition of the meaningful utterance either of (35.a) or (35.b). Hearers who recognize them as part of the common ground of the ongoing exchange simply *satisfy* this requirement by checking it, those who don't are invited to cooperate with the speaker by *accommodating* them into the common ground as long as there is no contradiction with what is already known and no reason to distrust the speaker.[35] When it is impossible either to satisfy or accommodate these presuppositions the utterance is *incongruous*, as it happens with the following examples:

(36)

(a) *Mickey Mouse *has resigned* from his position as an honorary member of the International Olympic Committee. (Mickey does not exist – existential presupposition failure)

(b) *The building of the Olympic Stadium *has resigned* from his position as an honorary member of the International Olympic Committee. (The building of the Olympic Stadium is not a person – categorial presupposition failure)

(c) *Lamine Diack *has resigned* from his position as First Commodore of the Klingon Fleet (The position of First Commodore of the Klingon Fleet is not a really existing office – existential presupposition failure).

(d) *Lamine Diack *has resigned* from his position as native of Dakar (Being native of Dakar is not an office one can resign from – categorial presupposition failure).

(e) *Vladimir Putin *has resigned* from his position as an honorary member of the International Olympic Committee. (Vladimir Putin never held such an office, and so could not resign from it – factive presupposition failure).

The source of the presuppositions in (35.a-b) and of the presupposition failures in (36.a-e) is to be located in the presuppositional preconditions attached to the predicate *to resign* (x_1, x_2) and, more specifically, to its two argument places: if the actual discourse referents filling these argument places do not respect these presuppositional requirements the utterance results in incongruity. Congruity Theory posits predicates, different layers of semantic hierarchy, as the source of all presuppositions and sees meaning construction as a process globally guided by the search for possible congruous application of predicates to potential argument terms, so that complex, discourse spanning, predicate-argument structures[36] are established.

While this view of predicates as the source of presuppositions plays a crucial role in Congruity Theory, it is by no means unique to it. Recently, a major work in lexical semantics, such as Asher (2011: ix) has put the contextual "mechanisms of presupposition justification" at the centre of an enriched view of compositional meaning construction processes. Asher (2011) sees "type presuppositions" – presuppositions

---

[35] On the notion of *accommodation* see Lewis (1979) and Thomason (1990). David Lewis' original "rule of accommodation for presupposition" is quite powerful and unrestricted: "If at time *t* something is said that requires presupposition *P* to be acceptable, and if *P* is not presupposed just before *t*, then – *ceteris paribus* and within certain limits – presupposition *P* comes into existence at *t*" (Lewis 1979: 340). Successive researchers were left to tackle with the arduous, and still not clearly resolved, task of listing the relevant *cetera* and delimiting the extent of the "certain limits".

[36] Let us remind one last time the reader of the ambiguity of the term *argument* between semantics and argumentation theory. Per se, *predicate-argument structures* in linguistic semantics have absolutely nothing to do with *argument (macro-)structure* in Freeman's sense in argumentation theory, which roughly corresponds to *argumentation structure* in Pragma-Dialectics. Yet, ultimately, also argumentation structure, as a part of the meaning of a discourse, can be brought back to predicate-argument structure. But that's another story.

that an argument term belongs to a given semantic type – as originating from the selectional restrictions of predicates and develops a sophisticated account of how predicates and arguments adapt to each other in context to guarantee that presuppositions are justified. Asher's (2011) notion of "type presupposition" extends to embrace various kinds of categorial and factual presuppositions, but does not go as far as to suggest that *all* presuppositions generate from predicates. Closer to this view and, consequently, to Congruity Theory is the analysis of Pieter Seuren who distinguishes two types of "predicate conditions": *satisfaction conditions* and *preconditions* as components of the semantic representation of any predicate. Seuren locates "the source of presuppositions", including existential presuppositions, in the preconditions of lexical predicates (cf., in particular, Seuren 1985: 244 and Seuren 2009: 341).[37]

In line with Seuren (1985, 2009), predicates are analysed by Congruity Theory in terms of the *presuppositions* they places (or n-uples of argument places) and of their semantic *entailments* (Seuren's *satisfaction conditions*). For instance, a rough attempt at analysing the semantics[38] of the predicate *to resign* according to the Congruity Theory template would go as follows:

(Def. 1) *to resign* $(x_1, x_2) =$
Presuppositions: $\exists x_1\ \exists x_2$: x_1 is a person, x_2 is an office, at some time t_{-1} just before the time of verification t x_1 held x_2.
Entailments: at time of verification t, x_1 performs a speech act which conventionally causes x_1 to cease to hold x_2.

Congruity Theory postulates that for the predicate to be *congruous* with respect to its arguments, the semantic features associated to the actual arguments have to be both *compatible* with and *more informative* than (that is to say hyponyms of) the presuppositions associated to the argument place (cf. Rigotti and Rocci 2001 for a more developed discussion). Rigotti (2005: 79) states the *principle of congruity* in the following terms:

> The requirement of congruity between predicate and argument can be formulated in terms of the following law: There is semantic congruity between a predicate term and the argument terms it is applied to when the characteristics imposed by the predicate on each argument place are hyperonyms of the characteristics of the respective arguments

---

[37] Seuren (1985: 244) offers an exemplarily perspicuous exposition of the relationship between predicates and presuppositions: "Presuppositions are derived from the preconditions associated with the predicates in the semantic lexicon. A sentence figuring a predicate with some precondition will have as a presupposition the application of the precondition in question to the term or the terms in question". Presuppositions are thus distinguished from "denotational entailments", which originate in the satisfaction conditions.

[38] Here we use a style of presentation of predicate presuppositions and entailments that is similar to those used for lexical sematic analysis by Charles Fillmore (2003), Igor A. Mel'čuk (2012) and Pieter Seuren (1985).

When a predicate appears as the rheme of a speech-act of the assertive type, its presuppositions have to be *satisfied* (or *accommodated*) in the common ground of the dialogue, while its entailments are *added* to the common ground. For this reason we can also call them *update conditions*.

While Congruity Theory can be used, and will be used here, as a semi-formal template for discussing the meanings of lexical and constructional predicates, its main raison d'être is providing a unitary account for speech acts and discourse relations based on the idea of *abstract predicates* imposing a requirement of congruity on the *discourse utterances*, on the *participants* of the communication event, as well as on the *discourse context* (Rigotti and Rocci 2001; Rigotti 2005; Rocci 2005). Such an account can incorporate insights from existing analyses of argumentatively relevant speech acts (cf. van Eemeren and Grootendorst 1984, Houtlosser 2002) and from linguistic theories of rhetorical relations and discourse coherence (cf. Mann and Thompson 1987, Asher and Lascarides 2003).

The key notion in this unified account is the notion of a *pragmatic predicate*, also called a *connective predicate* (cf. Rigotti 2005).

2.3.4.2 Pragmatic Predicates and Rhetorical Relations

A *connective predicate* is an *action* predicate saying what the *speaker* does to the *addressee* with the *utterance* to which the predicate relates. In other words, pragmatic predicates are performative predicates. Like other predicates, connective predicates are analyzed in Congruity Theory in terms of the *presuppositions* they impose on their argument frame and of the *entailments* (or *update* conditions) with which they update the common ground. Since these pragmatic predicates not only *represent* but also *realize* social actions, their presuppositions function as contextual *preparatory conditions* for the performance of the act ad their update conditions are *pragmatic effects*, involving the creation of commitments. These are stored in a specific area common ground, which we can call, with Hamblin (1971), *commitment slate*.

Connective predicates (CP) capture the pragmatic, illocutionary, level of discourse, covering both the area addressed by Searlean illocutionary acts and the level of rhetorical relations responsible for discourse coherence. Pragmatic predicates allow to seamlessly extend the principle of congruity to the analysis of the semantico-pragmatic coherence of whole discourses. The minimal schema of a connective predicate is the following:

(Def. 2) CP_{U0} (Spk, Hr, U_0)

Note that the third argument U_0 is identified with the *utterance* (or broader discourse sequence) whose pragmatic function is being analysed and not directly with its *propositional content*. This allows us to accommodate also the pragmatic predicates associated with utterances such as (26), which do not have a semantic propositional content proper, but only a purely pragmatic frame (e.g. *Spk* greets *Hr* by uttering U_0).[39]

(37) Hi!

A congruity theoretic analysis of (27) would go along the lines of (Def. 3), below.

(Def. 3) *Hi!*$_{U0}$ (Spk, Hr, U_0, U_{+1}) =
Presuppositions:
Spk and *Hr* are in an informal interpersonal relationship, at some time t_{-1} just before the time of utterance t_0 *Spk* and *Hr* are not engaging in interaction, U_0 is a conventionally appropriate utterance of *Spk* at t_0, U_{+1} is a possible utterance of *Hr* at a time $t_{+1}>t_0$
Pragmatic effects:
By uttering U_0 at time of utterance t_0, *Spk* recognizes *Hr* as a potential interaction partner, inviting *Hr* to reciprocate the recognition in U_{+1} and granting to *Hr* the faculty of initiating interaction in U_{+1}.

Since the earliest formulations of Congruity Theory, terms such as *connettivo sequenziale* ('sequence connective', Rigotti 1993) or *connective predicate* (Rigotti 2005) have been used to refer to the class of pragmatic predicates. As we can see from the analysis of the most simple example in (37), the notion of connective predicate makes it possible to treat discourse and dialogue coherence in terms of predicate-argument congruity. We can see that the utterance at issue is not the only utterance appearing as argument of the connective predicate. Thanks to the argument U_{+1}, the analysis in (Def. 3) makes manifest how the greeting projects the second turn (second-pair part) of an adjacency pair, or dialogical relation.

In fact, the domain of discursive action addressed by connective predicates includes countless predicates – such as, for instance, *answering*, *objecting*, *concluding*, *explaining*, *clarifying*, *correcting* and *self-correcting*, *elaborating*, *proposing*, *refusing* and *counter-proposing* – that are relational also in the sense that they are logically dependent from the content of another utterance. To represent this the argument frame of the pragmatic predicate routinely includes what Rigotti calls *phoric arguments:* that is *anaphorically* recovered utterances (U_{-n}) and *cataphorically* projected ones (U_{+n}). Finally, as it happens with tacitly understood arguments of lexical predicates, inferred contextual propositions (X) can be understood as being part of the frame. As a result, the generic frame of a connective predicate can be extended as in (Def. 4).

(Def. 4) CP_{U0} (Spk, Hr, X, U_{-n}, U_0, U_{+n})

In this framework the felicity conditions imposed by Searlian illocutions (cf. Searle 1969; Searle and Vanderveken 1985) – which typically involve the speaker and hearer – are reinterpreted either as presuppositions imposed by the connective

[39] This dispenses us of the special proviso that Searle and Vanderveken (1985: 215) formulate about the act of greeting: " 'Greet' is only marginally [italic is ours] an illocutionary act since it has no propositional content".

predicate on n-uples of argument places including for instance the speaker (*Spk*), the hearer (*Hr*) and the utterance U_0 or as pragmatic effects of the predicate.[40]

Rhetorical relations, as postulated by Rhetorical Structure Theory (RST, Mann and Thompson 1987, 1988), are even more straightforwardly rendered in terms of presuppositions and effects of the connective predicate. For instance, in RST *Evidence* is a two-place relation holding between two spans of text, the nucleus (N) and the satellite (S), which are pragmatically asymmetric, the nucleus being "more essential to the writer's purpose" (Mann and Thompson 1987: 31) than the satellite and enjoying a greater degree of autonomy. The relation is then defined in terms of the constraints associated with the two argument places and of its intended effect:

(Def. 5)
Evidence

> *Constraints on N*: The reader might not believe N to a degree satisfactory to the writer;
> *Constraints on S*: The reader believes S or will find it credible;
> *Constraints on N+S combination*: The reader's comprehending S increases the reader's belief of N;
> *Effect:* The reader's belief of N is increased.

A first, rough and "literal", conversion of the *Evidence* relation in Congruity Theory terms would look as follows:

(Def. 6)

> *Evidence* (Spk, Hr, U_0, $U_{+1/-1}$) =
> Presuppositions:
> *Spk* believes the propositional content p of U_0, *Hr* may not believe the propositional content p of U_0 to a degree satisfactory to *Spk*, the propositional content q of $U_{+1/-1}$ is part of the common ground of *Spk* and *Hr* or will be found credible by *Hr*, the propositional content p of U_0 can be inferred from q of $U_{+1/-1}$. (together with the appropriate context).
> Pragmatic effects:
> By uttering U_0 at time of utterance t_0, *Spk* invites *Hr* to accept the propositional content p of U_0 by making it follow inferentially from q of $U_{+1/-1}$ (together with the appropriate context).

One can observe that in the perspective of Congruity Theory different versions of the *Evidence* relations need to be specified according to the cataphoric or anaphoric nature of the evidence. Clearly, a relation like RST's *Evidence* is argumentative in nature as it concerns precisely the link between arguments and standpoints. The rough translation in terms of Congruity Theory proposed here already modifies the language of the original to put it in line with an inferential view of argumentation

[40] Consider, for instance, the preparatory conditions of the assertion as formulated by Searle (1969). One of them states that 'It is not obvious to the *Speaker* that the *Hearer* knows (does not need to be reminded of) *p*'. This type of condition is treated as a (relational) presupposition imposed by any 'assertive' connective predicate on the argument places characterized by the roles of the *Speaker*, the *Hearer* and the asserted proposition.

and to account for contextual, implicit premises. It, however, still falls short of a fully adequate representation of argumentative moves in discourse.

I will focus more closely on rendering argumentative speech acts and argumentative relations in the next section. Here, let us pause and take stock as regards the general notion of a *connective predicate*. To summarize, the congruity theoretic construct of connective predicate addresses the areas of meaning covered by notions such as *illocutionary act*, or *dialogue act*, as well as those which are usually treated in terms of discourse relations, *coherence relations* (Knott and Sanders 1998), *rhetorical relations*, or *rhetorical predicates* (Grimes 1975).

While the conflation of these two kinds of notions might seem surprising from the viewpoints of traditional speech-act theory, it appears quite a natural step to take in view of the analysis of discourse and, especially, dialogue. The need of such a conceptual merger is recognized, for instance, by Segmented Discourse Representation Theory (Asher and Lascarides 2003) one of the more influential and formally sophisticated theories of discourse and dialogue. Asher and Lascarides (2003: 305) observe:

> Searle and more recent AI work [...] typically take speech acts to be a property of an individual utterance. In contrast, SDRT shows that many types of speech acts must be understood relationally, because successfully performing the speech act is logically dependent on the content of an antecedent utterance. So technically speaking, the type must be (at least) a two place relation [...]. For example if one uses an utterance to conclude something, then that conclusion must be antecedent to hypothesis or argument. And this is relational because successfully performing this speech act is dependent on the content of the antecedent.

One limitation of SDRT, for the purposes of this book, is that the theory only addresses the truth-functional consequences of the performance of speech acts, that is the effect that their performance has on the truth conditions of a discourse representation progressively updated by the successive utterances of the discourse. The authors openly admit that they are not interested in accounting for Searlean illocutionary forces in terms of commitments. For this reason, this kind of approach cannot be easily interfaced with a commitment based normative theory of argumentation such a Pragma-Dialectics. In contrast, the congruity theoretic approach based on *connective predicates* outlined here, while much less formally developed than SDRT, adopts a basically Searlean notion of commitment and interfaces easily with Pragma-Dialectics.

Illocutions and discourse relations alike often receive little or no linguistic manifestation, and hearer have to infer them, in order to make sense of an utterance or discourse sequences. This remain true for connective predicates, which should not be identified with lexical or grammatical units, but the abstract predicates that are ultimately inferred by discourse interpreters in their effort of accounting for discourse coherence in term of the *congruity* of manifested utterances to pragmatic functions that justify their fit in the discourse as a whole and/or their relevance in the dialogical exchange. While connective predicates can be often tacit, there are still a number of linguistic items whose business is to impose quite detailed constraints on the connective predicates to be established by interpreters in their search for congruity. We can say that these items conventionally express *connective predicate tem-*

plates, which the hearer has to flesh out in the dialogue context (cf. also Ginzburg et al. 2003 on this point).

2.3.4.3 Discourse Relations and Argumentative Reconstruction

The role of modals as direct and indirect indicators of *argumentative* discourse relations will be a major focus in the following chapters of this book, and especially in the case studies presented in Chap. 4. It is therefore important to understand how the discourse relations captured by Congruity Theory's *connective predicates* fit in a normative pragmatic view of argumentation. What is the place of these argumentative discourse relations in a theory of argumentation?

The reconstruction of a discourse as a contribution to an argumentative discussion is greatly helped by a detailed *descriptive* analysis of discourse meanings specifying the performance a variety of speech-acts that can be reasonably considered to *count as* the performance of dialectically relevant speech acts. By *dialectically relevant* speech-acts here I mean specifically those speech acts defined as relevant to the resolution of the dispute in the Pragma-Dialectical model of the critical discussion (cf. van Eemeren and Grootendorst 2004: 67).

Argumentatively relevant speech acts include not only such acts as *advancing a standpoint*, *drawing a conclusion* from available premises or *presenting an argument* in support of a standpoint, but also a variety of other speech acts[41]: *casting doubt* on the interlocutor's standpoint, *demanding a clarification* of the content of the interlocutor's standpoint, *conceding* a proposition as part of the common ground, presenting a *refutation* (*rebuttal* or *defeater*) of an argument presented by the interlocutor. Virtually all these argumentatively significant speech-acts are *relational*, in the sense that they all establish a discourse relation – though not necessarily an inferential relation – between the utterance that realizes the speech-act and another utterance, anaphorically recovered or cataphorically anticipated. Many of these speech acts are also inherently *dialogical* as they relate to a previous or anticipated utterance of the other dialogue participant.

[41] See van Eemeren and Grootendorst (2004: 67) for a list of speech acts relevant in the context of a critical discussion. It should be observed that some of the examples that I mention here (e.g. *drawing a conclusion* and *presenting a refutation*) do not appear in van Eemeren's and Grootendorst's list. This happens because, in view of their contribution to a critical discussion, they can be seen as variants of more general speech acts. *Drawing a conclusion* can be a way of manifesting one's standpoint (or to establish the results of a discussion). *Refutations*, *rebuttals*, *defeaters* – the terminology and the relevant conceptual distinctions in this area are not well established – are special kinds of arguments, so that presenting a refutation can be construed as a subtype of the speech-act of presenting an argument. Speech-acts that are equivalent from the viewpoint of their dialectical import can have very different rhetorical effects. As classical rhetoricians teach us, drawing a previously unmentioned conclusion from already conceded premises can have an altogether different persuasive effect that declaring one's standpoint upfront and then proceed to present arguments in support of it, which the other party may or may not concede.

The Pragma-Dialectic theory of speech acts, as it was developed starting from van Eemeren and Grootendorst (1984), does itself recognize higher lever relational speech acts, starting from the speech act of argumentation itself. Monadic speech acts (e.g. assertion) are also present and are seen as constituents of those higher level configurations. There are a number of *technical* differences between Pragma-Dialectics and Congruity Theory in the treatment of relational speech acts, which I cannot discuss here.[42] Most are very minor, in my opinion, and easily overcome.

What is more important here is considering the basic division of labor between a descriptive model of discourse such as the one offered by Congruity Theory, and a normative, idealized model of argumentation. The types of speech acts that one and the other have to recognize is not the same. Many distinctions that are relevant for a descriptive account of discourse have no critical import in a reconstruction of arguments. At the same time, it should be clear when the performance of a given speech act in discourse *counts as* incurring an argumentatively relevant commitment in the context of a given stage of a critical discussion.

For instance, according to the Pragma-Dialectical model of the critical discussion, the proper reconstruction of the confrontation stage of a dispute requires that we determine the standpoint or standpoints being advanced by the participants. As shown by Houtlosser (2002) there is a variety of signals that can be observed in a descriptive analysis of argumentative discourse that can justify the analyst's conclusion that a participant incurred the commitments that characterize the act of advancing a standpoint. These include making manifest the anticipation of doubt in the speaker's presentation of an assertive speech-act, an interlocutor's reaction of non-acceptance, doubt or challenge to the speech-act as well as the presence of an argumentative follow-up provided by the speaker (cf. Houtlosser 2002: 173–182). Consider the following example used by Houtlosser (2002: 181) to illustrate how the presence of an argumentative follow-up can serve as an indirect indicator:

(38.a) She must be very brave, *because* she never complains.

Here, as observed by Houtlosser (2002: 180–181) the connective *because* has clearly an *inferential* rather than a causal reading – from the point of view of causation between events 'she never complains' is an effect – so that an interpretation of the clause it introduces as causal explanation is excluded and we must conclude that 'she never complains' is offered as an *argument.*

Houtlosser also discusses in detail other more complex cases where *because* does not refer to the *inferential* connection, but to a *causal* one and shows that they can be argumentative as well if certain conditions are met. In Chap. 5 as well as in the case studies in Chap. 6 it will be shown how non-epistemic readings of the

[42] I hope to write a paper very soon on this topic, easing out technical difficulties with the aim of providing a seamless interface between Pragma-Dialectics and Congruity Theory, so that Congruity Theory can be used as an option for the in-depth analysis of an argumentative discourse as a preliminary step for (i) its analytical reconstruction, (ii) the consideration of its strategic arrangement (*dispositio*) from a rhetorical viewpoint in the context of strategic maneuvering.

modals are pretty similar to causal readings of the connectives with regards to their capacity of conveying argumentative discourse relations indirectly through their evocation of causal loci.

Here I will leave out the more complex case of indirect indicators and dwell on the easy case of (38.a), to show what kind of distinction a descriptive theory of discourse relations may want to recognize and how they map onto the categories necessarily recognized by the normative model. First of all, it can be observed that (38.a) provides redundant marking of the inferential connection between 'her never complaining' and 'her being brave'. The presence of the modal *must* clearly marks the standpoint as the result of an inference, i.e. as a conclusion. The reasons why the epistemic interpretation of the necessity modal *must* has this potential as an argumentative indicator will be discussed in detail in Chaps. 5 and 6. For now, let us be content with observing that the indicative potential is there. The role of *must* as an argumentative indicator becomes apparent in (38.b), where the argumentative relation remains clear despite the absence of the discourse connective *because*:

(38.b) She *must* be very brave. She never complains.

Let us sketch how Congruity Theory would represent the speech-acts realized by the two utterances in (38.a) and (38.b) by assigning a connective predicate to each of them:

U_1: 'She must be very brave'	*Conclude_f* $_{U1}$ (Spk, Hr, U_1, U_X)	
U_2: '[because] she never complains'	*Argument*$_{U2}$ (Spk, Hr, U_2, U_1)	$U_x = U_2$

It can be observed that, contrary to RST and SDRT, each utterance in the developing discourse is assigned a connective predicate that specifies what kind of relational speech-act the utterance at issue is performing. In interpreting U_1 the hearer recognizes, also thanks to the presence of the modal *must*, that the utterance is performing an act of conclusion from some body of evidence and that this act is forward-looking (*cataphoric*) and it opens up a place, an argument$_S$ place (argument$_S$ in the semantic sense) to be saturated in the continuation of the discourse. The utterance U_2 is then understood as providing an argument for U_1 saturating the open argument place. It can be observed that the discourse is observed in its linear development, describing the pragmatic function of each step. As a result, discourse relations are often represented redundantly, contrary to what happens in other theories of discourse relations, where a compact representation of the resulting discourse structure is given priority over the representation of each unit as a speech-act in a linearly unfolding discourse. In RST, for instance, U_1 and U_2 would have been respectively the nucleus and the satellite of the same *Evidence* relation.

In fact, the "redundant" connective predicates *Conclude_f* and *Argument* are simply *converses* (like, say, *buying* and *selling*)[43] and represent the same relation from the viewpoint of different stages of the discursive development. This redun-

[43] On the semantic notion of *converse* or *conversive* predicates see Fillmore (2003: 255) and Rocci and Luciani (2016b: 71)

dancy is, to some extent, detectable at the surface of the text in the original example (38.a), where both *must* and *because* appear. In reality, redundant representation of converse relations is normal in discourse, also at the level of propositional contents, as demonstrated by (39), below.

(39) Finally, Peter **sold** his house. It was **bought** by a young woman from Zurich.

A different, very obvious, *dispositio* of the same argument of (38.a) is represented in (38.c), below:

(38.c) She never complains. She must be very brave.

This discourse will be analyzed in terms of connective predicates as follows:

U_1: 'She never complains' *Assert*$_{U1}$ (Spk, Hr, U_1)
U_2: 'She must be very brave' *Conclude_b*$_{U2}$ (Spk, Hr, U_2, U_1)

Here, at least in the absence of a richer real context, U_1 is not immediately recognized as a premise. So that the argumentative discourse relation is established non-redundantly when the backward-looking (*anaphoric*) variant of the *Conclude* connective predicate is recognized.

Now, clearly at the level of the commitments that are relevant for a dialectical and inferential reconstruction of the argument (38.a), (38.b) and (38.c) are perfectly equivalent. In the three cases the speaker's behavior *counts* as a performance of two dialectically relevant speech-acts: (i) *putting forward a standpoint* and (ii) *presenting argumentation* in support of said standpoint in view of the (manifested or anticipated) doubt of an antagonist. This mapping between connective predicates and dialectically relevant commitments is summarized in Table 2.2.

Interestingly, in (38.c) if the speech act of U_1 is recognized as counting as an argument it is *a posteriori* because U_2 has been recognized as an anaphoric act of conclusion. It should be stressed that these are just three of many possible dialectically equivalent discursive or dialogic configurations. For instance, a dialogue in which someone asserts "She is very brave" as an unproblematic statement, then is met with the doubt of the interlocutor ("Why do you say, so?") and then produces

Table 2.2 From descriptive discourse analysis to reconstruction as a critical discussion: the mapping between connective predicates and dialectically relevant commitments

Examples	Connective Predicates in the unfolding discourse	Dialectically relevant commitment
(38.a), (38.b)	*Conclude_f*$_{U1}$ (Spk, Hr, U_1, U_X)	The propositional content p of U_1 is put forward as a standpoint;
	Argument$_{U2}$ (Spk, Hr, U_2, U_1)	The propositional content q of U_2 is presented as an argument supporting the standpoint.
	$U_x = U_2$	
(38.c)	*Assert*$_{U1}$ (Spk, Hr, U_1)	The propositional content q of U_1 is presented as an argument supporting the standpoint;
	Conclude_b$_{U2}$ (Spk, Hr, U_2, U_1)	The propositional content p of U_1 is put forward as a standpoint;

an argument in support ("See, she never complains") would produce exactly the same dialectically relevant commitments on the part of the protagonist (cf. Houtlosser 2002: 175–178). Yet, this dialogue would certainly deserve its own descriptive analysis in terms of connective predicates.

In a good congruity theoretic analysis, the dialectical equivalence of different discourse structures should emerge clearly from the conditions (presuppositions and pragmatic effect) that define each of the connective predicates. These conditions, if properly formulated, will contain – sometimes redundantly, always sufficiently – all the commitments that make them *count as* the performance of a certain dialectically relevant speech act in the idealized reconstruction of the argument. They will contain them along with other dialectically irrelevant conditions that will be filtered out. The dialectically irrelevant information, for the most part, will turn up again to be relevant (i) in the analysis of features of the discourse genre and concrete activity type, (ii) in view of an analysis of rhetorically relevant choices (deep *dispositio*, in particular).

As it will gradually emerge in the following chapters, the mediation of a rich layer of descriptive pragmatic analysis of discourse in terms of Connective Predicates is important to describe *how exactly*, i.e. through what kind of interpretive path, modal expressions and other indicators contribute to the reconstruction of the dialectical, as well as of the inferential, properties of arguments.

2.4 The Sociopragmatic Dimension: Argumentation in Context

There is another set of considerations, of a *prima facie* very different nature, which appears as important as semantic and pragmatic considerations in constraining and guiding argumentative reconstruction: the social context of interaction in which the argument takes place.

The study of the social context of interaction in which arguments occur and, in particular, of the *activity types* (Levinson 1979) in which arguments are embedded is a fast growing area of research in argumentation theory (cf. Rigotti and Rocci 2006b; Rocci 2008; and van Eemeren 2010).

Arguments take place within broad domains or spheres of human activity – *justice, education, politics, finance*, etc. – each characterized by an overarching concern, a "point" or *raison d'être*. Each domain gives rise to (more or less) institutionalized *interaction fields* – e.g. the American judicial system, or the Swiss banking system – where sets of institutionalized commitments consistent with the domain's *raison d'être* give rise to social roles defined in terms of their prerogatives and obligations towards other roles in the field.

At the same time, argumentation emerges as part of very specific, situated, activity types: *jural interrogation, press conference, analysts and investors conference call*. And these specific activity types result from the application and adaptation of broad culturally shared *interaction schemes* (e.g. *questioning*) to specific spots of the institutional interaction field (e.g. the criminal trial in a specific judicial system, the communication between listed companies and investors under a specific national system of regulations for the financial markets, etc.).

Taking into account this twofold contextual dimension (cf. Rigotti and Rocci 2006b), which we could call *sociopragmatic* (Dascal 2003), is essential for both for reconstruction and for evaluation for several reasons (see also van Eemeren 2010, 2013).

First of all, the point or basic concern of a domain defines the *issues* that can be relevantly debated within it and the specific activity types arising with institutionalized fields further restrict the range of relevant issues and hence of the standpoints that can be advanced. In other words, the context contributes to define the possible types of issues, which are roughly corresponding the different *status causae* or *stases* of the rhetorical tradition. By restricting the type of issues it also restricts the possible semantic type of the standpoints. In turn, the semantic type contributes to define the possible means of proof, and hence the *loci* that can be invoked. Consider, for instance, the use of an *alibi*. The alibi is an argument based on the maxim *If a physical entity e is at place p at time t, then it is not at a different place q at the same time t,* a complex maxim, which draws from the *locus from alternatives* and the *locus from time and place*. Such a maxim is relevant at a specific stage of a criminal trial, which corresponds to the *conjectural* stasis in classical rhetoric. It would be completely out of place, for instance, in a deliberative activity type in the context of a practical issue. The reverse applies to an *argument from consequences* – a subtype of the *locus from the final cause*. An argument from consequences is relevant in the *practical* stasis but irrelevant in the conjectural. The application of the argument from consequences to a descriptive, conjectural, issue results in the fallacy *ad consequentiam*.

Secondly, *procedural restrictions* characterizing the activity type might affect the allocation of the burden of proof (e.g. the presumption of innocence in the criminal trial) as well as the kind of arguments that can be *de iure* brought to bear on the issues.

Thirdly, the materiality of the activity types (e.g. time constraints, knowledge limitations) might further restrict *de facto* the options that are available to the arguers.

Finally, the domain ontology – both in its material an in its social aspects – creates characteristic affordances for certain kinds of arguments, in particular the use of certain *loci* or argument schemes. The availability of relevant culturally accepted taxonomies and authoritative definitions will affect the availability of the definitional loci; while the culturally shared frames of the activity type will clearly affect the recourse to the *locus from correlates* and to other circumstantial loci. For each domain different experts will be recognized, affecting the recourse to the *locus from authority*.

It might seem at first sight that in an investigation bridging semantics and argumentation such as the present one these *sociopragmatic* factors are of secondary importance.

In fact, it has been shown (cf. Greco-Morasso 2010; Rigotti and Greco-Morasso 2010; Rigotti et al. 2006) that the analysis of arguments that support deliberation in a given context is greatly enhanced by the use of semantic analysis to elucidate the meaning of the terms that define the social ontology of the context. Relevant semantic analyses should involve both the *cultural keywords*[44] that give access to important *endoxa* of the chosen field of interaction and of *framing words*[45], which give access to the natural and social ontology of the field and provide, as suggested above, the domain ontology where the *loci*-relations are to be located. As observed in Rocci and Musi (2014), abstract, logical words like modals are not value laden, domain specific nor are they markers that specifically concern interaction. Yet, the skeletal meanings of the modals provide a fundamental scaffolding to the folk ontologies of social reality and interactions that emerge from discourse. I have already given a hint of this scaffolding in Sect. 2.2.3.2, in discussing how causal loci have a deep connection with certain flavors of modality that are most prominent in discourse on human actions.

An examination of modalities in contextualized arguments is particularly productive for their proper reconstruction if we pay attention at *how these modalities relate to the different aspects of the social context* of interaction that we have briefly listed above. As anticipated in the introduction, my chosen sphere of human activity for this examination will be that of finance. More precisely, by looking at newspaper articles in the Italian financial press (cf. Miecznikowski et al. 2012; Palmieri and Miecznikowski 2016; Rocci and Luciani 2016a), I will consider a particular "spot" where the complex interaction field of the financial markets (Rocci 2014), with its network of institutions, intersects the interaction field of the news media.

In Chap. 6 starting with from basic concern of the financial domain for investment decisions (*raison d'être*), I will examine how a number of constraints and affordances of different origin affect how argumentation unfolds in the discourse genre of financial news. These constraints originate from the basic ontology of the physical world (investment decisions concern the possibility *future* returns and the future is unsettled), from the social norms and commitments to which the involved actors are bound (e.g. financial regulations, contracts), from the strategic behavior of the social actors (e.g. information asymmetries between company insiders and investors), from social circumstances (e.g. the expertise gaps between journalists, financial experts and lay readers), from the material requirements of the activities (e.g. the urgency and time constraints of the journalists). The analysis will show how these contextual constraints on argumentation are reflected by modal structures of different "flavors" in discourse. These modal structures become relevant for argumentative reconstruction each time the arguments exploit or need to address these features of the context.

[44] Cf. Rigotti and Rocci (2005) on the logical and pragmatic functioning of keywords as pointers to *endoxa* in enthymematic argument and Rocci and Monteiro's (2009) analysis of *interactivity* as a concrete example of semantic analysis of a value laden keyword.

[45] Cf. Rocci and Luciani (2016b) on frames and semantic analysis and the analysis of *conflict* in Greco Morasso (2010: 79–112) as a concrete example of semantic analysis of a framing word.

The hypothesis of this close connection between modal structures and the ontology of fields of interaction becomes less surprising once we have considered how the most commonly encountered flavors of modality are closely related to the ontology of human *action* and *interaction*, as we have seen, at least in part, in discussing causal *loci* in § 2.2.3.2. For instance, consider that reasoning and talk about *action*[46] and deliberation involve envisaging a state of affairs corresponding to the *desires* of the agent (*bouletic* modality) and generally possible in the future (circumstantial *dynamic* modality) as well as considering what actually *can* be done by the agent (participant internal *dynamic* modality) and what is *needed* to bring about the desire (*deontico-practical* modality). Moreover, since our agent has to cope with her partial and fallible access to the facts, she will consider what *may* be or *must* be the case in view of the evidence at her disposal (*epistemic* modality). Finally, since people do not act in a social and cultural void, our agent's action will inevitably relate to what *may*, *should* or *must* be done according to the *law*, to culturally shared *ethical norms* or *ideals* and to the *commitments* that the agent has entered in her interactions with other people (that is different sorts of *deontic* modality). One can already surmise how these modalities become relevant in relation to the constraints and affordances of financial communication sketched just above.

Toulmin ([1958] 2003) envisaged a direct connection between the context-dependent part of modal expressions and the different *fields* of argument, a somewhat elusive notion that he connected both with "logical types" of propositions and with fields of expertise and contexts of discussion.[47] The present book abandons Toulmin's idea of such a direct connection, but shows how a context-dependent semantics of modality can be a precious tool for investigating how contextualized argumentation is constrained by and takes advantage of the scaffolding provided by the ontology of the interaction field.

References

Anscombre J-C, Ducrot O (1983) L'argumentation dans la langue. Mardaga, Bruxelles

Aristotle (2014) The complete works of Aristotle: the revised Oxford translation, one-volume digital edition. In: (ed) Barnes J. Princeton University Press, Princeton (originally published in 1984 as 2 volumes)

Asher N (2011) Lexical meaning in context. A web of words. Cambridge University Press, Cambridge

Asher N (2013) Temporal modification. In: de Saussure L, Jaszczolt KM (eds) Time: language, cognition and reality. Oxford University Press, Oxford, pp 15–36

[46] We cannot enter here the discussion on the theory of action, but see Rigotti (2005) for a concise presentation of a basic model applied to discourse analysis.

[47] The notion of *field* in Toulmin is a puzzling one. At a certain point *fields* are identified with *logical types* (Toulmin [1958] 2003: 13): propositions having different logical type belong to different fields. At the same time it is clear from the examples of field given in the book that they also correspond to concrete fields of expertise, in some case to disciplines or contexts of discussion.

Asher N, Lascarides A (2003) Logics of Conversation. Cambridge University Press, Cambridge
Bitzer LF (1959) Aristotle's Enthymeme Revisited. Q J Speech 45:399–408
Chilton P (2005) Missing Links in Mainstream CDA: Modules, Blends and the Critical Instinct. In: Wodak R, Chilton P (eds) A New Agenda in (Critical) Discourse Analysis. John Benjamins, Amsterdam/Philadelphia, pp 19–52
Chilton P (2009) What is Critical Discourse Analysis and What Can we Mean by "Critical" in a Cross- Cultural Context? J Chinese Sociolinguist 2(13):10–24
Cicero MT (2003) Cicero's "Topica", Reinhard T (ed). Oxford: Oxford University Press.
Clark HH (1996) Using language. Cambridge University Press, Cambridge
Crook J (2004) On covert communication in advertising. J Pragmat 36(4):715–738. doi:10.1016/S0378-2166(03)00039-0
Dascal M (2003) Interpretation and Understanding. John Benjamins, Amsterdam/Philadelphia
Devlin K, Rosenberg D (1996) Language at work. Analyzing Communication Breakdown in the Workplace to Inform Systems Design. CSLI Publications, Stanford CA
Ducrot O (1980) Le mots du discours. Le éditions de minuit, Paris
Fahnestock J, Secor M (1982) A rhetoric of argument. Random House, New York
Fairclough I, Fairclough N (2012) Political discourse analysis. A method for advanced students. Routledge, London/New York
Fauconnier G, Turner M (2006) Conceptual integration networks. In: Geeraerts D (ed) Cognitive linguistics: basic readings. Mouton De Gruyter, Berlin/New York, pp 303–371
Fillmore CJ (1984) Lexical semantics and text semantics. In: Copeland JE (ed) New directions in linguistics and semiotics. John Benjamins, Amsterdam/Philadelphia, pp 123–147
Fillmore CJ (2003) Form and meaning in language, vol 121. CSLI Publications, Center for the Study of Language and Information, Stanford
Fillmore CJ, Baker C (2009) A frames approach to semantic analysis. In: Heine B, Narrog H (eds) The Oxford handbook of linguistic analysis. Oxford University Press, Oxford, pp 313–340
Freeman JB (1991) Dialectics and the macrostructure of arguments. A theory of argument structure. Berlin/New York, Foris
Freeman JB (2000) What types of statements are there? Argumentation 14(2):135–157
Freeman JB (2005) Acceptable premises: an epistemic approach to an informal logic problem. Cambridge University Press, Cambridge
Freeman JB (2011) Argument structure: representation and theory. Springer, Amsterdam
Garssen B (2001) Argumentation schemes. In: van Eemeren FH (ed) Crucial concepts in argumentation theory. Amsterdam University Press, Amsterdam, pp 81–99
Ginzburg J, Sag IA, Purver M (2003) Integrating conversational move types in the grammar of conversation. In: Kühnlein P, Rieser H, Zeevat H (eds) Perspectives on dialogue in the new millennium. John Benjamins, Amsterdam/Philadelphia, pp 25–42
Greco Morasso S (2010) Argumentation in dispute mediation. John Benjamins, Amsterdam/Philadelphia
Grice HP (1975) Logic and conversation. In: Cole P, Morgan JL (eds) Speech acts. Academic Press, New York, pp 41–58
Grimes JE (1975) The thread of discourse. Mouton, The Hague
Hamblin CL (1970) Fallacies. Methuen, London
Hamblin CL (1971) Mathematical models of dialogue. Theoria 37(2):130–155
Hansen HV (2002) The straw thing of fallacy theory: the standard definition of fallacy. Argumentation 16(2):133–155
Hermogenes (2009) Les États de Cause. Patillon M (ed) Les Belles Lettres, Paris
Hitchcock D (1985) Enthymematic arguments. Informal Logic 7(2):83–97
Hobbs JR (2005) Toward a useful concept of causality for Lexical Semantics. J Semant:181–209. doi:10.1093/jos/ffh024
Hofweber T (2014) Logic and ontology. In: The Stanford encyclopedia of philosophy. Zalta EN (ed). http://plato.stanford.edu/archives/fall2014/entries/logic-ontology/ . Accessed 8 November 2016

Houtlosser P (2002) Indicators of point of view. In: van Eemeren FH (ed) Advances in Pragma-Dialectics. SicSat, Vale Press, Amsterdam, pp 169–184

Jackson S, Jacobs S (1980) Structure of conversational argument: pragmatic bases for the enthymeme. Q J Speech 66(3):251–265

Jacobs S (2000) Rhetoric and dialectic from the standpoint of normative pragmatics. Argumentation 14(3):261–286

Katzav J, Reed CA (2004) On argumentation schemes and the natural classification of arguments. Argumentation 18(2):239–259

Kay P (1997) Words and the grammar of context. CSLI Publications, Stanford

Kendall, Gavin. 2007. What is critical discourse analysis? Ruth Wodak in conversation with Gavin Kendall. Forum Qualitative Sozialforschung/Forum Qual Soc Res 8 (2) http://www.qualitative-research.net. Accessed 4 October 2016

Knott A, Sanders T (1998) The classification of coherence relations and their linguistic markers: an exploration of two languages. J Pragmat 30(2):135–175

Kratzer A (1981) The notional category of modality. In: Eikmeyer H-J, Rieser H (eds) Words, worlds, and contexts: new approaches in word semantics. Mouton de Gruyter, Berlin/New York, pp 38–75

Kratzer A (2012) Modals and Conditionals. Oxford University Press, Oxford

Kronning H (1996) Modalité, cognition et polysémie: Sémantique du verbe modal devoir. Uppsala University, Uppsala

Kronning H (2001) Pour une tripartition des emplois du modal devoir. In: Dendale P, van der Auwera J (eds) Les Verbes modaux. Cahiers Chronos 8. Rodopi, Amsterdam, pp 67–84

Kruger AN (1975) The nature of controversial statements. Philos Rhetor 8(3):137–158

Levinson SC (1979) Activity types and language. Linguistics 17:365–399

Lewiński M (2011) Towards a critique-friendly approach to the straw man fallacy evaluation. Argumentation 25(2):469–497

Lewis D (1979) Scorekeeping in a language game. J Philos Log 8(1):339–359

Mann WC (1988) Dialogue games: conventions of human interaction. Argumentation 2(4):511–532

Mann WC (2002) Dialogue macrogame theory. In: Proceedings of the third sigdial workshop on discourse and dialogue. Association for Computational Linguistics, Philadelphia, pp 129–141

Mann WC, Thompson SA (1987) Rhetorical structure theory: a theory of text organization, vol. Technical. University of Southern California (USC), Information Sciences Institute, Marina del Rey

Mann WC, Thompson SA (1988) Rhetorical structure theory: a functional theory of text organization. Text 8(3):243–281

McCawley JD (1993) Everything that linguists have always wanted to know about logic – but were ashamed to ask, 2nd edn. University of Chicago Press, Chicago/London

Mel'čuk IA (2004) Actants in semantics and syntax I: actants in semantics. Linguistics 42(1):1–66

Mel'čuk IA (2012) Semantics: from meaning to text. John Benjamins Publishing, Amsterdam/Philadelphia

Miecznikowski J, Rocci A, Zlatkova G (2012) L'argumentation dans la presse économique et financière italienne. In: Gautier L (ed) Les discours de la bourse et de la finance. Frank und Timme, Berlin, pp 65–83

Montague R (1970) English as a formal language. In: Visentini B (ed) Linguaggi nella società e nella tecnica. Edizioni di Comunità, Milan, pp 189–223

Musi E (2015) Dalle apparenze alle inferenze: i verbi sembrare ed apparire come indicatori argomentativi. Ph.D. dissertation, Università della Svizzera italiana

Musi E, Ghosh D, Muresan S (2016) Towards feasible guidelines for the annotation of argument schemes. In: Proceedings of the third workshop on argument mining (ArgMining2016):82–93. http://aclweb.org/anthology/W/W16/W16-2810.pdf

Musi E, Rocci A (2016) Obviously epistentials are argumentative indicators: evidence from an Italian and English corpus of newspaper articles. In: Proceedings of the workshop on the Foundations of the Language of Argument at COMMA (September 13, 2016, University of Potsdam)

Nølke H, Fløttum K, Norén C (2004) ScaPoLine. La Théorie Scandinave de la Polyphonie Linguistique. Paris, Kimé

O'Regan J (2006) This risky order of discourse: the normativity debate in critical discourse studies. Crit Discourse Stud 3(2):229–235

Palmer F (2001) Mood and modality, 2nd edn. Cambridge University Press, Cambridge

Palmieri R (2014) Corporate argumentation in takeover bids. John Benjamins, Amsterdam/ Philadelphia

Palmieri R, Miecznikowski J (2016) Predictions in economic-financial news: author's stance and argumentative loci. J Argument Context 5(1):48–73. doi:10.1075/jaic.5.1.03pal

Perelman C (1977) L'Empire Rhétorique: Rhétorique et Argumentation. Vrin, Paris

Perelman C, Olbrechts-Tyteca L (2008 [1958]) Traité de l'argumentation. La nouvelle rhétorique. Les Éditions de l'Université de Bruxelles, Brussels

Piazza F (2000) Il corpo della persuasione. Edizioni novecento, Palermo

Pollaroli C, Rocci A (2015) The argumentative relevance of pictorial and multimodal metaphor in advertising. J Argument Context 4(2):158–199

Pollock JL (1987) Defeasible reasoning. Cogn Sci 11:481–518

Pollock JL (2010) Defeasible reasoning and degrees of justification. Argument Comput 1:7–22. doi:10.1080/19462161003728921

Portner P (2009) Modality. Oxford University Press, Oxford

Recanati F (1989) The pragmatics of what is said. Mind Lang 4(4):295–329. doi:10.1111/j.1468-0017.1989.tb00258.x

Rescher N (2006) Presumption and the practices of tentative cognition. Cambridge University Press, Cambridge

Rigotti E (1993) La sequenza testuale. Definizione e procedimenti di analisi con esemplificazioni in lingue diverse. L'analisi Linguistica e Letteraria 1(1):43–148

Rigotti E (2005) Congruity theory and argumentation. Stud Commun Sci (Special Issue: Argumentation in Dialogic Interaction):75–96

Rigotti E (2007a) Relevance of context-bound loci to topical potential in the argumentation stage. Argumentation 20(4):519–540

Rigotti E (2007b) Can classical topics be revived within the contemporary theory of argumentation. In: van Eemeren FH, Blair A, Willard CA, Garssen B (eds) Proceedings of the 6th ISSA conference in argumentation. SicSat, Amsterdam, pp 1155–1163

Rigotti E (2008) Locus a causa finali. L'analisi Linguistica e Letteraria, XVI (Special Issue "Word meaning in argumentative dialogue"):559–576

Rigotti E (2009) Whether and how classical topics can be revived in the contemporary theory of argumentation. In: van Eemeren FH, Garssen B (eds) Pondering on problems of argumentation. Springer, New York, pp 157–178

Rigotti E (2011) Review of Frans H. van Eemeren: strategic maneuvering in argumentative discourse. Extending the pragma-dialectical theory of argumentation. Argumentation 25:261–270

Rigotti E (2013) The nature and functions of loci in Agricola's De Inuentione Dialectica. Argumentation 28(1):19–37

Rigotti E, Greco Morasso S (2010) Comparing the argumentum model of topics to other contemporary approaches to argument schemes: the procedural and material components. Argumentation 24(4):489–512

Rigotti E, Rocci A (2001) Sens–non-sens–contresens. Stud Commun Sci 1(2):45–80

Rigotti E, Rocci A (2005) From argument analysis to cultural keywords (and back again). In: van Eemeren FH, Houtlosser P (eds) Argumentation in practice. John Benjamins, Amsterdam/ Philadelphia, pp 125–142

Rigotti E, Rocci A (2006a) Tema-Rema e Connettivo: la congruità semantico-pragmatica del testo. In: Gobber G, Gatti MC, Cigada S (eds) Syndesmoi. Connettivi nella realtà dei testi. Vita e Pensiero, Milano, pp 1–27
Rigotti E, Rocci A (2006b) Towards a definition of communication context. Stud Commun Sci 2:155–180
Rigotti E, Rocci A, Greco S (2006) The semantics of reasonableness. In: Houtlosser P, van Rees A (eds) Considering pragma-dialectics. Lawrence Erlbaum, Mahwah, pp 257–274
Rocci A (2005) Are manipulative texts coherent? Manipulation, presuppositions, and (in-) congruity. In: Schulz P, de Saussure L (eds) Manipulation and ideologies in the twentieth century: discourse, language, mind, Discourse approaches to politics, society and culture, vol 17. John Benjamins, Amsterdam/Philadelphia, pp 85–112
Rocci A (2006) Pragmatic inference and argumentation in intercultural communication. Intercult Pragmat 3(4):409–442. doi:10.1515/IP.2006.026
Rocci A (2008) Analysing and evaluating persuasive media discourse in context. In: Burger M (ed) L'analyse linguistique des discours médiatiques. Entre sciences du langage et sciences de la communication. Nota Bene, Québec, pp 247–284
Rocci A (2014) The discourse system of financial communication. Cahiers de l'ILSL 34:201–221
Rocci A, Luciani M (2016a) Economic-financial journalists as argumentative intermediaries. J Argument Cont 5(1):88–111
Rocci A, Luciani M (2016b) Semantics and verbal communication. In: de Saussure L, Rocci A (eds) Verbal Communication, Handbooks of communication science, vol 3. Mouton de Gruyter, Berlin/New York, pp 57–75
Rocci A, Monteiro MW (2009) Cultural keywords in arguments. The case of "Interactivity". Cogency 1(2):65–100
Rocci A, Elena M (2014) From modal semantics to discourse: exploring the functioning of the adjective "possibile" in a corpus of Italian financial news. In: Blochowiak J, Grisot C, Laenzlinger C, Durrlemann-Tame S (eds) Papers dedicated to Jacques Moeschler, vol 1–21, Geneva
Sanford AJ, Graesser AC (2006) Shallow processing and underspecification. Discourse Process 42(2):99–108. doi:10.1207/s15326950dp4202_1
Searle JR (1969) Speech acts – an essay in the philosophy of language. Cambridge University Press, Cambridge
Searle JR, Vanderveken D (1985) Foundations of illocutionary logic. Cambridge University Press, New York
Seuren PAM (1985) Discourse semantics. B. Blackwell, Oxford
Seuren PAM (2009) Language in cognition. Language from within vol 1. Oxford University Press, Oxford/ New York
Snoeck Henkemans F (1992) Analysing complex argumentation. SicSat, Amsterdam
Snoeck Henkemans F (2001) Argumentation Structures. In: van Eemeren FH (ed) Crucial concepts in argumentation theory. Amsterdam University Press, Amsterdam, pp 101–134
Stati S (1998) Il lessico dell'argomentazione. In: Bondi M (ed) Forms of argumentative discourse. Per un'analisi linguistica dell'argomentazione. Atti del convegno, Bologna 12–13 dicembre 1996. CLUEB, Bologna, pp 51–56
Thomason R (1990) Accommodation, meaning and implicature: interdisciplinary foundations for pragmatics. In: Cohen PR, Morgan J, Pollack ME (eds) Intentions in communication. Cambridge University Press, Cambridge, MA, pp 325–363
Toulmin SE (2003) The uses of argument. Cambridge University Press, Cambridge
van Benthem J (1996) Logic and argumentation. In: van Benthem J, van Eemeren FH, Grootendorst R, Veltman F (eds) Logic and argumentation. North-Holland, Amsterdam, pp 27–42
van Benthem J (2009) One Logician's perspective on argumentation. Cog J Reason Argument 1(2):13–25
van Dijk TA (1993) Principles of critical discourse analysis. Discourse Soc 4(2):249–283

van Eemeren FH (2009) Strategic Maneuvering between Rhetorical Effectiveness and Dialectical Reasonableness. Stud Logic Grammar Rhetoric 16(29):69–91

van Eemeren FH (2010) Strategic maneuvering in argumentative discourse. John Benjamins, Amsterdam/Philadelphia

van Eemeren FH (2013) That's how it goes! Towards the identification of patterns in argumentative discourse. In: Sellami AL (ed) Argumentation, rhetoric, debate and pedagogy: proceedings of the 2013 4th international conference on argumentation, rhetoric, debate, and pedagogy. QatarDebate Center, Doha, pp 269–286

van Eemeren FH, Garssen B, Meuffels B (2003) I don't have anything to prove here. The (un) reasonableness of evading the burden of proof. In: van Eemeren FH et al (eds) Proceedings of the fifth conference of the International Society for the Study of Argumentation. Sic Sat, Amsterdam, pp 281–284

van Eemeren FH, Grootendorst R (1984) Speech acts in argumentative discussions. Foris Publications, Dordrecht

van Eemeren FH, Grootendorst R (2004) A systematic theory of argumentation: the pragma-dialectical approach. Cambridge University Press, Cambridge

van Eemeren FH, Grootendorst R, Jackson S, Jacobs S (1993) Reconstructing argumentative discourse. University of Alabama Press, Tuscaloosa

van Eemeren FH, Grootendorst R, Henkemans FS (2002) Argumentation. analysis, evaluation, presentation. Lawrence Erlbaum, Mahwah

van Eemeren FH, Houtlosser P, Henkemans FS (2007) Argumentative Indicators in Discourse a Pragma-dialectical Study. Springer, Amsterdam

Vorobej M (2006) A theory of argument. Cambridge University Press, Cambridge

Walton D (1996) Argumentation schemes for presumptive reasoning. Studies in argumentation series. Lawrence Erlbaum, Mahwah

Walton D (1998) The new dialectic. University of Toronto Press, Toronto

Walton D (2007) Dialog theory for critical argumentation. John Benjamins, Amsterdam/Philadelphia

Walton D (2013) Methods of argumentation. Cambridge University Press, Cambridge/New York

Walton D, Krabbe E (1995) Commitment in dialogue: basic concepts of interpersonal reasoning. State University of New York Press, Albany

Walton D, Macagno F (2010) Wrenching from context: the manipulation from commitments. Argumentation 24(3):283–317

Walton DN, Reed CA, Macagno F (2008) Argumentation schemes. Cambridge University Press, Cambridge

Webber B, Stone M, Joshi A, Knott A (2003) Anaphora and discourse structure. Comput Linguist 29(4):545–587

Wilson D, Sperber D (2012) Linguistic form and relevance. In: Wilson D, Sperber D (eds) Meaning and relevance. Cambridge University Press, Cambridge, pp 149–168

Woods J (2010) Defeasible reasoning. In: Reed C, Tindale CW (eds) Dialectics, dialogue and argumentation. An examination of Douglas Walton's theories of reasoning and argument. College Publications, London, pp 239–261

Chapter 3
Three Views of Modality in Toulmin

Abstract The chapter reads as an original contribution to the exegesis of Toulmin's *The Uses of Argument*. It examines this modern classic of argumentation theory from the scarcely explored perspective of modality unveiling some latent inconsistencies in the fabric of the book and proposing some new interpretations of the key elements of the Toulmin model. The chapter also considers the reception of Toulmin's ideas on modality in later studies, as well as their relation to current contributions to argumentation theory and linguistics.

3.1 Toulmin and Modality Beyond the Toulmin Model

In this chapter I examine how Stephen E. Toulmin dealt with modality in his foundational book on *The Uses of Argument* (2003 [1958])[1]. Toulmin devotes a surprising amount of space to modality. I say *surprising* because comparatively little of this discussion passed to subsequent work in argumentation theory by other authors and to "standard" textbook presentations of the "Toulmin model", including the one co-authored by the philosopher (Toulmin et al. 1984).

The works in contemporary argumentation theory that single out modality as an object of attention, or at least foreground it in the context of a broader discussion, are few and far between. While this small literature often employs the Toulminian term *modal qualifier* (as in Freeman 1988, 1991, 2011, 2013; Pinto 2007) or terms somewhat reminiscent of it, such as *standpoint qualification* (Tseronis 2009), this does not mean that the views of modality expounded in it draw on Toulmin in a very extensive or deep way. This is not the case, for instance, for the above cited contributions whose views of modality have little in common with those of the British philosopher except, as we will see, the choice of placing modality outside the propositional contents of the individual utterances composing an argumentative discourse.

Despite the relative lack of interest for this specific aspect of his work, Toulmin's ideas on modality remain a most convenient and revealing vantage point for examining issues of modality in argumentation.

[1] The first part of the present chapter develops ideas initially presented in Rocci (2008a). In particular, some passages from this work are reproduced in § 3.2.

A. Rocci, *Modality in Argumentation*, Argumentation Library 29,
DOI 10.1007/978-94-024-1063-1_3

3.1.1 The Place of Modality in Stephen Toulmin's The Uses of Argument

Of the five chapters, or "essays"[2], that make up *The Uses of Argument*, the first four deal prominently and extensively with modal matters.

The first essay, entitled *Fields of Arguments and Modals*, draws a close parallel between the context-dependent and context-invariant aspects of modal semantics and of argumentative moves in order to introduce the notion of *field of argument* and defend the thesis of the field-dependence of the criteria relevant for evaluating the soundness or goodness of argumentation.

The second is entirely devoted to probability and takes what we would now call a pragmatic analysis of the adverb *probably* as the starting point of an ordinary language-based critique of both subjectivist and objective chance views of probability. Toulmin takes particular care in reviewing what Kneale (1949) and Carnap (1950) have to say on ordinary discourse on probability in order to reject the idea that the mathematical theory of probability offers a viable account of what ordinary talk of probabilities is about or a way to make vague ordinary notions of probability precise.

In the third essay, Toulmin finds a special slot for modalities, the *qualifier* (often exemplified by *probably* or *presumably*) in his analysis of the fundamental "layout" of arguments. It is this chapter that exerts a significant, if somewhat superficial, influence on the subsequent development of theory and even more on the analytic and textbook writing practice of the argumentation discipline. Basically, it is from this chapter that argumentation theory inherits the term *modal qualifier* and the key structural idea that this modal element occupies a distinct place in the argument and is not properly part of the propositional content of the claim.

Finally, the fourth chapter consists largely in an attack on the notion of "logical necessity" and, in general, of "logical modalities". This discussion draws on the treatment of the context-dependency of modality in the first chapter trying to cash in its broader philosophical consequences regarding the scope of logic. As anticipated in the Introduction, the present book is not primarily about non-deductive reasoning and the criteria for its evaluation. It focuses, instead, on reconstruction. Consequently, I will not directly dive into Toulmin's critique of deductive validity as a standard of argument *evaluation* and I will confine myself on Toulmin's remarks on the nature of logical modalities and their alleged irrelevance in the *reconstruction* of the function of modal words in ordinary discourse.

---

[2] Toulmin uses the term "essay". This term may reflect the fact that earlier versions of some of the chapters had appeared independently – in particular, the second chapter had appeared as Toulmin (1956) – and that even in the context of the book the "essays" appear somewhat more loosely connected than chapters in a treatise. As a matter of fact, the third essay, detailing the "Toulmin model" had a fortune which is largely independent from the matters discussed elsewhere in the book which testifies the ease of selective reading.

3.1.2 *The Uses of Toulmin's Views in This Book*

Toulmin certainly does not offer a theory of modal semantics nor a complete guidebook to approach modality in the task of analyzing and critiquing arguments. Modality is more of an heuristic and argumentative resource to bring home Toulmin's main point about the task of logic. Visible gaps and rough edges can be spotted between the different chapters so that it is not easy to put together a complete and fully consistent picture of modality. Here I will go through what Toulmin says about the meaning of the modals, including his extensive and usually ignored discussion of their *polyfunctionality*[3] in the first essay, and the semantico-pragmatic analysis of *probably* in the second[4].

The aim is not to criticize Toulmin as a semanticist – with more than fifty years of research in semantics in between – nor to assess the role of considerations of modality in the overall argument of his book. Rather, both the insights and the unresolved problems that emerge from the essays of *The Uses of Argument* will help to provide a sharper framing of the key research questions that were mooted in the introduction. The first two research questions, in particular, are inspired by Toulmin's work and are obviously directly addressed there:

(Q1) How are modals similar to arguments? Is there any relevant relation of isomorphism between the semantic structure of modalities and the "deep structure" or layout of arguments?

(Q2) Is there a special place for modality in the essential structure of arguments? Is there a distinctive functional slot or role that is assigned to modal meanings in such a "deep structure".

As we will see, in the first essay, Toulmin answers positively to both questions. And does so with regards to all readings or flavors of the polyfunctional modals. His idea that all flavors of modality are argumentatively relevant goes together with a context dependent analysis of the modals that was way ahead of his time. From the insights and flaws of this analysis, I will take up what is perhaps the most original challenge

---

[3] I follow van der Auwera (1999) in adopting *polyfunctionality* as the most generic term covering all forms of "context-dependent meaning", including, but not limited to, the phenomena of vagueness, ambiguity, homonymy and polysemy. In the case of modal expressions, *polyfunctionality* comes in handy since the precise semantic and/or pragmatic account of the way in which the different flavors of modality is an important issue in the linguistic literature on modality and one that intersects an account of the role of modal expressions in arguments.

[4] In their presentation of Toulmin's contribution in the *Fundamentals of Argumentation Theory*, Van Eemeren, Grootendorst, Snoeck Henkemans *et al.* (1996: 137–138) do give a brief account of how modals are related to argument fields in the first essay, but then comment on their omissions in a footnote: "Some of the implications Toulmin attaches to this observation relate to semantic and philosophical questions that are not directly relevant here. They pertain to the development of an adequate semantic theory for modal words and to the vigorous philosophical controversy about probability in the 1960s. In that controversy, Toulmin opposed the views on probability put forward by Carnap (1950) and Kneale (1949)". These omissions are, of course, perfectly reasonable in the context of the *Fundamentals*, but at least some of these "semantic and philosophical questions", the semantic ones in particular, are relevant here.

for the present book: showing how all the readings of the modals can be relevant from an argumentative point of view.

The second and, especially, the third essay deal with question Q2 and do so in relation to flavors of modality that linguists and philosophers would nowadays call *epistemic* or *doxastic*. With the notable exceptions of Robert Pinto (2001) and Lilian Bermejo-Luque (2011), argumentation scholars after Toulmin appear to deal exclusively with these epistemic and doxastic modalities. Yet, the question of the place or proper slot of epistemic modalities is far from settled in argumentation theory, nor it is in linguistics.

The investigation of specific linguistic modal expressions as argumentative indicators, which relates to the third research question addressed in this book, is not specifically addressed by Toulmin.

(Q3) What does the use of modal expressions in an ordinary discourse tell us about the arguments being put forth in the discourse?

Yet, his views on the role of the modal qualifier did to some extent inform the views of analysts on *what kind of information* a modal expression can give us on the structure of an argument.

3.2 The First View: Modal Forces, Criteria and Fields of Argument

3.2.1 Modal Forces and the Phases of an Argument

The quote placed in epigraph to the introduction is worth repeating here:

> These terms— 'possible', 'necessary' and the like— are best understood, I shall argue, by examining the functions they have when we come to set out our arguments (Toulmin 2003 [1958]: 17).

"Understanding" the modals in terms of their argumentative functions along the lines of Toulmin's account will lead, in turn, to understand something about arguments, which is crucial for Toulmin, namely their *field dependence.* Toulmin establishes an exact parallelism between the semantics of modal words like *may*, *must*, *possible*, *cannot* and discourse moves, corresponding to different phases of an argument: taking an hypothesis into consideration (*may*), excluding an hypothesis (*cannot*), and concluding (*must*), as shown in Table 3.1 below.

Table 3.1 Modals and phases of an argument according to Toulmin (2003 [1958])

Modal markers	Phases of the argument
Possibility: *may / possible/* [*can*]	Putting forward an hypothesis as worth considering;
Impossibility: *cannot*	Ruling out an hypothesis;
Necessity: *necessarily/ must*	Having ruled out other hypotheses, presenting one particular conclusion as unequivocally the one to accept.

Toulmin (2003 [1958]: 28) claims that "the meaning of a modal term [...] has two aspects" an invariant force, to be understood in terms of the argumentative moves that correspond to the different phases of an argument, and variable "field dependent" *criteria*, which Toulmin also characterizes as *reasons*, *grounds, standards*. The phases identified by the different modals must be seen in relation with Toulmin's quest for a new "procedural" notion of *argument form*, based on a judicial metaphor, as an alternative to the notion of argument inherited from (formal) logic.

A few pages earlier, Toulmin hedges quite a bit on his commitment to providing a semantic analysis of the modals denying that by connecting modal words with phases of argument, he is presenting a "formal analysis" or "formal definition" of modal words like *possible*, of which, he says, "it would be difficult to give any strict dictionary equivalent" (Toulmin 2003 [1958]: 18). Speaking of *must,* Toulmin (2003 [1958]: 20) again repeats that "The connection between the meaning of the word and the sort of situation I have indicated is intimate, but not of a sort which could be adequately expressed in a dictionary definition". Despite these hedging maneuvers, I maintain here that what Toulmin presents in the first essay *is* a semantic analysis of the modal words in argumentative terms (albeit not a "strict", "formal" or dictionary-like one). He does want to characterize in argumentative terms the *meaning proper* of the modals, and, in particular, what he calls the "two aspects" of the meaning of the modals (*force* and *criteria*) and not to establish a somewhat looser association between modal meanings and arguments. The strict reading is, in the end, the most charitable as it is not quite clear what such a looser association would be[5].

If we translate this analysis in the terms of Speech Act Theory – which is something that Toulmin himself does with his Austin-inspired analysis of *probably* in the second essay – the different modals turn out to be *illocutionary markers* signaling

[5]There are several additional reasons for maintaining this strict reading: 1) The strict reading is required by Toulmin's pragmatic analysis of *probably*, where he rejects attempts to analyze probability statements as being *about* real-world (or subjective) *probability relations*. For Toulmin, *probably* does not denote an external or cognitive "probability relation" because it expresses performatively a guarded, tentative conclusion. The performative use is primitive, and seemingly objectified uses such as those of the abstract noun *probability* have to be explained as derived from the performative one. 2) Similarly, the strong reading is required by Toulmin's rejection of *logical possibility* (as lack of contradiction or consistency) in the fourth essay. For Toulmin (2003 [1958]: 157), being *possible* is never simply a matter of successfully avoiding "contradicting our data" as it involves being "a genuine candidate-solution whose backing we shall have to investigate and whose acceptability we shall have to evaluate". As in the case of probability, an externalized relation (consistency) is rejected as the meaning of the modal term in favor of the characterization of possibility in terms of an argumentative operation. 3) Toulmin (*Ibid.*, p. 33) declares that he wants to avoid the use of the term *meaning* because it is too imprecise, while his notions of force and criteria operate with "finer distinctions". Speaking of their *meaning* would suggest either the idea that modals are simply ambiguous or simply monosemous. In other words, he avoids *meaning* because he wants to be a more precise semanticist! 4) Finally, the whole hedging manoeuver performed by Toulmin needs to be put in its historical context on the backdrop of the accusations that were leveled against ordinary language philosophers of merely doing "lexicography". See Thomasson (2002: 123) on this allegation of "lexicography" against Toulmin's Oxford teacher Gilbert Ryle.

different kinds of illocutionary acts. In terms of Congruity Theory (Rigotti 2005a, b), they encode pragmatic *connective predicates* corresponding to different discursive actions relevant in the process of argumentation. This translation into speech act-theoretic terms appears legitimate also because the *force* of the modals is understood by Toulmin precisely in terms of the commitments they dynamically create in the follow-up of the argument. Saying that p *is possible* or *might be the case* commits the arguer to consider the proposition as a hypothesis worth investigating in the follow-up of the argument: "to call something 'possible' and then ignore it indefinitely without good reason is inconsistent" (Toulmin 2003 [1958]: 18). Similarly, saying that p *is impossible* commits the arguer to discounting p from the follow-up of the argument. If someone still behaved as if p were a suggestion worth of his "respectful consideration" after having ruled out the proposition as impossible, he would "lay himself open to a charge of inconsistency or perhaps of frivolity" (*Ibid.*).

Commenting on the force of *must*, Toulmin (2003 [1958]: 20–21) observes that not all arguments arrive to the point where all possibilities worth considering are ruled out except one that unequivocally *must* be the case. He thus briefly introduces two additional modal forces, which are not directly tied to the three basic phases: *probability* and *presumption* – two notions that will enjoy considerable fortune in the development of argumentation theory:

Probability (*probably*) is provisionally characterized as the situation where, in light of the evidence, one hypothesis appears more worthy than the others so that the arguer is entitled to take the step to the conclusion "with more confidence than the step to the others" (*Ibid.:* 20). The second essay, discussed at length in the following pages, will then provide a surprisingly *different* characterization of probability.

Presumption (*presumably*) is characterized as akin to necessity (*must*) but containing a conditional, counterfactual element. It corresponds to the situation where the arguer is able "to show that one particular answer would be *the* answer, supposing only we were confident that certain unusual or exceptional conditions did not apply in the particular case" (*Ibid.*: 20–21). Similar views of presumptive reasoning will re-emerge in the third essay, when Toulmin discusses the relationship between modal qualifier and rebuttal.

In themselves, the pragmatic forces Toulmin associates with the modals are quite interesting from the viewpoint of a descriptive pragmatic analysis of argumentative discourse of the kind envisaged in Chap. 2. The tools of Congruity Theory can serve as to make more explicit the kind of moves Toulmin envisages as associated with the various modal expressions in the firs essay of the *Uses of Argument*. For instance, the congruity theoretic rendering of the *modal force* of *may/ possible* could be as follows:

(Def. 1) *May/ Possible/ Can* $_{U0}$ (Spk, Hr, E, U_0, U_{+1}) =
Presuppositions:
The propositional content p of U_0 is not part of the common ground of *Spk* and *Hr*, E is a segment of the common ground of *Spk* and *Hr* bearing on the truth or falsity of p, U_{+1} is a projected utterance of either *Spk* and *Hr* relating p to E.
Pragmatic effects:
By uttering U_0, *Spk* presents p of U_0 as a hypothesis worth of the consideration of *Hr* in view of E and commits *Spk* and *Hr* to perform U_{+1}

Clearly, this kind of exercise of explicitation opens new questions that cannot be answered in the context of Toulmin's own presentation, such a as, for instance, the role of the evidence E and its relationships with the common ground and with the propositional content p. Before opening such discussion, it will be wiser to consider more carefully the proposal in its original context.

Opening the third essay, Toulmin (2003 [1958]: 87) discusses briefly how the phases presented in the first essay relate to the analysis of the layout of arguments that he is about to present, and which will become known as the "Toulmin model". Here we learn that the three phases make up the "gross, anatomical" structure of an argument, "marking the progress of the argument from the initial statement of an unsettled problem to the final presentation of a conclusion"[6]. Each of these phases corresponds to one or more *paragraphs*, while the units of the "Toulmin model" consider argument at "the level of individual sentences" within each paragraph, which corresponds to the "finer, as-it-were, physiological" structure. Thus, the three phases associated with the modals would be macro-illocutionary acts taking scope on discourse units of the size of a paragraph.

However, if we consider this distinction between these two levels of structure – let us call them *macro-* and *micro-*structure – things become slightly bizarre for two reasons:

First, it does not seem mandatory that speech acts like *advancing an hypothesis*, *ruling out an hypothesis* or *concluding* are realized by a complex, paragraph-length, manifestation. They can be realized very well by a simple sentence, as shown by most of the short examples that Toulmin uses in the first essay. Rather, they look like *relational* speech acts, presupposing the presence of other units of information in the discourse (e.g. the evidence on the basis of which I rule out an hypothesis). In this respect, they are not very different from the functional units that appear at the micro-level analysis.

[6] Clearly, Toulmin's macro-structural "phases" look as an attempt to capture the dialectical (as opposed to the inferential) organization of arguments and they somewhat remind us of the "stages" of the Pragma-Dialectical model (van Eemeren and Grootendorst 2004). However, the overall view of the argumentative dialogue that they suggest is different. Participants in Toulmin's phases as seen as putting forward (possibly multiple) hypotheses about an "unsettled problem", which are then tested so that they are ruled out by the evidence or retained as a conclusion. In the Pragma-Dialectical *critical discussion*, on the contrary, protagonists put forward a *standpoint* that they are committed to defend in the face of rational criticism from an antagonist. If the standpoint is successfully defended it is retained as a conclusion.

Second, the same modal expressions initially associated with the macro-structural phases also appear as a part of the micro-structure, corresponding to the slot of the modal *Qualifier*. This leaves us with a choice between two construals of the theory:

(A) The modals are *either* illocutionary markers at the macro-structural level *or* part of the micro-structure. As a matter of fact, Toulmin discusses certain modals (mostly) in relation with the macro-structure (*may, cannot, must*, in the first essay) and others (mostly) in relation to micro-structure (*plausibly, probably*, in second and third essay).

(B) Slightly more plausibly, all modals appearing in argumentative texts contribute at the same time *both* to the macro-structure (indicating the phase) and to the micro-structure (specifying a modal *Qualifier*).

Both reconstructions of Toulmin's theory of argument are highly speculative and I will not pursue them further. In fact, Toulmin, in the scant remarks opening the third essay, does not tell us anything about the precise way in which the macro-structure is supposed to be connected to the micro-structure. For instance, we do not know whether the first phase (putting forward an hypothesis) necessarily involves "arguments" describable at the micro-structural level with the Toulmin model and, if it does, whether the *Qualifiers* appearing in those arguments are always possibility modals (like *may* or *possible*). In the following pages I will put aside the distinction between the macro- ("anatomical") and the micro- ("physiological") level and I will more generally accept that the accounts of the modals in the different essays are not integrated, but rather autonomous. Whether they are fully compatible or not remains to be seen.

3.2.2 *From Modal Criteria to Fields of Argument and Logical Types*

With respect to the first essay I will concentrate my attention on the interplay of the two aspects of modal meaning: *force* and *criteria*. Toulmin illustrates the role of *must* in the speech act of *presenting one particular conclusion as unequivocally the one to accept* with the following examples:

(1) Under the circumstances, there is only one decision open to us; the child *must* be returned to the custody of its parent.

(2) Considering the dimensions of the sun, moon and earth and their relative positions at the time concerned, we see that the moon *must* be completely obscured at the moment. (Cf. Toulmin (2003 [1958]: 19)

According to Toulmin, in (1) and (2) the force of concluding conveyed by *must* remains the same, what changes are the *criteria* (reasons, grounds) for concluding, which vary according to what Toulmin calls the different "field of argument" to which the arguments belong.

Toulmin's choice of examples is striking because the *must* appearing in (1) and (2) manifest clearly distinct modal meanings or "flavors" that the philosophical and linguistic literature on modal semantics keeps separate. By the nomenclature that I have started introducing in Chap. 2, in (1) *must* manifests some shade of *deontic* or *deontic-practical* modality, indicating that in the given circumstances a course of action is morally or legally required or practically necessary. While in (2) the meaning of *must* is epistemic. With few notable exceptions (cf. Kronning 1996), in the linguistic literature it is only in the latter reading that necessity modals like *must* are analyzed as indicating the drawing of a conclusion from evidence.

Toulmin makes two hypotheses on the nature of the polyfunctionality of the modals that are far from trivial from a linguistic-semantic viewpoint – and were even less trivial at the time. In order to better appreciate them we need to introduce a little bit of the terminology that linguistic semantics uses to set apart different kinds of polyfunctionality.

The first choice made by Toulmin is the hypothesis that modal expressions like *must* or *cannot* (the only form for which Toulmin provides a detailed analysis) are not simply *ambiguous* between separate meanings, nor are they *polysemous* words manifesting a network of meanings sharing family resemblances and connected by motivational links to an etymologically basic sense. He sees them as basically *monosemous* words whose core meaning (the "field invariant" modal force) remains constant as it combines with contextual information (the "field dependent" criteria). In other words, modals are not polysemous but rather *context-dependent*. A polysemous word like, say, *ring* is associated with multiple linguistically encoded meanings ('circular piece of jewelry', 'circular entity', 'arena', 'group of people operating together clandestinely') that are bound together by a network of motivational links, involving sometimes abstraction and generalization, sometimes metaphor, sometimes other kinds of motivational links depending on historical accident[7]. In contrast, as context-dependent words, *must* and *cannot* are closer to deictic words, the indexicals (like *I, today, tomorrow*) or the demonstratives (such as *he, this, there*), whose invariant linguistically encoded meaning picks a certain type of information in the context and uses it to contribute to the meaning of the utterance (cf. Recanati 1989, 2009 and Kay 1997, 2004). The view of the modals as context-dependent words is now well represented in linguistic semantics (cf. Kratzer 2012, Portner 2009), but by no means prevalent (see *contra* Talmy 1985, Sweetser 1990, Nauze 2008, Boogart 2009). Following, in part, Toulmin's suggestion, in this book I will adopt a context dependent semantics for modal words as the best suited to account for their role in arguments and their potential as argumentative indicators.

The second choice made by Toulmin is to hypothesize that the context-invariant core meaning of the modals (the force) is the indication of an argumentatively relevant illocutionary act and that the contextual information that is picked-up to complement it (the criteria) is related to the contextually variable aspect of argumentation,

[7] I borrow the example of *ring* and the glosses from Langacker (1991: 3). The characterization of the different motivational links is mine, though. I am not sure Langacker would agree with it *in toto*.

the "fields of argument". The next section will follow closely the detailed analysis that Toulmin provided of the invariant and context-dependent aspects of the meaning of *cannot* across different "fields of argument". Before moving to this illustration of how "fields of argument" supply the contextual parameter of the modals, however, it is worth pausing a moment to consider what exactly these "fields" are supposed to be.

In their critical assessment of Toulmin's contribution for the *Fundamentals of Argumentation Theory*, Van Eemeren et al. (1996: 155) cite the notion of field of argument as one prime example of Toulmin's "vagueness, ambiguity, and sometimes even inconsistency in the use of key terms". Certainly Toulmin does not give a fully explicit definition of the term and largely relies on concrete examples of fields to convey this concept to the reader. The closest thing to a definition that we find is a partial characterization of *field of argument* in terms of the *logical type* of the statements appearing in the argument. Toulmin first introduces the notion of *logical type*, as follows:

(Def. 2) The statements of our assertions, and the statements of the facts adduced in their support, are, as philosophers would say, of many different 'logical types' – reports of present and past events, predictions about the future, verdicts of criminal guilt, aesthetic commendations, geometrical axioms and so on. The arguments which we put forward, and the steps which occur in them, will be correspondingly various: depending on the logical types of the facts adduced and of the conclusions drawn from them, the steps we take – the transition of logical type – will be different. (Toulmin (2003 [1958]: 13)

On this basis, he then introduces the "technical term" *field of argument*:

(Def. 3) Two arguments will be said to belong to the same field when the data and conclusions in each of the two arguments are, respectively, of the same logical type: they will be said to come from different fields when the backing or the conclusions in each of the two arguments are not of the same logical type. (Toulmin (2003 [1958]: 13)

The concept *logical type* itself is not defined by Toulmin, but rather treated as if it was of common currency ("as philosophers would say"). In a footnote, van Eemeren et al. (1996: 136) offer a precious pointer for determining which "philosophers" Toulmin had in mind, suggesting that his Oxford teacher Gilbert Ryle is the likely source of the concept. Looking at the role of logical types in Ryle's work contributes to illuminate this otherwise opaque Toulminian notion in a way that proves particularly interesting for a study of the relationship between argumentation and semantics. Ryle's definition of logical type, reproduced in (Def. 4) below,

(Def. 4) The logical type or category to which a concept belongs is the set of ways in which it is logically legitimate to operate with it (Ryle 2009 [1949]: LX)

appears in a book devoted to the philosophy of the mind that was intended as a demonstration of philosophical method at least as much as a critical contribution to the philosophical problem discussed (Cf. Thomasson 2002: 117). The concept of mind and other mental concepts were to function as a major case study to put to the test a method of *conceptual analysis* that Ryle was honing since his earlier major contributions (Cf. Ryle 1937) and to provide an example of the specific contribution of philosophy as opposed to that of the empirical sciences. As observed by Thomasson (2002), for Ryle (2009 [1949]) the task of conceptual analysis is to study the meaning of linguistic expressions to map their "logical geography" uncovering not their particular lexicographic definition but the proper logical types or categories to which they belong.

This kind of investigation rests on the assumption that there is a "logical grammar" that dictates how concepts can be combined to form meaningful wholes (e.g. simple statements or complex discourses). The laws of this logical grammar can be discovered by looking at the nonsense or incongruity that arises from their violation within a complex expression. Categories, or types, are discovered by substitution of expressions within a meaningful whole: those expressions that tolerate substitution without giving rise to incongruity (*type mismatch*). Thomasson (2002) shows that Ryle draws this method from Husserl's work on the idea of an a priori (or "pure") grammar, presented in the fourth of the *Logical Investigations*[8]. Ryle moves the investigation from the extremely general categories of Husserl's a priori grammar – which remain somewhat close to traditional grammatical categories – to the richer inventory of categories that is relevant for the construction of a *formal ontology*. In many ways, the methodology followed by Ryle anticipates what was to become the fundamental methodology of natural language semantics in philosophy and linguistics. Congruity Theory itself systematically employs a version of this methodology (cf. Rigotti and Rocci 2001).

Keeping in mind its likely ancestry, we can better make sense of how Toulmin uses the concept of logical type in the passages quoted in (Def. 2) and (Def. 3) above. In (Def. 2) Toulmin seems to view arguments as a kind of coherent whole that results from the applications of constituents of the appropriate type: the logical type of the "facts adduced" and of the "conclusion drawn" determines the logical type of what Toulmin provisionally calls "the step" that has to operate the transition. For instance, if we assume that 'statements of observable facts' are one logical type (A) and 'predictions of future events' are another logical type (B), arguments moving from A to B will require an adequate "transition of type" of a certain kind. Let us call it 'A → B'. Moving from observable facts (A) to conclusions of a different logical type, like verdicts of guilt (C) will require a different "transition of type" 'A → C'. The definition of *fields* in (Def. 3) can be interpreted accordingly. Arguments belong to the same field when they have the same configuration of types in the data and conclusion or, in other words, when they require the same "transition of type". Thus, 'A → B' and 'A → C' in the previous example define two distinct fields of argument. In the third essay, as we will see, the functioning of these "transitions" or "steps" will be analyzed in terms of *warrants* legitimating the move from data to conclusion and field-dependent *backing* legitimating the warrant.

Seen from this angle, Toulminian fields of argument do not seem too far removed from two closely connected ideas that were evoked in Chap. 2, namely, the idea that

[8] On the methodological role of nonsense in semantics and its Husserlian origin in the fourth logical investigation see Rigotti and Rocci (2001).

standpoints belonging to different semantic types require different kinds of evidence in relation to the issues they address, and the idea that the argument schemes on which the transition from evidence to conclusion is based should be defined in terms of semantico-ontological relations which impose certain congruity requirements on the semantic types involved in the data and in the conclusion. As mentioned in Chap. 2, both ideas have ancient origins: the first in the rhetorical doctrine of the *stases* (or *status causae*) and the second in the doctrine of *topoi* (or *loci*). As a simple example of the interplay of loci with the "logical type" of standpoint and evidence, consider again the prediction of future events from presently available evidence. Certain loci, such as the locus *From Authority* of the *testimony* sub-type and the application of locus *From efficient cause* with maxims *from effect to cause*, will be immediately ruled out as incongruous with the logical types of the standpoint and evidence. Others such as arguments using maxims *from cause to effect* and arguments exploiting observed past correlations via the locus *From Concomitances* will remain, in principle, possible.

It is clear, however, that, despite his use of "logical types" in defining the fields of argument, Toulmin's perspective is, in certain respects, very far from the one evoked above. To become aware of this distance it is sufficient to examine Toulmin's lush exemplification of logical types of statements, which includes pretty general categories such as "predictions about the future", but soon drifts to increasingly contextualized kinds like "aesthetic commendations", "verdicts of criminal guilt", "geometrical axioms". All along Toulmin's book the types of conclusions and data are always described in terms that pertain to specific disciplines or contexts of use.

It never looks as if the establishment of the relevant logical types is a product of a conceptual analysis, as it was in Ryle. While Toulmin deftly uses the analytical tools of ordinary language philosophy to provide semantic analyses of linguistic expressions like *probably* and *cannot,* which pertain to the field-invariant structure of arguments, he does not seem to be interested in applying the same tools to the task of properly setting apart the different logical types that determine the fields of argument. Toulmin seems to think that the philosopher has no business in determining logical types and that such a task is better left to the practitioner of the different disciplines or, more generally, to those acquainted with certain arenas of discussion. In several passages along the book and especially, as observed by van Eemeren et al. (1996: 155) in the third essay, Toulmin gives the impression – without explicitly stating it – that for all purposes fields of arguments coincide with disciplines or with informal areas of expertise. This interpretation of fields of arguments as fields of expertise certainly played a major role in the reception of Toulmin's work and has been somewhat authorized by Toulmin's later more socio-historical use of the term field of argument to indicate an "intellectual enterprise". It, however, immediately runs into two problems if it is taken jointly with the more formal definition of fields based on logical types given in Toulmin (1958).

Firstly, if establishing logical types is delegated to experts of the subject and the fields end up coinciding with the subject themselves we will have to admit, as observed by van Eemeren et al. (1996: 155), that a (meteorological) prediction about tomorrow's weather in southern Switzerland and a (financial) prediction about the US Dollar – Swiss Franc exchange rate do not belong to the same logical type of standpoint *qua* predictions. For sure, statements of financial predictions and of meteorological prediction *are not* semantically equivalent and do not have, to use Ryle's words, the same "logical geography". But this is not because of a lack of a common semantic type 'prediction of future events', but because the underlying semantic type of the event predicted is different. For instance, in one case we have natural events, while, in the other, we have the results of a multitude of deliberate human actions – a crucial difference which confers important characteristics to predicting in finance, which in lack in meteorology. This kind of detailed comparison is possible only if we distinguish "logical types" through conceptual (i.e. semantic) analysis. As anticipated in Chap. 2, § 2.4, semantic analysis is indeed a powerful tool for understanding how standpoints and arguments relate to the domain ontology of a context of argumentative discussion.

Secondly, if we identify fields with disciplines or areas of expertise, it follows (from II, above) that arguments within a discipline are characterized by the same configuration of types of data and types of conclusion, that is by only one kind of "transition of type". This is manifestly not true in many (if not all) disciplinary areas. If we remain in the area of applied finance and applied economics in general, for instance, we have both arguments that move from observed data concerning the current and past economic situation to support *predictions of future events* (forecasts on the basis of economic models) and arguments that move from similar data to support *economic policy recommendations* (on the basis of economic models and systems of social values and preferences). Furthermore, if we move from applied to theoretical economics we will have empirical arguments moving from observed data to support the descriptive adequacy and predictive power of economic models (on the basis of econometric methods), as well as purely theoretical arguments moving from certain formal properties of economic models do derive further formal properties[9]. It is therefore clear that a discipline or domain of expertise can involve several kinds of "transitions of type".

3.2.3 Cannot*: An Early Analysis of Modal Context-Dependency*

The most complete illustration of the field dependent functioning of the modals is provided by Toulmin's analysis of the "modal" *cannot*. I now turn to this analysis with three goals in mind:

[9] In her pioneering text-linguistic investigation of the illocutionary act of prediction in economics research papers, Merlini (1983) offers a semantico-illocutionary typology of the claims that are put forth and argued for by economists. While relatively old and not formulated from an explicitly argumentation-theoretic viewpoint this typology her work remains highly valuable for analyzing arguments in academic economics and about the economy in general.

I. assessing its descriptive adequacy as a theory of modal semantics and, more specifically, of the polyfunctionality of modal expressions;
II. comparing it with the better known views of Toulmin on probability expressions and on modal qualifiers;
III. introducing the style of analysis of modal expressions as relational and context-dependent, which will be fully deployed in Chap. 4 and will provide us the scaffolding to investigate how modality relates to arguments in the final part of the book.

Toulmin (2003 [1958]: 22–27) examines several uses of *cannot*, presenting them first in the form of a conversational utterance providing a situational context through a vividly narrated vignette, then in a sort of normalized sentence form, whose syntactic structure is meant to make explicit certain kinds of information – critical for the analysis that most natural conversational uses leave implicit. Except for the first example in (3), where I quote extensively, I will just provide the "normal form" together with a summary of the context.

(3) "A large piece of metal falls from a lorry on the road. The driver, a pale, seedy-looking young man, gets down from his cab and makes toward it as if to pick it up. We see this and say to him, 'You can't lift that weight single-handed: hang on a moment, while I get help or some lifting-tackle'. He replies, 'Bless you, I've done the like often enough', and going up to it hoists it deftly back on to the lorry again."
"Your physique being what it is, you can't lift that weight single-handed – to attempt to do so would be vain".

(4) Context: the addressee has organized a public meeting in Town Hall, sending invitations to ten thousand people. Most of them may well show up.
"The seating capacity of the Town Hall being what it is, you can't get ten thousand people into it –to attempt to do so would be vain".

(5) Context: the addressee has described a fox hunting scene speaking of the *tail* of the fox (instead of the *brush*).
"The terminology of hunting being as it is, you can't talk about a fox's tail – to do so is an offence against sporting usage"

(6) Context: The manuscript of the novel authored by the addressee contains an incongruence, referring to the same character as someone's *sister* and as *he*.
"The nomenclature of sexes and relationships being what it is, you can't have a male sister – even to talk of one would be unintelligible."

(7) Context: The addressee is caught smoking in a non-smoking compartment of the train.
"The by-laws being as they are, you can't smoke in this compartment, Sir – to do so would be a contravention of them."

(8) Context: The addressee has ejected his dissolute son from the house.
"Standing in the relationship you do to this lad, you can't turn him away without a shilling – to do so would be unfatherly and wrong"

The six examples are clearly intended to showcase three groups of uses: (3) and (4) are physical impossibilities, (5) and (6) are linguistic improprieties and, finally, (7) and (8) concern respectively *illegality* an *immorality* – they are thus *deontic* uses (but Toulmin does not employ this term).

Examples that would invite the label of *logical* or *formal* impossibility are "deliberatedly" avoided. Toulmin intends to turn later to a critique of logical modalities but first needs to put in place the analysis of the "simpler", unproblematic sorts of modality. He, however, anticipates some of his critical views saying that, typically, cases of formal impossibilities also involve one or more of the "simpler" sorts of impossibility. In particular, formal impossibilities in scientific theories can typically be reduced either to "solid, physical impossibilities", like (3) and (4) or to linguistic improprieties like (5) and (6) (*Ibid.* p. 26).

Toulmin's explanation of the polyfunctionality of *cannot* rejects both the ambiguity/polysemy hypothesis ("as many dictionary entries as there are sorts of possibility and impossibility") and the purely monosemic alternative ("to say that there is no difference in meaning between these varied uses"). While the rejection of ambiguity/polysemy is mostly due to reasons of descriptive adequacy and elegance of the semantic explanation, the rejection of monosemy is philosophically charged as it is directly connected with the critique of logical modalities.

Toulmin claims that in each of the examples discussed the *force* of the modal remains the same and consists in "an injunction to rule out something-or-other – to dismiss from consideration any course of action involving this something-or-other" (Toulmin 2003 [1958]: 27). All these uses can be reduced to a common pattern (Def. 5) expressing an the invariant force and a number of open variable slots (P, Q, R and S) that are filled differently according to the different fields of argument to which the grounds for ruling out the hypothesis belong.

(Def. 5) 'P being what it is, you must rule out anything involving Q: to do otherwise would be R and would invite S'

The variables correspond to the kind of course of action (Q) that is being ruled out, are the *grounds* (P) for dismissing it ('the driver's physique', 'fox hunter's jargon', 'a father's relationship with his son'), the nature of the offence involved (R) and the eventual consequences or sanctions resulting from the attempt (S). This latter variable, however, is not always relevant as moral examples involve no specific sanction.

With the formula in (Def. 5), Toulmin anticipates a style of analysis of the polyfunctionality of modal words that will appear in formal semantics around 1980 and will gain currency in various strands of linguistic semantics in the late 1980s and 1990s. Modals are not *one-place operators* taking a propositions in their scope, but *relational predicates* whose semantic analysis includes *covert argument slots* (in Toulmin P, R and S) to be saturated in the context of utterance with a specific kind of contextual information. Their apparently widely divergent uses can be reduced to a core meaning and to the different saturation of the variable slots with propositions indeed belonging to different semantic types. The semantic type of these propositions and their precise identification sometimes can be made explicit by the syntax of the modal sentence – as in Toulmin's examples – but in many uses it is either recovered from the preceding discourse or inferred by the hearer from contextual information. This is exactly the style of analysis of modal expressions that I will adopt in Chap. 4, drawing mostly from the theory of Relative Modality (Kratzer

2012, Portner 2009). Where Toulmin differs markedly from the analysis that will be presented in Chap. 4 is the choice of an argumentative illocutionary level relation (his *force*) as the invariant frame.

I will now consider whether this analysis works from a linguistic semantics viewpoint as an account of the polyfunctionality of *cannot* and of other modal expressions – considering in particular *must* in examples (1) and (2) and the other modal verbs, adjectives and adverbs mentioned by Toulmin when discussing the relationship between modal force and phases of an argument. Examining how well Toulmin's hypothesis fares with the semantic facts of English modal verbs and adjectives is not an idle exercise in checking how good at linguistic semantics a philosopher was in 1958. It can give critical indications on the directions that an account of the relationship of modality with argumentation can or cannot take. Even if this might not be immediately apparent, the evidence against Toulmin's idea of modal force is quite strong and it is revealing to see where and why the analysis falls apart. To see that we need to move beyond the beautifully crafted examples presented by Toulmin. For once, the fact that he chooses, unexpectedly, to focus his discussion on the impossibility manifested by *cannot* rather than is not devoid of consequences, as we will see presently.

Toulmin justifies the choice of *cannot* instead of expressions of "necessity" or of "probability" saying that he wanted to start by picking a word unhindered by the big philosophical baggage of the former two. However, if we think of the way in which *necessity* and *possibility*, since Aristotle, have been customarily discussed together in the company of their external and internal negations – in the so-called Modal Square of Opposition[10]– the idea of *impossibility* as an unexplored field feels slightly bizarre.

Surprisingly, Toulmin does not tell us much about *can*. Initially, *can* is not listed with *may* and *possible* among the possibility expressions whose business is to introduce an hypothesis worth considering. Later on, can is fleetingly mentioned in this role (Toulmin 2003 [1958]: 35), but does not appear in the examples, which typically use the adjective *possible*. Toulmin does not say anything about the exact relationship in which *can* stands with respect to *cannot*. Given the relatively transparent word formation of *cannot*, it would be normal to consider whether its meaning arises compositionally from the possibility modal and an external negation. Especially in view of the importance of the aforementioned Aristotelian square in the traditional treatment of the logic of the modals. In fact, in the illocutionary and argumentative analysis proposed by Toulmin where the forces of *can* and *cannot* are, respectively, "putting forward an hypothesis as worth considering" and "ruling out an hypothesis" the scope of negation and the compositionality of its contribution may become a less trivial matter[11].

---

[10] See § 1.6.1 of the Introduction for a short discussion of the Modal Square of Opposition.

[11] At first blush, it seems difficult to obtain the full meaning of "ruling out a hypothesis" from either the external or the internal negation of "putting forward a hypothesis as worth considering". But this is hardly surprising, as the interpretation of negations ostensibly scoping over the performance

Interestingly, if we look at *can* certain problems inherent in Toulmin's analysis become apparent. Consider a sentence like (9.a):

(9.a) John *can* lift 100 kg single-handed.

Here the *criteria* called forth by the modal appear to be the same of Toulmin's *cannot* example: "John's physique being what it is". The modal seems to indicate *what is* ***possible*** *to accomplish with respect to John's muscles, and physical build in general*. The paraphrase of *can* with *possible* seems fitting. But can we say that the above criteria coincide with the *kind of evidence* at our disposal in evaluating the event of 'John lifting 100 kg single-handed' *and* that this event is evaluated as a *hypothesis worth considering*? Let us consider different kinds of arguments that might be relevant in this case:

(9.b) John *can* lift 100 kg single-handed. He has an exceptionally powerful physique.
(9.c) John *can* lift 100 kg single-handed. I've seen him doing the like often enough.
(9.d) John *can* lift 100 kg single-handed. The trainer told me so.

In (9.b) we rely indeed on our knowledge of *John's physical build* to come to the conclusion via a causal locus (possibly the *Locus from Instrumental Cause*). At the same time, John's physical build is the criteria of the modal. This is what we expect from Toulmin's hypothesis. But, let us consider (9.c), where we base our argument and on an inductive version of the locus *From Analogy*, projecting past occurrences onto future similar situations. Clearly, the mode of the transition from the evidence to the claim is different. Here, following Toulmin, we should expect that the criteria of the modal should change as the grounds of our argument change. But there is a strong intuition that they don't: in (9.c) *can* continues to mean 'possible with respect to John's physical build'. This is even clearer in (9.d) where the argument is based on the locus *From authority – expert opinion* subtype – but *can* does not come to mean "possible with respect to what an authority said" – the modality is still about the physique. So, the propositions that we take as our arguments do change (observation of physical build vs. observation of similar action vs. expert opinion) but the facts that are relevant for the modality remains the same (physical build).

It has to be stressed that this does not mean that the criteria of the modal cannot change. They do change. Consider (9.e):

(9.e) John can lift 100 kg. The trainer has given his permission / is OK with it.

Here *can* is no longer referring to *physical conditions* but to the *authority of the trainer*. Toulmin was right in saying that the meaning of the modals involves a contextual component, but examples (9.b, c, d) showed us that this contextual component does not always change as we change the grounds of our conclusion. This suggests that the *criteria* of the modal (the variable slot that we need to saturate to get the complete meaning) and the *grounds* in the argumentative sense (the kind of

---

of speech acts (e.g. *I don't promise I will speak with the counselor*) is notoriously quirky. On the interpretation of negation in relation to speech acts see Moeschler (1992).

reasons that we put forward in support of our modal utterance) are not one and the same thing.

Let us now consider the second issue: are these arguments supporting 'John lifting 100 kg single-handed' as an hypothesis worth considering? We have successfully paraphrased can with *possible* in (9.b, c, d) and the arguments have to do with what is possible or impossible to do with a physique like John's. However it is not sure that this possibility that the arguments support corresponds to the act of "putting forward an hypothesis as worth considering" as Toulmin would have it. In fact, it seems that in the above arguments the arguer rather presents conclusive evidence to assert the general compatibility of the action with the agent's body, rather than presenting an uncertain hypothesis on this action taking place on any particular occasion[12]. Interestingly, we can convey the very same compatibility as being *the content of a hypothesis worth considering* if we add a second modal, of a different kind, which contemporary semanticists would call *epistemic* or *doxastic*:

(10) Maybe /Perhaps /It (may/might) be the case that John can lift 100 kg single-handed

As observed above, Toulmin rarely speaks of *can* as indicating the force of advancing a hypothesis as worth considering, preferring to use the adjective *possible.* It is then fair to check whether my criticism holds also for this form. And it does, as shown in (11):

(11) *Perhaps* it's *possible* to re-set the body's aging clock and *maybe* make someone live longer. (example retrieved through *Google*)

Here the two epistemic markers *perhaps* and *maybe* do refer to an hypothesis worth considering, but *possible* – which here takes "biological laws" as its criteria – is definitely part of the content of that hypothesis.

Similarly, Toulmin sees *must* as having the force of "presenting one particular conclusion as unequivocally the one to accept", while the grounds on which the conclusion is based may pertain to different fields, for instance legal/moral vs. astronomical/ physical, as illustrated by Toulmin's examples in (1) and (2) quoted earlier in this chapter. Yet, in (12) we find *must* embedded as the *content* of an hypothesis worth considering:

(12) If someone wants, for instance, to buy clothes, he must know where to buy them. He must go to different shops. *Maybe* he *must* negotiate with the sales-person. (example retrieved through *Google*)

Clearly, *must* in (12) does not indicate the performance an act of conclusion, nor any other kind of speech act. Let us consider a further example. The Italian necessity modal *deve* 'must' in example (13) below would have to be considered similar in kind to *must* and signal an act of drawing a conclusion on a required course of action from legal grounds:

[12] Compare examples (9.b, c and d) with the congruity theoretic explicitation in (Def. 1) to check this hypothesis.

(13) Un aspirante fotografo **dovrà** effettuare 55 passaggi burocratici e bussare alla porta di una ventina di uffici [...]
'A would-be photographer will have to perform 55 bureaucratic passages and knock at the door of about twenty offices' (*Il Sole 24 Ore*, April 4, 2006).

But certainly this is not the case in example (13). In this text the journalist is indeed *asserting* that the unnamed '55 bureaucratic passages' are *necessarily entailed* by Italian laws and regulations. We cannot say, however, that he is supporting the standpoint that a certain course of action is legally required for photographers by presenting Italian laws and regulations as an argument for drawing this conclusion. If anything because the specific laws and regulations remain unanalyzed and textually inaccessible, just as the modalized (prejacent) proposition is utterly uninformative about the specifics of the entailed actions.

We can conclude that the modals do indeed point to a variable set of "grounds" or "criteria", as Toulmin surmised, but the relation they establish between this often implicit background and the underlying modalized proposition is not always an argumentative one. More analytically, the above discussion shows three facts that contradict Toulmin's analysis of modal context-dependency:

(a) Modal "criteria" (the context-dependent part of modal meanings) do not always map on types of arguments for a conclusion.
(b) Modal "force" (the context-invariant part of modal meanings) does not always correspond to the force with which the standpoint is advanced or to a phase or type of speech-act in an argumentative discussion.
(c) There are certain modal flavors that seem closer to argumentative speech acts than others: the examples in (10), (11) and (12) show that there is one kind of modal flavor, the epistemic-doxastic, that can readily used to embed other modalities and make them the content of whatever argumentative move it signals. Correspondingly, these modals appear to be more "detached" from the specific grounds on which the conclusion is based.

This evaluation of the viability of Toulmin's views of modal semantics has pretty strong consequences for some of the major philosophical points that Toulmin was trying to make in his book. Both his critique of mathematical probability theory as an adequate source for a philosophical understanding of probability and his critique of logical possibility are based on the allegation of mistaking *one particular type of ground* on which the argumentative force of the modal is based for the *force* itself of the probability or possibility modal. This critique loses much of its weight once it is clear that the allegedly invariant argumentative force of the modal isn't quite apt to capture many ordinary uses of the modals, such as the ones examined above, where either (a) there is no argumentative speech-act to speak of, or if there is an argumentative move either (b.1) the *grounds* for the conclusion do not match the criteria of the modal or (b.2) the *force* of the conclusion does not correspond to the force of the modal. I will pursue this point a little further discussing Toulmin's critique of logical possibility.

3.2.4 *First Interlude: Salvaging the Pragmatic Account of the Modals Through the* **Polyphony** *and* **Delocutivity** *Route*

With a certain amount of ingenuity one can conceive ways in which the speech-act based account of modal forces can be technically salvaged. A possible route for explaining away at least some of the counter examples presented above is offered by the concepts of *polyphony* and *delocutivity*, which are extensively used in various versions of the Theory of Argumentation within Language (TAwL, cf. Anscombre and Ducrot 1983) and in related linguistic approaches in the French tradition.

The polyphonic theory of utterance (*théorie polyphonique de l'énonciation*), as understood in the broad TAwL tradition (Ducrot 1984; Nølke et al. 2004; Kronning 2014) brings Bakhtin's idea of the multi-voicedness of discourse into the formal micro-analysis of individual utterances, which are seen as staging a complex network of points-of-views or voices, to which the voice of the actual speaking subject relates in different ways, assenting to them, distancing itself, denying them. The semantics of linguistic forms is seen as setting up "instructions" or "traces" of the utterance process that guide the recovery and the fleshing out of this stratification of voices in the actual context of utterance (cf. Nølke, Fløttum and Norén 2004: 23–24).

Delocutivity refers to the emergence of derived *descriptive* uses of originally *performative* expressions. Originally introduced by the French linguist E. Benveniste, it became a keystone of the "radically ascriptivist" approach to semantics defended by TAwL (cf. Ducrot 1993), under the influence of the ascriptivist analyses of ethical terms in Richard Hare (1970). For Hare evaluative terms such as *good* do not have a denotative meaning, but express an "act of commendation". In order to account for the fact that *good* can occur also in embedded contexts (e.g. *If this hotel is* ***good****, it must be expensive*) and, as Ducrot (1993: 125) observes, in unembedded contexts where it does not perform an actual commendation (e.g. *This hotel is* ***good****, but it is too expensive*), Hare (1970) splits the illocutionary force of the utterance into a "sign of mood" or *tropic* and a "sign of subscription" or *neustic*. Embedded occurrences of *good* still express the tropic associated to the evaluative word, but lack a *neustic* of their own; that is: they lack the subscription of the speaking subject and are only "*potentially* commendatory" (Hare 1970: 23). Polyphony theory, with its radical detachment between the stances ascribed to the different voices or viewpoints and the actual speaking subject provides a natural environment to extend Hare's account also to unembedded contexts, such as discourse sequences connected by *but* (cf. the polyphonic analysis of *mais* 'but' in Nølke, Fløttum and Norén 2004).

This route would open up a scenario where the possibly embedded and generally non-argumentative uses of the modals are explained away as involving virtual, echoed or evoked arguments. A semantics of this kind might even seem germane to some of Toulmin's philosophical positions, but – to my knowledge – no one among

Toulmin's continuators has tried to pursue its construction for the whole set of modal forces and modal flavors touched in the first essay.

In contrast, the delocutivity route was clearly suggested by Toulmin to generalize his speech-act analysis of *probably* to the whole range of probability expressions, including the adjective *probable* and the noun *probability*. This suggestion is taken up in Ennis' (2006) defense of Toulmin's speech-act analysis.

I will have more to say on *probably* later in this chapter, but I will not inquire further on the general viability of the 'polyphony and delocutivity route' for the full range of modals. I will accept at face value the findings in (a), (b) and (c) above, on the lack of matching between modals and argumentative speech acts. These negative findings, however, must be taken together with the challenge posed by those surprising facts that Toulmin *does* show in his analysis. His *cannot* examples in (3)-(8) *do* show that different kinds of physical, linguistic, legal or moral impossibility expressed by this modal *can* have an intimate relationship with the ruling out of an hypothesis (of different logical types) on different kinds of grounds. In these examples the kinds of arguments supporting the conclusion and the kind of propositions (the criteria) determining the fully fledged meaning of the modal *do* coincide. The cases of mismatch I have examined later do not make these matching cases simply go away or look merely coincidental, what they do is to render the direct explanation in terms of argumentative speech acts that Toulmin proposes *prima facie* implausible. Rather than taking the polyphony and delocutivity route to salvage the hypothesis and explain the non-argumentative cases as derived I will adopt a different hypothesis on the core semantics of the modal, a semantics from which both the argumentative and non-argumentative uses can be derived.

Finding a more adequate framework of explanation of the argumentative relevance of all "flavors" of modality is the challenge I take up from Toulmin in this book. It is, I should stress, a challenge no one has considered up to now. All those that have elaborated upon Toulmin's treatment of modality in the subsequent argumentation literature have taken a different path. Instead of considering the potential argumentative significance of all kinds of modals and rejecting as descriptively inadequate the speech-act level analysis, most treatments in argumentation theory retain (aspects of) the speech-act analysis, but restrict their attention to the sole epistemic-doxastic modalities, such as *probably*, for which this analysis appears defendable. This restriction, in fact, happens to some extent already in Toulmin's book if we consider chiefly the most frequently quoted third essay on the layout of arguments and the essay on probability.

3.2.5 On the Irrelevance of Logical Modalities

In detailing the consequences of his context-dependent analysis of *cannot* and of the distinction between *force* and *criteria* that lies at its core, Toulmin tackles the cases of formal impossibilities (and possibilities) that were initially left out and anticipates aspects of a discussion of logical possibility that he will carry out with greater detail in the fourth essay of the book. Considering how the issue of logical modalities in Toulmin is intimately connected with the distinction between force and criteria it will be useful to briefly discuss it here, before moving to the pragmatic analysis of probability expressions and to the notion of modal qualifier in the second and third essay, which can be considered in relative independence.

The fourth essay is where Toulmin brings to task two millennia of logical theorizing – from Aristotle's syllogistic onwards – which allegedly lost sight of the practical task of evaluating real world arguments because it became enthralled by its fascination for one very particular kind of argument, namely the "analytical syllogism". Logicians disregarded the field dependency of arguments and tried to erect the *criteria* that are appropriate for "analytical arguments" as field independent standards of good reasoning with the result of casting most of the reasoning that is actually done in science and ordinary life outside the pale of a logic condemned to practical irrelevance. For Toulmin, the "logical modalities" – the notions of possibility and necessity employed by logicians – lie at the core of this unfortunate restriction of the scope of logic.

Toulmin (2003 [1958]: 137–138) claims that the exclusive interest of logicians for the "analytic syllogism" has brought them to abusively conflate a series of conceptually independent distinctions: *necessary* vs. *probable* arguments, *formal* vs. *informal* ones, *warrant-using* vs. *warrant-establishing* ones, arguments whose functioning can be accounted just in terms of the meaning of a few "logical words" (such as standard quantifiers and truth-functional connectives) vs. arguments that cannot be explained in terms of those words, and, finally, *analytic* vs. *substantial* arguments. These oppositions, and the alleged exclusive focus of logic for their first members, would require an extensive discussion both in light of the state of logic at the time of writing and in light of its successive development. As announced in § 1.6.2 in the Introduction, I will not dare to dive into this discussion. Instead, I will put on again the linguist's hat and consider the criticisms that Toulmin levels against "logical modalities" as if it was simply a matter of semantics.

Importantly, Toulmin uses the term "logical modalities" more liberally than one could expect, including, for instance, among the logically necessary statements not only properly analytical statements (e.g. *Necessarily, either it is raining or it is not raining*) but any non-analytical statement that is a necessary conclusion given "the data and backing on the basis of which we are arguing" (Cf. Toulmin (2003 [1958]: 140). In other words, he tacitly assumes a *relative* use of the notions of possibility and necessity also on the part of the logicians. Thus, his discussion could be better qualified as a criticism of the *logician's view of modalities in general* than as a criticism of *logical modalities* in the strict sense. In fact, for Toulmin this general view of modality is questionable precisely because it is informed by criteria that should legitimately pertain only to what I have just called "*logical modalities* in the strict sense"[13]. I will distil here, from the broader discussion, three distinct criticisms that Toulmin moves to the logician's treatment of modality.

[13] In reality, Toulmin does not trace explicitly this difference between *logical necessity* in the strict sense (analyticity) and *relative logical necessity* of a conclusion given certain premises. The lack

The first is a general criticism of the lack of attention of logician for the field-dependency of modal terms: "where possibility, necessity and the like are treated in practice in a field-dependent way logicians [...] at most concede, grudgingly, that there may be other looser senses of words like 'necessity' which are used in talking about causation, morality and the like" (Toulmin (2003 [1958]: 137). Certainly, at the time the interest for the logic of deontic and other types of modal concepts was just beginning to emerge (Cf. von Wright 1951). Even more clearly, Toulmin's awareness of the context-dependent functioning of the modals and his perception of the lush variety of "grounds" or "criteria" that can be invoked by the modals in their everyday uses are quite rare at the time. As we will see in Chap. 4, systematic treatments of modal meanings as context-dependent by philosophers and linguists start appearing more than a decade later than *The Uses of Argument*, even if it is always possible to find glimpses on this issue in the writings of earlier authors. It is, thus, difficult to deny the merit of Toulmin's most general critical point: philosophers have indeed been mostly concerned with a limited inventory of "flavors" of modality, considered of special philosophical significance, *and* have not considered worth investigating how the striking variety of modal "flavors" that are used in ordinary reasoning comes to be expressed by a limited inventory of modal words.

In Toulmin, the idea that logicians failed to consider the field-dependence of modal notions is intertwined with a more specific, twofold, criticism of the logician's view of modality. By putting notions of *inconsistency* (or contradiction) and *consistency* at the heart of their definitions of, respectively, impossibility and possibility, logicians have elevated "criteria of possibility and impossibility applicable to analytic arguments into positive definitions of the terms" (Toulmin (2003 [1958]: 140). The target of Toulmin's criticism is made explicit in the following definitions of the modal notions:

(Def. 6.a) 'P is logically impossible' means 'P is either self-contradictory, or contradicts the data and backing on the basis of which we are arguing',

(Def. 6.b) 'P is logically possible' means 'P is not logically impossible (as just defined)', and

(Def. 6.c) 'P is logically necessary' means 'the denial of P is logically impossible (as just defined). (Toulmin (2003 [1958]: 140)

Toulmin's criticism of this kind of definitions of modal notions is twofold.

On the one hand, the criteria of *consistency* and *inconsistency*, being specific of analytical reasoning, cannot apply to the vast majority of the ordinary uses of the

---

of an explicit distinction of this kind makes his commentary on certain key examples rather blurred and difficult to follow (among them is his favorite *Dwight D. Eisenhower as a possible member of the U.S. Davis Cup team* which I discuss here) because it is not always completely clear whether what is contended is the irrelevance of logical possibility/ impossibility strictly intended, or the irrelevance of logical possibility/ impossibility *relative to a body of premises*. On the other hand, one can understand how this distinction may have looked scarcely important in Toulmin's view: if formal logic only works in analytical arguments, then all relative necessities are, in the end, analytical necessities given certain minor premises (unless empirically derived major premises are spuriously introduced). The distinction is, however, of vital importance for building a working semantics of the modals.

modals, which would require their own field-specific criteria. This abusive extension of the analytic criteria has the unfortunate effect of making virtually impossible to qualify a conclusion as *necessary* unless as "a loose and imprecise *façon de parler*, resulting from sloppiness of thought" (*Ibid.:* 141) and, conversely, of extending indefinitely the domain of the *possible* far beyond what could be an hypothesis genuinely worth considering in the relevant field of argument.

On the other hand, *(in-)consistency*, even when it *is relevant* as a ground or criterion for ascribing possibility or impossibility in analytic reasoning (e.g. in mathematics), fails to fully define "the whole meaning of these terms" (*Ibid.:* 160). It is just a criterion, and remains distinct from the *force* of the modal notion. The latter corresponds to "the practical implications" (*Ibid.:* 28) of the use of the modal in a discussion; i.e. to the argumentative "moral" (*Ibid.*, pp. 30 and 161) one draws from the recognition of the criteria. For instance, recognizing a contradiction is one thing and ruling out an hypothesis as mathematically impossible because of that contradiction is another.

Let us examine more closely the first aspect – the irrelevance of (in-)consistency criteria – using one of Toulmin's examples, appearing both in the first and in the fourth essay:

(14.a) Dwight D. Eisenhower is a *possible* member of the team selected to represent the U.S.A. in the Davis Cup match against Australia.

(14.b) Dwight D. Eisenhower will be selected to represent the U.S.A. in the Davis Cup match against Australia. (Cf. Toulmin 2003 [1958]: 34–35 and 157)

Commenting on this example Toulmin observes that (14.b) is meaningful and does not involve any "demonstrable contradiction" (Toulmin 2003 [1958]: 34). Later, in connection with the same example, Toulmin refers to definitions (Def. 6.a) and (Def. 6.b), and speaks of a proposition that "successfully avoids contradicting our data" (*Ibid.:* 157). Yet, even if (14.a) satisfies the conditions of "logical possibility" in (ii) no one "outside the philosopher's study" would dare to submit it as a serious possibility, worth of consideration, when discussing the composition of the U.S. Davis Cup team. Toulmin concludes that, outside analytical reasoning, the logical criteria of possibility, impossibility and necessity "do nothing to show us that any conclusion we shall be concerned in practice is genuinely possible, impossible or necessary" (*Ibid.:* 157).

Should we agree with Toulmin that the consistency criteria invoked by the logician's definition of possibility in (Def. 6.a) fail to account for the meaning of *possible* in (14.a)? Actually, before pronouncing consistency considerations as irrelevant in the account of this particular flavor of possibility we should consider that the notion of consistency in (Def. 6.a) and (Def. 6.b) is relational, and different notions of possibility can arise by choosing different sets of propositions, different relevant "data", with which the modalized proposition represented by (14.b) is supposed to be consistent. Certainly, (14.b) is not self-contradictory, which rules out logical impossibility in the strict sense. Yet, there may still be a relevant set of propositions which plays the role of "data and backing" with which (14.b) enters into contradiction, making (14.a) patently false on a certain salient reading of the modal *possible*.

Let us reconstruct a plausible context of use for (14.a). Such a possibility assessment could be uttered either as part of a *deliberation* process, say, by one of the members of the committee in charge of selecting the U.S. Davis Cup team, or by a third party in a secondary discussion aiming to *predict* the outcome of the deliberation of said committee[14]. For the sake of simplicity, I will assume that (14.a) is uttered in the primary context of the deliberation. In order to specify the relevant bits of contextual information that play a role in determining the "flavor" of the modal it is useful to follow Toulmin's example and offer a syntactically more developed variant of our utterance, such as the one in (14.c).

(14.c) In view of *his tennis-playing form* and of *our selection criteria*, Dwight D. Eisenhower is a *possible* member of the team selected to represent the U.S.A. in the Davis Cup match against Australia.

In (14.c) the complements introduced by the preposition *in view of*, provide one possible meaningful specification of (14.a) that assumes that the committee shares a set of more or less explicit, more or less formal selection criteria, which determine among other things what sort of people satisfy the minimal requirements to be eligible as Davis Cup team members. What is as stake here is not possibility with respect to axioms of logic, or possibility with respect to physical laws, but possibility with respect to criteria belonging to two apparently heterogeneous logical types: certain relevant facts (tennis playing form), and certain necessary conditions (the minimal requirements) that need to obtain for a certain institutional fact to be the case. As we will see in Chap. 5, this kind of modal flavor – pervasive in our reasoning about society but rarely recognized in semantic treatments of natural language modal expressions – has been called *anankastic* (Cf. Kronning 1996, Conte 2010 [1995]).

This helps us to establish with respect to which relevant "data and backing" we should check the consistency of the modalized proposition. Thus, we can hypothesize that the use of *possible* in (14.a, c) is to be understood in terms of consistency of the proposition 'Dwight D. Eisenhower is selected to play against Australia' with the relevant facts about his tennis playing form and with the committee's criteria. circumstances. Can this analysis account for the oddity of (14.a, c) by providing a suitable inconsistency? If we imagine, quite realistically, that the criteria should at least include something like (15),

(15) The members of the U.S.A Davis Cup team must be professional tennis players who have won international matches during the past 12 months

and we add the fact that Dwight D. Eisenhower, who in 1958 was 68 years old and suffering of a heart condition, has not won any international professional match in the past 12 months, the trick is done and we get an inconsistency that explains the

[14] The context of financial argumentation that will be discussed later in this book provides a wealth of examples of predictive arguments that take an underlying process of rational deliberation as their object.

oddity of *possible* in (14.a). At this point someone could object that we have made too easy our refutation of the alleged irrelevance of consistency by selecting a clear cut criterion such as (15). But, let us suppose that the committee has no other criterion than (16):

(16) The members of the U.S.A Davis Cup team need to be the best tennis players in the country.

we can still have our inconsistency as long as we take (17) as a truthful characterization of Dwight D. Eisenhower's tennis playing form:

(17) Dwight D. Eisenhower is not one of the best tennis players in the country.

Obviously, if we consider (17) from an epistemic point of view, we see very well that the reasoning through which people arrive to hold true this proposition might well include all sorts of complex (and perhaps messy) inferences moving from more basic facts about Dwight D. Eisenhower. None of this, however, is directly the business of the modal *possible* in (14.a, c). The point is that as long as someone recognizes (17) as true (which most sane people will do), they will recognize the inconsistency and, in doing so, the falsity of (14.a, c). Treating the semantics of the possibility/impossibility modals in terms of consistency/ inconsistency does not require that we consider them logical modalities in the strict sense, nor does require that the "data and backing" with respect to which we check the consistency/inconsistency of the prejacent proposition be *a priori* truths derived from an uninterrupted deductive chain. Chapter 4 will present a tradition of analysis of modal meanings that considers logical relations of *consistency* and *inconsistency* as part of the context-invariant core meaning of the modals and works out the different contextual meanings of the modals in terms of the different kinds of propositions that are invoked as the relevant "data and backing". In this view, the context-invariant core is not pragmatic (as in Toulmin), but logical.

The second criticism that Toulmin moves to the logical definitions of modalities is very subtle: *(in-)consistency,* even when it is a relevant ground, does not tell us what possibility and impossibility are really about. It's not the whole story. And the missing part of the story has to be accounted in terms of the consequences for the development of the discussion: a possibility is *admitted to further consideration*, while impossibilities are *ruled out of the discussion*. The objection has a strong intuitive appeal: that there should be more to the notion of impossibility than the simple recognition of an inconsistency fits our intuition.

We should, however, consider more carefully whether Toulmin's pragmatic account of the "missing part of the story" is adequate and can be applied to all the uses of the modals. The discussion of examples such as (10), (11), (12) and (13) in § 3.2.3 above seems to indicate that modals, in many of their everyday uses, do not correspond to the kind of argumentatively relevant speech acts that Toulmin envisages. If this is the case, the pragmatic account of the missing part of the story cannot be generalized.

In fact, Toulmin's discussions of possibility and impossibility, for instance in relation to the Eisenhower example in (14.a), contain certain revealing turns of phrase that gently shift the modality under discussion towards the pragmatic story.

For instance, when Toulmin uses expressions such as (18.a) and (18.b) he makes a very specific use of the modals, which can be characterized as *meta-pragmatic* and *meta-argumentative*.

(18.a) Is this a *possible* conclusion? (Toulmin 2003 [1958]: 157)
(18.b) Have I a *possible* case? (*Ibid.*: 158)

Of course, in Toulmin's pragmatic and argumentative account of the modals these explicitly meta-argumetative uses are *ex hypothesi* to be considered as a stylistic variant of the functionally equivalent simple uses. Saying that 'p *is possible*' amounts to say that 'p *is a possible conclusion*'. However, if we explore the alternative hypothesis that these meta-pragmatic and meta-argumentative uses are indeed a special case we obtain an interesting glimpse into the reasons why Toulmin's pragmatic account of the consequences of modality functions for a limited class of uses of the modals.

A meta-argumentative modality can be effectively characterized with the aid of the components of a normative-pragmatic model of argumentation, such as Pragma-Dialectics (van Eemeren 2010). For instance, a meta-argumentative possibility can be conceived as an argumentative speech-act whose performance is consistent with the *set of dialectical rules* governing the discussion, with the *material starting point* of the discussion and with the *store recoding the preceding moves* of the discussants. Examining the consistency of the performance of the argumentative speech act with respect to the above mentioned sets of information means, at the same time, characterizing it with respect to the development of the discussion. At every moment in the development of an argumentative discussion a range of developments consistent with the dialectical rules and with the state of the discussion remains open. Argumentation theorists have sometimes envisaged dynamically this set of argumentative possibilities with the aid of *dialogue profiles* (Walton and Krabbe 1995) or *dialectical profiles* (van Eemeren 2010). Saying, meta-argumentatively, that one '*must* conclude that p' means characterizing the speech-act of 'concluding that p' as the only alternative left that is consistent with the rules and the state of the discussion.

Seen from this angle, the pragmatic consequences of the meta-argumentative uses of the modals appear as a special case of a much more general phenomenon: the fact that modalities quantify over alternative situations ("possible worlds") of some kind.

3.3 The Second View: Probability and Speech Acts

3.3.1 **Probably** *and Guarded Commitment*

As mentioned earlier, Toulmin's discussion of probability concepts and probability idioms combines a pragmatic analysis of the modal adverb *probably* with a critique of the ideas of Kneale (1949) and Carnap (1950) on the nature of probability concepts and the meaning of probability expressions in ordinary discourse. Here I limit myself to considering the speech-act account of *probably*, leaving the discussion of the different conceptions of probability in the background. Toulmin (2003 [1958]: 49) proposes the following speech-act based analysis of *probably*:

(Def. 7) When I say 'S is probably P', I commit myself guardedly, tentatively or with reservations to the view that S is P, and (likewise guardedly) lend my authority to that view.

This pragmatic analysis is directly inspired by an early analysis of *I promise* and *I know* as performative formulas by J.L. Austin, the founder of speech-act theory (Austin 1946). Toulmin uses two examples to argue for the pragmatic analysis in terms of *guarded commitment*, reproduced here as (19) and (20).

(19) There comes a moment in time in the life of a well-brought-up little boy when he finds himself in a quandary. For the last week he has come every day after tea to play with the little girl who lives in the next street, and he has begun to value her esteem. Now bed-time is near, Mother has come to fetch him away, and his companion says with bright eyes, 'You *will* come tomorrow, won't you?' Ordinarily he would have answered 'Yes' without a qualm, for every other evening he has fully intended to come next day, and known of nothing to stand his way. But… but there was some talk at home of a visit to the Zoo to-morrow; and what if that, and tea in a tea-shop afterwards, and the crowds in the Tube, meant that they were late getting home, and that he was to fail after saying 'Yes'?... How difficult life is! If he says 'Yes' and then cannot come, she will be entitled to feel that he has let her down. If he says 'No', and then is back in time after all, she will not be expecting him and he will not be able, decently, to come; and so he will have deprived himself, by his own word, of his chief pleasure. What to say? He turns to his mother for help. She, understanding the dilemma, smiles and presents him with a way out: 'Tell her that you'll *probably* come, darling. Explain that you can't *promise,* since it depends on what time we get home, but say that you'll come if you possibly can'. Thankful for the relief, he turns back and utters the magic word: 'Probably'. (Toulmin 2003 [1958]: 44–45).

(20) A complex disturbance at present over Iceland is moving in an easterly direction. Cloudy condition now affecting Northern Ireland will spread to N.W. England during the day, *probably* extending to the rest of the country in the course of the evening and night. (Toulmin 2003 [1958]: 47)

The little boy in (19) is performing what Toulmin calls a "guarded and restricted undertaking" (*Ibid.* p. 48) or "qualified declaration of one's intentions" (*Ibid.* p. 47) aimed to shield, or insure, him against "some of the consequences of failure" (*Ibid.* p. 46). Had the boy uttered the performative formula *I promise* and failed to show up, he would have given reason to be strongly reproached by the girl. But even uttering an unqualified *I shall come* would have laid him open to a reproach as, Toulmin argues, it would have given reason to *expect* his arrival, perhaps making preparations. Similarly, the weather forecasters producing (20) are making a "guarded and restricted prediction" (*Ibid.* p. 48) avoiding to put at stake in full the authority of their expertise on that item and so avoiding, to a limited extent, the consequences of failure. As Toulmin observes, the need of qualifying their prediction with *probably*, is felt more keenly by the forecasters of the meteorological office than by the little boy in (19). Such is the authority of expertise of the institution that any non qualified

Table 3.2 Assertive and commissive scales (Cf. Toulmin 2003 [1958]: 46)

	Assertive scale	Commissive scale	Meaning (or force)
Introduced by explicit performative formula:	"'*I know* that S is P'"	"*I promise* to do A"	"I expressly commit myself"
Unqualified:	"'S is P'"	"'I shall do A'"	I commit myself "to a lesser degree and only by implication"
Qualified by *probably*:	"'S is *probably* P'"	"'I shall *probably* do A'"	"I expressly avoid unreservedly committing myself. I insure myself thereby against some of the consequences of failure"

prediction would be taken contextually as if it was prefaced by the assertive performative formula *We know*.

For Toulmin, *commissive* and *assertive* utterances qualified by *probably*, such as those in (19) and (20), are closely related to those prefaced by *I promise* and *I know* studied by Austin (1946: 143–144) and form with them two parallel series, or scales, reconstructed in Table 3.2, above.

The term *assertive*, used in Table 3.2, is a Searlean term with no equivalent in Austin[15]. Using it in relation to Toulmin's account constitutes an anachronism and involves a modicum of distortion[16]. If we were to follow the later *William James Lectures* (Austin 1962), the meteorological forecast in (20) would fall either under the *expositives*, which are acts of expounding of views, conducting of arguments and clarifying, or, given the institutional authority of the meteorological office, under the *verdictives*, which correspond to the delivery of a finding or verdict of fact or value on the basis of evidence or reasons. The promises and declarations of intention in (19) would be *commissive* also in Austin (1962). Interestingly, when Austin discusses commissives in lecture XII, he observes that there is a difference between true *commitments* and simple *declarations of intention*, so that one could be tempted to split the class of commissives. Austin exemplifies the latter with expressions qualified by *probably*.

Looking at how Toulmin constructs the two series of Table 3.2 starting from Austin's (1946) remarks on the parallelism between *I know* and *I promise*, it seems fair to say that he sees *probably* as a *modifier of the illocutionary force* of different kinds of speech acts (at least of *assertive* and *commissive* ones). The nature of the modification of the act is characterized, rather vaguely, by Toulmin using the adver-

[15] For the sake of simplicity, in this book I will, as a rule, normalize speech-act nomenclature to the Searlean terminology (Searle 1969), which is the one adopted by Pragma-Dialectics (van Eemeren and Grootendorst 2004) and by Congruity Theory (Rigotti 2005) – two theories that play an important background role in this investigation.

[16] Toulmin refers Austin's ideas as first expounded in his 1946 essay, while Austin's "classic" formulation of speech-act theory is given in his *William James Lectures* held at Harvard in 1955 and later published as Austin (1962). Austin had given courses on *Words and Deeds* at Oxford in the years 1952-1954 and Toulmin might well have been exposed to his theory from other sources than the 1946 essay.

bials *guardedly, tentatively, with reservations* and the rather convoluted expression *expressly avoiding unreservedly committing*. At any rate, Toulmin makes clear that this modification has an interactional point, which consists in limiting the answerability of the speaker and shielding him from "some of the consequences of failure" (Toulmin 2003 [1958]: 51).

In relation to this interactional point Toulmin also uses the expression *hedging* (*Ibid.* p. 48), which conveys precisely this idea of a strategy aimed at avoiding unwanted and unpleasant consequences. This is a non-technical use of the ordinary idiom and refers in a derogatory way to the abuse of the shield provided by *probably*. Yet, the function of *probably* according to Toulmin matches very well the technical semantic notion of *hedge* (Lakoff 1973) as an operator that makes the attribution of a predicate to an argument term fuzzy or partial. Here, what is made fuzzy or partial is the performance of a certain kind of speech act by the speaker, with its associated commitments. Toulmin's *probably* also matches the related but broader discourse-analytic notion of *hedge* (cf. Markkanen and Schröder 1997), which covers a range of strategies aimed at limiting the responsibility of the speaker towards its utterance and thus mitigating its interactional consequences.

The pragmatic analysis allows Toulmin to avoid what he believes to be the fundamental error plaguing both objectivist and subjectivist theories of probability: the search of a "designatum", of a "referent", of a "thing" denoted by the word *probability* (Cf. Toulmin 2003 [1958]: 61). According to Toulmin, the pragmatic approach differs not only from objectivist approaches that seek a thing "out-there" to which probability statements correspond (e.g. observed relative frequencies), but also from subjectivist ones that treat probability statements as statements "about the present strength of the speaker's beliefs" (*Ibid.*, p. 78).

The latter point deserves to be underscored as it points to one of the major issues in the treatment of epistemic modalities not only in the analysis of arguments, but more broadly in natural language semantics; namely, the treatment of those epistemic modal expressions that do not appear to contribute to the basic truth-conditions, or propositional content, of the utterance in which they appear. In rejecting the idea that utterances modalized by *probably* are *about* the strength of speaker's beliefs, Toulmin appeals to a distinction between what an utterance *states* and what is *implied in* it: even if we are legitimated to draw inferences on the strength of the speaker's beliefs from the kind of modal qualifier used the utterance remains about the qualified state of affairs and not about the belief. Toulmin insists that the modal does not contribute to the asserted content of the utterance and maintains that '*Probably* p' corresponds to 'asserting guardedly that *p*' and not to 'asserting that you are tentatively prepared to assert that p'. According to Toulmin (2003 [1958]: 79), utterances of the form '*Probably* p' are no more about the speaker beliefs than bare assertions of p are: in both cases the utterance states something about the world and enables us to infer something about the speaker's belief.

While Toulmin maintains that his treatment of *probably* carries further the analysis of modal terms presented in the first essay (Cf. Toulmin 2003 [1958]: 41), we can see that, in fact, the two pragmatic analyses bear only a superficial resemblance. In the first essay, modals like *possible, cannot* and *must* are *illocutionary force indicators* rather than *illocutionary force modifiers*[17]. In terms of Congruity Theory, they are full blown *connective predicates* corresponding to specific argumentative moves that mark different phases of an ideal argumentative dialogue: 'advancing an hypothesis as worthy of consideration', 'ruling out an hypothesis' and 'concluding'.

Probably, on the other hand, intervenes to modify the commitments of different kinds of speech acts and is characterized in terms that do make reference to argumentative moves or to an argumentation process. Consistently, the analysis of *probably* is illustrated, in (19) and (20), with examples that do not involve inter-subjective argumentation: both the little girl and the audience of the weather forecast have simply to trust the speaker. As we have seen, in the first chapter, Toulmin had given a completely different characterization of *probably*, couched in argumentative terms as an indicator of a situation where, in light of the evidence, one hypothesis appears more worthy than the others (*Ibid.*, p. 20). In the second chapter, the idea that *probably* has to do with the relationship between the evidence and a conclusion is not part of the definition of the meaning of the modal. The connection between the meaning of *probably* and the evidence-conclusion relationship is only *indirect*.

It is only after having put in place the independent analysis of *probably* as an illocutionary modifier hedging the responsibility of the speaker that Toulmin observes that the reason for qualifying the commitment can be "the sub-standard quality of the *evidence* and *argument* at the speaker's disposal" (*Ibid.*, p. 48). Concluding the chapter Toulmin generalizes this remark, stating that "it is the quality of the argument or evidence at the speaker's disposal which determines what sort of qualifier he is entitled to include in his statements" (*Ibid.*, p. 83).

Toulmin echoes his analysis of the modals in the first essay when he insists that *probable* like *cannot* "is a term that keeps an invariant force throughout a wide variety of applications" (*Ibid.*, p. 77). Yet, if we look at the actual analysis of *probably* we find that idea that the meaning of the modal is composed by invariant *force* and variable *criteria*, which was embodied by the relational template used to elucidate the meaning of *cannot* is largely, if not completely, abandoned. The criteria variables migrate, so-to-say, outside the meaning of the modal, which is reduced to the sole *force*. The quality of the *evidence* may be what motivates the hedging of a commitment using *probably*, but there is no open slot for the *evidence* in the semantic analysis of *probably*.

In fact, Toulmin, in criticizing the works of Kneale (1949) and Carnap (1950), explicitly rejects the relational conceptions of probability terms according to which *probability* is relative to the *evidence*. It is important to stress that his criticism is not limited to the view that reduces probability to observed relative frequencies (Cf. Toulmin 2003 [1958]: 72), but invests the very idea that the meaning of *probably* should be elucidated in terms of a relation between a body of evidence and a hypothesis. Toulmin presents several lengthy arguments aimed at rejecting different versions of the relational conception of probability. Here, I shall be concerned only

[17] Referred to Toulmin (2003 [1958]) this terminology is obviously an anachronism.

with the most general ones, which apply to the very idea of probability as a relationship between evidence (*e*) and a hypothesis (*h*).

Firstly, for Toulmin probability expressions cannot properly point to bodies of evidence other than the actual evidence at the disposal of the speaker at the moment of utterance. Something that the speaker knows to be false cannot be literally said to be *probable* with respect to a different body of evidence: one can only say that it *seems* probable to other people, or that it *seemed* probable to the speaker in the past. For Toulmin (*Ibid.*, p. 50) examples referring to counterfactual past probability, such as (21) below, must be seen as involving some sort of "*oratio obliqua*".

(21) Alonso was leading the race with one lap to go. *It was very probable* that he would win. Then suddenly, going into the first corner, he lost the back end and spun off into the gravel trap and then hit the tyre wall.

With a little bit of terminological actualization, one could reconstruct Toulmin's position as follows: the use of *probable* in utterances such as (21) should not be explained by simply saying that the modal manifests a relation *probable* (*e*, *h*) and in this particular occurrence the variable *e* is identified with the evidence available to the spectators at the beginning of the last lap (rather than with the evidence available to the spectator); it should be considered an instance of free indirect speech reporting a real utterance (or staging a fictional one) by a (real or fictional) subject for whom it is permissible to use *probable/probably* in its original function of illocutionary modifier indicating guarded assertion. In sum, for Toulmin *epistemic probability* – that is probability in view of all the known relevant evidence (cf. Freeman 1991: 123–124) – is basic, while other kinds of probability are derived through what I have called above (cf. § 3.2.4) the *polyphony and delocutivity route*; which consists in setting up "voices" other than the speaker and deriving secondary descriptive properties from expressions whose original meaning is performative rather than descriptive.

A second general criticism of relational approaches to the semantics of probability expressions is that they confuse the qualification of a conclusion as probable with the "support relation" between the evidence and the conclusion that authorizes such a qualification. According to Toulmin, this amounts to confusing the categorical statement of a conclusion with the hypothetical statement of the bearing of the evidence on the conclusion (cf. Toulmin 2003 [1958]: 74–75). This distinction corresponds to the distinction between the conclusion of an argument and its backing (*Ibid.*, p. 77). Table 3.3, below, summarizes how Toulmin draws the distinction between the two categories.

Table 3.3 Probability and support

Backing (hypothetical):	**Conclusion** (categorical)**:**
'If A, then B'	'A, so B'
Evidential support:	**Probability:**
'Evidence *e*, if available, would suggest very strongly that *h*'	'*e*; so very likely *h*'

Cf. Toulmin (2003 [1958]: 74–75, 77)

The distinction to which Toulmin points is a fundamental one. In this book we will repeatedly point to the close connections that modal expressions have with conditionals on the one hand, but also with argumentative connectives (*so*, *therefore*) and causal ones (*because*). On their own, the argumentative connections expressed by concluding connectives, causal relations and hypothetical ones are subtly related[18]. As we will see in the following chapters, modal expressions of different "flavors", according to very specific semantic conditions of their context of use can function as indicator either of the relations in the left column of Table 3.3, being thus closer to conditionals, or of those of the right column, being thus closer to argumentative, sometimes to causal, connectives[19].

However, saying that *probably* falls firmly in the second camp, being used to conclude (guardedly) for a certain hypothesis does not entail that it should not be regarded as relational, and more specifically as a relation between evidence and an hypothesis. In fact, the notion of *conclusion* illustrated in the right column of Table 3.3 is itself a *relational* notion, just like the hypothetical one that forms its backing. The very fact that Toulmin uses the same *couple* of variable letters (*A*, *B* or *e*, *h*) to illustrate both kinds of notions attests to that. Commenting on this distinction, Toulmin compares *probable* to the predicate *true*, saying that what applies to the latter (which is not relative to evidence), applies also to the former (*Ibid.*, p. 76). It is surprising, however, that – having evoked the distinction between a conclusion and its backing – Toulmin does not compare *probably* to *must*, which appeared as the sign of the speech-act level relation of *concluding* in his first essay. Contrary to what Toulmin seems to suggest here, his choice of seeing *probably* as unary rather than as relational has nothing to do with the distinction between the act of 'concluding B from A' and the conditional 'If A then B' that forms the backing of the conclusion. Both are relational.

It should be clear, at this point, that despite the fact that Toulmin presents his analysis of *probably* as a direct continuation and refinement of the analysis of modal terms offered in the first essay, in fact, he starts afresh with a different approach, informed by other "local" preoccupations and produces something very different. This is consistent with the textual history of the second essay, which appeared earlier independently as Toulmin (1956). Whether the two analyses are squarely inconsistent or just badly in need of clarifications on their missing links is hard to tell, given the vagueness of certain areas of both proposals.

What can be said, however, is that the connections that Toulmin draws between the two analyses are unconvincing. In particular, it is hard to see how *probably* could be said to exhibit field dependent criteria in the same sense of *cannot*, as Toulmin repeatedly suggests (Cf., for instance, Toulmin 2003 [1958]: 77). Certainly, Toulmin does shows that *probably* can apply to a wide variety of judgments, includ-

[18] This is, of course, something of which Toulmin was keenly aware given the importance he gives to the reflections of Ryle (1950) on the topic.

[19] Moreover, as we will see in Chap. 6 natural language conditional expressions, which interact closely with modals, are not purely hypothetical as they do not remain neutral with respect to the truth of the antecedent and consequent they embed.

ing those, like the judgments of artistic excellence (22), where an assignment of numerical probabilities would be meaningless.

(22) This painting is probably the finest product of the whole Paduan school (*Ibid.*, p. 83)

Yet, this variety of application is quite another thing than the different kinds of impossibilities and improprieties that were examined in the case of *cannot* in the first essay, which spanned physical impossibility, linguistic improperty, illegality and immorality. It is true that an adverb like *probably* can be applied to conclusions about all sorts of issues, but it's functioning is not affected by the context in the same way in which the functioning of *cannot* is. Using the terminology currently employed in dealing with the semantics of modality, we can say that *probably* always remains an *epistemic* modal independently of the kind of judgment it modifies. As we have seen above in § 3.2.3, examples (10), (11) and (12), epistemic modals can embed other kinds of modals irrespectively of their, force, with which they do not interact. Consider the following examples:

(23) *Probably*, you *can* smoke on the terrace.
(24) *Probably*, he *can* lift 100 kg single-handed.

In (23) and (24), the presence of *probably* does not directly hedge or otherwise affect the deontic permissibility or physical possibility of the action in question. *Probably* has to do, in both cases, with the amount of evidence that the speaker has to assert this possibility or permissibility. It does not *by itself* indicate a *marginal* capacity of lifting that weight or the *borderline* status of the action of smoking on the terrace (although it can be used to indirectly convey these ideas by implicature in a conversation). It only indicates the lack of fully compelling evidence. One could also observe that, in itself, the embedding of the possibility modals by *probably* sits singularly badly with Toulmin's idea that the invariant force of possibility is to present a hypothesis as worth considering.

3.3.2 Ennis vs. Freeman on Toulmin's Probably

In a contribution to a 2006 collection of essays devoted to *Arguing on the Toulmin Model*, Robert Ennis provides a defense of what he calls "Toulmin's speech-act contextual definition of *probably*" (Ennis 2006: 146). The externalities of this short contribution are notable: it was presented at a 2005 OSSA conference devoted to discussing the different aspects of the Toulmin model and published in the subsequent volume, where it is the only essay devoted to the *qualifier* component. It will be interesting to look at certain difficulties his defense runs into by contrasting them with some remarks by James Freeman, another scholar interested in Toulmin's legacy. In a book that takes the "Toulmin model" as its starting point, Freeman (1991) devotes a chapter to modalities, where he rejects Toulmin's view of *probably* not because it's a pragmatic view, but firstly and foremostly because it's an *unary* rather than relational view.

Ennis (2006: 146) limits his defense of Toulmin's speech-act hypothesis to the sentential adverb *probably,* leaving aside "the closely-related *probable* and the distantly-related *probability*" and excluding the embedded occurrences of the adverb, for which he will need a separate explanation. Before introducing his main argument in support of the hypothesis, Ennis (2006: 150) briefly comments on the initial plausibility of Toulmin's examples of the little boy and of the weather forecast. He does not, however, take notice of the difference between the commissive act of the boy and the assertive act of the forecaster. Likewise, he does not comment on Toulmin's *probably* being a hedge or qualification of different types of act – an illocutionary modifier rather than a fully-fledged connective predicate. Ennis (2006: 148) refers to the "act of guardedly committing" as if it were one speech act, rather than as a way of performing different kinds of acts.

The main positive argument in favor of the speech-act hypothesis is based on a comparison with competing hypotheses via a substitution test[20]. Different glosses of *probably* are substituted for the adverb in a constructed argumentative text – the so-called "raccoon argument" (an example he already used in Ennis 2004):

(25) Raccoons rarely attack a human when they do not feel threatened and do not feel that their young are threatened. The raccoon that is ambling across the yard does not feel threatened by us, and it does not feel that its young are threatened-its young are not around. So the raccoon will *probably* not bother you.

Four are the alternatives taken into consideration:

(Def. 8.1) an *objective specific numerical* definition where *probably* stands for a specific probability value or a specific probability range or distribution (e.g. probability is 0.97);

(Def. 8.2) an *objective non-specific numerical* probability (e.g. probability is less than 1.0, but substantially above 0.5);

(Def. 8.3) an *objective non-numerical definition* consisting in a paraphrase in terms of the adjective *probable* (e.g. it is probable that);

(Def. 8.4) a *subjective specific numerical definition* construed in terms of a numerical speaker's "degree of belief" (e.g. my degree of belief is 0.97)

I have no idea why the two additional possibilities of a *subjective non-specific numerical* (e.g. speaker's degree of belief is lesser than 1.0 but significantly higher than 0.5) and of a *subjective non-numerical* definition (e.g. speaker does not fully believe *p*, but believes it more than *non-p*) are not considered by Ennis. They are hardly more implausible than the bizarre definition in (Def. 8.4).

---

[20] A very curious trait of Ennis' paper is that the author sets out to accomplish a typically linguistic endeavor – the semantic analysis of a lexical unit – using the quintessentially linguistic method of substitution tests, but does not show any awareness of the existence of lexical semantics as a field. He prefers to speak of an "ordinary language argument" and mentions individual differences in the beliefs about 'what we would say' as one of the methodological problems of the "ordinary language movement". Allusion to the ordinary language *movement*, makes it look as if the paper was written in the 1950s rather than in 2006.

In the discussion, alternatives (Def. 8.1) and (Def. 8.4) are easily dismissed as plausible substitutes of *probably* in the "raccoon argument", simply because they require numbers "that are just not available" (Ennis 2006: 151). The Toulminian analysis and alternatives (Def. 8.2) and (Def. 8.3) pass the Raccoon substitution test.

The second substitution test involves what Ennis (2006: 152) calls the retention of a "speaker's inconsistency", exemplified in (26). As Ennis correctly recognizes the inconsistency concerns commitments, rather than propositional content. In fact, it looks like a variant of Moore's paradox (cf. Sorensen 2014) with *probably* in lieu of a belief predicate. Here is how the substitution test goes according to Ennis:

(26) I *probably* did not roll boxcars, but I did roll boxcars. (stimulus, inconsistent)

(26.0) I guardedly commit to not having rolled boxcars, but I did roll boxcars (retains inconsistency, passes test)

(26.1) The probability that I did not roll boxcars was 0.97222…, but I but I did roll boxcars (consistent, fails test)

(26.2) The probability that I did not roll boxcars was substantially above 0.5 but below 1.0…, but I did roll boxcars (consistent, fails test)

(26.3) It was probable that I did not roll boxcars, but I did roll boxcars (consistent, fails test)

(26.4) My degree of belief that I did not roll boxcars was 0.97, but I did roll boxcars (consistent, fails test)

Ennis (2006: 52) concludes that all the alternatives to Toulmin's hypothesis fail to retain the inconsistency. Unfortunately, Ennis' application of this semantic test is compromised by a problem with the respective scope of the modal and of the past tense and the result has everything to do with which operator takes the larger scope in the sentences constructed by Ennis, and nothing to do with what it is supposed to test. In (26.0), where the paraphrase of the modal (*I guardedly commit..*) lies outside the scope of the past tense, the inconsistency is retained, while in all the other examples, where the paraphrase of the modal falls under a past tense operator the inconsistency is lost. As simple as that. But, there is no reason to make an unembedded *probably* fall under past tense and thus refer to *past* beliefs, *past* objective numerical probabilities, *past* objective non-numerical probabilities, etc. Take example (26.3), which is supposed represent the objective non-numerical view of probability, it suffices to turn the adjectival construction in the present as in (26.3.bis) and the inconsistency suddenly reappears!

(26.3.bis) It *is* probable that I did not roll boxcars, but I did roll boxcars (inconsistent, passes test)

Clearly, the time is important because the probability is relative two different states of the evidence. With (26.3) we can imagine a situation like the following: I roll 2 six-sided dice using a dice cup, then I lift the cup to check the outcome. Before lifting the cup it *was* probable that I did not roll boxcars, but now, having lifted it, I can assert that I did roll boxcars. There is no inconsistency because the ascription of probability and the assertion refer to two distinct states of the evidence. In (26.3.bis) the ascription of probability and the assertion appear to refer to the same information state, creating a contradiction between the ascription of probability and the epistemic status required by the felicity conditions of assertion (you should not assert propositions that you deem probably false). Examples (26.1) and (26.2) work in the same way and become inconsistent if turned to present tense.

The above remarks acquire a fuller significance in light of Freeman's (1991), rejection of Toulmin's unary analysis of *probably*. Following Max Black's (1967) analysis of probability expressions, Freeman maintains that the underlying semantic frame of *probably* is always relational and has the following structure:

(Def. 9.1) Given evidence D, probably p

This relational structure is muted where *probably* is used in a seemingly absolute sense. Yet, even in these "absolute" uses *probably* should not be understood as unary but rather as "elliptical". In these utterance D is made to correspond to an unanalyzed or partially unidentified background of relevant aspects of the "state of the world". Freeman continues to follow Black in observing that the relational nature of probability is most evident in expressions using the noun *probability* where the "assertive force" associated with the adverb is absent:

(Def. 9.2) Given evidence D, the *probability* of p is such-and-such

When *probably* is used "in the context of arguments", its relational structure re-emerges:

(Def. 9.3) Given $p_1, p_2, \ldots, p_n$, *probably* q

According to Freeman, in these occurrences the ascription of relative probability is combined with the assertion of q, "albeit in a hedged manner" (Freeman 1991: 115). Interestingly, Freeman maintains that the unarticulated background that is presupposed by the apparently unary occurrences of *probably* remains there also in argumentative occurrences, so that a more precise formulation of the semantic frame of *probably* in arguments should be the following:

(Def. 9.4) Given $p_1, p_2, \ldots, p_n$, together with background B *probably* q

Even if the arguer, in many cases may not have "any salient view of such background information" (*Ibid.*: 116) the B slot remains crucial, in Freeman's conception to distinguish the purely relative sense of probability from the "assertive" uses of *probably*. In the non-assertive sense (also called *inductive probability*) we speak of the probability of a proposition "given a specific set of evidence statements" (*Ibid.* 123), while in the case of *epistemic probability*, "given all the known relevant evidence". The "assertiveness" or performativity of unembedded *probably* can be derived from the fact that the adverb always refers to *epistemic probability* that is probability in view of all the relevant evidence available to the speaker at the moment of utterance. When I use *probably* in an argument I can make explicit a part (or the entirety) of the propositions making up the relevant set of evidence B, but it is understood that I do not abstract from any known proposition impinging on the probability of q.

Going back to the problematic test proposed by Ennis, the past tense ascriptions of probability of non-q relativizes the evidence to another moment in time (arguably to a moment where the only evidence available is that the chance of rolling boxcars on any given roll is 1/36) and is therefore not incompatible with the assertion of q. For this very reason ascriptions of probability under the scope of a past tense operator are not a good paraphrase of *probably*, which always expresses *epistemic probability*, as defined above, and it is therefore always, so to say, in the present tense.

For the subjective definitions of *probably* such as (26.4), which are based on the idea of the subjective strength of a belief, things are even simpler. Consider the following, negatively phrased, instance of Moore's paradox:

(27) I believe that I did not roll boxcars, but I did roll boxcars

The "paradox" is due to a contradiction between the belief explicitly expressed in the first part of the utterance and the belief[21] state that is required for asserting felicitously the second part of the utterance One of the original reasons of the interest for this "paradox" was the fact that it cannot be accounted for in a purely propositional analysis and requires a pragmatic analysis of the act of assertion. If we put the belief predicate under the scope of a past tense operator the paradox, quite trivially, goes away:

(27.bis) I believed that I did not roll boxcars, but I did roll boxcars

This is exactly what happens in (26.4) in Ennis' inadequate substitution test. Finally, if we were to put in the past tense Toulmin's paraphrase in terms of "guarded commitment" (26.0) the expression would lose its performativity and the incongruity would, again trivially, go away:

(26.0.bis) I had guardedly committed to not having rolled boxcars, but I did roll boxcars

To summarize: the test proves nothing as regards the superiority of Toulmin's paraphrase over the four alternative analyses discussed by Ennis. On the contrary, the deconstruction of the test with the help of Freeman (1991) has suggested that, in order to account for the apparent performativity of unembedded uses of *probably*, we are not obliged to have recourse to Toulmin's unary analysis and sever the link with conditional notions of probability, but rather we need to distinguish between the *epistemic* uses – understood as indexical uses that refer to all the relevant evidence available to the speaker at the moment of utterance – and uses that abstract only certain kinds of facts.

I will return to Freeman's relational notion of modalities in § 3.4.2 in the context of the discussion of the modal qualifier as a distinct component of the "Toulmin model" in the third essay. As for the remaining arguments mustered by Ennis in defense of Toulmin's analysis of *probably*, two of them deserve a brief discussion.

There is a third substitution test that Ennis uses to compare Toulmin's analysis with the four competing views of *probably*. This concerns the retention of inconsistency in a situation of dialogical contradiction as the following:

---

[21] In Searlean terms this belief is the sincerity condition of the illocutionary act of assertion.

(28.0) A: The raccoon will probably not bother you
B: The raccoon will probably bother you

According to Ennis (2006: 153) Toulmin's analysis retains the inconsistency because the two participants "are expressing opposing commitments". This is a sui generis inconsistency, however. Ennis (2006: 155) maintains that since committing oneself and guarding one's commitment are actions they do not have truth-conditions, which entails that the *probably* utterances, considered in their entirety, are not propositions and do not have truth-conditions, even if they do contain propositions. So, the utterances are in some sense inconsistent, but since "neither statement is a proposition because each includes an action" the pair is not "contradictory in the technical sense" (Ennis 2006: 156).

The objective hypotheses (Def. 8.1, 2 and 3) also retain the inconsistency and pass the test. And, it must be noted, do so through a plain relationship of propositional contradiction rather than via the somewhat murky pragmatic relation of opposition sketched above. On the other hand, according to Ennis (2006: 153) the test is useful because it allows us to eliminate the subjective analysis (Def. 8.4) from the range of viable paraphrases of *probably*. The reason is that the two beliefs expressed in (28.4) are consistent as "statements about A's and B's degrees of belief".

(28.4) A: My degree of belief that the raccoon will not bother you is 0.95
B: My degree of belief that the raccoon will bother you is 0.95

Again, Ennis' construction and interpretation of the tests leave something to be desired. The critical question here is the following: does interpreting *probably* in terms of subjective belief entail saying that we have a *statement about* 'degrees of belief' any more than interpreting it in terms of guarded commitment entails saying that there is a *statement about* 'the action of guardedly committing'? Obviously, should we read the *guardedly committing* part as a part of the propositional content rather than as a non-propositional illocutionary marker, the commitment reading would as well fail the test. Conversely, if there were a possibility to view the expression of a degree of subjective belief not as part of the propositional content, but as an extra-propositional commentary, such a reformulation of the subjective belief hypothesis would pass the test.

In fact, by choosing a case of dialogical contradiction with *probably* for the test, Ennis chooses a kind of example that, during the last few decades (see for, instance, Papafragou 2006, Portner 2009 and several of the essays in Egan and Weatherson 2011), has been hotly debated in the linguistic semantic literature on epistemic modal expressions – including adverbs such as *probably*, but also the epistemic readings of modal verbs like *might* or *must*. Dialogical contradiction examples lie at crossroads of two debated issues in the semantics of these epistemic expressions: the first is whether these expressions should be seen as referring to the subjective individual beliefs of the utterer, the second is whether they should be considered part of the propositional content of the utterance (i.e. contributing to its truth conditions) and, if they are not, to what level of meaning should they be seen as contributing to. Consider the following two examples of dialogical contradiction:

(29.a) A: This *must* be John's car entering the driveway.
B: No. John's Trabant does not sound like a V12 Ferrari!

(29.b) A: It *might* rain tomorrow.
B: No. There is no way it is going to happen this season.

If the epistemic modals *must* and *might* in A's statements refer to A's subjective beliefs at the moment of utterance *and* they are part of the propositional content there should be no contradiction with B's statements. A state of affairs can well be necessary (29.a) or admissible (29.b) in view of A's subjective beliefs and still be false and ruled out by evidence in B's possession. There would not be any inconsistency. Yet, B's reactions imply that A is *wrong* and that these statements should be revised in light of the new evidence provided.

Several different ways out from this quandary have been sketched in the literature (see, in particular, Papafragou 2000, 2006 and Portner 2009). One possibility is to say an epistemic modal like *must* in (29.a) is indeed relating to the utterer's subjective belief state, but is not part of the propositional content being asserted. In this view A is doing two things: (I) asserting that 'John's car is entering the driveway', (II) commenting on the relationship between this proposition and A's subjectively available evidence. While the asserted content falls under the scope of the dialogical negation, the non-propositional commentary is not affected by it. If we want to apply a similar explanation to *might* in (29.b) we run into an additional difficulty: while the prejacent proposition of *must* (or *probably*) can be thought as an asserted propositional content, albeit perhaps a somewhat weakened one, the epistemic attitude conveyed with *might* appears too weak to be compatible with the felicity conditions of the act of assertion, even in a weakened version.

Thus, a non-propositional view of *might* appears to require that a special type of speech act, distinct from assertion, is postulated. This is the route recently taken by Swanson (2006, 2011, see also the discussion in Portner 2009: 155–158) who proposes a sophisticated pragmatic account of the epistemic modal *might* that ends up deeply revising the very notion of what an assertion is. "Assertions" in Swanson's broadened sense do not require knowledge nor reasonable full belief. They can be performed with different degrees of *authority,* which reflects on the range of subjective probabilities associated with the proposition. By asserting a proposition I "advise the addressee on how to update her subjective probabilities" (Portner 2009: 158). Epistemic modal expressions are seen as modulating the degree of authority of the assertion. The speaker who utters '*might* p' takes the position of someone who is "communicating from a position of ignorance", admitting an extremely wide range of subjective probabilities for p. What is the communicative point of performing such an "assertion"? Swanson (2011: 261) illustrates it with the following example:

(30) Suppose I have no idea of where my car keys are, and neither has my housemate. He gets home from work – and so he has no good sense of where I have looked – and I ask him if he knows where my keys are. He says: "The keys *might* be on the kitchen table".

The housemate, Swanson observes (2011: 265), "is attempting to ensure only that I not *inadvertently* rule out or overlook the possibility that my keys are on the kitchen table". As Portner (2009: 158) summarizes, by uttering the *might* sentence the speaker performs the act of "rasing a possibility of which the addressee might have previously been unaware". Interestingly, Swanson's speech act of *raising a possibility* is not very far from Toulmin's original speech-act view of possibility as "putting forward an hypothesis as worth considering", which I have reconstructed in terms of Congruity Theory in the opening sections of this chapter.

The inverse way out from the quandary posed by the dialogical contradictions with epistemic modals such as *must* and *might* in (29.a) and (29.b) is to say that they *do* contribute to the truth conditions (propositional content) of the utterances in which they appear, but they are not about subjective individual beliefs. They are instead about the evidence available to a community of epistemic agents. The exact construal of this community knowledge is a tricky step: understanding it simply as the *common ground* between speaker and addressee does not lend the right results in certain contexts, so that several alternative configurations have been proposed (e.g. the sum of the knowledge of speaker and addressee pooled together).

There is, finally, a third, intermediate, way of escaping the problems posed by dialogical contradiction of epistemic modals, which is adopted by Papafragou (2000, 2006). Papafragou distinguishes between the overall truth-conditions or propositional content of an utterance and the proposition that carries the main communicative relevance in such an utterance. Typically, negation and relations of dialogical contradiction do not entail the falsity of the whole propositional content of the target utterance, but only of the predicate-argument nexus that constitutes the *rheme* or focus of the utterance. According to Papafragou, in the case of epistemic modals the modalities may be part of the propositional content – and this irrespectively from their construal as individual speaker's knowledge or communal knowledge – but they are rarely the predicate that represents point of the utterance. Thus they are not properly psychological statements *about* beliefs. Their main relevance lies in the embedded prejacent. In the end, they are typically uttered in order to invite the addressee to adopt a certain epistemic stance towards the prejacent, not to inform her about the speaker's beliefs. This is why dialogical contradiction only targets the prejacent.

In the final part of his paper on *probably* Ennis discusses an issue closely related to the presumed performativity and non-propositionality of epistemic modals. Namely, the analysis of embedded occurrences of the adverb, which obviously fall outside the explanatory reach of the speech-act hypothesis strictly conceived. Ennis examines *reported speech* (31.a), *conditionals* (31.b), *imperatives* (31.c), *questions* and *wonderings* (31.d):

(31.a) Ennis said that the raccoon would probably not bother me.

(31.b) If the raccoon probably will not bother her, then she can probably safely turn her back to the raccoon.

(31.c) Please, make sure that the raccoon probably would not bother her

(31.d) Will the raccoon probably not bother her?

Clearly, these embedded contexts cannot involve the performance of a speech act of "guarded commitment". Ennis, however, taking a version of what I have called the "delocutivity route" (see § 3.2.4), sees them as derived from the properly performative ones: (31.a) reports the performance of the act by someone else, while the other examples are explained as predicating the reasonableness or justification of performing the act, so that, for instance, the imperative sentence in (31.c) is analyzed as (32):

(32) Please, make sure that a guarded commitment to the proposition that the raccoon will not bother her is justified.

Thus, embedded occurrences are reduced either to reports of performance of the actions or to situations where "the action is lurking in the institutional background, so to speak, waiting to be performed". In all cases the embedded meanings are "a function of the combination of the delimited Toulmin hypothesis and the context in which the term *probably* appears" (Ennis 2006: 162). This is probably the most successful part of Ennis' defense of Toulmin's hypothesis. It requires to somewhat relax the requirements of meaning compositionality but it does so in a systematic way that is hardly uncommon in natural language[22]. The explanation offered appears reasonable as long as the embedded examples appear natural.

In fact, what might raise questions it is the naturalness of the explanandum – the embedded occurrences of *probably* – rather than the viability of the explanation, as we will see in the next section.

3.3.3 Second Interlude: Modality as Detached from the Propositional Content

Both the first and the second theory of modality in the *Uses of Argument* locate modals in the performative, pragmatic, component of utterance meaning; modals being respectively reconstructed as pragmatic predicates (illocutionary force indicators) and as modifiers of pragmatic predicates (illocutionary force modifiers). As a consequence, in these theories modals are not part of the propositional content of the utterance. In the previous section we have seen how the issue of the (non-)propositionality of the modals emerges repeatedly in connection with the arguments that Ennis (2006) uses to defend Toulmin's theories. As the final part of this chapter will show, Toulmin's third view of modality also places modals outside the propositional microstructure, by making the modal qualifier a distinct part of the macro-structural "layout of arguments". It is then worth pausing on what we know about the (non-)propositionality from research in linguistic semantics and pragmatics and assessing the implications for argumentation theory.

---

[22] This of Ennis' defense is presented in a discussion of Searle's (1969) criticism of speech-act theories of the meaning of *good* and *probably* and refers to principles of meaning composition discussed by Searle. I decided not to discuss Searle's views on Toulmin's *probably*.

Over the last four decades several linguists have considered epistemic modal adverbs such as *probably* as unembeddable (Doherty 1987a, b), or, at least, severely limited in their ability to occur in the scope of other operators (cf. Portner 2009 and the literature cited therein). These views are associated to a strand of linguistic thought that maintains that at least some epistemic modal expressions, including adverbs, but also modal verbs and certain uses of belief predicates, are not properly part of the propositional content of the utterances in which they occur and are somewhat akin to performative expressions signaling illocutionary force.

The linguistic evidence that is considered when discussing the idea of epistemic expressions not contributing to truth-conditions is varied, but always quite subtle and difficult to handle. We have already encountered it in the previous section, discussing examples of dialogical contradiction of epistemically modalized utterances. We have also seen that *probably* takes scope over tense operators and refers to what Freeman calls *epistemic probability*. Other data (cf. Doherty 1987b, Hengeveld 1988: 236–237, Papafragou 2006, Portner 2009: 145–151) called forth to support of this hypothesis typically include the apparent unacceptability of these modals in the antecedent of conditionals, in questions, and the fact that they cannot occur under the scope of a sentential negation[23]:

(33.1.a) *If Alice perhaps/ probably won, they will be preparing a party at home.
(33.1.b) *If Alice may have won, they will be preparing a party at home.
(33.2.a) *Has Alice probably won? / *Did Alice probably win?
(33.2.b) *May Alice have won?
(33.3.a) *Alice has not perhaps won / *Alice has not probably won
(33.3.b) Alice may not have won (*may* takes scope over *not*)
(33.3.c) Alice cannot have won (*not* takes scope over *can*)

One of the philosophical sources of this strand of linguistic research, in itself quite separate from issues of argumentation, is very close to Toulmin's *probably*. In the same 1956 collection of essays on conceptual analysis where Toulmin's piece on *probably* originally appeared, James O. Urmson published an essay[24] delimiting a class of "parenthetical verbs", comprising mental predicates such as *think*, *believe*, and *suppose*. Parenthetical verbs are delimited "grammatically" by the possibility of occurring in the first person present either in a matrix clause governing a *that*-clause (34) or parenthetically (35), as implied by their name.

(34) I believe that he is at home
(35) He is at home, I believe / He is, I believe, at home.

[23] Some authors (e.g. Doherty 1987b) also consider whether the modal expression can take a contrastive stress and function as the (contrastive) focus of the utterance. These data bring in a further layer of complexity to an already messy situation, because they have to be considered in combination with the other parameters (e.g. the possibility of carrying a contrastive stress in a question) and also because the intuitions of acceptability concerning them are highly dependent on the discourse context and on the intended interpretation (e.g. whether we have to do with an echo question).

[24] The paper was first published in *Mind* in 1952.

Functionally, occurrences in the matrix clause (34) can play the exact same role of parenthetical occurrences (35), but they are not bound to this "purely parenthetical" function and are capable of what Urmson sees as derived, quasi-parenthetical uses. According to Urmson, the functional distinctive trait of parenthetical verbs is that the proposition they embed (34) or modify parenthetically (35) – their prejacent – counts as asserted, that is retains the "implied claim to truth" of an indicative sentence. In fact, according to Urmson (1956: 197), the prejacent is the only proposition asserted as parenthetical verbs are not asserted, they "show rather than state". They function as "warning, priming or orientating signals" (*Ibid.*) that guide the hearer in understanding different aspects of the context of utterance of the asserted prejacent: (i) its "emotional significance" (e.g. *regret*, *rejoice*), (ii) its "logical relevance" in a discussion (e.g. *admit, deduce, conclude, suppose*)[25], (iii) the reliability of the evidence that the speaker has for his statement (e.g. *guess, think, believe, know*).

Urmson (1956: 196–197) observes, in apparent reference to group (iii), that some parenthetical verbs have the function "to modify or to weaken the claim to truth that would be implied by a simple assertion" and later discusses verbs of group (iii) as modifying a "claim to reasonableness" that is also implied by naked assertion.

Sentence adverbials, which Urmson sees as syntactically similar to parenthetical verbs in their positional mobility and loose attachment to the sentence, are considered to play essentially the same function as the verbs, with adverbs covering the three areas identified for the verbs: (i) *happily, unfortunately*; (ii) *admittedly, consequently, presumably*; (iii) *possibly, probably, certainly*. For Urmson (1956: 200) the main difference between the verbs and the adverbs is that the latter are "more impersonal", a feature that he exemplifies with a contrast between *admittedly* and *I admit*: while the former signals that the prejacent "would be regarded by anyone as an admission", the latter only signal that the statement plays this logical role in the current exchange.

The main philosophical point that Urmson is making with his analysis of parenthetical verbs and the closely matched sentence adverbs is to analyze the philosophically important verb *to believe* as functionally equivalent of *probably* in its parenthetical uses. The non-parenthetical uses of the verb are then analyzed as derived from the parenthetical ones, according to a variant of what I have earlier called the "delocutivity and polyphony route" (see Sect. 3.2.4). Both verb and adverb are analyzed in terms that, unsurprisingly, are very close to Toulmin's *probably*: both indicate that "evidence justifies the guarded claim to truth" that the speaker makes. As we have seen, Toulmin insists in severing the guarding of the commitment from the evidential situation that justifies it, while Urmson keeps the

---

[25] It is interesting to observe that the function of indicating "how the statement is to be taken as fitting logically into the discussion" (Urmson 1956: 198) covers both the logical aspects *stricto sensu*, illustrated by *deduce* and *conclude*, and the dialectical ones as illustrated by *admit*. According to Urmson, by saying *I admit* p the speaker signals that p is either "support" for the opponent's position or "part of" the opponent's position, forestalling an opponent's reaction such as *But, don't you see that it's part of my point?*

two moves together as two aspects of the indexing of how the assertion fits in its discourse context[26]. This is related to the fact that the critical target of the two philosophers is not exactly the same: psychological and dispositional views of belief predicates for Urmson, objectivist and subjectivist theories of probability for Toulmin. Yet, the strategy of philosophical argumentation they adopt is clearly similar in that it invokes notions from a then fledging Austinian[27] speech-act theory to deny that the terms analyzed, in their primary occurrence, denote anything outside the speech event.

In linguistics, the impact of Urmson's paper was significant and twofold. On the one hand, the idea that parenthetical verbs do not contribute to propositional contents and instead "show rather than state" how the propositional content fits its context of utterance is discussed by scholars that see non-truth conditional meanings as "procedural" instructions for discourse interpretation and look into the relation between the lack of syntactic integration and the mode of contributing to discourse interpretation (Blakemore 1990, 1997, 2006, Ifantidou Trouki 1993). On the other hand, the idea that the complement or prejacent of parenthetical verbs counts as asserted was taken up by an early and influential paper by J. Hooper (1975). Refining and extending Urmson's class of parentheticals Hooper defines a class of *assertive predicates* and further subdivides into *strong assertives* and *weak assertives* on linguistic grounds[28]. Hooper's notion of assertive predicates influences a number of scholars in linguistics (cf. Venier 1991 for a review) who broaden the range of epistemic expressions considered to be non-propositional and further develop the speech-act analysis. The paper is also noteworthy as the main linguistic reference of van Eemeren et al. (2007) when discussing the role of epistemic expressions as indicators of the speech act of putting forward a standpoint in an argumentative discussion in a Pragma-Dialectical framework.

Let us now briefly examine some typical non-propositional analyses in linguistics. These analyses hypothesize two distinct layers at which prima facie synonymous expressions of possibility, probability, necessity or certainty belong. The labels proposed for the two layers in three "classic" proposals of the 1970s and

---

[26] Urmson suggests that the subdivision of parentheticals in three functional groups is not clear cut or without overlap, "with verbs that might with equal reason be placed in either of two groups" (Urmson 1956: 200). I suspect that what Urmson had in mind here was chiefly the contiguity between group (ii), comprising the parenthetical that place the assertion in its logical context and group (iii), which includes those evaluating the reliability of the statement in view of the evidence, as he makes this comment immediately after putting *presumably* in group (ii) and the closely related *probably* in group (iii).

[27] Urmson insists that no mental acts or activities are described by parentheticals. For instance, there is no mental act of *concluding* or of *guessing*, only assertions occurring in specific logical contexts or evidential situations. He sees parenthetical verbs as closely related but distinct from Austinian performatives (Urmson 1956: 205). In the paper, however, he does not explicitly elaborate on the distinction between these two classes.

[28] Hooper (1975) looks at mood choice (indicative vs. subjunctive) in complement clauses in Romance languages.

Table 3.4 Two layers of epistemic modality

Author:	*Possibly* – layer (non-propositional)	*Possible* – layer (propositional)
Lyons (1977)	*subjective epistemic modality*	*objective epistemic modality*
	is part of Hare's (1970) *neustic* level, that is of the speaker's subscription to the assertive mood of the utterance.	is part of Hare's *phrastic* level, which corresponds to the propositional content.
Doherty (1987a, b)	*modalized attitude* (= epistemic modality)	*modalized propositional meaning* (= alethic modality)
	"the attitude about the truth of p follows possibly or necessarily from whatever knowledge the inference happens to be based on" (Doherty 1987b: 49)	"the truth of p itself follows (possibly or necessarily) [...] from whatever circumstances may be relevant" (Doherty 1987b: 49)
Hengeveld (1988)	*epistemological modality*, which includes two subtypes:	*objective modality*,
	subjective modality, which concerns the speaker's "expression of the degree of his commitment with regard to the truth of the content" (Hengeveld 1988: 236) (Hengeveld (1988: 240) further subdivides subjective modality into *epistemic* proper and *boulomaic*. The latter includes complex attitudes such as *wish* and *hope*, which have both a desire and a knowledge component. For instance, hope entails (or presupposes) epistemic possibility. I will ignore this latter subdivision in the present discussion.)	which concerns the speaker's "evaluation of a state of affairs in terms of his knowledge" (Hengeveld 1988: 236)
	evidentiality, which concerns the indication of the source of the knowledge of the content.	

1980s can be illustrated in Table 3.4 by the contrast between the adverb *possibly* and the adjective *possible* in examples (36.a) and (36.b).

(36.a) Possibly it will rain tomorrow
(36.b) It is possible that it will rain tomorrow

Some approaches shown in the table, including Lyons (1977) and Hengeveld (1988), associate the occurrence of the modality in an extra-propositional layer of the utterance with the *subjective* nature of the evaluation.

The issue of the *subjectivity* vs. *objectivity* of epistemic expressions, which we have encountered in analyzing dialogical contradiction in the previous section, is raised repeatedly in the literature without a clear consensus on the definition of the contrast (cf. Portner 2009: 108–110, 137–138, 158–167 for a discussion). In contrast with Lyons, Kratzer (1981) does not see it as connected with extra-propositionality – a notion that is not part of her framework – but rather with the *quality of the evidence*. According to Kratzer, subjective epistemic modals involve a recourse to lower quality evidence, including superstitions and stereotype. Others see the issue of subjectivity in terms of the contrast between evidence that is accessible only to the speaker and evidence that is intersubjectively accessible to a community of epistemic agents. In this vein, Nuyts (2001a, b) proposes to separate two distinctions that Lyons (1977) appears to conflate: the *performativity* and the *subjectivity* of epistemic expressions.

Epistemic modal expressions are said to be used *performatively* when they refer to an actual epistemic evaluation of the speaker at the time of utterance, *descriptively* when they refer to another knowing subject, to another time of evaluation, or to a conditional, potential or negated act of evaluation. As Nuyts (2001a: 40–41) explains, this notion of performativity is weaker than the one involved by illocutionary indicators in speech-act theories. While both notions involve a linguistic expression referring to an act of the speaker performed here and now, modals signal a mental act, while illocutionary indicators a communicative one; the linguistic expression is *constitutive* of the communicative act, while the mental act occurs in principle independently from its linguistic representation. To avoid confusion, Faller (2002: 212) christens Nuyts' notion of performativity *m-performativity* (where 'm' stands for *modal* or *mental*). If certain epistemic modal expressions cannot be embedded in the antecedent of conditionals, cannot be questioned, negated and cannot fall within the scope of tense operators, as examples (33.1.a)-(33.3.c) seem to suggest, the idea that these expressions are constrained to appear exclusively as m-performatives offers an underlying pragmatic rationale for this behavior[29]. It is easy to see that the idea of m-performativity closely matches the idea of epistemic probability that Freeman associated with the meaning of *probably*.

According to Nuyts (2001b: 399), *subjectivity* is clearly distinct both from performativity and from the idea of "quality of evidence":

> The dimension of subjectivity should not be defined in the traditional terms of the quality of the evidence for an (epistemically qualified) state of affairs, but rather in terms of the question whether the evidence (and the conclusion drawn from it) is only available to the speaker or is rather more widely known (including to the hearer).

Subjectivity is thus contrasted with *intersubjectivity* rather than with objectivity. Not all m-performative modal expressions are subjective in this sense. In the previous section we have seen that *probably*, when occurring in the context of an argument, re-activates the relational semantic structure of probability, as the argument

[29] Portner's (2009) sees this notion of performativity as mere indexicality and therefore unable to fully explain non-propositionality, observing that indexicals can be embedded without any special problem. It is true that M-performatives are a very special type of indexicals as they simultaneously index the *speaker*, the *time of utterance* and the *actual world*. Yet, Portner's objection has a certain bite, because indexicals such as, for instance, *I* and *now* can combine compositionally with negations and counterfactuals, shifting their reference as a result (*If I were you, I would have resigned by now*). For an exclusively m-performative epistemic modal like *perhaps* an attempt at negative or counterfactual embedding does not result in a shift, but simply fails. A sentence like *If Alice had perhaps won the race, they would be preparing a party at home* cannot be interpreted as 'If the speaker's evidence at the moment of utterance were compatible with Alice winning the race, they would be preparing a party at home'. The only possible interpretation has the unembedded epistemic modal taking scope over the entire conditional sentence.

makes explicit a part or the entirety of the set of propositions that constitutes the relevant evidence. It is clear, at this point, how Nuyts' decoupling of subjectivity and performativity can be relevant for a study of modality in argumentation.

A corpus study conducted by Nuyts (2001b) on German and Dutch data shows that the adverbial uses of *wahrscheinlich* and *waarschijnlijk* ('probably'/ 'probable')[30] can occur in contexts where there is clear reference to a body of evidence available to an epistemic community (e.g. a study conducted by a group of researchers, a body of disciplinary knowledge) and, at least in principle, shareable also with the addressee, but they can occur also in contexts where there are no contextual pointers to intersubjectively available evidence.

It is worth observing that the examples of intersubjectivity reported in Nuyts (2001b) include contexts where the intersubjective evidence is mentioned (e.g. a particular scientific study is named in the text), contexts where a body of intersubjective evidence is implied (e.g. a popular science author is assumed to report the state of knowledge in a field rather than her personal opinions and hunches) and, finally, "contexts involving explicit logical reasoning", including both scientific reasoning and "common sense reasoning in non-scientific discourse" (Nuyts 2001b: 389). Only the third type of context is argumentative. Unfortunately, the study in Nuyts (2001b) does not focus on argumentation and is not very systematic in reporting the occurrence of the different epistemic expressions in the three types of intersubjective contexts. In contrast, for the purposes of the present study it will be important to distinguish between *mentioning* intersubjective evidence, *implying* or presupposing its availability and *displaying* it in an argument.

Nuyts concludes that the epistemic adverbs are strictly neutral with regards to the dimension of subjectivity. In contrast, the adjectival occurrences of *wahrscheinlich* and *waarschijnlijk* are overwhelmingly tied to contexts where intersubjective evidence is available, unless they occur as part of explicitly subjectivizing constructions (e.g. *it seems probable to me*).

Finally, Nuyts examines the m-performative uses of mental state predicates corresponding to English *to think* and *to believe* and finds that they "systematically express subjectivity", occurring predominantly "in contexts in which the speaker voices personal opinions, very often about topics in the realm of strictly individual experiences concerns" or in contexts "involving antagonism between the views of speaker and hearer" (Nuyts 2001b: 390–391). The landscape emerging from the study thus suggests that subjectivity, as defined by Nuyts, overlaps only in part with performativity.

As repeatedly noted by Portner (2009), a common problem afflicting the whole intricate literature on the non-propositionality / non-truth-conditionality of epistemic modals is the slipperiness of the data used to support the analyses. Corpus based studies such as Nuyts (2001a, b) are a relatively recent innovation in this kind of semantic discussions and nearly all the earlier studies (before 2000) and still

[30] Note that, contrary to English and to Romance languages, German and Dutch adverbs do not have a distinguishing morphological mark and are set apart from their adjectival counterparts by syntax only.

many of the more recent ones rely exclusively on constructed examples, such as (33.1.a)-(33.3.c). Individual intuitions about these examples can vary significantly. We have seen that Ennis, who is not a linguist but certainly is a native speaker of English, considers embedded uses of *probably* as something that needs to be explained – which presupposes that he consider them something that people say. Furthermore, the battery of embedding tests used to establish non-propositionality is quite diverse and the application of different tests does not result in the same partition of the tested modal expressions. The modal expressions that are supposed to be non-truth-conditional are themselves a diverse bunch, including lexically and syntactically diverse items such as modal adverbs, modal verbs and parenthetical uses of belief predicates. This diversity can make certain tests non-applicable, hard to interpret, uninformative or trivial for certain classes of items[31].

A recent corpus study by Hacquard and Wellwood (2012) represents a first step forward towards a picture of the embedding restrictions of epistemic modals based on authentic data. Hacquard and Wellwood (2012) focus on epistemic vs. non-epistemic interpretations of modal verbs and examine *might*, *can*, and *must* in antecedents of conditionals, questions, and complements of attitude predicates. The resulting picture is much more nuanced than what earlier studies based on semantic tests on invented examples would suggest, with differences in the distribution of *might* and *must*. In general, epistemics turn out to be significantly rarer in antecedents of conditionals and in matrix questions, but they do occur in embedded questions introduced by "inquisitive predicates" (*ask*, *wonder*), with *must* limited to contexts where the "solipsistic inquisitive" *wonder* introduces someone's thinking (Hacquard and Wellwood 2012: 20). Epistemic modal verbs do appear in the complements of some attitude verbs (e.g. those expressing belief) but they are excluded from complements of attitudes expressing desires or commands. Hacquard and Wellwood (2012) conclude that the data they examined appears to support the hypothesis that epistemic modals do contribute to propositional content. One could also say, in Nuyts' terms, that they are not restricted to m-performative occurrences. As Hacquard and Wellwood (2012) observe, several findings remain in need of an explanation. For instance, the limited distribution in conditional antecedents and matrix questions seems to suggest that what Hacquard and Wellwood (2012: 19) call an *objective interpretation* "with the modal anchored either to the addressee's or some collective knowledge state" is less accessible than a "subjective" one, anchored to the speaker, and requires special contextual conditions.

Hacquard and Wellwood (2012), focus on modal verbs, investigating only a subset of the spectrum of epistemic expressions interested by earlier hypotheses on non-propositional epistemic modality, and uncover a nuanced behavior with respect to the proposed tests for propositionality, where pragmatics and discourse context seem to play an important role in determining the outcome of the test (Hacquard and Wellwood 2012: 20–22). Sentential adverbs and structurally parenthetical epistemic

[31] I refrain from discussing here these issues, which strictly pertain to the methodology of linguistic semantics. Certain difficulties in applying tests for propositionality are discussed in Rocci (2005: 111–115).

verbs may be truly "syntactic orphans" (cf. Blakemore 2006 for a discussion) and have their non-propositional status more structurally ingrained. Yet, it is very likely that a closer look to these classes would also give a nuanced picture, with the behavior of individual lexical items requiring special interpretation. Consider the epistemic adverb *perhaps* in (37):

(37) Has Alice *perhaps* won?

In contrast with the debated examples with *probably* in (31.d) and (33.2.a), there is little doubt that the occurrence of *perhaps* is perfectly natural in yes-no questions such as (37)[32]. Yet, the acceptability of the adverb in a question context does not indicate that it contributes to truth-conditions. In fact, *perhaps* is not properly part of the questioned content: (37) is not asking whether it is epistemically possible that Alice has won. In fact, by uttering (37) the speaker performs a complex speech-act conventionally associated to the constructional template formed by the adverb and the interrogative form of the sentence[33]. In terms of Congruity Theory, it can be tentatively analyzed as a connective predicate with the following presuppositions and twofold pragmatic effect[34]:

(Def. 10) *Perhaps*_Question$_{U0}$ (Spk, Hr, E, U_0, U_{+1}) =
Presuppositions:
The propositional content p of U_0 is not part of the common ground of *Spk* and *Hr*, E is a segment of the common ground of *Spk* and *Hr* bearing on the truth or falsity of p, U_{+1} is a projected utterance of *Hr* relating p to E.
Pragmatic effects:
By uttering U_0,
Spk puts forth p of U_0 as a cautious hypothesis in view of E, worth of the consideration of *Hr*
Spk proposes to *Hr* to contribute with a utterance U_{+1} to determine whether p or $\neg$ p by considering all the evidence available to *Hr* (including E).

We can observe that this connective predicate also involves the presupposition that a body of evidence E is intersubjectively available (e.g. *Has Alice perhaps won? I hear people cheering downstairs*) and can be activated as a premise in an argument. If this analysis is correct, the contribution of *perhaps* in (37) is that of a *non-truth-conditional, performative, intersubjective* expression. The complex constraints

---

[32] Doherty (1987b: 53), from which I take the example sentence, also recognizes the possibility for the non-propositional modalizer *perhaps* to occur in questions. Doherty proposes an analysis of this phenomenon that is substantially different from the one I propose here.

[33] Here I use "constructional template" to refer informally to what is more properly called a *construction* in the technical sense of Construction Grammars, that is of 'a form-meaning pair such that some aspect of its form or some aspect of its meaning is not strictly predictable from its component parts or from other constructions" (cf. Goldberg 1995). While non-compositional internally, constructions are productive and combine compositionally with linguistic units that saturate their open slots. See Rocci (2007) for the analysis of a similar construction in Italian.

[34] I adapt the analysis of (37) from Rocci (2007: 146–151) which proposed a Congruity Theoretic analysis of an Italian language construction which performs a modalization of yes-no questions that is very similar to what *perhaps* realizes in (37).

imposed on the pragmatic effects of the utterance and on the common ground by the use of *perhaps* in questions such as (37) shows that non-propositional epistemic modalities cannot be accounted for simply by Hooper's notion of assertive predicate, nor by any simple model of the stratification of the utterance where the epistemic evaluation can be attached at different levels of the speech act, but require a rich analysis of the discourse-pragmatic level of meaning. Swanson's (2006, 2011) analysis of non-propositional *might* in terms of a speech act of "raising a possibility" is broadly in the same vein.

Possibly, the broad lines of a solution to the issue of non-propositionality and performativity of epistemic modalities can be sketched in terms of Congruity Theory with non-propositional elements seen as contributing to the presuppositions and pragmatic effects of the pragmatic predicate at issue. I use this kind of approach in Rocci (2007) to deal with an Italian epistemic construction occurring in interrogative contexts. In outlining such a proposal, however, one should draw a sharp distinction between

(a) linguistic constraints preventing individual lexical items from occurring in embedded contexts and related scope restrictions, which need to be studied in a case by case fashion using rich natural data,

(b) the non-propositionality and performativity of an epistemic modality as a *discourse phenomenon*.

While its recognition by the addressee may be guided by the choice of modal expression, non-propositionality as a discourse phenomenon does not ultimately depend on linguistic constraints, but rather on whether the *entire modal proposition* or just the *prejacent proposition* is *made relevant* as content in a given discourse move (cf. also Papafragou 2000: 110–123)[35].

Argumentation is a case in point. Whenever arguers present a claim as a conclusion from arguments, and whenever such a claim or its status as a conclusion is put

[35] There is no contradiction between recognizing (non-)propositionality as a discourse phenomenon rather than a linguistic one and adopting pragmatic predicates as a way to represent their contributions. Note that according to Congruity Theory, pragmatic predicates are not linguistic units, not even abstract, tacit, phonologically null linguistic units. They are a construct aiming to reflect the *logico-semantic nature* of the pragmatic organization of discourse, not to reduce the pragmatic organization of discourse to a level of linguistic structure. In this respect Congruity Theory differs subtly from a theory of rhetorical relations like SDRT (Asher and Lascarides 2003), with which it has many similarities. SDRT's ambition is to model precisely the construction of discourse structures as a level of linguistic competence, albeit a very peculiar one: building "a competence model of the construction of logical form" for discourse (Asher and Lascarides 2003: 76), where these logical forms are seen as an intermediate step between the surface of discourse and "*deep* interpretation"(*Ibid.*). Congruity Theory is not fully formalized and does not provide any "mechanical" model of discourse interpretation. Congruity Theory, instead, offers to analysts a way of representing the *results* of (deep) discourse interpretation in a (more) explicit fashion, which is something that can be useful for discussing individual linguistic-pragmatic hypotheses, for discourse criticism of various sorts (including argumentative criticism), for drawing typologies of discourse genres or of kinds of discourse moves in specific social contexts of interaction, and, more generally, for working in the fields of "socio-pragmatics" and "onto-pragmatics" – as opposed to "psycho-pragmatics" (cf. Dascal 2003: 412).

into question by a critic, as analysts we should ask ourselves whether any epistemic modalities appearing in the utterance of the claim are part of what is concluded or questioned or not. The answer, ultimately, is not a just a matter of linguistic semantics, but of reconstructing in a pragmatically realistic way the dialectically relevant commitments in which the arguers incur. Certainly, the linguistic constraints and proclivities associated with an expression contribute to guiding the hearer and, as a consequence they need to be taken into account also by the analyst engaging in argumentative reconstruction[36], but considerations of pragmatic and logical relevance in the context of the discussion are ultimately far more decisive in determining whether a modal is part of the propositional content of a particular standpoint or premise or, instead, the prejacent only should be considered. Consider the following examples:

(38.a) *Perhaps*, Alice is too outspoken
(38.b) Alice is, *I think*, too outspoken
(38.c) *I think* that Alice is too outspoken

Imagine a discussion where arguers are trying to establish whether Alice has the social skills required for discussing some delicate matter without compromising the interpersonal relationship with her counterpart. Here the discussion is about Alice's personality and social skills, not about the beliefs of the arguers. In (38.a) and (38.b) this contextual information is harmonic and, so-to-say, redundant with the linguistic constraints associated with *perhaps* and with the parenthetical position of *I think*. In (38.c) the same contextual information becomes *decisive* for identifying the relevant proposition with the prejacent 'Alice is too outspoken' – as an argument supporting the negative standpoint 'Alice does not possess the required skills to carry out that delicate discussion'. The reconstruction will be different if the same form occurs in a local context that makes the speaker's beliefs a necessary part of a premise (or a necessary part of the standpoint). Suppose that in the same discussion someone says (39):

(39) I've known Alice for many years and *I think* that Alice is too outspoken

Here the belief predicate is made relevant as part of the propositional content of a premise because it is necessary to "hook" the locus *From authority* activated in the argument: it's because it is *the belief of someone who has experience* that (38.d) can function as an argument.

In sum, we need to decide whether, in a precise local context, it makes sense to reconstruct the exchange as a meta-level debate about belief/ knowledge states, or as ultimately concerning the only the prejacent propositions. Certainly, the limitations and proclivities of the linguistic forms we encounter can function as argumentative indicators help us in this choice – or, alternatively, turn out to be relevant at a different level, to account for the rhetorical maneuvering for the participants.

---

[36] Of the closeness between the perspective of the addressee, interpreter of the utterance, and of the analyst engaging in rational reconstruction when we deal with *direct argumentative indicators* I have spoken in the Introduction (§ 1.3).

What is, then, *in the context of an argument*, the role of a modality that is not part of the content of a standpoint or argument? Re-reading Toulmin's third essay and the "Toulmin model", with an eye to the slot occupied by the modal qualifier, will help us to better frame this question. It will help us to see the detachment of the modality from the propositional content also in another light, in dialogue with the logical rather than with the linguistic tradition and with a focus on the relationship between the microstructure (proposition/ statement level) and macrostructure (argument level).

3.4 The Third View: Modal Qualifiers and the Layout of Arguments

3.4.1 The Toulmin Model

In the third essay of his book, Toulmin finally presents the eponymous "Toulmin model" of argument structure. Its early critics may have been quick to dismiss the proposal as merely "a syllogism lying on its side"[37], but the "layout" that Toulmin proposes for the finer, micro-level of argumentation is certainly the aspect of Toulmin's work that had the greatest influence on the development of argumentation studies in rhetoric, communication, computer science and other disciplines, starting with the early appreciation by Brockriede and Ehninger (1960). One could be tempted to say that it was the *only* part of the *Uses of Argument* that had a substantial influence on argumentation studies. As van Eemeren et al. (1996: 153) observe, in many presentations and especially in textbooks – including the one Toulmin himself co-authored with Richard Rieke and Allan Janik (1984) – "the model is completely isolated" from the broader philosophical argument about the nature and scope of logic that Toulmin develops in the book. Recent publications (cf. in particular Hitchcock and Verheij 2006) have inspected more thoroughly the implications of Toulmin's thought for argumentation, but have kept the model as the starting point and center of interest of their reflections, including those concerning the place of modality in arguments (Ennis 2006).

The components of the "Toulmin model", presented in Fig. 3.1 below, are well known, even if not always well understood.

The functioning of the "model" is famously illustrated by the "Bermuda" example in Fig. 3.2:

The conclusion or standpoint of the argument is called *claim (C)*, while the facts appealed to in support of the claim are the *data (D)*. Toulmin, in line with his "procedural" view of argument structure, introduces the *data* as the answer to a challenger's question: "What have you got to go on?". A third component, the *warrant*

[37] I find this gibe anonymously quoted in the appreciative paper of Brockriede and Ehninger (1960: 47). See also van Eemeren *et al.* (1996: 149), who observe that "in the reviews [...] little or no attention is paid to the model".

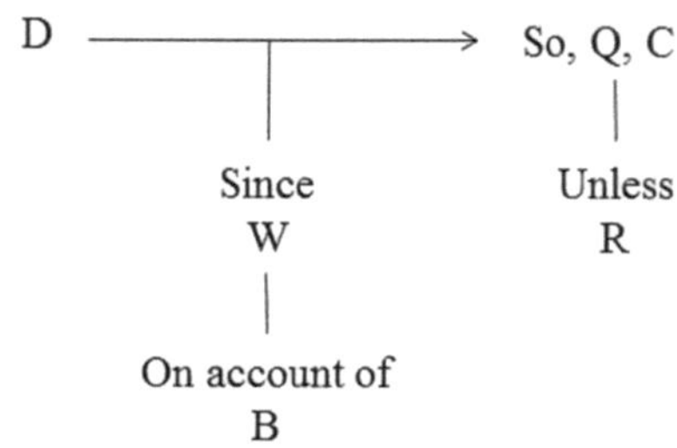

Fig. 3.1 The "Toulmin model" (Adapted from Toulmin 2003 [1958]: 97)

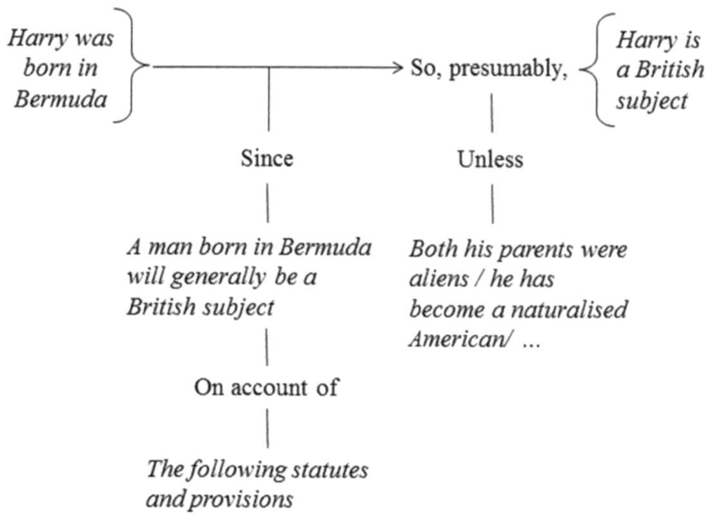

Fig. 3.2 The "Bermuda" example illustrating the functioning of the "Toulmin model" (Adapted from Toulmin 2003 [1958]: 97)

(*W*) is represented by "general, hypothetical statements, which can act as bridges" (Toulmin 2003 [1958]: 91) and authorize the inferential step from the *data* to the *claim*. Warrants correspond to answers to a different challenger's question: "How do you get there?". Toulmin expounds the Warrant either as a simple conditional, as in (Def. 11), or with a more developed, meta-argumentative, phrasing as in (Def. 12):

(Def. 11) If *D* then *C*

(Def. 12) Data such as *D* entitle one to draw conclusions, or make claims such as *C*.

Toulmin believes the meta-argumentative formulation in (Def. 12) to be more "candid" (*Ibid.*). This is because the second formulation more clearly separates the Warrant, which Toulmin sees simply as an inference rule rather than as a universal factual statement about the world, from its *backing* (*B*), which corresponds to the facts in virtue of which the warrant has authority. According to Toulmin (*Ibid.*, p. 100), the distinction between *warrant* and *backing* is "comparable" to that between the field-independent force of the modals and their field-dependent criteria.

Thus, statements such as (40.a), (41.a) and (42.a), for Toulmin, have the same force when we see them as warrants, but when we look at their backing, in virtue of which their force as inference warrants holds, we can see how they will vary greatly moving from one field of argument to the other, as shown in (40.b), (41.b) and (42.b).

(40.a) A whale will be a mammal.
(41.a) A Bermudan will be a Briton.
(42.a) A Saudi Arabian will be a Muslim.
(40.b) A whale *is classifiable as* a mammal
(41.b) A Bermudan *in the eyes of the law* is a Briton
(42.b) Every/ Nearly every Saudi Arabian *has been found to be* a Muslim

Examples (40–42.b), which I adapted from Toulmin, showcase the three kinds of backing for which Toulmin provides examples[38]. In (40.b) the factual backing of the warrant is "a system of taxonomical classification", in (41.b) "the statutes governing the nationality of people born in the British colonies" and, finally, in (42.b) the warrant appeals to a "statistical record" (*Ibid.*, p. 96). On hindsight, these different factual bases somewhat remind of the different loci (cf. Rigotti and Greco-Morasso 2010) underlying the different argument schemes, the *habitudines* ('relations' from Lat. *se habere ad* 'to relate to') of the Medieval and Renaissance theories of dialectic and rhetoric. For instance, in (40.a, b) the inference warrant appears to be backed according to the locus *From definition.* In the retrospective preface to the 2003 edition of the *The Uses of Argument*, Toulmin observes that in 1958 he was not aware of the relationship between his views and the Aristotelian tradition of the *Topics* and adds that if he were to rewrite the book, he "would point to Aristotle's contrast between 'general' and 'special' topics as way to throwing clearer light on the varied kinds of 'backing' relied on in different fields of practice and argument" (Toulmin 2003 [1958]: viii). It is impossible to guess how exactly he would have gone about developing such a parallel, and what role the *common topics* would have had in it vis-à-vis the *special* ones.

The *modal qualifier* (*Q*) is a distinct category in the argument layout, separate from the claim. It is closely related to the field-independent force of the warrant. It provides an "explicit reference to the degree of force which our data confer to our claim in virtue of our warrant" (Toulmin 2003 [1958]: 93).

The *modal qualifier* is often seen in the light of Toulmin's pragmatic analysis of *probably*: in this perspective modal qualifiers modify the claim of the argument expressing its *force* at a pragmatic level. Yet, Toulmin says (2003 [1958]: 94) that

[38] In fact, Toulmin (2003 [1958]: 105) alludes to a "fourth" kind of backing with the following example: *The practice of doing A leads to the following intolerable consequences, etc.* This is mentioned together with the taxonomical, statistical and legal ones as one of the possible backings of a warrant of the form 'An A can be certainly taken to be a B'. The example seems a fragment of moral reasoning based on an argument from consequences. But it is hard to see how it can provide a fitting backing for a warrant with the form 'An A can be certainly taken to be a B'. I leave the question at that, but not without commenting on the way in which Toulmin's *backings* (and fields of argument) seem to hint to the necessity of addressing the level of *topics* (*loci*, argument schemes) without actually developing it.

the qualifier comments "on the bearing of the Warrant on the step from Data to Claim". This can suggest a different interpretation of the qualifier as a higher-level operator qualifying the relation between Data and Claim stated by the Warrant.

The ambiguity between the two readings of the qualifier has even a visual side. Considering that in the graphical presentation of the model the arrow is used to "symbolize the relation between the Data and the Claim in support of which they are produced" and that the Warrant is written "immediately below the arrow" to indicate the "authority for taking the step" from D to C (Toulmin 2003 [1958]: 92), we might well think that Q should appear as taking scope over the arrow. Yet, the actual graphical representation, with *Q* slotted between *So* and C, rather gives credence to a reconstruction of the *modal qualifier* as taking scope on the claim itself.

Under the second reading – that is Q commenting the step/arrow – Toulmin's qualifier would match the view of "hedges" or "qualifications" as "operator-operators" discussed in Woods (2010: 246):

> On one reading "probably" qualifies the conclusion operator "so". It maps a conclusion operator of whatever strength to one that is weaker than it. It does this in a way that signifies that the underlying consequence relation has a corresponding lack of strictness.

Woods (2010: 247) contrasts this view of hedges with another view where hedges like *probably* are sentence operators "signifying part of the content of the argument's terminal proposition". As mentioned by Woods, this is the approach followed by standard modal logic, including epistemic logic. In such a reading, p, '*probably* p', '*perhaps* p' are different propositions that can appear as the conclusion of an argument. This allows – Woods observes – "for the consequence relation to be both unqualified and as strong as it gets". This latter construal of the modal qualifier, however, is clearly not among those found in Toulmin. When Toulmin presents *probably* as unary, he clearly does not sees it as forming a complex proposition, but as modifying the assertive speech-act. In both reconstructions of Toulmin's qualifier the modal is *separate* from the propositional content of the claim.

The issue of the interpretation of the modal qualifier is discussed in depth in Freeman's (1991) reconstruction and revision of the Toulmin model. Freeman choses the second view for *his own* approach to representing modalities in arguments, but clearly attributes the first, pragmatic, view to Toulmin to save the consistency with Toulmin's analysis of *probably*. Freeman's remarks on modality – which have been briefly examined in Sect. 3.3.2 – deserve further detailed examination, but before moving to that, it is worth pausing a moment to consider the idea of *modality as a separate component* of the layout of an argument in view of the broader landscape of the logical tradition.

3.4.2 *Third Interlude: The Modal Qualifier and the Place of Modality in the Logical Tradition*

The modal qualifier is one of those areas where the "Toulmin model" introduces a sharp change of perspective with respect both to modern formal logic and classical syllogistic logic. It is easy to characterize this shift by saying that modalities in logic are part of the *microstructure* of the statements that make up the argument, while Toulmin's modalities are a separate component of the argument's *macrostructure*[39]. As observed by Woods (2010), cited above, modal logic treats modalities as unary operators that require application to one propositional argument in order to obtain another proposition, a *modal proposition*. As stated in the Introduction, modal logic is concerned with investigating the validity of arguments *containing* modal propositions. We have seen that the same is true, with the necessary adjustments, for Ancient and Medieval contributions to modal logic.

In contrast, Toulminian qualifiers are either illocutionary modifiers modifying the claim of the argument by specifying its force, or modifiers qualifying the relation between Data and Claim stated by the Warrant. Neither interpretation sees the result of modal qualification as contributing to the propositional content of the Claim nor of any other component of the model.

Yet, this simple characterization runs the risk of being misleading, both in the case of formal logic and in the case of syllogistic. Things become more complicated because the modalities that appear in the macrostructure in Toulmin's account *are not exactly the same* modalities that we find in logic in the microstructure of propositions.

3.4.2.1 The Modality of Apodeictic Judgements

In fact, both classical syllogistic and formal logic become often quite elusive when it comes to representing the kind of modality expressed by Toulmin's *qualifier*. This elusiveness is favored by the fact that logic traditionally looked at deductive arguments where the force of the qualifier is always that of *necessity*. The fact that the conclusion of a syllogism is *apodeictic, i.e.* necessary, *in view of the premises* (as per Aristotle, *Topics*, I, 1, 100a25–26) does not find any explicit representation in the canonical form of the syllogism. Consider the conclusion of (43.a) as opposed to (43.b):

(43.a) Animals that feed their newborn with milk are (classified as) mammals
Whales feed their newborn with milk
Whales are (classified as) mammals

(43.b) *Necessarily*, whales are (classified as) mammals

[39] The terms microstructure and macrostructure are used here according to Freeman (1991, 2011).

Not only (43.b) is not the customary way of presenting the conclusion of a syllogism, but it would be seen as making the syllogism invalid if the necessity modal is interpreted *absolutely*, that is independently from the premises as it is the case in modal syllogistic.

In fact, Aristotle's modal syllogistic, to the extent to which it can be reconstructed and understood (cf. Hintikka 1973: 135), deals with syllogisms in which one or both premises are *apodeictic* (necessary) or *problematic* (possible) and thus distinguished from the non-modal (*assertoric*) premises. These distinctions make sense only when we have to do with independently necessary (*apodeictic*) propositions that make up the content of a conclusion, not just with propositions that are necessary *qua* conclusions.

In contrast, it is only the latter that Gottlob Frege had in mind when he liquidated the distinction between *assertoric* and *apodeictic* judgments, thus decreeing the invisibility of the "modal qualifier" in the "formalized language of pure thought", while manifesting, at the same time, a keen sense of the use of certain necessity modals in discourse:

> What distinguishes the apodeictic from the assertoric judgment is that it indicates the existence of general judgments from which the proposition may be inferred – an indication that is absent in the assertoric judgment. By saying that a proposition is necessary I give a hint about the grounds for my judgment. But, since this does not affect the conceptual content of the judgment, the form of the apodeictic judgment has no significance for us (Frege 1960 [1879]: 4).

When modals were finally re-admitted in the formal language it was, in fact, not the same modals that Frege had expunged. Formal systems of modal logic are concerned with modeling valid inferences from statements containing modal operators of necessity ($\Box$) and possibility ($\diamond$). Within a specific system of modal logic (e.g. the propositional system KT) these modal operators are defined *syntactically* by the *axioms* of the system (e.g. axiom T: $\Box\ \alpha \rightarrow \alpha$), which are mirrored in the *possible worlds semantics* by the formal properties of the *accessibility relation*[40] R which restricts the set of worlds over which the modal quantifies (e.g. the property of *reflexivity* corresponding to T, that says that each world *w* is related to itself by R). If the modal logic is to have philosophical or practical significance, R also receives a concrete interpretation as corresponding to a kind or "flavor" of modality. According to its axioms (and the corresponding formal properties of R) a system can be a more or less sensible choice for the task of modeling a particular modal "flavor". For instance, KT may be a reasonable choice to model various kinds of *alethic* necessity and possibility, but – because of axiom T and the corresponding property of reflexivity – is certainly not capable to model reasoning with *deontic* modalities, as clearly not all that is obligatory is also the case in the world where the obligation holds. About these matters more will be said in Chaps. 4 and 5.

Now, when we apply a system of modal logic to the concrete task of deducing a conclusion from a set of assumptions (contingent propositions, which are neither axioms nor theorems in the system) we cannot simply employ the modal operator of

[40] Accessibility relations and their properties will be presented in more detail in Chap. 4.

necessity *of* the system at issue to indicate the relative necessity of the conclusion *qua* deduced from those assumptions. All (normal) propositional modal systems have a *necessitation rule* according to which if α is *provable* within the system, then it is also necessarily true ($\Box\alpha$). However, this obviously concerns only the propositions that are *theorems* within the system and not those that require additional assumptions. The meaning of a necessity operator – which is a unary operator – within a system of modal logic is fixed once for all by the accessibility relation. Thus, this operator cannot be used to indicate the *relative* "necessity" of a conclusion which depends on propositions that are contingent with respect to that accessibility relation.

3.4.2.2 Aristotle on the Necessity to Be and the Necessity to Say

Introducing the conclusion of a syllogistic formula containing letter variables, Aristotle often uses the word ἀνάγκη ('necessity') or the phrase ἐξ ἀνάγκης ('of necessity', 'necessarily') even if the syllogism's premises are simply assertoric rather than apodeictic. So, after all, the kind of modality expressed by Toulmin's *qualifier* is there also in Aristotelian logic.

This sign of "syllogistic necessity", as Lukasiewicz (1957: 10) calls it, disappears from the most usual modern presentations of syllogistics, which operate purely in extensional terms. Lukasiewicz (1957) tries to reduce syllogistic necessity to extensional universal quantification: ἐξ ἀνάγκης would simply indicate that the syllogistic formula's implication holds *for any value assignment* of the variables (cf. Blanché 1970: 65)[41].

The necessary nature of the conclusion of the syllogism is also stated in Aristotle's definitions of the syllogism in the *Prior Analytics* (I 1, 24b18–20), in the *Sophistical Refutations* (1, 165a1–2), and in the *Topics* (I 1, 100a25–26). As we have seen in the Introduction, the latter passage introduces the distinction between *demonstrative syllogism* and *dialectical syllogism*:

> Now a deduction is an argument in which, certain things being laid down something other than these necessarily comes about through them. It is a demonstration, when the premisses from which the deduction starts are true and primitive, or are such that our knowledge of them has originally come through premisses which are primitive and true; and it is a dialectical deduction, if it reasons from reputable opinions (*Topics*, I 1, 100a25–26, transl. by W.A. Pickard-Cambridge, Revised Oxford Translation in Aristotle 2014).

It is clear from this passage that syllogistic necessity characterizes syllogisms generally and it is therefore not limited to those which are drawn from necessary

---

[41] It has to be said, for the sake of historical exactitude, that Lukasiewicz's (1957) own interpretation of Aristotelian syllogism "from the standpoint of modern formal logic" is very far from the modern textbook vulgate, which Lukasiewicz calls the "false form". For Lukasiewicz syllogisms in their true form are not *inference schemes* but *implications* (e.g. *If A is predicated of all B and B is predicated of all C then (it follows of necessity that) A is predicated of all C*): they are, in other words, dominated by an *if...then*, not by a *therefore*. This difference has no bearing on the issue of the necessity of syllogistic conclusions that concerns us here.

premises. It appears that Aristotle makes a distinction between the necessary nature of the conclusion of a syllogism *qua conclusion* and the necessity that characterizes a conclusion being drawn from necessary premises. In view of this distinction, when Aristotle stresses the non-necessary nature of most of the conclusions of *rhetorical syllogisms* (*Rhetoric* I, 1357a) he must refer to the second kind of necessity and not to the syllogistic necessity of the conclusion:

> There are few facts of the necessary type that can form the basis of rhetorical deductions. Most of the things about which we make decisions, and into which we inquire, present us with alternative possibilities. For it is about our actions that we deliberate and inquire, and all our actions have a contingent character; hardly any of them are determined by necessity. Again, conclusions that state what holds for the most part and is possible must be drawn from premisses that do the same, just as necessary conclusions must be drawn from necessary premisses; this too is clear to us from the Analytics. It is evident, therefore, that the propositions forming the basis of enthymemes, though some of them may be necessary, will in the main hold for the most part. (*Rhetoric* 1357a, transl. by W.R. Roberts, Revised Oxford Translation in Aristotle 2014)

Yet, even if Aristotle cannot be taken to say that the non-necessary nature of the premises compromises the syllogistic necessity of the conclusion, the passage remains surprising. It only takes into account inferences from necessary premises (that is, for Aristotle, premises that are *always* true)[42], which lend necessary conclusions or inferences from *usually* true premises, which lend conclusions "that state what is merely usual". Inferences from contingently true premises that allow to deduce, and hence to infer with certainty, other contingently true propositions are left completely out of the picture[43].

While Aristotle adheres to the distinction between the two kinds of necessity in practice, there is reason to think, with Fait (2004: 101), that theoretically he considers it "a troublesome point". Fait (2004) points to a passage of the *Posterior Analytics* (I 6, 75a18–27) where this troublesome issue is explicitly presented as a paradox and a solution is sketched out. In fact, not a very convincing one. Leaving out an irrelevant complication, the passage runs as follows:

---

[42] As convincingly argued by Hintikka (1973), Aristotle subscribes to the so-called "principle of plenitude" according to which all genuine possibilities are sooner or later realized in time. Therefore "whatever is always true is true necessarily according to Aristotle" (Hintikka 1973: 136).

[43] As an example of deduction of contingent conclusions from contingent premises consider the following: *All the participants in the experiment experienced dizziness and nausea. Some participants were men. Some men experienced dizziness and nausea.* The premises are observational facts and the reasoning seems to proceed in an entirely extensional manner through quantification. The fact that in the passage of the *Rhetoric* Aristotle explicitly considers *deliberative* contexts and thus mostly *practical*, and, subsidiarily, *predictive,* standpoints concerning *future* human actions may explain, but only in part, the absence of deduction from contingent truths. Future events cannot be simply inferred from contingent propositions without the mediation of an intensional nomic premise, which can be either necessary – with respect to a given domain of quantification – or probabilistic. However, even in predictive arguments *minor premises* can be purely contingent observational facts. Predictive standpoints will be discussed in-depth in Chap. 6.

> Of [contingent] accidents [...] there is no demonstrative knowledge. For one cannot prove the conclusion of necessity, since it is possible for an accident not to belong (this is the sort of accident I am speaking about). Yet one might perhaps puzzle about what purpose there could be in asking these questions about them, if it is not necessary for the conclusion to be the case – one might as well ask chance questions and then say the conclusion. But we should ask questions not because the conclusion will be necessary in virtue of the points asked, but rather because it is necessary for anyone who says them to say the conclusion, and to say it truly if they hold truly (As translated by Fait 2004: 102[44])

The passage, as observed by Fait (2004: 103), has to be read in a dialectical questioning context. This is not surprising as it is with dialectical syllogism that the problem of contingent premises surfaces. What is the point for an arguer, wonders Aristotle, of asking the opponent to concede contingent premises? Since they are not necessarily true, we cannot derive necessary conclusions from them. One could as well ask random premises and then just add the conclusion. The difficulty for Aristotle here is that it is difficult to see how a real necessary (always holding) connection can be established between propositions that are not themselves necessary (always true). The solution suggested by Aristotle is a distinction between a *necessity to be* (ἀνάγκη εἶναι) and a *necessity to say* (ἀνάγκη λέγειν)[45]: the contingent premises do not make the conclusion a necessary state of affairs, but it is *necessary for the arguers to assert* the conclusion if they accept the premises, and to assert it truthfully if the premises hold true. Unfortunately, the necessity to say remains a somewhat mysterious concept. As observed by Fait (2004: 104), the notion seems to have a deontic rather than alethic flavor, similar to the obligation to obey "the rules of a game". If the reading suggested by Fait is correct Aristotle's necessity to say is a meta-pragmatic deontic necessity to assert the conclusion relative to the rules of the dialectical dialogue game. This reading, however, does not explain why this necessity is not simply a necessity to assert given certain premises, but also a *necessity to assert truthfully when the premises are true*. It remains to be explained, in other words, what makes the rules of the game truth-preserving in the absence of a necessary connection between facts. The problem is simply shifted.

3.4.2.3 The Schoolmen's *Necessitas Consequentiae* and *Necessitas Consequentis*

The well-known distinction between *necessitas consequentiae* and *necessitas consequentis* that is associated with the work of the Scholastic philosophers of late the Middle Ages is more transparent than Aristotle's apparent distinction between *necessity to be* and *necessity to say*. At the same time, this distinction bring us farther from Toulmin's qualifier because the attention here is not on the meaning of necessity in the conclusion of an inference, but rather on the related problem of the

[44] The translation differs only slightly from Jonathan Barnes' translation in the Revised Oxford Translation.

[45] These two Greek formulas are not Aristotelian technical terms and first appear in the writings of Oxford philosopher and Aristotelian exegete H.W.B. Joseph (cf. Fait: 2004: 105).

meaning of necessity in the consequent of a conditional. It is well worth digressing, however, because the latter distinction plays a crucial role in the modern contextual semantic theories of modality used in Chap. 4.

The distinction is developed in the context of the discussions on God's foreknowledge and future contingent propositions (cf. Knuuttila 1982, 2013) and is, in fact, already found in Boethius (*Consolatio philosophiae* V, 6.27–30)[46], who employs the terms *necessitas conditionata* vs. *necessitas simplex*. Later St. Thomas Aquinas introduces in his own discussion of God's foreknowledge the terminological couple *necessitas consequentiae* vs. *necessitas consequentis*[47] that will stick in the logical vocabulary. In the *Summa contra gentiles* (lib. 1 cap. 67 n. 10) Aquinas discusses whether God's foreknowledge entails that every action happens *necessarily*, and therefore excludes human freedom. Following Boethius, Aquinas argues that God's eternal vision does not make what happens necessary with respect to their proximal causes (cf. Knuuttila 2013). Furthermore, he adds that there is a necessity of the consequence from God's foreknowledge of an action to the future happening of said action but this *necessity of the consequence* does not mean that the action becomes *absolutely* necessary. Aquinas uses perceptual evidence as an analogy. If I see that Socrates is sitting, then I must necessarily conclude that he is sitting, but my seeing does not make Socrate's sitting an *absolute* necessity:

> Praeterea, si unumquodque a Deo cognoscitur sicut praesentialiter visum, sic necessarium erit esse quod Deus cognoscit, sicut necessarium est Socratem sedere ex hoc quod sedere videtur. Hoc autem non necessarium est absolute, vel, ut a quibusdam dicitur, necessitate consequentis: sed sub conditione, vel necessitate consequentiae. Haec enim conditionalis est necessaria: si videtur sedere, sedet (*Summa contra gentiles* lib. 1 cap. 67 n. 10 in Aquinas 2006).

> 'Furthermore, if something is known by God as seen in presence, then what God knows will be necessary, in the same manner as as it is necessary that *Socrates is sitting* from that *he is seen sitting*. This, however, is not necessary absolutely, or, as some say, with a necessity of the consequent: but conditionally, or with a necessity of the consequence. In fact, what is necessary is the following conditional *If he is seen sitting, then he is sitting*.'

The two *necessitates* can be formally represented as follows, using a propositional modal logic notation:

(Def. 13) Necessitas consequentiae: $\Box(p \rightarrow q)$

(Def. 14) Necessitas consequentis: $p \rightarrow \Box q$

---

[46] "Duae sunt etenim necessitates, simplex una, ueluti quod necesse est omnes homines esse mortales, altera condicionis, ut si aliquem ambulare scias eum ambulare necesse est. Quod enim quisque nouit id esse aliter ac notum est nequit, sed haec condicio minime se cum illam simplicem trahit. Hanc enim necessitatem non propria facit natura sed condicionis adiectio; nulla enim necessitas cogit incedere uoluntate gradientem, quamuis eum tum cum graditur incedere necessarium sit. Eodem igitur modo, si quid prouidentia praesens uidet, id esse necesse est tametsi nullam naturae habeat necessitatem." (*Consolatio philosophiae* V, 6.27–30 in Boethius 2005).

[47] The terms used by Boethius were *necessitas condicionis* 'necessity of the conditional' and *necessitas simplex* 'simple necessity'.

In the passage quoted above Aquinas implicitly recognizes that the actual scope of a modal expression can be different from what the syntactic form of the sentence seems to suggest. While the modal appears in the consequent of the conditional it takes scope over the entire conditional.

There is clearly a relationship between the *necessitas consequentiae* and the necessity of the step from premises to conclusions: we can say that the step from premises to conclusion is necessary when a conditional having the premises as antecedent and the conclusion as consequent holds necessarily, as illustrated by the following example stating the necessary (tautological) status of the conditional of the *modus ponens*:

(Def. 15) $\Box (((p \rightarrow q) \wedge p) \rightarrow q)$

The latter notion of necessity of the step from premises to conclusion is one and the same with the notion of *validity* of the argument that licenses the conclusion: an argument is valid when it is not logically possible for its conclusion to be false while all of its premises are true.

3.4.2.4 Modality as Strength of Support in James Freeman

Students of argumentation interested in arguments that are not deductively valid have sought to extend the idea of strength of support to modalities other than necessity. James Freeman's (1991, 2011) work on argument structure is an example.

Freeman (1991: 112) agrees with Toulmin's idea that "modalities have distinct function from premises and conclusions". Freeman gives a dialectical justification for this independence. The modal qualifier corresponds to a distinctive challenger's question: "How sure do your reasons make you of the claim?". Freeman models the question after those used by Toulmin to introduce the data and warrant, but in Toulmin there is no question for the qualifier. At the same time, as we have seen, Freeman rejects Toulmin's purely pragmatic and unary view of *probably* in favor of a relational view of modalities.

Claiming validity, i.e. the specification of a deductive modality is one possible answer to Freeman's qualifier question, with weaker modalities corresponding to other possible answers:

> Deductive modalities claim that the premises give complete support to the conclusion, that they necessitate it, while inductive modalities make the weaker claim that the premises give sufficient evidence to justify accepting the conclusion, but not necessarily conclusive evidence (Freeman 1991: 112).

The job of modalities, for Freeman (1991: 117), is to "ascribe a certain degree of probability to the conclusion *vis-à-vis* the premises". That's why, from a structural viewpoint, modalities are attached to the arrow from premises to conclusion, as in Fig. 3.3.

Fig. 3.3 The modality slot in the macrostructure of an argument (Adapted from Freeman 1991: 128)

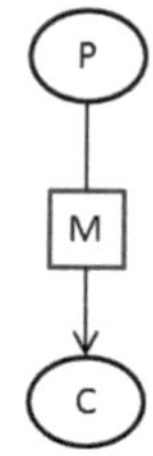

Developing Freeman's approach to modality in argument diagramming[48], Mark Vorobej (2006) proposes a straightforward treatment of the strength of the link between premises and conclusion in terms of conditional probability: an invalid argument is said to be *reliable* when "the probability that the conclusion is true, given that each of the argument's premises is true, is greater than half" (Vorobej 2006: 54). Finer distinctions among modalities in terms of degree of reliability can be drawn assigning a precise numeric probability value to the modality *m* corresponding to the conditional probability that the conclusion C is true given that each of the *n* premises P is true. In a formula:

(Def. 16) $\text{prob}\ (C/P_n) = m$

It worth reminding here that, by itself, the *conditional necessity* or the *conditional probability* of a conclusion given the truth of its premises tells us nothing about its *epistemic force*, that is about the degree of certainty one should attribute to such a conclusion. Epistemic force will also depend on the epistemic status of the premises. Only if the premises are known to be true the conditional necessity and the conditional probability would directly translate into epistemic necessity and epistemic probability.

Vorobej (2006), and Freeman (1991) before him, seem to conceive modality in their diagrams in purely conditional terms, as validity (conditional necessity) or as conditional probability of the step from premises to conclusions. For them modality is separate both from premises and conclusion because it characterizes their connection.

When Toulmin, somewhat obscurely, characterizes the modal *qualifier* as an "explicit reference to the degree of force which our data confer to our claim in virtue of our warrant"? (Toulmin 2003 [1958]: 93) he appears to take a similar position. But, as Freeman (1991: 126) observes, this position is incompatible with the analysis of *probably* as a unary pragmatic modifier "guarding" the assertion, which Freeman (1991:119–127) criticizes. According to the latter view, Toulmin's qualifiers "qualify the conclusion directly, rather than the arrow from premises to conclusion" (Freeman 1991: 121).

For Freeman (1991: 119), the pragmatic effect of guarding or hedging the assertion of the conclusion, which Toulmin takes as the primary meaning of *probably*, is

[48] In one of the very few citations of the book, Vorobej (2006: 174) acknowledges that his treatment of modalities in argument diagramming derives from Freeman's.

a derived "emotive meaning"[49]. For Freeman, the pragmatic meaning of *probably*, modifying the assertive commitment, derives from the fact that the adverb is restricted to indicating epistemic probability (see above § 3.3.2), that is probability given all available relevant evidence.

The discussion of Toulmin's "pragmatic" view of modality further clarifies how Freeman views modalities in arguments. As observed above, the sign M sitting on the arrow has a purely relational, conditional, meaning. Yet, the arguer's commitment towards the content of the premises together with the arguer's commitment to consider all the relevant evidence make M an epistemic modality. In turn, the epistemic nature of the modality signals a certain qualification of the assertive commitment towards the content of the conclusion.

I have hitherto reconstructed Freeman's view of the modals as relational, in contrast with Toulmin's unary pragmatic view of *probably*. This is not entirely correct. Discussing the relationship between modality and argumentative connectives. Freeman, initially invites the reader to view the modals "as functioning as logical indicator words" (Freeman 1991: 117), like *therefore* or *because,* as both the modals and these connectives "constitute part of the material connecting premises and conclusions". Immediately thereafter, though, Freeman makes clear that connectives and modals play distinct roles: the former signal that the premises support the conclusion, the latter indicate how strong this support is. They work as "illation-sign modifiers" (*Ibid.*).

This means that, in the end, Freeman's modalities, are not truly relations as implied by his discussion of *probably*, but unary operators that require a relation between propositions as their semantic argument. This clear-cut distinction, so-to-say between the arrow and the M modifying the arrow, is drawn at the level of the functionally distinct slots that we should recognize in reconstructing the macro-structure of an argument and is not always exactly mirrored by the surface structure of discourse. Modal words, Freeman (1991: 119) recognizes, often "do double duty as both logical indicators and modifiers of logical indicators": when proper connectives are not present, the modals can "indicate a conclusion" as well as signal how strongly the premises support it.

Freeman's account here is certainly candid in recognizing the phenomena. Yet, in the following chapters I will try to demonstrate that by treating the modals as modifiers on relations rather than as relational predicates we lose precious insights on their linguistic semantics and on their argumentative functioning. Considering the occasions where they act as indicators of an argumentative relations as a case of *indirect* indication of a related but distinct semantic category rather than as an exploitation of their inherently relational semantic structure we risk focusing *only*

[49] I reckon from the context of the passage that by using Stevenson's term "emotive meaning" Freeman intends a pragmatic meaning, indicative rather than denotative of a speaker's attitude, which arises besides the denotative meaning in the use of a word and which is derived from the denotative meaning. The choice of term is not entirely felicitous because it evokes the arousal of emotional responses as well as aspects of Stevenson's notion of emotive meanings that are not really relevant here.

on one aspect of the meaning of modal expressions: their quantificational force. In his later work on argument structure, Freeman (2011: 120) is explicit in indicating the modals' quantificational force as pretty much the only aspect worth considering in view of argumentative reconstruction:

> [...] adverbial expressions such as *obviously, evidently, surely, apparently* all have force but [...] their cognitive meaning is sufficiently vague as to render their force the only aspect of their meaning worthy of interest. Where their cognitive meaning is clear it is synonymous with *necessarily*

Leaving aside a discussion of the specific adverbs mentioned in this passage (but see, for instance, Merlini-Barbaresi 1987 on *obviously* in arguments), the general problem in considering only the force of a modal expression is that the constraints that each expression puts on the selection of the *grounds* or *criteria* of the modality fall out of the picture. And these grounds or criteria are often what is most important to understand what the modal does in an argument. I have loosely used here the Toulminian terms *grounds* and *criteria*, which, as we have seen, come with the baggage of conceptual problems characterizing Toulmin's theory that we have amply dissected. In the next chapter a more precise framework will be put in place to deal with this aspect of modal meaning as part of a relational view of modality.

Having discussed the cases where modals "do double duty" as connectives, Freeman (1991: 119) feels also compelled to mention another situation where modal expressions do not correspond to the M slot: necessity modals pertaining to logical or mathematical necessity do not claim that "relative to certain conditions" a statement must be true, but that "the statement must be true in itself or considered just by itself" (i.e. it's a logical or mathematical tautology). Such claims are part of the content of the propositions that occur in an argument as premises and conclusion rather than part of the macrostructure of the argument. The remark is correct and important in view of argument reconstruction. Yet, at the same time, it highlights that the view of modalities as illative force modifiers cannot give us an unified semantics of modal words and does not make it clear how the notion of *necessity* as deductive validity of an argument relates to the seemingly absolute notions of necessity of logical and mathematical tautologies. The unified relational view of modal semantics adopted in the next chapter will hopefully make this relation a bit more transparent.

3.4.2.5 Taking Stock

This limited excursus on different structural slots occupied by modalities in different reconstructions of reasoning in logic and argumentation theory allows us to identify three types of modality occupying three structurally different slots:

(A) Modalities appearing as part of the propositional content of a conclusion (or of a premise). We have this kind of modality in the conclusion when we want to argue that something is, in some sense (e.g. logically, mathematically, physically, legally), necessary, possible or impossible in its own right (and not just *given that the premises are true*).

(B) Modalities characterizing a move from premises to conclusion in an argument, in terms of conditional necessity (*necessitas consequentiae*) or conditional probability.

(C) Epistemic modalities characterizing a proposition, e.g. the conclusion of an argument, as more or less certain in view of the evidence known to the arguer.

In the next chapter, by introducing the Relative Modality approach to modal semantics I will set up a basic and versatile tool for dealing with these three structural levels. As the name suggests in this kind of approach the notions of relative, conditional, modality, which we have encountered in dealing with (B) level modalities, plays a pivotal role at the three levels.

3.4.3 Modal Qualifiers and Warrants

In connection with the above discussion of the different structural slot of modality in argument, the close relationship that Toulmin sees between qualifier and warrant deserves further scrutiny.

Toulmin (2003 [1958]: 93) states that "warrants are of different kinds, and may confer different degrees of force on the conclusions they justify" that "explicit reference to the degree of force which our data confer to our claim in virtue of our warrant". Warrants that authorize us to accept a claim "unequivocally" entitle the arguer to qualify the claim with *necessarily*, while other warrants are subject to "conditions, exceptions or qualifications", which can be manifested by qualifiers such as *probably* or *presumably*. Toulmin sees qualifiers as commenting "implicitly"[50] on the bearing of the warrant on the step from data to claim.

The close relationship between warrant and modal qualifiers emerges indirectly when Toulmin examines universally quantified statements setting apart their reading as *warrants* from their underlying *backings*. Not only Toulmin (2003 [1958]: 104–105) considers this distinction as parallel to that between *force* and *criteria* in the analysis of modal meanings, he also renders the *force* of the warrant in modal terms. For instance, a plain universally quantified statement with the form of (44.a) manifests more transparently its function of warrant when it is translated as (44.b):

(44.a) All A's are B's
(44.b) An A can *certainly* be taken to be a B

[50] One might wonder why Toulmin sees linguistically manifest elements such as the modal qualifiers as being *implicit* comments. A plausible interpretation could be that Toulmin refers to the fact that modal qualifiers are not *asserted* comments. According to this interpretation the remark is obviously consistent with Toulmin's non-propositional analysis of *probably* as an illocutionary modifier, and reminds even more closely of Urmson's (1956) analysis of parentheticals.

If, on the contrary, we take the universal statement as expressing a backing, a more transparent translation would depend on the field, as shown in (44.c, d, e):

(44.c) The proportion of A's found to be B's is 100% (statistical records)
(44.d) A's are ruled by statutes to count unconditionally as B's (statutes)
(44.e) The class of B's includes taxonomically the entire class of A's (taxonomy)

As we can see from (44.b), the "expansion" of an universal premise "as a warrant" (*Ibid.* : 110) involves the inclusion of a modal word (*certainly*, in this case), which appears to mirror the modal qualifier accompanying the claim. This parallelism between warrant and qualifier is clearly shown when Toulmin (2003 [1958]: 110) presents the following argument form:

(45) X is an A
An A is *certainly* a B
So X is *certainly* a B

Also for warrants that do not have the force of a universal quantifier (e.g. Nearly all A's are B's), the 'modal' formulation is considered the most perspicuous: "these differ from 'All A's are B's' in only one respect, that where before we wrote 'certainly' we must now write 'almost certainly', 'almost certainly not' or 'certainly not'" (Toulmin 2003 [1958]: 105).

Despite their central role in setting apart warrants from backings, these modal expressions appearing in the explicit formulation of the warrants are not given any special name by Toulmin, nor their function is compared explicitly with earlier discussions of the function of the modals in general. Even their clearly intended connection with the qualifier is left implicit and can be gleaned only from the examples.

Thus, the obvious question of whether and how the earlier pragmatic analyses of the modals as relational illocutionary markers (in the first essay) or as unary illocutionary modifiers (in the second essay) could also be applied to these *modals-in-the-warrant* is to remain unanswered.

Considering how Toulmin hints at a close parallelism between warrant and qualifier, one might be tempted to conclude that qualifiers simply correspond to the force of the warrant, that is on the degree of support that it offers to the claim given the data. As we have seen in the previous sections (§§ 3.4.2.3 and 3.4.2.4), this force or degree of support, is a conditional notion, treatable in terms of conditional necessity (*necessitas consequentiae*) or conditional probability. As such, it is clearly distinct from the *overall degree of confidence* that the arguer may have towards the claim. As we have observed above in § 3.4.2.4, the latter would depend not only on the support offered by the warrant, but also on the degree of confidence that the arguer has towards the data, or, to put it in another way, on the reliability of the data.

Toulmin does not discuss this distinction in the third essay, nor he mentions the issue of data reliability. However, if we look back at the overall argument of the essay on probability and at the strength with which he rejected the identification of probability with support and the relational analysis of probability terms, we must conclude that in his view the qualifier must have been distinct from the force of the warrant and conveying, in some way, an 'absolute' rather than relative indication of the confidence of the arguer towards the claim.

There is, however, one interesting passage in the essay on probability where the relationship between the unnamed modalities in the warrant and the modal qualifier is hinted at. Toulmin (2003 [1958]: 83) says that "probability-terms" do not only qualify assertions, promises or evaluations in themselves but also provide "an indication of the strength of the backing" that the speaker has for the speech act" and that it is "the quality of the *evidence* or *argument* at the speaker's disposal which determines what sort of qualifier he is entitled to include in his statement". Freeman (2011: 298) finds this remark hard to reconcile with Toulmin's pragmatic view of *probably*:

> Toulmin does not seem to realize that this indication gives modal qualifiers a semantic/literal meaning besides their pragmatic/ emotive force. It is by virtue of this literal meaning that modal qualifiers make a claim about the strength of support premises give to their conclusion.

3.4.4 *Modal Qualifiers and Rebuttals*

The Toulmin model also includes a sixth element, the *Rebuttal* (R), which has been left out of the picture up to this point. The visual presentation of the model suggests that rebuttals have an especially close relationship with modal qualifiers. Both the qualifier and the rebuttal stem from the fact that many warrants do not authorize to draw a conclusion "unequivocally" but only subject to qualifications and conditions, or exceptions (Toulmin 2003 [1958]: 93). They both "comment implicitly" on the bearing of the warrant on the step from data to claim. While the qualifier indicates the strength or force that the warrant provides to this step, the rebuttal indicates "the circumstances in which the general authority of the warrant would have to be set aside" (*Ibid.*: 94). Commenting on Toulmin's diagram, Freeman (2011: 11) notes the attachment of the rebuttal to the modality and sees this graphical layout as due to the rebuttal indicating "the conditions when the warrant would have to be set aside and so the force of the claim invalidated".

Clearly, Toulmin's *rebuttal* is not an *actual* rebuttal. By "actual rebuttal" I mean is the assertion of a proposition that constitutes an argument against the claim, as one could find, for instance in (46):

(46)	Arguer:	Harry was born in Bermuda. So, presumably, he is British.
	Critic:	*But*, he has become a naturalized American.

An actual rebuttal, recognizing the factuality of a proposition capable of ruling out the claim, has no place in the discourse of the arguer. Its inclusion would completely change the nature of the discourse, as it happens in (47) where the speaker cannot be any longer said to be arguing for the claim that Harry is British:

(47) Harry was born in Bermuda. He should be a British subject. *But*, in fact, he has become a naturalized American.

Toulmin's rebuttal does not seem a *concession* either. By concession I mean, broadly, the recognition by the arguer of the factuality or possibility of a proposition that the critic considers or may consider an actual rebuttal, that is an argument against the claim, combined with the implicit or explicit indication that the arguer considers this proposition either outright irrelevant or, in any case, insufficient to rule out the claim, as it happens in example (48).

(48) *Yes*, Harry has lived in the USA for several years, *but* he was born in Bermuda. So, presumably, he is a British subject.

In (48) the arguer concedes the factuality of the proposition 'Harry lived in the USA for several years' but considers it *irrelevant* or, at least, *insufficient*, to rebut the tentative conclusion that Harry's citizenship is British. In Toulmin's example the proposition 'Harry has become a naturalized American' would be relevant and sufficient to rebut the claim, *should it be the case.*

In fact, Toulmin's category corresponds not to an actual act of rebutting, but, to use Toulmin's own words, to "conditions of exception or rebuttal" (*Ibid.*: 93). This notion of a *condition of exception* is conveyed by the choice of the conjunction *unless* to introduce the rebuttal in the standard presentation of the model.

Freeman (1991, 2011) has examined in detail how such a category may fit into the overall architecture of the Toulmin model, offering a *dialectical* interpretation both of the *rebuttal* in tune with his dialectical view of the Toulmin model as a whole[51]. Here I will briefly examine his interpretation, arguing that, despite its independent merits as a general theory of rebuttals, it fails to capture an aspect of *Toulmin's* rebuttals, which is relevant for the present investigation. This aspect emerges more clearly if we examine the notion of condition of exception and the meaning of *unless* from a *semantic* point of view. Such an examination can cast light on the interaction of modalities with conditions of exception and with other kinds of conditions in the structure of arguments.

In Freeman's (2011) dialectical reading, the Toulmin model is seen not as a model of an argumentative text, but as a model of a dialogical exchange between a proponent (the *arguer*, in this book) and a challenger (here usually called *critic*). Conditions of exception find their place in such a model because they could be the object of a critical question of the challenger:

> The challenger might be quite willing to admit that, all things being equal, the premises provide a sufficient case for accepting the conclusion. The problem is that from her perspective not all things are equal. There is a question in her mind whether some condition or state of affairs hold which, should it hold, would either constitute evidence against the proponent's conclusion or call the reliability of the proponent's inference to that conclusion into question. It is not necessary that the challenger accept on evidence that this condition holds. It is sufficient that there be a realistic possibility, from her perspective that it does. (Freeman 2011: 20)

[51] Freeman's (1991, 2011) dialectical justification of the modal qualifier was briefly presented in 7.2.3.

According to Freeman (2011: 22) it is sufficient that the question of a potential rebuttal arise to make the proponent's argument insufficient, that is no longer able to create a presumption for the claim or no longer justified in using a certain modality to describe the force of the conclusion. Thus, according to Freeman, in the presence of a rebuttal, be it "recognized as a fact" or "simply mooted as a significant question", the proponent must take it into consideration either by addressing it with a counter-rebuttal, as in (49), or simply by signaling that the pretended rebuttal is ineffective in defeating the argument, as in (50).

(49) Arguer: John can vote in the Westwood municipal elections, as he is a resident.
Critic: How can be so sure that he can vote? Might Jones not be an American citizen?
Arguer: Jones has an American passport.
(Adapted from Freeman 2011: 25)

(50) Even if Mrs. Wilson and her daughter had a reconciliation, the fact is that Mrs. Wilson's will leaves all her estate to charity. Therefore her daughter has no share in her estate. (Adapted from Freeman 2011: 28)

In the latter case, as Freeman (2011: 27) recognizes, the conjunction *unless*, indicating the exceptive condition, is no longer appropriate and will be substituted either with *even if*, when conceding the possibility of the pretended rebuttal, or with *even though*, when conceding its factuality. Rebuttals addressed in the latter way are a kind of *concession*. Inasmuch rebuttals are addressed by the proponent (in either way), they also become part of arguments as products and need to be accounted for in the macrostructure of the argument.

It is impossible to do justice here to Freeman's sophisticated and detailed theory of rebuttals. It includes the fundamental distinction between evidence against the proponent's conclusion (rebutting defeaters) and evidence against the reliability of the inference in the proponent's argument supporting the conclusion (undercutting defeaters); it discusses several kinds of counter-rebuttals addressing the two fundamental types of rebuttal and examines fine points of argument diagramming. I have little to say in this book about these important contributions to the analysis of refutation and counter-refutation (but. What has to be observed here is that Freeman's (2011) reconstruction of rebuttals, and its rationale for their inclusion in the argument's macrostructure, does not sit well with Toulmin's examples of rebuttal and arguably does not cover what in Toulmin's view appears to be the most straightforward reason for including "conditions of exception or rebuttal" in the layout of an argument.

Consider the following examples adapted from Toulmin:

(51) Harry was born in Bermuda. A man born in Bermuda will be a British subject. So, *presumably*, Harry is a British subject, *unless he has become a naturalized American*.

(52) Anne is one of Jack's sisters. All Jack's sisters have red hair. So, *presumably*, Anne now has red hair, *unless she has already turned white*.

It seems reasonable, or at least consistent with Toulmin's text, to think that the exceptive conditions that appear in examples such as (51) and (52) are to be interpreted as epistemic possibilities, and thus could potentially be "mooted as a significant question" by a challenger. In Freeman's view, this should be sufficient to make the argument unable to create a presumption for the claim or no longer "justified in describing the force of the argument with the modality he used" (Freeman 2011: 22). In order to make the argument viable again the arguer should address the possible rebuttal with an appropriate kind of counter-rebuttal or dismiss it as irrelevant. As observed above, the Toulminian rebuttals exemplified in (51) and (52) are certainly not dismissed: if they were, the use of *unless* would be semantically incongruous. They are not addressed with counter-rebuttals, either. Yet, there is no sign that Toulmin's examples are meant to be taken as defective arguments in need of further support. In fact, (51) and (52) look reasonably natural texts that an arguer could put forth, in the appropriate conversational context, without the risk of looking self-defeating.

In view of the above, the most straightforward account of the role of rebuttals in the Toulmin model is to consider that the rebuttals in examples (51) and (52) do not need to be addressed because they are already completely "covered" – so to say – by the modal qualifier. The conclusion is qualified as presumptive, exactly because the possibility subsists of special circumstances "in which the general authority of the warrant would have to be set aside" (Toulmin 2003 [1958]: 94). In his dialectical reading of the Toulmin model, Freeman (2011) sees the rebuttal arising as a question of the challenger and targeting an argument already equipped with its modal qualifier: hence the idea of the modality becoming no longer justified once the rebuttal has been mooted. In my reading, Toulmin's modal qualifiers are not prior to rebuttals: they emerge jointly with them and also *in view* of them.

Beside avoiding the problems expounded above, this interpretation of Toulmin's rebuttals lends interesting insights into the close relationship between modal qualifier and rebuttal.

First of all, in a dialectical perspective, it suggests a way of dealing with rebuttals that was overlooked in Freeman's (2011) detailed treatment of rebuttals. When a rebuttal is mooted as a significant question by the challenger, the proponent may react by specifying a suitable modal qualifier if the force of the modality was previously left unspecified. If a questioned rebuttal makes a previously specified modal qualifier unjustified, the proponent may deal with it by weakening the modal qualifier without completely retracting the standpoint.

But what makes a modal qualifier justified or unjustified in view of a mooted rebuttal? It is the force of a modal explicitly or implicitly assigned to the rebutting proposition. Loosely speaking, the force of the modal qualifier of the argument has to be *complementary* to the explicit or implicit modality of the rebuttal.If we had to do with numerical probabilities we could say, for instance, that a modal qualifier with a force of 0.7 would be justified with rebuttal whose implicit or explicit modality is 0.3, and so on. Of course, even in a framework that sees the modals as a kind of quantifiers – such as the variant Relative Modality analysis presented in the following chapters – a translation of modal meanings into numerical probabilities is not necessarily feasible or desirable[52]. However, even in the absence of numerical probability values, this idea of complementarity can be applied in a more qualitative manner. If we go back to the first essay of the *Uses of Argument*, Toulmin (2003 [1958]: 20–21) says that *presumably* corresponds to a situation where the arguer is able "to show that one particular answer would be *the* answer, supposing only we were confident that certain unusual or exceptional conditions did not apply in the particular case" (*Ibid.*, p. 20–21). If we adopt Toulmin's own definition of *presumably*, then, we can conclude that in examples such as (51) and (52), the use of the modal qualifier is justified as long as their rebuttals are considered "unusual or exceptional". Now, both in (51) and in (52), a rebuttal is mentioned by the proponent using an *unless* clause, which does not make explicit the modality of the exceptive condition. Even if there is no explicit linguistic marking of modality in the rebuttal, an interpreter minimally charitable towards the proponent can pragmatically infer the intended modality of the rebuttal from the very fact that the proponent uses *presumably* as a modal qualifier. Assuming that the proponent *believes his qualifier is justified*, one can conclude that the exceptive conditions that he mentions are believed to be an exceptional circumstance.

The complementarity between the qualifier and the epistemic evaluation of the rebuttal has to be considered in connection with the semantics of the *exceptive conditions*, which are introduced by *unless* clauses. *Exceptive conditions* can be considered to be the semantic facet of Toulminian rebuttals. Following Higginbotham (2003: 65), I will consider the exceptions introduced by *unless* clauses to be, for all relevant purposes, equivalent to *if+not*-clauses and thus to introduce a negative conditional antecedent. If we provisionally assume, with Higginbotham, that *if* clauses express the material implication[53], *p unless q* would translate in the propositional calculus as $\neg q \rightarrow p$. Note that, according to the semantics of the material implication, the falsity of the antecedent does not rule out the truth of the consequent. In other words, if q is true, p can be either true of false. This semantics may seem too minimal, and the fact that the truth of q does not rule out p seems in contrast with the concept of *exceptive condition* associated with *unless* in the Toulminian examples considered above. In reality, however, one very frequently finds occurrences of

[52] Cf. Portner (2009: 73–79) on detailed arguments for rejecting an analysis based on numerical probability for the core modals of English. The main argument is that such an analysis would apply to epistemic modals only an would preclude a unified synchronic and diachronic account of the different flavors of modality (*Ibid*: 77). Furthermore, Portner believes that even in the case of English probability expressions a numerical probability analysis is not justified outside of "mathematically sophisticated language which explicitly builds on probability theory" (*Ibid.*).

[53] As it is well known, the material implication fails to provide an adequate account of *lato sensu* counterfactual conditionals (sometimes called 'subjunctive conditionals'). Lewis (1973) proposed a modal analysis of counterfactual conditionals based on possible world semantics. Grice (1989) famously argued that a material implication analysis can be maintained for *indicative conditionals* as long as it is supplemented by pragmatic inferences based on the cooperative principle. In contrast to the Gricean analysis, which is still adopted by several students of pragmatics, formal semanticists such as Kratzer have proposed modal analyses of indicative conditionals in line with those of counterfactuals (Cf. Kratzer 2012, esp. Chap. 5).

unless where the truth of q would not rule out p and the minimal interpretation ¬q → p is pretty much the one intended:

(53) I'm leaving *unless* I get a pay rise. (WebCorp)
(54) He won't go to sleep *unless* you tell him a story. (WebCorp)
(55) They can't vote *unless* they have French citizenship (WebCorp)
(56) Spain And Italy Are Toast *Unless* Germany Allows The ECB To Print Trillions Of Euros (WebCorp)

In (53) the point is that not getting a pay raise would bring about the leaving, but even with a pay rise the employee may leave nevertheless. Similarly, in (54) the child may still refuse to go to sleep even if the addressee tells him a story and, in (55), the people concerned may turn out to be barred from voting from some reason even if they have French citizenship (e.g. criminal conviction). Finally, in (56), Spain and Italy may still become (financially) "toast" even if the EBC does print trillions of Euros.

There are other occurrences where *unless* seems to imply also that q→ ¬p. Example (57) below clearly implies that, when they find food, black bears *do* pose a threat. In this stronger reading, p *unless* q would be equivalent to ¬q ↔ p.

(57) Black bears pose little threat, unless they find food (WebCorp)

I will not discuss here whether this stronger, less frequent, reading corresponds to a polysemy of the connective *unless* or to contextual pragmatic inferences. What is more interesting, in the context of the present discussion, is to examine whether the weaker reading of *unless* as ¬q → p is sufficient to render the Toulminian idea of rebuttal.

Let us set aside the doubt that this kind of formalization might be contrary to the overall spirit of Toulmin's enterprise and try to apply the formula ¬q → p as a translation of the connective *unless* in the Toulmin model. Clearly, the proposition q ought to correspond to the Rebuttal R. As for the proposition p, the correspondence is much less straightforward. Should we identify p with the Claim C? Probably not. Here Freeman's (2011) distinction between *rebutting defeaters* and *undercutting defeaters* comes in handy. *Rebutting defeaters* are propositions that, should they hold, would "constitute evidence against the proponent's conclusion" – they are directly aimed at striking down the claim, independently of the argument. *Undercuting defeaters*, on the other hand, "call the reliability of the proponent's inference to that conclusion into question". If we look at how Toulmin presents the Rebuttal we find that Rs appear much closer to the latter: they correspond to "special facts" that "may make the case an exception to the rule or one in which the law can be applied only subject to certain qualifications", "circumstances in which the general authority of the warrant would have to be set aside" (Toulmin 2003 [1958]: 93, 94). Thus, in the Rebuttal the connective *p unless q* should take scope over the whole inference. The argument p of the connective *unless* should be identified with the inferential step from the Data to the Claim authorized by the Warrant. More precisely, the scope *q* of *p unless q* is a conditional having the premises (Data, Warrant) as *antecedent* and the conclusion (Claim) in the consequent. The attentive reader will have already noticed that this is basically the same conditional that falls under the scope of modality in the analysis of the necessity of the move from premises to conclusions (cf. Def. 15 in 4.2.2 above).

For the sake of simplicity, let us ignore that such a conditional will have to be represented either as a necessary conditional or as conditional probability. If we are to remain within the boundaries of the extensional propositional calculus, we should come up with something like this:

(Def. 17) $\neg R \rightarrow ((D \wedge W) \rightarrow C)$

Interestingly, on the material implication reading of $\rightarrow$, this formula is equivalent[54] to

(Def. 18) $(D \wedge W \wedge \neg R) \rightarrow C$

This suggests a view of the Rebuttal as a negative condition on the inference. The warrant authorizes the inference from the Datum to the Claim *provided* that the Rebuttal is false. Once the condition R has been epistemically evaluated as *improbable* or *implausible*, it acts as a sort of premise, authorizing a *probable* or *plausible* conclusion in favor of the Claim. Should the Rebuttal be proven false, this falsity would count as a premise allowing to support the Claim without qualification.

This reading of Rebuttal as a *negative conditions* uncovers an interesting gap in the Toulmin model. The model includes *negative conditions* on the inferential step taken by the argument (i.e. the Rebuttals) as one of the essential components of the "finer, physiological" structure of an argument. Yet, comparable *affirmative* (non-negated) *conditions* do not receive the same treatment. Toulmin briefly mentions that "the step from data to conclusion" may be "subject to conditions" (Toulmin 2003 [1958]: 93), but the more general notion of *condition* does not make it to the Toulmin model, where we find instead the more specific notion of Rebuttal.

Considering that the exclusive focus on the negative conditions is a gap in the Toulmin model may seem bizarre at first sight. Yet, if it seems so, it is only because the settled picture offered by the model has canonized certain moves of argumentative discourse and rendered invisible others that are equally pervasive in argumentative practice. Presenting an argument as based on certain background assumptions, as valid only within certain parameters, specifying that an inferential step can be taken *provided that* certain expected conditions hold is not such an uncommon move. For instance, as shown in Rocci (2008b) and Miecznikowski et al. (2012), such a move is frequently found in the arguments supporting predictions in economic-financial news. The arguments supporting predictions are routinely presented as holding *provided that* certain more or less strongly expected conditions hold, such as in (45) below:

[54] On a modal reading of the conditional this equivalence is not generalizable (Cf. Kratzer 2012, Chap. 4).

(58.a) Firm's fortunes may rise as commodity prices fall (Headline)

(58.b) Major airlines around the globe continue to see strong passenger demand, so profit could climb *if they are able to raise prices* while their own costs drop as fuel prices fall. (From the body text) (*Wall Journal Europe* (WSJE) September 14, 2006)

The headline in (58.a) consists of a modalized claim supported by an argument based on a form of causal argumentation, while the passage in (58.b) presents a more developed form of the same argument where the modalized conclusion introduced by *could* holds only within a conditional frame (*if they are able to raise prices*). The epistemic stance of the arguer towards the condition is partially expressed by the form of the *if* clause: we have an indicative conditional rather than a counterfactual and thus we know that it is not regarded as impossible or only marginally possible. The presence of a weak modal qualifier, such as *could*, on the other hand, might contribute indirectly to signalling that the arguer's epistemic stance towards the condition is not one of certainty or near certainty either. In other cases it is the content itself of the condition that allows to infer a certain degree of expectation towards its realization. This happens when the condition corresponds to the continuation of a trend or the stability of a state of affairs, the correctness of an estimate, the truthfulness of a testimony or of an expert opinion, things going according to plan and so on. Examples of this kind will be discussed further in the final chapters of this book, which draw from a large corpus of Italian financial news their data.

It is worth noting that often these *if* clauses can be trivially converted in Rebuttal-like *unless* clauses which play an analogous function:

(59) Major airlines around the globe continue to see strong passenger demand, so profit will climb *unless they are unable to raise prices* while their own costs drop as fuel prices fall.

Saying that an argument holds *unless* a (more or less) unlikely condition happen to be the case and saying that said argument holds *provided that* its contradictory (more or less) likely condition is realized should be, in principle, equivalent.

In fact, these positive and negative conditions can be seen as functioning as a kind of *presumption* and the presence of similar presumed propositions plays a key role in certain accounts of the basic inferential structure of presumptive reasoning, in particular by the "deductive"[55] account of presumptive reasoning offered by Nicholas Rescher (2006).

In the final chapter of this book I will combine Rescher's view of presumptive reasoning with the integrated semantics of modals and conditionals offered by the Relative Modality analyses to present an account of the interaction of specific modal markers with positive presumptions as well as negative ones (equivalent to Toulminian rebuttals) in argumentation. . For the moment, the matter of Rebuttals is

[55] Cf. Rescher's account of presumptive reasoning is characterized as essentially deductive by Woods (2010: 247).

better left as it stands as the detailed examination of Toulmin's views draws to a close.

3.5 Toulmin's Three Takes on Modality and Their Legacy in Contemporary Argumentation Theory

3.5.1 Addressing the Toulminian Research Questions

This long "excavation" of the three theories of modality presented (and implied) in *The Uses of Argument* was undertaken with two of our main research questions in mind – those that were directly inspired by Toulmin:

(Q1) How are modals similar to arguments? Is there any relevant relation of isomorphism between the semantic structure of modalities and the "deep structure" or layout of arguments?

(Q2) Is there a special place for modality in the essential structure of arguments? Is there a distinctive functional slot or role that is assigned to modal meanings in such a "deep structure".

The two questions are independent. Yet, the answer that one gives to the first may help to shape the answer to the second. If it is found that modals and arguments are similar or parallel in some respect, this similarity might have to do with the modals' propensity for taking up certain functions in arguments.

Additionally, a positive (or partially positive) answer to the second question makes it reasonable to ask the third research question (Q3), about the role of modal expressions as *argumentative indicators*, and inevitably shapes its answer, without determining it completely.

(Q3) What does the use of modal expressions in an ordinary discourse tell us about the arguments being put forth in the discourse?

Adopting a certain view of the place of modality as a category in the context of an argument will largely shape my expectations about the kind of argumentatively relevant information that can be gleaned from concrete modal expressions, even if it may turn out that these concrete expressions are not limited to expressing modality per se – to use Freeman's expression they "do double duty" – and have also other stories to tell.

Toulmin answers positively to the first two questions, but, as shown in the previous pages, the answers he gives, taken together, do not form a cohesive nor an entirely consistent theory of modality in arguments. The different essays imply distinct and somewhat disconnected theories of modality, and the close comparison between them reveals several critical joints where apparent inconsistencies arise[56].

[56] As it should be clear from the previous discussion, many of these critical passages are affected by considerable vagueness so that one cannot characterize them as outright inconsistent without a considerable baggage of assumptions on how they ought to be made precise. Whether there could

Taken individually, some of Toulmin's proposals also encounter seemingly fatal empirical objections.

Consider, for instance, the key idea of the first essay that all flavors of modality directly play an argumentative function and that their context dependent criteria and context invariant force can be understood directly in argumentative terms. In § 3.2.3, this was shown not to be the case: criteria do not always map on types of arguments for a conclusion and force does not always signal a move or speech-act in an argumentative discussion. There is a clear rift between the epistemic or doxastic flavor of modality, which are inevitably tied to evidence and inferences (if not with argumentation) and other flavors of modality. The former, at first blush, do not appear to be related to a specific type of criteria of their own, as they refer to whatever counts as evidence in the context of utterance[57]. As for the other flavors of modality, their criteria may or may not be relevant as evidence for a claim.

3.5.2 *The Narrow View: Argumentatively Relevant Modals Are Epistemic Qualifiers*

When dealing with modality, most subsequent work on argumentation – including work co-authored by Toulmin – has ignored non-epistemic modalities, restricting (deliberately[58] or *de facto*) its focus to epistemic qualifiers.

On the one hand, this restriction of focus is fully justified in view of what emerged from the critical examination of the first essay.

On the other hand, the examples offered by Toulmin where the *criteria* of the non-epistemic modals *do* match the arguments supporting the claim should not be brushed away as *ad hoc* or merely coincidental. There is clearly a natural connection between the physical, semantic, stylistic, legal and moral impossibilities appearing in Toulmin's examples and the act of ruling out an hypothesis previously considered, even if these impossibilities do not semantically coincide with the act of

---

be ways to reconcile the different views of modality that emerge from the *Uses of Argument* remains a moot question that will not be pursued further in this book.

[57] In Chaps. 4 and 5 I will have to moderate this bald statement, taking into account the fact that individual epistemic expressions may be subject to preferences or outright restrictions concerning the kind of evidence.

[58] The topic of non-epistemic modalities is rarely the object of explicit attention in the argumentation literature. We have seen that Freeman (1991) briefly discusses the necessity of logical and mathematical tautologies and separates it from the necessity of concluding in view of a certain body of evidence. Snoeck Henkemans (1997: 109, also n. 7) is unusual in that she makes the restriction to "epistemic modalities" explicit, distinguishing them from deontic modalities. Only the former can be used to "indicate the extent to which the speaker is prepared to commit himself to the truth or acceptability of the propositional content of his standpoint" and, as such are of interest as argumentative indicators, while the latter are "part of the predicate of the proposition" and thus cannot play the role of illocutionary force indicators or illocutionary force modifiers.

ruling out an hypothesis. Capturing this natural fit will be an central concern of the following chapters.

A limited exception to the restriction of interest to epistemic modals in argumentation studies is represented by Bermejo Luque (2011), who includes two different types of modal qualifiers in her revision of the Toulmin model: the *epistemic modal qualifier* – which indicates the degree of pragmatic force of the act of concluding and is attached to the claim – and *ontological modal qualifiers*, which potentially characterize the content of all the components of the macrostructure of the argument (Bermejo Luque 2011: 115). Ontological modalities do not concern the relationship of a claim with evidence but characterize the status of a propositional representation with respect to the world by graduating its truth-value[59].

Of particlular significance are the ontological modalities of the *reason* (roughly Toulmin's datum) and of the *inference-claim* (roughly Toulmin's warrant), which, according to Bermejo Luque (2011: 171) determine the correctness of the epistemic qualifier of the conclusion. With this distinction, Bermejo Luque seems to make more explicit Toulmin's casual observation that probability expressions also serve to indicate the "strength of the backing" of a certain claim. She also appears to recognize that the two kinds of modality are of a fundamentally different nature and their role in arguments is different. In the next chapter I will briefly return to the way in which Bermejo –Luque draws the distinction between the two flavors of modality, which sharply diverges from the practice in linguistic semantics. I will also comment on problems that arise from her treatment of the interaction of modals and conditionals in the *inference-claim* (cf. Freeman 2011).

More generally, Bermejo-Luque's discussion of different structural slots for different kinds of modality in argument brings in the need of accounting for the structurally different roles that seem to emerge for the modals in arguments. In the previous sections we have encountered (a) seemingly absolute modals appearing as part of the propositional content in premises and conclusions, (b) purely conditional indications of necessity or probability and (c) epistemic qualifications of a propositions in view of all the relevant evidence. A semantics of modality useful for argumentation theory needs to tell us how these three types of modality relate to each other and how they differ. I will argue that it is possible to maintain a relational analysis of the modals without effacing the distinction between hypothetical relations and the relational act of concluding from evidence. Also seemingly 'absolute' uses of modal qualifiers will be shown to be better understood in relational terms.

Leaving aside Bermejo – Luque's idea that different kinds of modality are relevant in argumentation, the views of (epistemic) modalities espoused by argumentation scholars appear to be divided in two camps. The *semantic view*, which sees the

---

[59] Cf. Bermejo-Luque (2011: 171): "In acknowledging ontological probability values I am endorsing the view that propositional contents can have values other than true or false". As a consequence, the values of ontological possibility and ontological probability are incompatible with truth: "a representation that is, in fact, or false, cannot be ontologically probable" (*Ibid.*). So, for instance, realized ontological possibilities are not possibilities anymore. It is not clear what Bermejo-Luque is going to do with *necessity* in her modality-as-truth-value framework. Is ontological necessity less or more than truth?

modals as expressing the degree of support that the premises offer to the conclusion and the *pragmatic view*, which sees the modal qualifiers as extra-propositional illocutionary force modifiers working on the degree of commitment of the speaker towards a proposition.

As we have seen, Freeman (1991, 2011) and Vorobej (2006) fall firmly in the first camp. Pinto (2007) espouses a complex variant of the semantic view[60], where probability expressions denote the reasonableness of the adoption of a doxastic attitude of "expectation" or "partial belief" in view of the evidence. More generally, modal qualifiers signal that a certain doxastic attitude (certainty, expectation, presumption) is appropriate, or proportionate, in view of "the evidence or premisses on which the inference is based" (Pinto 2007: 1). They have both a doxastic[61] and an evaluative (or deontic[62]) component as they say that a certain doxastic attitude is proportionate, *right* so-to say, in view of the evidence. With respect to Freeman's and Vorobej's view, Pinto's conception introduces a further layer of complexity by making explicit reference to a series, or scale, of doxastic attitudes towards the conclusion that are evaluated as proportionate to the quality of the evidence instead of simply referring to the degree of support. In contrast with the pragmatic views of Toulmin and Ennis, all these semantic approaches the hedging or qualification of the assertive commitment is seen as derived from the semantic, denotative, content of the modal (cf. also Pinto 2007: 3).

Pragma-Dialectical scholars deal with modal qualifiers in terms of qualification of the arguer's commitment towards the propositional content of a standpoint. They thus appear to subscribe to the pragmatic view. However, this does not necessarily mean that they subscribe to the idea – which we encountered in Toulmin and Ennis – that epistemic modal qualifiers do not have a basic denotative meanings. Rather, this pragmatic orientation seems to reflect an interest for those aspects of the functioning of epistemic modals that directly interact with speech-acts and commitment, given the prevailingly pragmatic and dialectical focus of the theory.

---

[60] Pinto (2007: 1) marks explicitly the distance between his proposal and the pragmatic aspects of Toulmin's view: "Toulmin's idea that the function of *probably* and *probable* is to express guarded commitment or assertion plays little or no role in this paper". On the other hand, Pinto's account is clearly related to Toulmin's idea that the quality of the evidence at the speaker's disposal determines what sort of qualifier he is entitled to include in his statement (cf. § 7.3).

[61] In the logical tradition *epistemic modalities* concern what is possible/ probable/ necessary in view of *what is known* by a subject, while *doxastic* ones what is possible/ probable/ necessary, given the *beliefs* or opinions of a subject (cf. Hintikka 1962). Linguists rarely consider this distinction and tend to use only the term *epistemic modality*. As for Pinto, it is not surprising that he elects to use *doxastic* to refer to a cognitive attitude that has yet to be evaluated in view of the evidence.

[62] The closeness of the evaluative and deontic dimensions is well known. There is a small step between saying that adopting a certain cognitive attitude towards a proposition in view of the evidence is *good, right, correct* and saying that such an attitude *should* be adopted.

3.5.3 *The Epistemic Qualification of Standpoints in Pragma-Dialectics*

Francisca Snoeck Henkemans (1992: 108–117) addresses linguistic expressions of modality as *indicators* in the task of argumentative reconstruction. In Pragma-Dialectics argumentative indicators are linguistic expressions, textual features or behavioral cues that point to speech acts that are instrumental in the various stages of dispute resolution. For Snoeck Henkemans (1992: 109) modal words "can be used to indicate the extent to which the speaker is prepared to commit himself to the truth or acceptability of the propositional content of his standpoint". This expression of the degree of commitment "enables the analyst to determine what degree of justificatory or refutatory potential the argumentation should have, in order to lend sufficient support to the standpoint" (Snoeck Henkemans 1992: 113).

An equivalence between the expression of the "degree of certainty with which the standpoint is advanced" and the indication of the "degree of justificatory potential" can be established if we analyze the pragmatics of argumentation using the speech-act theoretic framework provided by Pragma-Dialectics. By advancing a standpoint an arguer commits himself to presenting adequate evidence in support of it in face of the expressed or implicit doubt of another party. By advancing a standpoint *with a certain degree of confidence*, the arguer *ipso facto* commits himself to providing evidence supporting the standpoint with a *matching degree of force or weight* (cf. Snoeck Henkemans 1992: 112–113). In fact, this commitment to the proportionate matching of certainty and weight of evidence, can be considered part and parcel of the commitment to a critical discussion. This is reasonably implied by the ideal model of the critical discussion, even if it is not made entirely explicit in the rules and commandments for the critical discussion (cf. van Eemeren and Grootendorst 2004). *Commandment 7*, in fact, only deals with one limit case of this matching, which concerns arguments presented as deductively conclusive, which need to be formally valid, without proposing an analogous matching justificatory force for lower levels of commitment.

The determination of the degree of justificatory potential that is required to "lend sufficient support to the standpoint" is not the only kind of indication that epistemic modal expressions can provide to the analyst in a Pragma-Dialectical framework. Van Eemeren et al. (2007: 28–45) discuss the role of epistemic expressions as indicators in the confrontation stage of a critical discussion as signs that a standpoint has actually been advanced. The discussion is particularly interesting because deals with a case of *indirect indication*[63], and because it nicely showcases the chiefly pragmatic and dialectical focus of the Amsterdam school scholars in dealing with modality.

[63] Cf. Rocci (2008a: 167): "let us establish a distinction between *direct indicators*—which are signs that, in some of their uses, *refer* to argumentative moves (e.g. the connective because) and *indirect indicators*, which refer to semantic or pragmatic categories *correlated* to argumentative moves. This distinction will help us in dealing with the modals' role as indicators".

According to the Amsterdam scholars, an analyst, in order to recognize that a standpoint has been advanced in real discourse, should consider whether the conditions of this speech act have been met. Advancing a standpoint, in the pragma-dialectical speech-act analysis (cf. Houtlosser 1994, 2002), means taking responsibility for a positive or negative position with regards to a proposition p. Taking responsibility entails, first of all, a commitment to defend the position with arguments. It also entails, as a sincerity condition, the belief[64] that p or non-p is the case and the intention to provide justification if necessary. The commitment to defend a point of view also presupposes, as a preparatory condition that the speaker supposes that "the assertive is not immediately acceptable to the listener" (van Eemeren et al. 2007: 29). The authors suggest that the analyst should adopt the satisfaction of this preparatory condition as a criterion for recognizing the performance of the act. The satisfaction of the preparatory condition can be detected in discourse either from actual opposition from the interlocutor or from a mode of presentation of the assertive that "makes an issue of the acceptability of the performed assertion (van Eemeren et al. 2007: 33).

The authors argue that certain epistemic expressions, in certain uses, can be used to convey by pragmatic implicature the idea that the assertion is not immediately acceptable to the interlocutor. Two types of expressions are considered: "propositional attitude indicators" (PAI), which basically correspond to first person present indicative epistemic attitude predicates (e.g. *I'm sure that, I think that*) and "force modifying expressions" (FME), which are impersonal adjectival or adverbial expressions (e.g. *It is certain that, supposedly*)[65]. Epistemic uses of modal verbs are not considered. According to van Eemeren and colleagues, both PAIs and FMEs can function as indicators when they are used parenthetically, in Urmson's (1956) and Hooper's (1975) sense and are therefore *assertive* in Hooper's sense. Epistemic expressions that are not parenthetical and become part of the asserted propositional content rather than qualifying the assertion of the prejacent cannot work as indicators of standpoints because they "cannot serve to modify the force of the assertive" (Cf. van Eemeren et al. 2007: 30). For instance, they argue, the epistemic predicate in (60a) is not parenthetical – given that (60.b) is not possible – , and asserts that it is probable that John will come this afternoon, rather than modifying the force of the assertion that John will come this afternoon.

(60.a) It is likely that John will come this afternoon

(60.b) *John will come this afternoon, it is likely.

---

[64] Clearly, as Houtlosser (2002: 171) observes, the speaker may well privately not believe the standpoint he is defending, but he is publicly committed to this belief. He cannot, for instance, advance a standpoint and assert he does not believe the opinion advanced as a standpoint.

[65] I induce this characterisation from the examples, as van Eemeren, Houtlosser and Snoeck Henkemans (2007) do not provide a definition of the two classes.

Those PAIs and FMEs that are parenthetical include *weak assertives* (e.g. *I think that, supposedly*) and *strong assertives* (e.g. *I'm sure that, It is certain that*)[66]. Both can be used to implicitly[67] draw the attention on the issue of the acceptability of the performed assertion, albeit in different ways.

- *Weak assertives* "prevent the listener from thinking that what he has said is founded on strong evidence" implying that "the listener who accepts this does it at his own risk" (*Ibid.* 33).
- *Strong assertives* make explicit the belief of the speaker towards the asserted propositional content. Given that this belief is part of the sincerity conditions of the assertion, its expression is unnecessary unless the speaker assumes that the assertion will encounter opposition and by stressing that the proposition is "obvious, clear, beyond doubt" (*Ibid.*) he wants to invite acceptance from the listener.

In both cases the epistemic expressions function as indicators of a dialectically relevant move (advancing a standpoint) not because their semantic meaning or pragmatic force *consists* in denoting or signaling this move, but *indirectly* through a mechanism of implicature.

3.5.4 Epistemic Qualifiers as Presentational Devices

A significantly revised view of epistemic expression in a pragma-dialectical framework is offered by Tseronis (2009) in an investigation that intersects only partially the domain of modality. Tseronis considers three kinds of qualifications that can accompany the advancing of a standpoint: epistemic (e.g. *probably*), evaluative (e.g. *unfortunately*) and "stylistic" (e.g. *frankly*)[68]. I consider here only the first category. The most striking feature of Tseronis' account of qualifiers is that he completely abandons the idea that epistemic qualifiers modify the force of an assertive speech act, or even "clarify the exact force of the assertion" (cf. van Eemeren et al. 2007: 32) prompting its recognition as an act of advancing a standpoint. He squarely says is that they are not part of the dialectically relevant components of an argument and only function as presentational devices[69]:

---

[66] For the sake of brevity I omit the category of *semi-assertives* which is also used by van Eemeren, Houtlosser and Snoeck Henkemans (2007) as it raises problems of semantic analysis that would require a rather lengthy discussion.

[67] For a detailed Gricean reconstruction of the two pragmatic implicatures see van Eemeren, Houtlosser and Snoeck Henkemans (2007: 32–34)

[68] The ground covered by Tseronis has a larger, yet not exact, overlap with the categories of parenthetical verbs and corresponding adverbs discussed by Urmson (1956, see § 3.3.3 above). The parentheticals that fall outside Tseronis' view of qualifiers are those that comment on the "logical relevance" of a proposition in a discussion (e.g. *admittedly, consequently*).

[69] To be clear: as defined by Extended Pragma-Dialectics (van Eemeren 2010), presentational devices play an important role at the rhetorical level as a means of strategic maneuvering. Yet, at the dialectical level, they are, by definition, not relevant.

> Qualification is one choice from the presentational devices at the protagonist's disposal when designing the move of advancing a standpoint. This is so because qualification does not affect the argumentative interpretation of the utterance in which it appears as a standpoint and does not receive support in the ensuing argumentative discussion either. (Tseronis 2009: 31)

To put it in a Toulminesque way, there is no special slot for qualifiers in the macrostructural layout of arguments, nor are they part of the microstructural propositional content of the standpoint that arguments need to support. According to Tseronis (2009: 32), qualifications that accompany standpoints are not illocutiorary force modifiers of the *act of advancing a standpoint*. If they were they would change the nature of the illocutionary force: suggesting "a different degree of strength of the illocutionary force ends up defining a different illocutionary act"[70].

Given that the qualifiers are not illocutionary force modifiers, according to Tseronis, the object to which they adds a comment can be either the propositional content or "the performance of an illocutionary act as a whole" (Tseronis 2009: 33). The latter is commented in the case of "style" adjuncts, while epistemic qualifiers add *a comment conveying commitment to the truth of the propositional content of the standpoint* (cf. Tseronis 2009: 34), without interacting with the illocution of the act of advancing the standpoint.

In the case of the epistemic qualification, keeping apart the comment from the assertive illocution of the act appears troublesome as they do provide information about the commitment of the speaker towards the belief in the truth of the propositional content and, as we have seen, such a commitment features among the sincerity conditions of the assertive speech act of advancing a standpoint (cf. Houtlosser 2002: 171). More specifically, it appears as the sincerity felicity condition. This would bring us back to considering epistemic qualifications as modifying or, at least, clarifying the sincerity condition.

Tseronis rejects this view and postulates a strict separation between the commitment towards the truth of the propositional content – which is commented upon – and the commitment to defend the correctness of an expressed opinion towards this same propositional content – which is what the act of advancing a standpoint is about. This means that, supposing that I advance (61) as a standpoint,

(61) John will come this afternoon

I am not asserting the propositional content 'John will come this afternoon' but I am asserting my "positive position on the question" (Tseronis 2009: 35), thereby taking responsibility for defending it. That's why, in Tseronis' view, modulating the commitment towards the original proposition does not affect the act of advancing a standpoint.

---

[70] Unfortunately, Tseronis (2009) does not give examples of linguistic devices that, unlike qualifiers, *do* work as illocutionary modifiers and *do* change the nature of the speech-act. The only example given is the contrast between *requesting* and *insisting*, but while there is certainly a difference of strength between the two verbs (among other semantic differences), the example does not really amount to a case of application of a recognizable modifying device.

In this connection Tseronis also observes (*Ibid.*), quoting van Eemeren and Grootendorst (1984), that the positive *belief* towards a position is a public social responsibility and not a psychological state. I believe, however, that this remark, while correct, is irrelevant as an argument for separating the two commitments. The social nature of the "belief" in the sincerity conditions holds for all assertives, not just for standpoints. And one cannot *socially commit* to believing "a positive position on the question" of p, without *ipso facto socially committing* to believing that p is the case. As a consequence, even if I conceive advancing a standpoint as a meta-level speech act (as proposed already by Houtlosser 1994) its sincerity condition and the commitment to believing the truth of the underlying proposition cannot be segregated as Tseronis suggests[71].

According to Tseronis, the commitment that characterizes the speech act of advancing a standpoint is not gradable in nature and simply involves a binary choice between positive or negative standpoints: one is either ready to defend an expressed opinion with arguments or not. As a consequence, the arguments that are presented in support of a standpoint qualified with epistemic adverbs such as *perhaps* or *clearly* do not represent "support for the degree of probability" of the proposition, but simply "support for the positive or negative position" (Tseronis 2009: 36). This is an important corollary as it directly contradicts the idea, expressed earlier by Snoeck Henkemans (1992) that, by advancing a standpoint with a certain degree of confidence, the arguer commits himself to providing evidence supporting the standpoint with a *matching degree of force or weight* (cf.: 112–113)[72]. According Tseronis, when the protagonist advances a standpoint using utterances such as

(62.a) Clearly, John is sick
(62.b) Perhaps, John is sick

he is signaling two separate commitments, as evidenced by the paraphrases reproduced in (63.a) and (63.b) below:

(63.a) John is sick. I am sure.
(63.b) John is sick. I am not sure.

[71] Tseronis stresses that in an argumentative discussion *not believing* the propositional content of the expressed opinion does not constitute cheating or uncooperativeness. Correct, but only as long as we refer to a *private* mental state, which is irrelevant here. When we consider a *public manifestation* of lack of belief in the propositional content, we find that it makes the act of advancing a standpoint infelicitous, as in the following example: **I am ready to defend the thesis that John is not coming with us tonight – even if I don't actually believe it.* We find ourselves again in a variant of Moore's pragmatic paradox of belief. If we exclude the non-serious scenario of people playing a dialectical game for the sake of argumentative sparring, and consider real argumentative situations where participants are oriented towards a *problem-valid* resolution, the utterance cannot be felicitous. We must either conclude that the utterer is argumentatively uncooperative or that he is deeply irrational.

[72] It goes without saying that the perspective introduced in Tseronis (2009) is also squarely incompatible with Pinto's ideas of arguments needing to be proportionate to the doxastic attitude adopted as well as with Toulmin's (admittedly vague) remarks on the matching between quality of the backing, force of the warrant and force of the qualifier.

In both cases the speaker commits to argue for the *unqualified standpoint*. – the additional comment being aimed at "emphasizing the quality of evidence that he is ready to forward in support of the standpoint" (Tseronis 2009: 37).

I have argued above that, pace Tseronis, the two operations are not independent. Let us examine now the claim that the epistemic qualifications *are not* the target of the argumentation supporting the standpoint. With evaluative and stylistic qualifications it is rather straightforward to see that the arguments that support the qualified propositional content do not target the qualifier, as shown in (64) and (65):

(64) He's taken his temperature and it's 102 °F / 39°C. So, *unfortunately*, John is sick.

(65) *Honestly*, John is sick. He's taken his temperature and it's 102 °F / 39°C.

The data about John's fever cannot be taken as an argument supporting the evaluation that the circumstance is unfortunate in (64) or supporting the remark that the speaker is being honest in communicating this fact in (65). These would be entirely different standpoints requiring entirely different arguments. In both examples the comment is separate from the standpoint in a way that can be rendered with a formula similar to that proposed by Tseronis for (63.a,b). The possible introduction of the discourse marker *by the way* makes this separation explicit, as in (66), below.

(66) He's taken his temperature and it's 102 °F / 39°C. So, John is sick. (By the way,) this unfortunate.

(67) He's taken his temperature and it's 102 °F / 39°C. So, John is sick. (By the way), I'm being quite honest in telling you this.

Adverbs like *probably* and *clearly*, which Tseronis considers epistemic, are another story altogether:

(68) He's taken his temperature and it's 100 °F / 37.7 °C. So, John is sick. **By the way,* this is *probable.*

(69) He's taken his temperature and it's 104 °F / 40 °C. So, John is sick. **By the way,* this is *clear.*

It seems that in (68) and (69) presenting the epistemic evaluation as detached and independent from the act of putting forth a standpoint creates a kind of nonsense that did not arise with the evaluative and stylistic qualifications. One could hypothesize that the reason why the epistemic comments refuse the insertion of *by the way* is that *probably* and *clearly* do not comment on the propositional content in isolation expressing a degree of commitment, but relate this propositional content with the evidence that supports it[73]. If this is the case, the very notion of qualification as *comment* would prove inadequate for them. Epistemic expressions would have to be associated to a wholly different category of non-propositional meanings: the one comprising also argumentative connectives.

[73] This is not the end of the story about *clearly* and *probably*. The two adverbs and their adjectival counterparts (*probable* and *clear*) differ in noticeable and yet mysterious ways that deserve further attention. The point made here was limited to showing that neither supports the idea of a modal qualifier expressing a comment on the speaker's commitment towards the proposition independently from the act of advancing a standpoint.

3.5.5 *Beyond Epistemic Scales*

Leaving aside the specific problems encountered by the view of qualifiers proposed in Tseronis (2009), it can be observed that all the approaches reviewed in § 3.5 – both the pragmatic and the semantic approaches – tend to see modal expressions as essentially conveying a degree on a scale. The nature of the scale varies greatly – it could be degrees of force of commitment, degrees of support offered by the evidence, discrete steps in a scale of doxastic attitudes. What remains constant is the focus on the scale as the primary dimensions along which modal expressions should be distinguished one from the other for the purposes of argumentation analysis. Certainly, quantificational force is one constitutive dimension of the semantics of modal terms, but an exclusive focus on scales and degrees risks to overshadow other differences among modals that are equally important for understanding argument. One can detect a hint of this attitude in Freeman (2011: 120), when he accept the idea that the cognitive meaning of modals "is sufficiently vague as to render their force the only aspect of their meaning worthy of interest", or in Tseronis (2009: 54) when he contends that "there is no difference in the discourse effect" of using a modal or an *evidential* expression when qualifying a standpoint and it is so because the question he is interested in is not the "lexical meaning" of the expressions but "the language user's choice to mark a high or low degree of probability/ certainty in argumentative discourse".

Actually, a key thesis of this book is that this question of degree is too narrow and that central aspects of modal meanings relevant for argumentative reconstruction remain hidden if we preemptively apply this straightjacket. As anticipated in Chap. 1, considering how epistemic modal expressions also encode evidential notions (such as the indication of the source of knowledge, or the mode of knowledge) is important to develop a semantic account of these expressions capable to capture their argumentatively relevant aspects and, in particular, to understand how they function as indicators of the inferential relations between a standpoints and arguments.

Van Eemeren and Henkemans (2007) have shown, with their Gricean analyses of the implicatures arising from the use of strong and weak assertives, how a pragmatic analysis can enrich the potential of an item as an argumentative indicator in discourse. This book takes a complementary route, trying to show how semantics can lend complementary insights on argumentatively discourse meaning, not only because semantic analyses can be fruitfully integrated with a reconstruction of discourse meanings by implicature – for instance, Simon-Vandenbergen and Aijmer (2007) largely adopt this strategy in their study of English certainty expressions – but also because implicatures are not the only source of discourse meanings. As it will be argued, certain discourse meanings of importance in reconstructing arguments directly arise from the way in which the incomplete semantics of these expressions deictically integrates aspects of the context.

Toulmin's first essay elects to focus on *cannot*, as we have seen, instead of looking at expressions of *necessity* or of *probability* in order to avoid the big "philosophical baggage" of these two notions. A remark that may seem bizarre, or even disingenuous, thinking of the way in which impossibility has been part of the – square – picture of modalities and their interrelations since Aristotle. A bit more charitably, one can see how Toulmin wanted to avoid attacking from the outset his two main polemical targets, namely the notion of logical validity and the theory of probability.

This essay remains an exception. For the most part, the argumentation literature not only restricts itself to looking at epistemics, and to looking at their semantics in terms of degrees of force on a scale, but also prevalently dwells on the middle values of the scale, represented by *probably* and by expressions that are seen as occupying the same area of the scale (e.g. *presumably*, parenthetical *I think*) and are therefore dealt with as functionally synonymous in the mono-dimensional scalar view. Possibility terms receive cursory attention[74], while necessity modals are evoked mostly when speaking of logical validity as a counterpart of probabilistic or presumptive reasoning. In contrast, the following chapters will not tackle directly the (formidable) issue of the semantics of probability expressions, focusing instead on the exploration of an ample gamut of modal meanings whose basic quantificational force can be analyzed either as possibility or necessity.

References

Anscombre J-C, Ducrot O (1983) L'argumentation dans la langue. Mardaga, Bruxelles

Aquinas T St (2006) Summa contra gentiles. In: Alarcón E (ed) Corpus Thomisticum. St. Thomae de Aquino Opera omnia. Fundación Tomás de Aquino 2000–2013. http://www.corpusthomisticum.org/. Accessed 5 Oct 2016

Aristotle (2014). The complete works of Aristotle: the revised Oxford translation, one-volume digital edition (ed) Barnes J. Princeton University Press, Princeton (originally published in 1984 as 2 volumes)

Asher N, Lascarides A (2003) Logics of conversation. Cambridge University Press, Cambridge

Austin JL (1946) Other minds. Suppl Proc Aristotelian Soc 20:148–187

Austin JL (1962) How to do things with words. Harvard University Press, Cambridge, MA

Bermejo-Luque L (2011) Giving reasons. A linguistic-pragmatic approach to argumentation theory. Springer, Dordrecht

Black M (1967) Probability. In: Edwards P (ed) The encyclopedia of philosophy. Macmillan and Free Press, New York, pp 464–479

Blakemore D (1990) Performatives and parentheticals. In: Proceedings of the Aristotelian Society, vol 91. Aristotelian Society, Wiley, pp 197–213

Blakemore D (1997) On non-truth conditional meaning. In: Rolf E (ed) Pragmatik. VS Verlag für Sozialwissenschaften, Wiesbaden, pp 92–102

Blakemore D (2006) Divisions of labour: the analysis of parentheticals. Lingua 116(10):1670–1687. doi:10.1016/j.lingua.2005.04.007

Blanché R (1970) La logique et son histoire: d'Aristote à Russell. Armand Colin, Paris

Boethius S (2005) De consolatione philosophiae. Opuscula theologica. Moreschini C (ed). Saur Verlag, München/Leipzig

[74] But see the brief interesting remarks of Pinto (2007) on the role of possibility terms in argumentative discussions that can be compared with Toulmin's views on presenting a hypothesis as worth considering and Swanson's (2006, 2011) speech act of raising a possibility.

Brockriede W, Ehninger D (1960) Toulmin on argument: an interpretation and application. Q J Speech 46(1):44–54
Carnap R (1950) Logical foundations of probability. University of Chicago Press, Chicago
Conte, M-E (2010[1995]) Epistemico, deontico, anankastico. In: Venier F, Proietti D (eds) Vettori del testo. Carocci, Roma, pp 151–159 (Originally published in From pragmatics to syntax. Modality in second language acquisition, eds. Anna Giacalone Ramat and Grazia Crocco Galèas, 3–9. Tübingen: Gunter Narr)
Dascal M (2003) Interpretation and understanding. John Benjamins, Amsterdam/Philadelphia
Doherty M (1987a) Epistemic meaning, vol 21. Springer, Berlin
Doherty M (1987b) Perhaps. Folia Linguistica 21(1):45–65
Ducrot O (1984) Le dire et le dit. Les éditions de minuit, Paris
Ducrot O (1993) A quoi sert le concept de modalité? In Dittmar N, Reich A (eds) Modality in language acquisition/Modalité et Acquisition des Langues, ed. Mouton de Gruyter, Berlin/New York, pp 111–129
Egan A, Weatherson B (2011) Epistemic modality. Oxford University Press, Oxford
Ennis RH (2006) Probably. In: Hitchcock D, Verheij B (eds) Arguing on the Toulmin model. Springer, Dordrecht, pp 145–164
Fait P (2004) Aristotle on a puzzle about logical consequence: necessity of being vs. necessity of saying. Topoi 23(1):101–112
Faller MT (2002) Semantics and pragmatics of evidentials in Cuzco Quechua. Ph.D dissertation. Stanford University, Stanford
Freeman JB (1988) Thinking logically. Basic concepts for reasoning. Prentice Hall, Englewood Cliffs
Freeman JB (1991) Dialectics and the macrostructure of arguments. A theory of argument structure. Foris Publications, Berlin
Freeman JB (2011) Argument structure: representation and theory. Springer, Amsterdam
Freeman JB (2013) What types of arguments are there?. In: Mohammed D, Lewiński M (eds) Virtues of argumentation. Proceedings of the 10th international conference of the Ontario Society for the Study of Argumentation (OSSA), Windsor (ON), pp 1–15
Frege, Gottlob. 1960. Translations from the philosophical writings of Gottlob Frege. In: Geach P, Black M (eds). Blackwell, Oxford
Goldberg AE (1995) Constructions: a construction grammar approach to argument structure. University of Chicago Press, Chicago
Grice HP (1989) Studies in the way of words. Harvard University Press, Cambridge, MA
Hacquard V, Wellwood A (2012) Embedding epistemic modals in English: a corpus based study. Semant Pragmatics 5(4):1–29
Hare RM (1970) Meaning and speech acts. Philos Rev 79(1):3–24
Hengeveld K (1988) Illocution, mood and modality in a functional grammar of Spanish. J Semant 6:227–269
Higginbotham J (2003) A note on conditionals and compositionality. USC Work Pap Linguist 1:63–69
Hintikka J (1973) Time and necessity. Oxford University Press, Oxford
Hitchcock D, Verheij B (2006) Arguing on the Toulmin model: new essays in argument analysis and evaluation, vol 10. Springer, Berlin
Hooper JB (1975) On assertive predicates. In: Kimball J (ed) Syntax and semantics. Academic, New York, pp 91–124
Houtlosser P (1994) The speech act "advancing a standpoint." In: van Eemeren FH, Grootendorst R (eds) Studies in Pragma-Dialectics. SicSat, Amsterdam, pp 166–171
Houtlosser P (2002) Indicators of point of view. In: ed. van Eemeren FH (ed) Advances in Pragma-Dialectics. SicSat, Vale Press, Amsterdam, pp 169-184
Ifantidou-Trouki E (1993) Sentential adverbs and relevance. Lingua 90(1-2):69–90
Kay P (1997) Words and the grammar of context. CSLI Publications, Stanford
Kay P (2004) Pragmatic aspects of grammatical constructions. In: Horn L, Ward G (eds) The handbook of pragmatics. Blackwell, Oxford, pp 675–700

Kneale WC (1949) Probability and induction. Oxford University Press, Oxford
Knuuttila S (1982) Modal logic. In: Kretzmann N, Kenny A, Pinborg J, Stump E (eds) The Cambridge history of later medieval philosophy. Cambridge University Press, Cambridge, pp 342–357
Knuuttila, Simo. 2013. Medieval theories of modality. In: Zalta EN et al. (ed) The Stanford encyclopedia of philosophy. Standford University Press, Stanford. http://plato.stanford.edu/archives/fall2015/entries/modality-medieval. Accessed 5 Oct 2016
Kratzer A (1981) The notional category of modality. In: Eikmeyer H-J, Rieser H (eds) Words, worlds, and contexts: new approaches in word semantics. Mouton de Gruyter, Berlin/New York, pp 38–75
Kratzer A (2012) Modals and conditionals. Oxford University Press, Oxford
Kronning H (1996) Modalité, Cognition et Polysémie: Sémantique du Verbe Modal Devoir; Acta Universitatis Upsaliensis, Studia Romanica Upsaliensia. Uppsala Univ, Uppsala
Kronning H (2014) La théorie modale de la polyphonie et les constructions conditionnelles en si. Langages 193(1):17–31
Lakoff G (1973) Hedges: a study in meaning criteria and the logic of fuzzy concepts. J Philos Logic 2(4):458–508
Langacker RW (1991) Concept, image, and symbol: the cognitive basis of grammar. Mouton de Gruyter, Berlin/New York
Lewis D (1973) Counterfactuals. Blackwell, Oxford
Lukasiewicz J (1957) Aristotle's syllogistic from the standpoint of modern formal logic, 2nd edn. Clarendon, Oxford
Lyons J (1977) Semantics. Cambridge University Press, Cambridge/London
Merlini L (1983) Gli atti del discorso economico: la previsione. Status illocutorio e modelli linguistici del testo inglese. Zara, Parma
Merlini Barbaresi L (1987) 'Obviously' and 'certainly': two different functions in argumentative discourse. Folia Linguistica 21(1):3–24. doi:10.1515/flin.1987.21.1.3
Miecznikowski J, Rocci A, Zlatkova G (2012) L'argumentation dans la presse économique et financière italienne. In: Gautier L (ed) Les discours de la bourse et de la finance. Frank und Timme, Berlin, pp 65–83
Mœschler J (1992) The pragmatic aspects of linguistic negation: speech act, argumentation and pragmatic inference. Argumentation 6(1):51–76
Nauze FD (2008) Modality in typological perspective. Institute for Logic, Language and Computation, Amsterdam
Nølke H, Fløttum K, Norén C (2004) ScaPoLine. La Théorie Scandinave de la Polyphonie Linguistique. Kimé, Paris
Nuyts J (2001a) Epistemic modality, language, and conceptualization: a cognitive-pragmatic perspective. John Benjamins, Amsterdam/Philadelphia
Nuyts J (2001b) Subjectivity as an evidential dimension in epistemic modal expressions. J Pragmatics 33(3):383–400
Papafragou A (ed) (2000) Modality: issues in the semantics-pragmatics interface. Current research in the semantics/pragmatics interface. Elsevier, Amsterdam
Papafragou A (2006) Epistemic modality and truth conditions. Lingua 116(10):1688–1702
Pinto RC (2001) Generalizing the notion of argument. In: Argument, inference and dialectic. Collected papers on informal logic. Kluwer, Dordrecht, pp 10–20
Pinto RC (2007) Evaluating inferences: the nature and function of warrants. In: Hitchcock D, Verheij B (eds) Arguing on the Toulmin model. Springer, Dordrecht, pp 115–143
Portner P (2009) Modality. Oxford University Press, Oxford
Recanati F (1989) The pragmatics of what is said. Mind Lang 4(4):295–329
Recanati F (2009) Compositionality, flexibility, and context-dependence. In: Markus Werning, Wolfram Hinzen, and Edouard Machery (eds) Oxford handbook of compositionality. Oxford University Press, Oxford, pp 1–29
Rescher N (2006) Presumption and the practices of tentative cognition. Cambridge University Press, Cambridge/New York

Rigotti E (2005a) Congruity theory and argumentation. Argumentation Dialogic Interact Spec Issue Stud Commun Sci 5:75–96
Rigotti E. (2005b). Towards a typology of manipulative processes. In de Saussure L, Schulz P (eds) New perspectives on manipulative and ideological discourse in pragmatics and discourse analysis. 'Discourse approaches to politics, society and culture' John Benjamins, Amsterdam
Rigotti E, Greco Morasso S (2010) Comparing the Argumentum Model of Topics to other contemporary approaches to argument schemes: the procedural and material components. Argumentation 24(4):489–512
Rigotti E, Rocci A (2001) Sens – non-sens – contresens. Studies in Communication Sciences 2:45–48
Rocci A (2005) La modalità epistemica tra semantica e argomentazione. I.S.U Università cattolica, Milano
Rocci A (2007) Epistemic modality and questions in dialogue. The case of Italian interrogative constructions in the subjunctive mood. In: de Saussure L, Puskas G, Moeschler J (eds) Tense, mood and aspect: theoretical and descriptive issues. Cahiers Chronos vol 17. Rodopi, Amsterdam/New York, pp 129–153
Rocci A (2008a) Modality and its conversational backgrounds in the reconstruction of argumentation. Argumentation 22:165–189. doi:10.1007/s10503-007-9065-8
Rocci A (2008b) Modals as lexical indicators of argumentation: a study of Italian economic-financial news. L'analisi Linguistica e Letteraria. Spec Issue Word Mean Argumentative Dialogue 16(2):577–619
Ryle G (1937) Categories. Proceedings of the Aristotelian Society 38:189–206
Ryle G (2009) The concept of mind. Routledge, New York
Searle JR (1969) Speech acts – an essay in the philosophy of language. Cambridge University Press, Cambridge
Simon-Vandenbergen A-M, Aijmer K (2007) The semantic field of modal certainty. Berlin/New York, Mouton de Gruyter
Snoeck Henkemans F (1992) Analysing complex argumentation. SicSat, Amsterdam
Snoeck Henkemans F (1997) Analysing complex argumentation: the reconstruction of multiple and coordinatively compound argumentation in a critical discussion. SicSat, Amsterdam
Sorensen R (2014). Epistemic paradoxes. In: Zalta EN (ed) The Stanford encyclopedia of philosophy. http://plato.stanford.edu/archives/spr2014/entries/epistemic-paradoxes/
Swanson E (2006) Interactions with context. MIT, Cambridge, MA
Sweetser E (1990) From etymology to pragmatics: metaphorical and cultural aspects of semantic structure. Cambridge University Press, Cambridge/New York
Talmy L (1985) Force dynamics in language and thought. In: Eilfort WH, Kroeber PD, Peterson KL (eds) Papers from the parasession on causatives and agentivity. Chicago Linguistic Society, Chicago, pp 293–337
Thomasson AL (2002) Phenomenology and the development of analytic philosophy. Southern J Philos 40(S1):115–142. doi:10.1111/j.2041-6962.2002.tb01926.x
Toulmin SE (1956) Probability. In: Flew A (ed) Essays in conceptual analysis. Macmillan, London, pp 157–191
Toulmin SE (2003) The uses of argument. Cambridge University Press, Cambridge
Toulmin SE, Rieke R, Janik A (1984) An introduction to reasoning. Macmillan, New York
Tseronis A (2009) Qualifying Standpoints: Stance adverbs as a presentational device for managing the burden of proof. LOT, Utrecht
Urmson JO (1956) Parenthetical verbs. In: Flew A (ed) Essays in conceptual analysis. Macmillan, London, pp 192–212
van der Auwera J (1999) On the semantic and pragmatic polyfunctionality of modal verbs. In: Turner K (ed) The semantics/pragmatics interface from different points of view. Elsevier, Amsterdam, pp 49–64
van Eemeren FH (2010) Strategic maneuvering in argumentative discourse. John Benjamins, Amsterdam/Philadelphia

van Eemeren FH, Grootendorst R (1984) Speech acts in argumentative discussions. Foris Publications, Dordrecht

van Eemeren FH, Grootendorst R (2004) A systematic theory of argumentation: the Pragma-Dialectical approach. Cambridge University Press, Cambridge

van Eemeren FH, Grootendorst R, Snoeck Henkemans F (1996) Fundamentels of argumentation theory. a handbook of historical backgrounds and contemporary developments. Erlbaum, Mahwah

van Eemeren FH, Houtlosser P, Snoeck Henkemans F (2007) Argumentative indicators in discourse a pragma-dialectical study. Springer, Amsterdam

von Wright GH (1951) Deontic logic. Mind 60(237):1–15

Venier F (1991) La modalizzazione assertiva: avverbi modali e verbi parentetici. Materiali linguistici. F. Angeli, Milano

Vorobej M (2006) A theory of argument. Cambridge University Press, Cambridge

Walton D, Krabbe E (1995) Commitment in dialogue: basic concepts of interpersonal reasoning. State University of New York Press, New York

Woods J (2010) Defeasible reasoning. In: Reed C, Tindale CW (eds) Dialectics, dialogue and argumentation. An examination of Douglas Walton's theories of reasoning and argument. College Publications, London, pp 239–261

Chapter 4
Relative Modality and Argumentation

Abstract The chapter draws on studies of the semantics of modality in the formal semantics tradition. Moving from the need of restricting the domain of quantification in a possible worlds semantics of modality, it presents a relational context-dependent semantics of modality inspired by the theory of Relative Modality. For Relative Modality modal expressions are *relational* and *context-dependent*. This approach is immediately applied to the analysis of conditionals and to the modeling of anaphora and rhetorical relations in discourse. The chapter demonstrates the potential of this approach for shedding light on the role of modal constructions as argumentative indicators. The vagueness and ad hoc-ness of modal conversational backgrounds plays an important role in the basic task of *determining the propositional commitments of the arguers* and for the potential for fallacy inherent in that vagueness. The contextual specification of modal conversational backgrounds is comparable to specifying the domain of *quantifiers*. Specifying the exact contribution of context dependent modal meanings to propositional content can impact, in turn, further aspects of the reconstruction of an argumentative discussion, notably including the semantic type of the propositions functioning as standpoints, the *loci*-relations supporting the inference from the premises to the standpoint, and the macrostructure of the argument.

4.1 Modal Semantics and Its Argumentative Implications

This chapter presents an approach to the semantics of modality and demonstrates through the discussion of illustrative examples its potential for shedding light on the role of modal constructions in arguments. This general semantic approach will be the foundation on which I will build the typology of modalities presented in Chap. 5, as well as the specific semantic analyses of Italian modal constructions of Chap. 6.

The basic ideas of this semantic approach are not original; they were introduced by Angelika Kratzer in the late 1970s and developed over the decades in a series of publications (Kratzer 1977, 2012). This approach is known as the theory of *Relative Modality* and remains one of the most influential logic-based approaches to modal meanings. Since its inception, Relative Modality has given prominence to the role

A. Rocci, *Modality in Argumentation*, Argumentation Library 29,
DOI 10.1007/978-94-024-1063-1_4

of context and discourse[1] and, over the years, its key ideas have contributed to formal approaches to the meaning of modality in discourse context (cf. Roberts 1989; Frank 1996; Geurts 1999).

Crucially, Relative Modality treats modal expressions as *relational* and *context-dependent*. This contributes to explaining how context enters into the determination of the myriad subtly different "flavors" of modality that can be manifested by *polyfunctional* modal terms (cf. van der Auwera 1999) such as *possible, necessary*, *can* or *must*. So that we manage to avoid postulating what would be a "monstrous" ambiguity in the semantics of these words (cf. Lycan 1994: 196).

This strategy of semantic analysis can deal satisfactorily with a number of issues that have emerged in Chap. 3 regarding the nature of modality, its place in arguments and the indicative potential of modal expressions. The key construct provided by Relative Modality to account for the contextual determination of modal 'flavors' is the *conversational background*. This notion provides a satisfying model of Toulmin's different modal "criteria" or "grounds" (cf. Chap. 3, § 3.2.2). The major flavors of modality recognized in the philosophical, logical and linguistic traditions can be accounted in terms of types of conversational backgrounds or combinations of conversational backgrounds. Relative Modality offers a *conditional* view of the modals, in the tradition of the Medieval notion of *necessitas consequentiae* (cf. Chap. 3, § 3.4.2.3). The same theoretical machinery that accounts for the variety of flavors, provides a solid account of the relationship of modality with *conditional structures*, which has been one of Kratzer's goals since the inception of the theory (cf. Kratzer 1978). The close relationship between modals and warrants, which emerged in Chap. 3 can be provided a better elucidation thanks to this unified approach to the semantics of modals and conditionals.

I will present, in § 4.2, a very basic and informal version[2] of the theory. I have little to contribute to its formal machinery, but the presentation prepares the terrain to the discussion of the implications of modal context dependency for argumentative reconstruction (§§ 4.3.1, 4.3.2, 4.3.3, 4.3.4, 4.3.5) and the view of relational modals as indicators of argumentative discourse relations (§ 4.3.6), which are the two core contributions of this chapter from the viewpoint of argumentation theory.

In Chap. 5, this informal version of Relative Modality will be the starting point of a procedural account of the emergence of the different interpretations of context-dependent modals from discursive and contextual saturation. The conceptual counterpart of this process of contextual saturation of conversational backgrounds is a typology of conversational backgrounds balancing logic and linguistic-pragmatic criteria. The resulting cartography of modal "flavors" will allow to investigate the interaction of modality and argumentation at a new level, with a special focus on the inferential aspects of argumentation. For the major nodes of the typology, I will also

---

[1] It is telling that Kratzer's doctoral thesis, published in German in 1978, was entitled *Semantik der Rede: Kontexttheorie, Modalwörter, Konditionalsätze* ("Discourse semantics: theory of context, modal words and conditional sentences").

[2] On the stance taken in this book with respect to formalization see Chap. 2, § 2.3.3.

propose a relational semantic analysis of possibility and necessity modals, inspired by Relative Modality, but recast into the semi-formal language of Congruity Theory.

4.2 The Theory of Relative Modality: From Possible Worlds to Discourse Structure

In this section I will present and critically discuss the theory of Relative Modality. The main purpose of this exercise is to lay down the conceptual groundwork not only for the discussions of the role of modality in arguments in this chapter and in the following, but also for the specific semantic analyses of Italian modal constructions presented in Chap. 6.

I will start with basic ideas about the structure of modal meanings which Relative Modality shares with the standard Kripkean approach to the semantics of modal operators as quantifiers over possible worlds (which is also called *frame semantics*). Then I will examine how the construct of *conversational background* differs from the key construct of the standard frame semantics (i.e. the *accessibility relation*) and why the Relative Modality approach has been characterized as a *premise semantics*, providing the general definitions of possibility and necessity relative to a conversational background (§ 4.2.7), on which all the semantic analyses of Chap. 6 will be also based. Furthermore, I will focus on Relative Modality's unified analysis of modals and conditionals (§ 4.2.8), which will turn out to be essential for dealing with the argumentative functioning of Italian conditional mood modal verbs in Chap. 6.

Finally, I will examine what the discourse semantic extensions of Relative Modality have to say on the anaphoric relations that are established in discourse in the process of saturation of the conversational background (§ 4.2.9). As it will become progressively clear in this and in the following chapter, this mechanism of anaphoric saturation of the conversational background turns out to play a crucial role in the functioning of modal expressions as indicators of argumentative discourse relations.

4.2.1 Alternatives and Domains of Quantification

In Chap. 2, modality was intuitively characterized through its connection with alternative situations. The standard way of treating alternative situations in modal logic and formal linguistic semantics is adopting the construct of *possible world*. A possible world is a complete situation, determined in every respect. We may think of a possible world as a maximal consistent set of propositions, the propositions that are true in that world (Peregrin 1993). For any proposition, either the proposition is true in the world or it is false (and its negation is true). Since the world is fully determined, adding any further propositions would make the set inconsistent. Other approaches do not treat alternatives as complete, but rather as sketchy, partial or schematic representations of situations[3], which can be incremented as more information is added in discourse. This book will not enter a discussion of the merits of the standard possible worlds approach versus these alternatives.

More precisely, modal meanings have been characterized as involving the *quantification* over some set of relevant alternative situations. Consider the following examples:

(1) Sarah Palin *could* have been vice-president of the United States of America

(2) Silvio Berlusconi *had* to resign

In (1) the prejacent proposition is presented as a strictly counterfactual (contrary to fact) state of affairs. This, however, is not the main point of the utterance. The point of the utterance is to communicate that there are – contrary to fact but envisageable – situations where this state of affairs is realized. Example (2), in contrast, presents a factual prejacent. And, again, factuality is not the point of the utterance, the point being the indication that there were no relevant alternatives to Berlusconi's resignation. In other words: all the relevant alternative scenarios involved Berlusconi resigning. Thus, the modals can be compared to *quantifiers*. The modal verb *could* in (1) is analogous to an *existential* quantifier, while *had to* in (2) is a *universal* quantifier. As it is well known, the structural parallelism between modalities and quantifiers clearly appears from their behavior with respect to negation, beautifully captured by the Square of Opposition (cf. *Introduction* § 1.6.1).

As it happens with other kinds of quantifiers, the quantification operated by the modals risks being utterly meaningless until we specify a *domain* of quantification. An uttered token of a sentence like (3)

(3) *Everybody* is having a good time (Stalnaker 1991)

cannot be really understood and given truth-conditions until we somehow restrict the domain of the quantification: the utterance could mean that everybody in a specific party or everybody in a specific group of hikers on a hiking trip is having a good time, while it is extremely unlikely that the utterance is taken to mean that everybody in the universe is having a good time (cf. Stalnaker 1991). Similar consideration hold for our modal examples. What is the relevant set of alternatives that I have in mind when I say that there is an alternative (albeit a counterfactual one) where Sarah Palin is vice-president? Selecting the relevant set is necessary to make my utterance meaningful and evaluable as true or false. Suppose that my interlocutor reacts to my statement and the following bizarre dialogue ensues:

---

[3]The notions of *text world* (Werth 1999) and *mental space* (Fauconnier 1994) in Cognitive Linguistics and the *situations* of Situation Semantics (Barwise and Perry 1983) are some examples of these partial, incomplete, worlds or situations.

(4) A: Sarah Palin *could* have been vice-president of the United States of America. Just think of it.
B: Tom Cruise and John Travolta could as well.
A: Well, no. Neither of them was nominated.
B: I mean: they're natural born American citizens and they're over 35.
A: OK. That's another matter.

It is clear that A and B implicitly restrict the set of relevant alternatives in different ways. A considers only the alternatives that are compatible with how the major parties nominated the tickets that would run for the presidency (the alternatives that could turn out to be true given the actual tickets for the presidency). For B, the range of alternatives is much broader and includes all the legally eligible individuals that could hypothetically have run for the vice-presidency. Once the two participants make their domain selection partially explicit their apparent disagreement disappears.[4] Both A's and B's construal of counterfactual possibility elect to consider only certain facts and to ignore other facts that would rule out the prejacent as a viable alternative once added to our information base. Both A and B choose to ignore the actual outcome of the election, but B's construal ignores much more than that, focusing on the legal precondition for eligibility.

The situation with Berlusconi's resignation is similar. We cannot tell whether the resignation was actually the only course of action available to him until we select the range of alternatives that we are prepared to take into consideration. The following example provides a context that makes the domain selection partially explicit:

(5) When the BTP-Bund spread reached 574 base points, Silvio Berlusconi *had* to resign. Otherwise, the financial doom of the country would have been sealed.

According to the reading suggested in (5), the modal quantifies over the courses of action compatible with Italy's financial stability – or, at least, those that do not involve Italy's imminent bankruptcy – and tells us that all these alternative courses of action involved Berlusconi's resignation. This is different than saying that resignation was, say, the only *legally* available option. To this someone could retort that as long as Berlusconi had a parliamentary majority, he *could* have remained in office.

These examples illustrate how modals, like quantifiers, need a context-dependent semantics to restrict their domain of quantification. This basic fact already has significant consequences for argumentative reconstruction, which are discussed below in § 4.3.1.

[4] Such divergences and shifts in modal domain selection are discussed in David Lewis' famous essay on "scorekeeping" in a language game. I will discuss the consequences of this phenomenon for argumentative confrontations in § 4.3.5.

4.2.2 *Modals Are Relational*

Both quantifiers and modals rely on the context of utterance to restrict their domain of quantification. In formal semantics it has been proposed (Partee 1991a, b) that a *domain restrictor* slot forms an essential component of the semantic structure of modal expressions, noun phrase quantifiers (e.g. *almost every*), adverbial quantifiers (e.g. *usually*) and other kinds of structures, including topic-focus articulation. Barbara Hall Partee proposed a general scheme, called a *tripartite structure*, to capture the parallelism between all these semantic structures (Fig. 4.1):

Consider the following Italian example – from the corpus of Italian business and financial news that I have exploited for the case studies expounded in Chap. 6:

(6) I rifiuti solidi urbani – secondo il piano del Governatore Vendola – **dovranno** [*dovere*.IND.FUT.3PL] essere riciclati al 55% entro il 2010 [...] (*Il Sole 24 Ore*, April 14, 2006)

'Urban solid waste – according to Governor Vendola's plan – **will have to** be recycled to 55% by 2010.'

The phrase *according to the governor's plan* in (6) explicitly provides the restriction of the quantification operated by the Italian necessity modal *dovere* ('must', 'have to'): the only alternatives considered are those that are in accordance with the measures contained in the governor's plan. Suppose that the actual situation of the recycling of urban solid waste turned out to be very different by 2010, with only 15% of

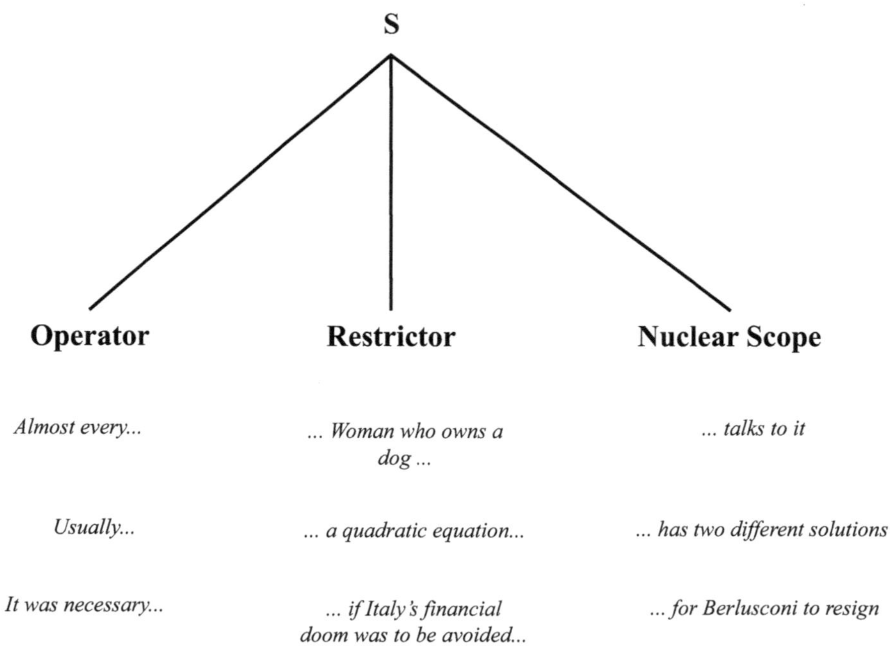

Fig. 4.1 Tripartite structures (adapted from Partee 1991a, b)

recycling. This would not affect the truth of (6) if it is due to *things not going according to plan*. This would place these alternatives outside the domain of quantification.[5] Kratzer (2012 [1977])[6] pays special attention to phrases such as *in view of the law*, *in view of what is known*, *in view of what is good for us*, which provide a syntactic manifestation of the domain restrictor of modal verbs such as *must* or *can*. Kratzer's *in view of* phrases are semantically analogous to the adverbial phrases with the present participle (e.g. *the by-laws being as they are*) which Toulmin used to make explicit what he calls the *criteria* of the modal, one of the two essential components of modal meanings (cf. Chap. 3, § 3.2.2, 3.2.3, 3.2.4, 3.2.5, 3.3).

Also for Kratzer these phrases do not amount to an optional modification of modal sentences, but rather manifest a constitutive component of their semantic structure. When this component is not manifested syntactically, which is most of the times, in order to interpret the modal we need to supply it from contextual information. This means that

> […] relative modal phrases like *must in view of* and *can in view of* should be considered as representing the semantic core of the modals *must* and *can* respectively. Modals are inherently relational. To be semantically complete, a modal requires two arguments: a restriction and a scope. The restriction may be represented overtly or may be provided by the context of utterance. (Kratzer 2012 [1977]: 8).

As we will see, the simple idea that modal expressions should not be understood as monadic operators on propositions but as relational predicates with the structure illustrated in Fig. 4.2 is extremely rich in consequences for our understanding of the role of modal expressions in arguments and their potential as argumentative indicators.

4.2.3 *Modals Are Context-Dependent Expressions Whose Valence Is Saturated in Context*

The relational hypothesis also has interesting consequences for another kind of questions, concerning the so-called *semantics-pragmatics interface*. In fact, Kratzer's hypothesis implicates a particular way of understanding the way in which

---

[5] Just like the fact that 'my aunt, who does not own dogs, does not speak to dogs' does not falsify the statement *Every woman that owns a dog talks to it*.

[6] Angelika Kratzer first presented what became later known as the theory of relative modality in Kratzer (1977) and further developed it in Kratzer (1981). Substantially revised versions of these two classic papers have been recently republished in Kratzer (2012), together with other essays that expand her theory of modality and conditionals. Since I'm not interested here in reconstructing the history of the theory of relative modality I will normally cite and quote from Kratzer (2012), specifying in square brackets the date of the original article. Occasionally, I will refer to the original when I need to quote a passage that has been omitted from the revised version. Another presentation of the theory of relative modality is Kratzer (1991), not reproduced in Kratzer (2012). Despite being largely a summary of Kratzer (1981, 1991) contains some passages that are illuminating in view of our purposes and is therefore quoted in this Chapter.

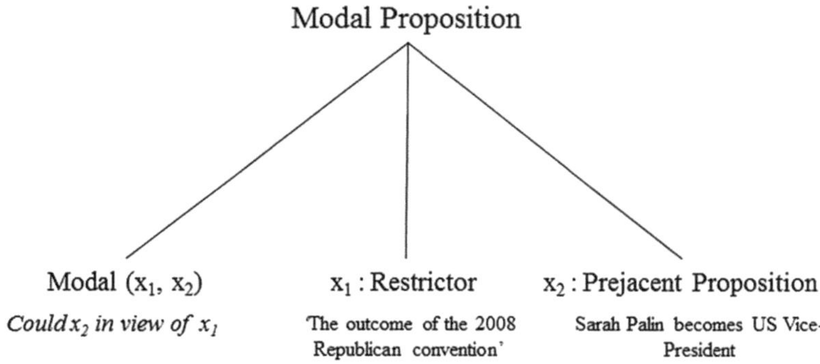

Fig. 4.2 The relational semantics of a modal viewed as a tripartite structure. The relational modal predicate appears as operator, while the nuclear scope corresponds to the prejacent proposition (Note that the line diagram here indicates *constituency* relations at the level of logical form, and not semantic *dependency* relations as customary in Congruity Theory graphs)

linguistically encoded and contextual information interact in the interpretation of the modals.

The restriction of a modal is not tacit or implicit in the same way as a *conversational implicature* is. Rather it can be roughly compared to the process by which, in a context like (7), the linguistically manifested content is enriched by supplying a contextually suitable entity to saturate the second argument place in the semantic frame of the predicate 'to eat' (x_1, x_2).

(7) A: Would you like a sandwich?
B: No, thanks. I've *eaten.*

While we don't know exactly what B has eaten we first of all establish that she has ingested something and that this something is edible solid food – not fresh water, or, say, the pebbles of the garden path (cf. also Groefsema 1995b). In Congruity Theory, this bit simply comes out of the need of saturating the semantic valence of the predicate *to eat* and to saturate it *congruently,* that is, in accordance to the presuppositions that characterize its x_2 argument slot. The discourse context of the utterance further restricts the way in which the slot is saturated. In Congruity Theory one can represent the influence of the discourse context in terms of congruity requirements imposed by the *connective predicates*[7] corresponding to the illocutionary functions and rhetorical relations characterizing the utterance. In this case the connective predicate involved is an argumentative one. For B's second sentence to function as an *argument* justifying her refusal x_2 has to be seen as something approximating quantitatively and qualitatively a full meal. We don't know whether it was a tuna sandwich, a salad or a five course gourmet meal, but we exclude, for instance, the eventuality that it was one cherry, or even several mint flavored candies. We can summarize the congruity requirements that guide the interpreter's inferences in (7) through the schema in Fig. 4.3.

[7] The congruity theoretic notion of a *connective predicate* is introduced in Chap. 2, § 2.2.3.2.

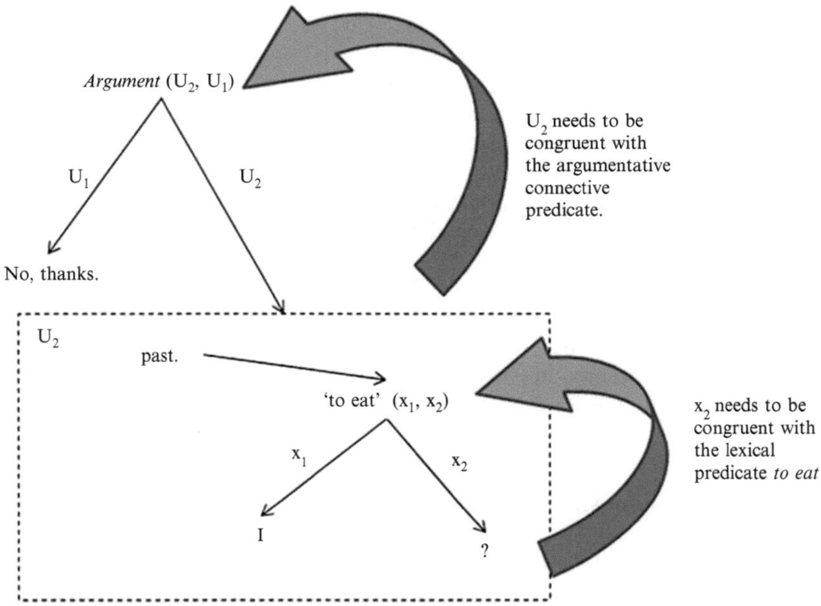

Fig. 4.3 Congruity requirements guiding the saturation of an unexpressed argument. The simplified representation of the argumentative connective predicate does not include the *Spk* and *Hr* argument places, which are not directly relevant here

Adopting Kratzer's hypothesis on the semantic structure of the modals leads us to see them as constitutively context-dependent and to see the narrowing down of their interpretation as a process of *saturation* guided by the presuppositions of their argument frames taking place in discourse context.

I will later return to discuss more in detail the process saturation of context dependent expressions in discourse as I see it as the main discourse-pragmatic phenomenon through which modal expressions become argumentative indicators.

4.2.4 Modal Restrictors Are ad hoc *and Vague*

Pausing on the inferential recovery of modal restrictors in discourse context leads us also to appreciate how *specific* and how *vague* they can be *at the same time*. The philosopher William G. Lycan offers a poignant illustration of this apparently paradoxical remark. According to Lycan (1994: 171) not only "everyday modalities" are contextually restricted, but "few of even those street-level restriction classes themselves correspond to recognizable philosophical categories". For instance, Lycan observes, "the concept of *natural* or *physical* necessity is hardly better known to everyday English than are the purely alethic modalities" (*Ibid.*). Lycan then proceeds to exemplify "real restriction" classes through an informal analysis of modal expressions extracted from a short dialogue passage in a novel. Here is a part of his sample (Lycan 1994: 173):

(8) *I have* to see you Ira. As soon as *possible*. I'm going out to the beach house this evening at six o' clock to talk to a man who's building a new jetty. I'll surely finish with him in a half-hour or so. *Could* you meet me out there at seven o' clock? [...]

(9) Ira, I *can't* talk any more. There are people in the house.

(10) Wait, wait, wait! – he insisted. I just want to clarify this one point. It boils down to 'have to'. *Could* you *possibly* be a little more precise?

In examining the passage Lycan ignores epistemic modalities like *surely* in (8) and concentrates on expressions such as those put in italic in the above examples, which he qualifies as "squarely alethic" (*Ibid.*). In view of the cartography of modal 'flavors' that will be presented in Chap. 5, § 5.2.4 it is worth taking good notice of the seemingly confusing terminology here.

4.2.4.1 An Aside on Alethic Modality

When he suggests that "purely alethic modalities" are extremely rare in conversational English, Lycan appears to refer here to modalities that are unrestricted or, better, that are restricted only by logical axioms. So that *necessity* is attributed to the axiom themselves as well as to the derived logical truths[8], and to tautologies, *impossibility* corresponds to contradiction and *possibility* is attributed to all non-self-contradictory statements.

A slightly broader view of "purely alethic necessity" could involve also the semantic axioms, so that a *ten minutes day* and Toulmin's *male sister* are recognized as inherently contradictory impossible entities. According to this view, alethic necessity and alethic impossibility corresponds to the *broadly logically determined* statement in Freeman's typology (cf. Chap. 2, § 2.3.3).[9] When dealing with natural language the boundaries between logical and semantic necessity is hardly clear cut, because, as insightfully observed by McCawley (1993: xviii), the distinction between logical words, deserving special analysis, and content words that can be safely substituted with letters, has often been largely a matter of habit.[10]

[8] In formal systems of modal logic the necessary status of logical theorems is captured by the *necessitation rule*, which establishes that if α is theorem of the system $\Box\alpha$ is also a theorem in the system (Garson 2014).

[9] Cf. Freeman (2005: 114): "by necessary statements [...] we mean alethically or broadly logically necessary statements only. I believe we can identify five classes of necessary statements: formally true statements – statements true by virtue of their logical form, semantically true statements, conceptually true statements, mathematical statements and metaphysical statements".

[10] In introducing his well-known and highly original logic textbook for linguists James McCawley states: "I hold that all elements of meaning can play a role in inference and in truth conditions and that is only by historical accident that logicians have largely confined themselves to the study of the logical properties of comparatively few elements (those expressed by *and, or, not, if, all, some, may* and *must*). I thus reject the distinction that is usually drawn between "logical" and "nonlogical" vocabulary. [...] I have realized that my rejection of that distinction forces me to reject another

It is clear, however, that when Lycan (1994) just speaks of "alethic" modalities, he refers to a much broader class, encompassing *natural* or *physical* modalities, and the plethora of idiosyncratically restricted modalities he finds in his fictional dialogue excerpt. What all these alethic modalities have in common is that they deal with necessity and possibility in view of *facts* of a certain kind, states of affairs that are the case in the world. Following Lycan (1994) as well as the more linguistically and cognitively oriented work of Kronning (1996) I adopt here this large view of alethic modality. Another term to refer to these modalities could be *ontological*. Taken in this broad sense alethic modalities are one of the fundamental classes in a cartography of modalities in natural language (cf. Kronning 2001a, b, c).

4.2.4.2 The Circumstances of the Affair

The facts that restrict the modalities examined by Lycan (1994) are diverse. Readers of the novel would infer that in (8) the speaker's pregnancy is among the facts that make seeing her lover necessary (*have to*). It is, however, more complex than that. As Lycan (1994: 174) observes, it is not just the pregnancy but the character's reasons for informing her lover about it, and the modality is "(further) relativized to whatever psychological or social norms apply". The possibility of *as soon as possible* and *could you meet me* in (8), on the other hand, centers mostly on "both the norms and the circumstances" of the addressee (*Ibid.*), but in the first one cannot exclude that also the circumstances of the speaker are taken into account. In (9) *cannot* refers to the speaker's immediate circumstances (there are people in the house): "her talking any more would be (as things are) incompatible with privacy, or with discretion on her part, or with prudence" (*Ibid.*). As for the *could you possibly* in (10), it is clear that the question is about the addressee's "presumed constraints" but it is quite hard to figure out what is envisaged exactly.

After examining the above and several other occurrences Lycan (1994: 176) concludes:

> My purpose is only to indicate that everyday English is shot through with restricted alethic modalities whose restrictions are almost capriciously diverse, rarely aligned with any easily specifiable modal concept known to logicians, and irreparably vague – yet calculated on the spot by ordinary human speakers/hearers with hardly a conscious thought.

I agree with Lycan's remarks on the *ad hoc*-ness and vagueness of the restrictions on modal quantification in ordinary discourse. I have, however, a substantially more optimistic view on the possibility to capture more precisely the general functioning, if not the individual calculations, of these restrictions and on the utility of a broad network of modal concepts for this task. In this latter respect, some of the modalities that Lycan comments on are better characterized as *partially* alethic, rather than

---

standard distinction, namely, that between axioms and rules of inference on the one hand and meaning postulates on the other: meaning postulates [...] are axioms and rules of inference for "nonlogical vocabulary", and there is no more grounds for singling them out for special status than there is for distinguishing "logical" and "nonlogical" vocabulary" (McCawley 1993: xviii-xix).

"squarely" so. Their restriction is based in part on factual circumstances but also includes norms and agents' goals that are better kept distinct from facts in the analysis of the restrictor.

In order to have a better grasp of the functioning of these *ad hoc* and vague modalities and to relate them with a cartography of broad modal categories we need to look more in detail at how the restriction of quantification works.

4.2.5 Interlude: Context Dependency in Language

It is important to stress that the way in which the domain restrictors of the modal expressions are determined in context, "calculated on the spot" as Lycan (1994: 176) puts it, is not a unique, idiosyncratic, phenomenon, but corresponds to one of the most basic dynamics of the semantics-pragmatics interface, albeit one whose pervasiveness is often underestimated by scholars who do not focus on the interaction of linguistically encoded and contextual information in language understanding.

It is well known that the contribution of indexical words to content and extension is mediated by the context of utterance. Take, for instance, the personal pronouns *I* and *you*. Their linguistic meaning can be seen as a simple *instruction* or *procedure* (cf. Wilson and Sperber 2012: 165–166) referring to the context of utterance: something like 'Find the speaker of the utterance u_0' or 'Find the addressee of the utterance u_0'. Their linguistic meaning is *token-reflexive* that is it cannot be made explicit without referring to the occurrence itself of the expression in an utterance (cf. Recanati 2008: 55). To determine the truth-conditions expressed by an utterance of (11) I need to resolve the instructions with respect to the context of utterance. These instructions appears to be part of the encoded linguistic meaning of the pronouns.

(11) *I* love *you*

Indexical words like *I* and *you* are one tiny subset of the broader class of context-dependent linguistic expressions. For most context dependent expressions, the resolution of the instruction in context requires more than picking up the basic contextual parameters of the utterance. Demonstratives are a case in point:

(12) *This* is very good.

The instruction associated with *this* could be very roughly paraphrased as

'Find a *salient* entity in the *proximity* of the *origo* of u_0',

where saliency can be determined by pointing or by non-verbal ostensive behavior of the speaker, by subtler means of coordinating attention in interaction, or simply by the hearer's inferences based on aspects of the common ground of the interaction, and, in particular, on the interaction's joint goals. Proximity also could be specified along multiple dimensions, especially if we aim at a unified account of the properly deictic and of the anaphoric uses of the demonstrative.

Interestingly, the plural first person pronoun *we* is similar to the demonstrative in that it calls on the pragmatic inferences of the hearer to resolve its instruction. Kay (2004: 675) cites the following example from Nunberg (1993):

(13) *We*$_{\mathrm{i}}$ do not know much about this part of the brain, which plays such an important role in *our*$_{\mathrm{ii}}$ lives, but *we*$_{\mathrm{iii}}$ will see in the next chapter.

The instruction encoded by *we* ask the hearer to pick out a set of people containing the speaker, whose identity has to be inferred "on the basis of the common conversational background and the rest of the utterance" (Kay 2004: 674). Each token of the pronoun in (13) picks out a different set of people:

(i) the writer + the relevant scientific community
(ii) the writer + the rest of humanity
(iii) the writer + the reader

Following Nunberg, Kay observes that the parameter of context that is addressed by the instruction is the speaker, but the referent that is to be worked out by inference is something else: a certain set of people containing the speaker. Kay glosses the instruction encoded by *we* as "I and you-know-who else". It is important to stress that what we have here is an instruction *demanding* to identify "you-know-who else" and not a vague minimal meaning of the type "I and somebody else" which can be *optionally* made more precise in context. This is shown by the infelicitous substitutions in (14). On the other hand, the felicitous substitutions in (15) show that the point is precisely *finding* the relevant sets.

(14) ?? *I and somebody else* do not know much about this part of the brain, which plays such an important role in *my and somebody else's* life, but *I and somebody else* will see in the next chapter.

(15) *Scientists* do not know much about this part of the brain, which plays such an important role in *human* lives, but *the reader* will see in the next chapter.

Wilson and Sperber (2012: 166) see pronouns as procedural expressions constraining "the inferential phase of comprehension by reducing the hypothesis space that has to be searched in arriving at the interpretation" and, more specifically, guiding "the search for the intended referent" thus contributing to the propositional content expressed by the utterance.

Now, the class of context-dependent, or *lato sensu* deictic, linguistic expressions is much larger than personal pronouns and demonstratives. Kay (2004: 676) provides an excellent general characterization of the broad class of instructional expressions:

> A number of grammatical constructions have been described in which part or all of the meaning of the construction is analogous to the "you-know-who" part of the meaning of *we*, a virtual instruction to the addressee to examine the common ground of the conversation (along with other interpretive content of the sentence) to fill in some partially specified part of the intended interpretation.

This kind of "virtual instructions" are hardly limited to a small number of grammatical structures. As observed by Recanati (1989, 2009), there is a relatively large range of lexical expressions whose linguistic meaning is "gappy" and can be characterized as involving a free variable whose value has to be assigned contextually. These variables may correspond to individuals, to eventualities, but also to properties or relations. In short, the gap to be filled in context may concern any aspect of semantic content. According to Recanati (2009), adjectives like *large* or *small* involve a free higher order variable corresponding to the class providing a covert standard or term of comparison.

(16.a) Dumbo is a *small* elephant (= *small for an elephant*)

(16.b) Mickey is a *big* mouse (= *big for a mouse*)

(16.c) Dumbo doesn't stand a chance against Godzilla. He's only a small elephant (= *small compared to the radioactive monsters who came from outer space*) (adapted from Heim and Kratzer 1998).

In examples (16.a) and (16.b) the value of the variable is provided by the noun which the adjective modifies. Yet, this process cannot be simply analyzed in terms of syntactically guided compositionality within the noun phrase. Following Heim and Kratzer (1998), Recanati (2009) remarks that the utterance of (16.c) would be appropriate also in a context where Dumbo is pretty big for elephant standards, because the relevant standard has been shifted to that of the giant *kaiju* monsters of Japanese B-movies. At best, I might add, the noun provides a *default* context for the saturation of the variable, or maybe just one among several highly accessible contexts for saturation.

The noun *foreigner* provides a less well-known example of contextually saturated variable, which is quite striking in its simplicity:

(17) The survey also revealed that 27 per cent of locals did say they saw value in working with *foreigners*. They also agreed that *foreigners* tend to be less demanding at work, and better skilled (WebCorp).

We can see that the content and extension of the common noun *foreigners* in (17) cannot be determined until we find out the country with respect to which we are speaking – in the example it's Singapore. *Foreigner* is clearly to be analyzed as two-place relation. Its lexical semantics could be specified roughly as follows in the style of Congruity Theory:

(Def. 1) *foreigner* (x_1, x_2) =
Presuppositions: $\exists x_1\ \exists x_2$: x_1 is a person, x_2 is a country.
Entailments: $\neg$ citizen (x_1, x_2)

The second argument x_2 is "covert" as it is normally *not* saturated by the syntax. This argument place *needs* to be saturated contextually, setting up this argument place for contextual saturation is sufficient to make *foreigner* a context dependent expression. The presuppositional selectional restriction associated with the x_2 argument place functions like an instruction guiding the search of a *definite* contextually salient saturation. This context-dependent functioning of the noun *foreigner* can be contrasted with that of verbs like *to eat*, whose second argument (internal argument) can be optionally left implicit (18.a) and understood as an indefinite (18.b).

(18.a) John has eaten.
(18.b) John has eaten some food (= There is some food that John has eaten).

In contrast, there is no way a sentence like (19.a) could be understood as expressing a minimal proposition with an indefinite tacit argument like (19.b).

(19.a) John is a foreigner.
(19.b) *There is a country* of which John is not a citizen.

This shows that the meaning of *foreigner*, like the meaning of *we*, is not generic but truly "gappy". The instructional, deictic, part of the meaning of *foreigner* really tells the hearer to search the common ground to find a *specific* country salient in the context of utterance.[11] So, the workings of *foreigner* are more akin to the functioning of verbs like *to follow* in (20), which admit *anaphoric* implicit arguments (cf. Groefsema 1995a). The difference is that with predicates like *to follow* the context-dependent anaphoric saturation is optional, as their argument can be expressed also syntactically.

(20) The guide left but the tourists didn't *follow* (adapted from Groefsema 1995a).

Having determined the "gappy" nature of modal meanings, we need to figure out what type of variable is responsible of the intrusion of context in them. Two different answers to this question are provided by the classic version of possible worlds semantics – prototypically represented by Saul Kripke's *frame semantics* – and by Relative Modality's own take on possible world semantics.

4.2.6 The Analysis of Restrictors: From Frame Semantics to Premise Semantics

In Kripke's pioneering work on the semantics of modal logic the quantification of the modals is restricted via a relation R between possible worlds (cf. Kripke 1963), now usually called an *accessibility relation*. Given any two worlds w and u ∈ W, a

[11] To be precise, there is one alternative to the deictic or anaphoric saturation of the argument with a definite contextually relevant country. The alternative consists in the binding of the covert argument variable to a quantifier. Consider the following sentence: *Everywhere in the world, foreigners feel like "second-class" citizens*. Here the covert "country" argument in *foreigners* (and in *citizens*!) is understood as bound to a universal quantifier. This behavior actually confirms the content dependent analysis of *foreigner*. It shows that the noun *foreigner* belongs to the sub-class of context dependent expressions that admit *deictic*, *anaphoric* as well as *bound* uses. Third person pronouns are the typical example of this class (cf. Recanati 2002). In the sentence *Every boy*$_a$ *thinks that John*$_b$ *loves his*$_{a,b,c}$ *mother* the possessive *his* can be interpreted as (a) bound, where *every boy* = *his*, (b) as anaphoric, where *John* = *his* and also as (c) deictic, where *his* = a contextually salient person. The context dependent expressions admitting the three uses can be contrasted with the strict indexicals, such as *you* and *I*, which only admit deictic uses. Interestingly, it will turn out that modals, like the noun *foreigner* have a covert argument and that they too admit *bound*, *anaphoric* and *contextual* resolutions of this variable.

relation wRu indicates, according to the intuitive interpretation originally offered in Kripke (1963: 70), that w′ is "possible relative to" w, or "possible in" w, so that instead of an absolute notion of possible world we have a relative notion, "of one world being possible relative to another". According to the currently prevalent terminology wRu is instead read as "world w has access to world u". While, for pedagogical purposes, having *access* is sometimes presented in terms of *seeing into* a world from another world (cf. Girle 2003: 37) – an obscure metaphor, which adds little to Kripke's lucidly terse presentation. The pair ⟨W, R⟩ is called a frame, sometimes a "Kripke frame", from which comes the name *frame semantics* that is often used to refer to the theory.[12]

The semantics of possibility and necessity expressions restricted by an accessibility relation can be given with the following definitions:

(Def. 2) $\Diamond \varphi$ is true in world *w* in W iff φ is true in at least one world, say *u*, in W and *w*R*u*

(Def. 3) $\Box \varphi$ is true in world *w* in W iff φ is true in all the worlds *u* in W so that *w*R*u*

Logicians building systems of modal logic have been chiefly, if not exclusively, interested in the formal properties of accessibility relations, which provide a semantic correlate to the axioms characterizing different systems of modal logic. Given a set of worlds W and an accessibility relation R, "certain axioms may be guaranteed to hold solely in virtue of the properties of the accessibility relation" (Kaufmann et al. 2006: 80) irrespectively of the truth values that the interpretation function assigns to expressions at individual worlds. Formal properties of accessibility relations that are of significance for distinguishing systems of modal logic include, for instance, *seriality*, *reflexivity*, *transitivity*, *symmetry* and *euclidity* (cf. Garson 2014, Kaufmann et al. 2006:). While these properties, also called "conditions on accessibility relations", are purely formal, their broader philosophical and practical significance – just like the significance of the axioms – depends on the fitness of the different systems so defined for modeling reasoning with particular types of modality. Let us consider, for instance, *seriality* (Def. 4) and *reflexivity* (Def. 5):

(Def. 4)	*Seriality*:	$\exists w'$: wRw′	*Axiom D*:	$\Box\varphi \rightarrow \Diamond\varphi$
(Def. 5)	*Reflexivity:*	wRw	*Axiom T:*	$\Box\varphi \rightarrow \varphi$

A serial accessibility relation guarantees that there is at least a world that is accessible from the world of evaluation so that necessity does not quantify vacuously

[12] To avoid all risk of misunderstanding, let me emphasize that Kripke's *frame semantics* has absolutely nothing to do with the homonymous *Frame Semantics* approach introduced by Charles Fillmore in cognitive linguistics. Likewise, *Kripke frames* have nothing to do with the notion of *argument frame* or *argument structure* of a predicate that is common in linguistic semantics. For instance, when we say that the argument frame of a modal has two argument slots, respectively for the restrictor and for the prejacent proposition in the scope of the modal we refer to *argument frame/ argument structure*. *Argument frames* are, to some extent, connected to Fillmore's *Frame Semantics* which generalizes the notion of an argument frame to provide a common format for the representation of semantic-conceptual information about types of situations or scenes that is variously indexed by linguistic expressions.

over an empty set of worlds. It offers a semantic counterpart to *Axiom D*, which is thusly called because it appears to be required in a system of logic to properly model *deontic* concepts: what is obligatory or morally necessary has to be also permissible or morally acceptable, lest we are put in an inescapable double bind. A reflexive accessibility relation guarantees that every world in W has access to itself so that what is necessary is also true in the world of evaluation, as stated in the T axiom.

This axiom characterizes *alethic* modalities in the broadest sense, even if the interaction of these everyday modalities with time and causality complicates the picture. In most ordinary alethic modalities the necessity of a state of affairs entails that it will be inevitably the case in due course, not that it necessarily is the case at the time of evaluation. Thus, T holds for these modalities provided that the prejacent proposition is viewed as including the reference to a time that is posterior to the time of evaluation of the modal proposition.[13] In contrast, both *doxastic* and *deontic* modalities correspond to an accessibility relation that is serial but not reflexive: one can be certain of things that are false and not all that is morally or legally obligatory actually happens.

The set of worlds W in a frame and the corresponding accessibility relation R can receive concrete interpretations consistent with the formal conditions on the accessibility relation. For instance, a deontic frame could be construed by identifying W with the set of worlds conceivable by human beings, and R with a relation that holds between two worlds w and w′ if and only if what is the norm or law for w is followed in w′ (cf. Portner 2009: 23). Different deontic frames can be conceived for different systems of norms. For instance, in one frame R could give access to *legally* compliant worlds, while in another to *morally* irreprehensible ones. As we have seen above, it is commonly assumed that such a concrete interpretation of the frame should require that R is serial, but not reflexive. In fact, a logico-philosophical investigation could proceed either from the formal properties of R to the possible concrete interpretation that such a frame can receive or from a concrete interpretation to the formal properties of R that are required to properly model valid reasoning with that kind of modality.

In the standard frame semantic view, the context dependency of the restrictor of modal quantification could be modeled by assuming that people effortlessly calculate on the spot a variety of accessibility relations "that are themselves unphilosophical and vague" (Lycan 1994:176) holding for equally vague and ad hoc sets of worlds. In view of the goals of the present study, this is not particularly appealing, because of two related problems.

Firstly, sets of possible worlds and accessibility relations are not entities that are easily and immediately relatable with arguments and with reasoning in natural lan-

[13] Lycan (1994: 177) even contends that, contrary to what "one would naturally think", in everyday discourse "some alethic modalities are not reflexive". I am not entirely convinced by his examples: some involve desires seen as emotional compulsions, which I would not treat as alethic, others have a conditional, counterfactual, component that changes the nature of their accessibility relation, or (according to the terms that I am going to introduce in the following pages) of their conversational background.

guage. To go back for a moment to Toulmin's original insight on the isomorphism between the structure of modal meanings and the structure of arguments, a set of possible worlds W let alone an accessibility relation R are not something that can be easily identified with the field-dependent *grounds* for the modal that Toulmin had in mind.

It is not necessary here to raise the broader issue of the psychological plausibility of possible worlds semantics (cf. Johnson-Laird 1978; Bell and Johnson-Laird 1998; Papafragou 2000: 33).[14] It is sufficient to consider the goal of finding out how different kinds of modality relate to premises, standpoints and inference licenses, as they are commonly understood, to see that frame semantic constructs such as W and R do not have immediate correspondence in the procedural, discursive, view of human reasoning that is typical of argumentation theory, despite their utility in the construction of a truth-preserving semantics for modal logics.

Secondly and relatedly, if we consider modals as relational expressions that need to saturate one of their arguments in discourse context, a frame ⟨W, R⟩ does not seem a particularly promising candidate as an argument slot to be saturated in discourse context.

In short, frame semantics does not have obvious handles to connect it to the discursive development of argumentation on the one side and of pragmatic interpretation on the other. This is a weak argument in itself, but it acquires significance when we consider comparatively what the alternative approach devised by Angelika Kratzer has to offer in both these respects. Kratzer (2012 [1977]) remains in a possible worlds semantics framework, but instead of basing the restrictor directly on a relation between possible words, conceives of it as a set of propositions or *premises* – hence the name *premise semantics* with which Lewis (1981) christens the approach.

A precursor to this approach, according to Kratzer (2012 [1977]: 8), can be found in Charles S. Peirce. The passage of the 1893 essay on *The nature of inference* from which Kratzer quotes is particularly lucid and deserves extended consideration. The

[14] The basic criticism to possible worlds semantics coming from cognitive psychology is based on experimental results suggesting that "human beings show little aptitude in considering the complete set of alternatives to a given situation, even where set is small" (Johnson-Laird 1978: 17). This criticism is originally framed in a psychologically realistic interpretation of the accessibility relation in terms of *conceivability*: "To make sense of certain modal logics, however, one need consider only certain possible worlds, namely, those that are 'accessible' to a given world. This notion may be taken to mean that there are only certain possible worlds that can be conceived of in a particular world, and hence for that world a necessary proposition is one that is true in all the worlds conceivable from within it, i.e., true in all the possible worlds that are accessible to it." (*Ibid.*). Clearly, such an interpretation forces upon possible worlds semantics a claim of psychological realism that was never intended by its proponents. The reading of accessibility as conceivability, in particular, appears a hasty identification. At best, conceivability can be one specific concrete interpretation of R. As for the human (in-)ability to entertain even relatively small sets of alternatives, it has been replied that in order to conceive modal concepts in terms of quantification over alternatives it is not necessary to have a discrete mental representation of each of them. This is true, in general, of our ability to conceive sets, which does not require individual representation of their members and thus applies also to ordinary quantifiers. For instance, in order to understand the sentence *Every Chinese knows something about Confucius* I do not need to store separate individual mental representations of 1,357,000,000 citizens of the People's Republic of China.

American philosopher had introduced notions of possibility and necessity as relative to *information*, where this term stands for "a state of knowledge that may range from total ignorance of everything except the meaning of words up to omniscience" (Peirce 1998: 43). *Informational* notions of possibility and necessity are defined as follows:

(Def. 6) *Informationally possible* is that which in a given information is not known to be false.

(Def. 7) *Informationally necessary* is that which in a given information is perfectly known to be true.

The information state can either correspond to our *actual* state of information – which gives what would nowadays be called epistemic and doxastic modalities or to a *hypothetical* information. *Physical necessity* and *possibility*, for instance, are based on a hypothetical information state where we imagine ourselves "to be thoroughly acquainted with all the laws of nature and their consequences but to be ignorant of the particular facts" (*Ibid.*). *Practical possibility* is another case in point, where "we imagine ourselves to know what the resources of men are but not what their dispositions and desires"; here possible is "what we do not know it won't be done. Peirce (1998: 43) devotes special attention to two limit cases that have philosophical significance: *essential possibility/necessity* and *substantial possibility/necessity*. Suppose there is a person "who knows no facts, though perfectly *au fait* at reasoning and well acquainted with the words involved", the essentially or logically necessary is what such a person would know to be true:

> For instance, he would not know whether there was or was not such an animal as a *basilisk* [...] but he would know that every basilisk there may be has been hatched by a serpent from a cock's egg. That is essentially necessary; because that is what the word *basilisk* means. (Peirce 1998: 43–44)

In contrast, *substantial* necessity and possibility refer to "supposed information of the present in the present, including among the objects known all existing laws as well as special facts" (*Ibid.*: 44). Peirce observes that all present substantially possible states of affairs are also substantially necessary, while he admits future contingents, avoiding determinism.

It can be observed how the empty *essential* modality and the full *substantial* modality form the two poles of a cline of alethic modality along which physical, practical and countless other alethic notions in everyday use can be situated, each basing its hypothetical information state on the selection of facts of a certain kind and the abstraction from facts of another kind.

4.2.7 Possibility and Necessity as Relative to a Conversational Background

Let us then present how possibility and necessity can be defined in a basic implementation of the premise semantics idea of Relative Modality. This basic version of Relative Modality, when implemented with the tools of possible worlds semantics is completely equivalent to the Kripkean frame semantics discussed above. In her works Kratzer proposes considerably more sophisticated versions of the premise semantics idea. I will present some aspects of the full theory later in this chapter. The basic version will however suffice to illustrate the main reason for adopting this kind of model, which is not (just) its power in providing correct truth conditions for modal expressions, but, rather, its ability to capture how modal expressions relate to their discourse context, in argumentative discourse in particular.

The key notion of the premise semantic approach of relative modality theory is the notion of a *conversational background*. The conversational background, which I will routinely indicate as B in the following pages, is a set of propositions, which plays the role of the restrictor in the argument frame of the modal. In short, modal expressions of necessity can be understood in terms of the *logical consequence* of the prejacent from the conversational background, while possibility expressions are to be conceived in terms of the *logical compatibility* between the prejacent and the background:

(Def. 8) [[*must/ necessarily* (B, φ)]] = {w ∈ W: [[φ]] is a logical consequence of B}
(Def. 9) [[*Can/May/Possibly* (B,φ)]] = {w ∈ W: [[φ]] is logically compatible with B}

4.2.7.1 Premises and General Principles of Rational Inquiry

The semantic clauses in (Def. 8) and (Def. 9) define the semantics of the modals through the relations of *logical consequence* and *logical compatibility*, for which a possible world semantics can be given as in Kratzer (1981: 43, 1991: 641):

(Def. 10) a proposition p is a logical consequence of a set of propositions A, if and only if p is true in all the worlds of the "universe" W in which all the propositions belonging to A are true;

(Def. 11) a proposition p is logically compatible with A, if and only if there is at least a world in W, where all the propositions of A *and* the proposition p are true

In other words, (Def. 11) tells us a proposition is a possibility relative to a given conversational background B, if and only if $B \cup \{\varphi\}$ is a *consistent* set of propositions.

More precisely, the conversational backgrounds denoted by phrases such as *in view of the law*, in *view of what is known*, which I have up to this point represented as a set of propositions B, is seen by Kratzer as a function *f* from worlds to sets of propositions. Let us consider, for instance, her presentation of the *deontic* conversational background, which Kratzer (1981: 45) identifies with the phrase *in view of what is commanded*:

(Def. 12) A deontic conversational background is a function f which assigns sets of propositions to members of W such that for any w ∈ W:
f(w) contains all those propositions p such that it is commanded in w that p – by someone, by the Law, etc.

What is commanded is a set of propositions that varies according to the world we are situated in. Should we think that Kratzer believes that in, say, our world there is one big f(w) containing all that is commanded and that we can derive all deontic necessities from it? For one, that would be an inconsistent set. Whe she revised her classic 1981 paper in Kratzer (2012), she made it clear that more localized sets of propositions are envisaged, speaking instead of "the content of a body of laws or regulations in w" (Kratzer 2012 [1981]: 37). It remains true, however, that the theory retains a somewhat holistic flavor in envisaging the conversational backgrounds.

A critical, and somewhat tricky, aspect of the theory is the notion of *proposition*, which is not intended to mean a linguistic or mental representation, not even a premise, despite the theory's nickname of "premise semantics". As it is customary in possible world semantics, propositions are simply seen as sets of possible worlds (Portner 2009: 51, Kratzer 2012 [1977]: 10), each proposition corresponding to the set of worlds in which the proposition holds true, in which it is a *fact*.

The proposition is seen ontologically, in relation to facts. From this point of view it is rather natural to say that a sentence expresses a proposition when it expresses a *possible fact*: something that can be a fact in a *world*, or perhaps in a *situation*. Knowing the meaning of this expression means knowing the situations in which the proposition is a fact. Possible world semantics, however, construes its ontology in a less intuitively obvious way, saying that for a linguistic expression, say, a sentence, to express a proposition means to denote a set of possible situations. And the notion of *possible world* is what possible world semantics has at its disposal for modeling situations.[15]

A consequence of defining a proposition as a set of possible worlds is that a set of propositions f(w) can immediately be translated in to a set of worlds ∩f(w), the intersection of all the propositions in f(w). Consider that, with respect to a conversational background f and a world w,

> a proposition p is necessary iff it is a logical consequence of f(w)

In view of the definition of logical consequence in (Def. 10) this means

> a proposition p is necessary iff p is true in all the worlds of W in which all the propositions of f(w) are true.

Thus, this necessity can be equivalently expressed as

> with respect to a conversational background f and a world w a proposition p is necessary iff p is true in all the worlds of ∩f(w).

[15] Some would say it is a very peculiar model of situation, again, counterintuitive and unwieldy in some respects. Worlds, in particular, are complete (every proposition or its negation holds in a world), while we would have the tendency to view situations as more localized, not completely determined and, perhaps, as having fuzzy margins. Semantic theories alternative to possible world semantics have proposed different views of situations and propositions. Situation Semantics (Barwise and Perry 1983), in particular, features situations that, unlike worlds, are incomplete. More recently Kratzer (2012 [2002]) proposed an approach that bridges the two, having *possible situations* as parts of *possible worlds*.

We can see how conversational backgrounds, in a possible world semantics implementation, do the same job of the accessibility relation in restricting quantification over a set of worlds. The equivalence between an accessibility relation R and a conversational background f can be established as follows:

For all w, w′ ∈ W: wRw′ iff w′ ∈ ∩f(w)

What could be the point, then, of achieving the results of a frame semantics of modality in a more circuitous fashion?

The answer is twofold. On the one hand, there is more in the full theory of relative modality expounded in Kratzer (2012 [1977, 1981]) than the fragment presented here. The full theory is not equivalent to a standard frame semantics. On the other hand, Kratzer (2012 [1977]: 10) makes it clear that the significance of the "premise semantics" idea does not exclusively lie in the power of the formalism and *is not bound to a possible worlds implementation*:

> All the analyses in this book are cast within a possible worlds framework [...] But the guiding idea of a premise semantics for modals can be implemented in any framework that provides suitable notions of logical consequence and compatibility. In fact, one of the main virtues of a premise semantics for modality is that it links the semantics of modals to general principles of rational inquiry that apply whenever we reason from a set of premises.

This is, in fact, one of two main related reasons of the appeal of premise semantics and relative modality theory in view of the present investigation of modality in argumentation. The other main reason is that the same tools that connect modality to "principles of rational inquiry" also provide the basis for connecting modal expressions to its dynamically developing discourse context, and, in particular, to argumentative relations in discourse.

4.2.7.2 How Premise Semantics Addresses the Contextual Construction of Conversational Backgrounds

The various "flavors" of the modals and their finer nuances can be expressed in terms of the different *conversational backgrounds* restricting the modal operator. One advantage of this kind of approach – the one that immediately caught the attention of linguists – is that the variety of modal "flavors" results from an invariant modal force and a variety of conversational backgrounds reconstructed in the context of utterance. One of the appeals of the theory is precisely that it allows for an indefinite number of conversational backgrounds, while at the same time it makes it possible to characterize broad classes of uses of the modals on the basis of the kinds of propositions that enter the background.

If we go back to example (6), for instance, we see that the necessity modal selects a very restricted deontic B, identified with the measures of governor's plan and the prejacent p will be true in every possible future development of the world w where the governor's plan is fully and successfully implemented.

By providing contextually situated meanings, relative modality addresses not only the choice between broad categories of "flavors" such as *deontic*, *alethic* and *epistemic* (cf. Kronning 1996) but also the fact that a category such as *deontic necessity* can be further subdivided in an indefinite number of sub-types, relative to *moral values, desires, goals, interpersonal commitments, social norms, laws, plans,* and so on. The same could be said for *alethic necessities*, spanning the full range from *logical* and *physical* necessity, down to what is necessary in view of the internal or external circumstances of an agent, according to what kind of *factual* propositions are allowed in the B.

Phrases like *according to the governor's plan, in view of the law* etc. (Cf. Kratzer 1981: 44–46) can narrow down the contextual selection of the B, which remains nevertheless a "black-box": as readers we don't know *exactly* what propositions make up the governor's plan, unless we have independent knowledge of this document. How can we become really sure of the propositions that are included in a conversational background?

Kratzer (1977: 342–343)[16] recounts us the (true?) story of one Professor Schielrecht – a third generation offspring of the Vienna circle whose main interest in philosophy is to show that "most of what most people say most of the time does not make sense". The good professor comments on a question a judge raises in the context of a murder case, namely *whether a murderer could have acted otherwise than he eventually did.*

> Professor Schielrecht said that the judge asks himself a question which does not make sense. Why not? Professor Schielrecht's answer was: Given the whole situation of the crime, which includes of course all the dispositions of the murderer, this man could not have acted otherwise than as he did. If he could have acted otherwise than he eventually did, he would have acted otherwise than he eventually did. So the answer to the question is trivial; there is no need to spend a single second on the problem. There is really no problem. And it does not make sense if one spends any time on a problem which only seems to be a problem but is not really one.(Kratzer 1977: 343).

The good professor interpreted the conversational background as containing *the whole* situation, something like Peirce's *substantial* modality as referred to a moment immediately preceding the murder. With one difference, though, Peirce still admits future contingents, while Kratzer's Schielrecht is squarely deterministic. Determinism, however, is not the problem here. The problem is, as Kratzer tells us, that Schielrecht misunderstands what the judge meant, which was whether the murderer could have acted differently *in view of such and such aspect* of the situation, which – I say – should have been prominent in the context of utterance.

It is at this point, however, that Kratzer (1977: 343) backs off, observing that is not sure whether people can be explicit enough to avoid this kind of misunderstanding and that she does not "want to get into the mess of context theory". A cautious attitude that was certainly excusable at the time – even if the very name chosen for *conversational backgrounds* intriguingly suggests that the relevant premises that serve as background for judgments of possibility and necessity depends on the con-

[16] As observed above, I refer to the original text of Kratzer's paper for s that have been expunged or altered in the 2012 edition.

versation we are having, from its jointly accepted purpose or direction and, lastly, from what has transpired in it.

Kratzer has been fully aware from the beginning of what we can call, stealing Lycan's words, the capricious diversity and irreparable vagueness of everyday conversational backgrounds. In the revised 2012 papers this is more explicitly said: "more often than not conversational background remain genuinely underdetermined and what speakers intend to convey is compatible with several choices of conversational background" (Kratzer 2012 [1981]: 32). Kratzer, however, devotes her attention to discussing the properties of broad classes of conversational backgrounds rather than to working out how vague backgrounds are *calculated on the spot* by speakers and hearers.

Re-examining Kratzer's theory from the standpoint of cognitive pragmatics, Anna Papafragou (2000: 34) observes that "one might get the idea from this exposition that there is a fixed inventory of conversational backgrounds favouring one modal interpretation over others", an idea that "presupposes a rather inflexible conception of the role of context in comprehension", while according to Papafragou's pragmatic view "constructing the context for understanding an utterance is part of the interpretation process, constrained by general pragmatic principles". Maybe this impression is not entirely warranted and depends, in part, on presentational choices of the original papers (see above in Sect. 4.2.7.1 the different presentations of the deontic background) that were later amended in the 2012 revisions. It remains true, however, that the context dynamics that affect the on-line construction of conversational backgrounds are not technically addressed in Kratzer's version of Relative Modality. Technically, the propositions in f(w) only depend on the world of evaluation w.

Papafragou, taking the pragmatic viewpoint of Relevance Theory, sees the contextual construction of conversational backgrounds as an inferential process guided by general pragmatic principles that account for "context selection" (Papafragou 2000: 35) . That this kind of "context theory" should be beyond the purview of a semantic theories such as Kratzer's is hardly surprising.

There are, however, aspects of context dynamics whose treatment does not require a fully-fledged pragmatic theory constraining the selection of the relevant context for utterance interpretation and can be modeled by simpler approaches to context update such as those developed by discourse semantics and dynamic semantics.[17] The Relative Modality analysis and, in particular, the notion of a conversational background is particularly germane to these discourse oriented theories of meaning. So, it is not surprising that some of the contextual implications of the construct of the *conversational background* have been explored by these approaches.

[17] On the relationship between these semantic theories and the development of argumentation theory see Chap. 2, § 2.1.2.

4.2.7.3 Conversational Backgrounds and Discourse Dynamics

In contrast with pragmatic theories that see utterance context as selected, or even constructed, dynamic semantic and discourse semantic approaches adopt a simpler model of context as a store of information (as well as of commitments and obligations) that is updated by successive utterances in discourse. As different cases and drawers in the store are populated and changed, so are the sets of propositions making up modal conversational backgrounds. In these versions of the Relative Modality analysis, modals function as *anaphorical pointers* towards the utterances where the propositions had been first introduced. As it will be shown later in this chapter, such an anaphorical analysis is crucial for understanding how modal expressions contribute to establishing relations among utterances and, in turn, for fully understanding their potential as argumentative indicators.

In fact, the *guiding ideas* of a premise semantics are particularly well adapted for accounting for this kind of discursive dynamics. This will become even more apparent once I have briefly introduced Kratzer's views on the similarities between modals and conditionals and their interaction. In contrast, the *implementation* of the theory of Relative Modality in Kratzer is not dynamic (cf. Portner 2009: 86): the value of the function f depends on the world of evaluation w, which is a static parameter. Additional contextual parameters that can and should be added to define certain kinds of background will be static as well.

In contrast to this static implementation, Lewis, who first coined the term *premise semantics*, in a famous paper on "scorekeeping in a language game" (Lewis 1979) explicitly includes Kratzer's conversational backrounds among those contextual variables whose value depends on a constantly updated *conversational score*, thus offering the first hint of what a dynamic treatment of relative modality might look like. Later in this chapter (§ 4.3.5) I will come back to Lewis' idea of conversational *scorekeeping* and its significance for argumentation theory.

In part, a dynamic development of Kratzer's premise semantics idea has been carried out by scholars working on modality within the *discourse semantic* approach (cf. Seuren 1985; Roberts 1989; Frank 1996; Frank and Kamp 1997; Geurts 1999: 175–200) and the related but separate *dynamic semantic* approach (cf. Veltman 1986; Beaver 2001). These two strands of research correspond to a general turn in formal semantics which, over the last three decades, has begun to look at meaning in a *dynamic* way, which reflects how discourse participants, by exchanging utterances in a monologue or dialogue, progressively establish a shared *discourse representation*, or *information base* – which one could also call the "world according to the text".

Within this general turn, the development of the *discourse semantic* approaches – introduced independently by Hans Kamp (2013 [1981]) and Pieter Seuren (1985) – has been influenced (cf. Portner 2009: 89) by the earlier work by Robert Stalnaker (1973, 1978), which, in order to model the dynamics of assertion and presuppositions introduced a notion of *common ground* cast in a possible worlds framework which closely parallels Kratzer's own *conversational backgrounds* (cf. Portner 2009: 89–90).

According to this view the semantic representation is built gradually through successive increments, or "updates" of a *common ground,* that is of that fragment of the world which is already shared by the discourse participants. This is what participants take for granted at the current stage of their interaction. According to Stalnaker (1973), we can think of the common ground as a set of propositions, which *partially* define a world that the participants are sharing through communication. New utterances add new propositions to the common ground extending the world fragment shared by the participants.

The common ground is underspecified with respect to the actual world, or any other single possible world. Possible worlds are worlds in a rather strong sense: they are complete, fully specified, and only a maximal consistent set[18] of propositions would define a single possible world. Instead, the propositions in the common ground only define a set of possible worlds that are consistent with them. Stalnaker calls this set of possible worlds the *context set*. As new utterances are added to discourse the *common ground* is increased and the *context set* of alternative worlds is reduced. That a given common ground is compatible with many possible worlds is quite natural. Consider the participants in a discussion: they will share a certain common ground, but where they disagree their views of the world will diverge.

Given this structural similarity between *common ground* and *context set* and, respectively, *conversational background* and the *set of accessible worlds*, it is tempting to envisage modal meanings in terms of operations on a developing discourse context. I will have something to say on such a perspective shortly (§ 4.2.9). Before approaching these developments again, however, it is necessary to dwell further on the basics of Relative Modality and, in particular, on its unified analysis of modality and conditionals.

4.2.8 Relative Modality and Conditionals

4.2.8.1 Relative Modality and *Necessitas Consequentiae*

It is possible to provide the semantics of the relative modal operators indirectly, by translating them in terms of an "absolute" modal quantifier roughly corresponding to Peirce's *essential modality* (let us use $\Box$ and $\Diamond$ as symbols of this variety of necessity and possibility), for which a standard frame semantics is assumed, and of a truth-functional connective ($\rightarrow$ and $\wedge$ respectively). This would give us the following indirect definitions:

(Def. 13) *Must/ Necessarily* $(B, \varphi) \Leftrightarrow \Box (B \rightarrow \varphi)$

(Def. 14) *May/ Can/ Possibly* $(B, \varphi) \Leftrightarrow \neg \Box (B \rightarrow \neg\varphi) \Leftrightarrow \Diamond (B \wedge \varphi)$

This inelegant *semantics by translation* is meant to make more apparent a single simple point: relative necessity in a premise semantics is always, in a certain sense, a kind of *necessitas consequentiae*. Let us consider the use of the English modal

[18] A set such that adding any single proposition to it would result in an inconsistency.

must in (21) and (22) in the light of the discussion of *necessitas consequentiae* in Chap. 3, § 3.4.2.3):

(21) If Alfred is a bachelor, he *must* be unmarried.

(22) Alfred is a bachelor. He *must* be unmarried.

Superficially in (21) the modal is syntactically embedded in the consequent of the conditional, but its semantic interpretation does not correspond to the logical form of the *necessitas consequentis* shown in (Def. 15).

(Def. 15) Necessitas consequentis: $p \rightarrow \Box q$

(Def. 16) Necessitas consequentiae: $\Box(p \rightarrow q)$

In other words (21) does not mean that if Alfred happens to be a bachelor *in the actual world* then he will be unmarried *no matter what the circumstances turn out to be like*, that is in all Peircean essentially possible worlds.[19] That would be, by the way, a bizarre thing to say because the set of essentially possible worlds includes also worlds where Alfred gets married, worlds where there is no Alfred, etc. In fact, the interpretation of (21) corresponds to the logical form in (Def. 16) where the necessity operator takes scope over the conditional (*necessitas consequentiae*): 'no matter what the world turns out to be like, if Alfred is a bachelor he will be unmarried'. Interestingly, in (22), where instead of a syntactic conditional we have two syntactically autonomous discourse units, we obtain the same interpretation corresponding to the *necessitas consequentiae*.

From these examples we can conclude that the modal expressions appearing in the consequent of a conditional appear to be *relative to the antecedent*. The antecedent and the conversational background appear to play the same role. The antecedent, however, cannot function as a fully-fledged conversational background. In our *bachelor* example, to make the consequence necessary, the antecedent *Alfred is a bachelor* needs to be conjoined with the essential conversational background–which includes, among other things the semantic axiom that *bachelor refers to unmarried adult males*. Thus conditional antecedents are *added* to pre-existing conversational backgrounds.

4.2.8.2 The Unified Analysis of Modals and Conditionals

This is the core of the integrated analysis of modality and conditionals proposed by Kratzer (1991: 648–649, 2012 [1981]: 64–68). According to this analysis the job of *if* clauses is to restrict the conversational background of the modal appearing in the consequent. The antecedent clause creates a new conversational background by adding its propositional content to the original background. Assuming that f(w) is the essential conversational background in (21), the clause *If Alfred is a bachelor* creates a new conversational background $f^{+}(w)$, as per (Def. 17):

[19] The set of Peircean essentially possible worlds will contain all logically consistent possible worlds where the semantic axioms of the English language also hold.

(Def. 17) $f^{+}(w) = f(w) \cup \{w: \text{Alfred is a bachelor in } w\}$

Kratzer's analysis goes further than saying that *if* clauses restrict the conversational background of modals occurring in the consequent. She hypothesizes that *all* conditional constructions are overtly or covertly modalized and that consequents where we do not find any overt modal marking contain a covert modal, which, by default, has the force of necessity. Thus, the conditional construction in (23) has roughly[20] the same semantic analysis of (21).

(23) If Alfred is a bachelor, he is unmarried.

In indicative conditionals without overt modal operators, the f(w) to which the content of the *if*-clause is added is often identified with normal, or expected conditions. A sentence like (24)

(24) If you show up on time, you will find plenty of room to park near the hotel.

is usually uttered on an assumed and *vague* backdrop of *normal conditions*, and it is not meant as stating that the condition of showing up on time is sufficient for finding parking *no matter what else could happen in the world*, but only *under normal conditions*. Truly exceptional situations – like those where an old Russian space station has just crashed on the hotel's parking lot destroying it – are not considered in evaluating the conditionals. Subjunctive, counterfactual, conditionals pose more complex problems, but they also involve conversational backgrounds *approximating* normal conditions.

4.2.8.3 Comparing Conditional Constructions with Anaphora in Discourse

Looking at the two *bachelor* examples we can observe that in (22) the proposition in the preceding utterance also gets added to the conversational background much like the conditional antecedent in (21). Here the further restriction of the necessity operator by the antecedent proposition seems to be realized *anaphorically* in discourse by the *premise* presented in the preceding unit.

As the reader might now surmise, this analysis of conditional and anaphoric antecedents as being added to an underlying conversational background is relevant for an analysis of arguments, where the force of the inferential step signaled by the modal relates to textually manifest premises on a backdrop of tacit, diffuse, and sometimes elusive background assumptions.

There are, however two differences between (21) and (22).

In (21) the antecedent of the conditional is not asserted. Its epistemic status may vary: the grammatical form of the conditional and the conversational context of the

---

[20] Only roughly though. The English modal verb *must*, like all modal lexical units, imposes specific constraints on the make-up of its conversational background that are not imposed by the phonologically null necessity modal operator. In Chap. 6 I will examine some of these fine grained lexical constraints on conversational backgrounds in relation to a few Italian modal expressions.

utterance may help to determine it with varying degrees of precision. As we add the antecedent to the conversational background the resulting conversational background ceases to be, in a broad sense, alethic. It contains facts and an assumption that may or may not be a fact. In contrast, in (22), the anaphorically recovered antecedent has been asserted and the conversational background remains alethic in Lycan's broad sense of the world. We could also say that the resulting conversational background is epistemic and includes the knowledge of the fact that Alfred is a bachelor and the knowledge of logical and semantic axioms. *Stricto sensu* epistemic backgrounds are always alethic: by definition only *facts* can be *known*.

The second, related, difference pertains the status of the prejacent: in (22) the prejacent is asserted and presented as the conclusion of an inference moving from the premise. In (21) it remains non asserted or only "conditionally asserted".

4.2.8.4 Epistemic Conditionals and Arguments

There are situations where conditionals cease to be non-asserted. Pragmatics can efface the difference between (21) and (22) when, in discourse context, (21) is understood to imply (22). In this cases the conditional antecedent ceases to be simply a presumptive condition and becomes a full-fledged premise. These contextual interpretations of the indicative conditional go by the name of *epistemic conditionals* (Sweetser 1990: 116–117; Dancygier and Sweetser 2006: 117–121).

According to Dancygier and Sweetser (2006: 117), who work in the cognitive framework of Mental Space Theory (Fauconnier 1994 [1985]), epistemic conditionals are used performatively to draw a conclusion and distinguish themselves from what the authors call *content conditionals* because they do not set up alternative mental spaces: their function being "simply to give background to the addressee, by invoking the relevant parts of the cognitive context which brought about this conclusion".

As far as I know, Kratzer has nothing to say on this notion of epistemic conditional. As we will see, this silence is not too surprising. They are, however, important for a fuller understanding of the functioning of conversational backgrounds and their manifestation in argumentative discourse. I will therefore devote some space to them, trying to bridge the Mental Spaces and Relative Modality frameworks.

Roughly, setting up alternative mental spaces means to prompt the addressee to imagine alternative enrichments of a certain state of the world or state of the information (base space) and consider the results or causal consequences of the enrichment. This activity, Dancygier and Sweetser (2006) tell us, is crucial in reasoning about alternative courses of action on the basis of their predicted effects. For instance in (15),

(25) If it snows too heavily during the night we won't be able to leave the hut tomorrow morning.

the speaker sets up two alternative stories: in one the group of friends remains in the hut blocked by the heavy snow, in the other the snowfall was not to heavy and (perhaps) they happily descend towards the valley floor.[21]

It is not easy to translate into the Relative Modality framework the idea of '(not) setting up alternative mental spaces', as it is an idea bred from a theory programmatically designed to capture non-truth-conditional aspects of meaning.

We can try to tell the story up to a certain (useful) point: a mental space is a partial characterization of a situation through a bundle of features.[22] It therefore corresponds to a set of possible worlds: the set of worlds at which all the features of the space are realized. Maybe we are not too astray if we say that mental spaces are very much like conversational backgrounds. When I use a *content conditional*, as in (25), by adding the conditional antecedent to the base conversational background of *(not) be able*, I obtain a qualitatively different, shifted, conversational background, which is no longer alethic. Maybe, the invitation to consider also the alternative extension of the base conversational background can be reconstructed as a conversational implicature. I'm not going to pursue this speculation here.

In contrast, in *epistemic conditionals* there is no real shift in the conversational background. Adding the antecedent does not change the nature nor the actual composition of the conversational background. Consider example (26), adapted from Dancygier and Sweetser (2006: 117):

(26) Rebecca: And who helped Mary typing her thesis?
Carl: JOHN typed her thesis!
Rebecca: Ah! If he typed her thesis, he *must* love her!

Dancygier and Sweetser say that in examples like (26) antecedents do not enrich the base space by setting up alternative spaces, they instead invoke the parts of the cognitive context that are relevant for concluding the consequent. Let us see how things are in terms of conversational backgrounds. Let us suppose that the basic conversational background of *must* in (26) is epistemic and corresponds to the common knowledge of Rebecca and Carl, what they know together. When the conditional is uttered, the proposition 'John typed Mary's thesis' is already part of this conversational background, even if it has only recently been added to it. We can still interpret the effect of adding the antecedent as the set-theoretic union of the base conversational background with the antecedent proposition, but it will be the special case (Def. 18) of the union of a set with one of its subsets:

(Def. 18) If $A \subset B$, then $B \cup A = B$

[21] From a logical point of view these natural language conditional structures are often interpreted as bi-conditional. (see Dancygier and Sweetser 2006: 35–42). I will not discuss this point here.

[22] Fauconnier and Turner (2006 [1998]: 307) give the following preliminary characterization of mental spaces: "Mental spaces are small conceptual packets constructed as we think and talk, for purposes of local understanding and action. Mental spaces are very partial assemblies containing elements, and structured by frames and cognitive models. They are interconnected, and can be modified as thought and discourse unfold." On the relation of mental spaces to domains and discourse worlds see Fauconnier (1994 [1985]: xxxvi-xxxvii).

There is no change in the conversational background and the conditional antecedent only serves to *highlight* which proposition in the conversational background is relevant as a premise to draw the conclusion. In example (26) the proposition had just been added in the previous exchange, but the same would hold for any context where it is reasonable to assume that the propositional content of the antecedent is part of the common ground of the interlocutors. Note that, on an *epistemic conditional* interpretation also example (21) would function exactly as (26) and would not feature a shift in the conversational background. Finally, the anaphorical recovery of the premise in (22) turns out to be of the same kind if we take the common ground at the moment of utterance as the original conversational background f(w). In all these cases $f^{+}(w)$ turns out to be identical with f(w) due to (Def. 18).

As observed above, Kratzer, in extending the Relative Modality framework to conditionals does not devote special attention to epistemic conditionals. This is not too surprising: from the point of view of a semantic analysis concerned with modelling correctly the contribution of modals and conditionals to the truth conditions these cases are trivial. From this viewpoint, the antecedent does not add really anything. Kratzer (2012 [1977, 1981]) is interested in cases where antecedents do change the conversational background and focuses on the treatment of the inconsistencies that might arise as a result of the extension of this set of propositions. These inconsistencies are *bound to arise* when we have subjunctive conditionals with a truly counterfactual interpretation (Kratzer 2012 [1981]: 66), as in (27), and *can arise* also in *deontic conditionals*, exemplified by (28).

(27) If Ayrton Senna had survived his crash at the 1994 San Marino Grand Prix, he *would* have won another championship

(28) If a murder occurs, the jurors must convene (in view of what the law provides) (Kratzer 1991: 648)

In (27) the presence of a contrary to fact antecedent is obviously incompatible with the underlying alethic f(w), giving rise to an empty $f^{+}(w)$ which wreaks havoc to the truth-conditions of the conditionals. In (28) the fact that *murders are illegal* renders the antecedent incompatible with the deontic f(w) *what the law provides in* w. The extensions of the basic Relative Modality approach that deal with these inconsistencies will be briefly discussed in § 4.3.4.

From the viewpoint of argumentation theory, epistemic conditionals are interesting also because with them we can have *an entire argument* whose only linguistically explicit part is a conditional. The expression of these arguments seems to be based on a particular type of implicature where the *communicative pragmatic inference* that guarantees the recovery of the implicit information follows the steps of the *argumentative communicated inference* which is recovered.[23] Pustejovsky (1995: 236–238) calls this pattern "enthymemic deduction". It is very tempting – but most

[23] Cf. Rocci (2006: 421) on this distinction: "Since these argumentative inferences represent the inferential path that is conveyed through the argumentative discourse, they are *communicated inferences*. […] Not surprisingly, the inferential paths communicated by argumentative discourse are not decoded, but are conveyed by the speaker relying on the pragmatic inferences of the hearer. The latter inferences closely interact with argumentative inferences, and yet are distinct from them;

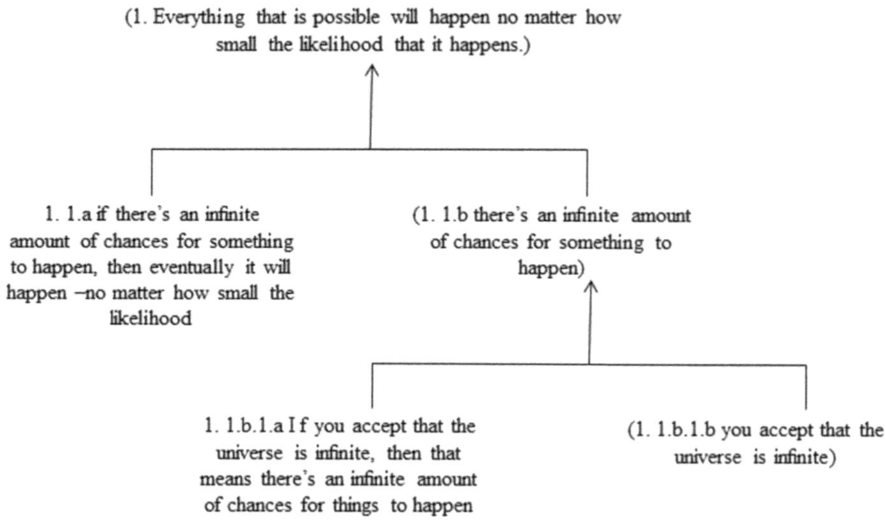

Fig. 4.4 Argumentation structure of example (29). Implicit parts are in *parentheses*

probably too simplistic – to reconstruct as an enthymematic *hypothetical syllogism* the "enthymemic deduction" underlying the interpretation of (26):

i.	Major Premise:	If he typed her thesis, he loves her	(explicitly asserted)
ii.	Minor Premise:	He typed her thesis	(anaphorically recovered)
iii.	Conclusion:	He loves her	(deductively inferred from *i* and *ii*)

In the following authentic example from Dancygier and Sweetser (2006: 118) a subordinative argumentation structure is manifested by juxtaposing two epistemic conditionals:

(29) *If you accept that* the universe is infinite, then *that means* there's an infinite amount of chances for things to happen, right? ... Well, if there's an infinite amount of chances for something to happen, then eventually it will happen –no matter how small the likelihood."

The argumentation structure of (29) can be represented as in Fig. 4.4 above, where the major premises (1.1.a) and (1.1.b.1.a) are the only elements that are explicitly manifested. The standpoint (1.) and the premise (1.1.b) are derived by "enthymemic deduction". Interestingly, (1.1.b.1.b) is not exactly presupposed as part of the common ground of the exchange. Rather the addressee is *invited* to accept it as a premise, which makes the whole argument conditional on this acceptance.

In commenting this example, Dancygier and Sweetser (2006: 118) – who do not pay special attention to argumentation — say that (19) "enlists the interlocutor as supporter for the assumption that the *if*-clauses hold, before engaging in conditional

they are *communicative inferences* rather than *communicated* since they are instrumental to recovering the communicated inferential path."

reasoning". These authors also observe (*Ibid.*: 117) that expressions such as *If you accept that* and *that means*, while not necessary to obtain an epistemic interpretation of the conditional, make its epistemic nature presenting the conditional relations as holding at the meta-level of the reasoning process involved rather than at the level of the ontological relations holding between the possible facts in the world ontology.

These comment are in line with a well-known proposal by Sweetser (1990: 80) that interprets modal verbs, conditionals and causal connectives in discourse as expressing causality at different levels: at the world level, at the epistemic level and at the speech act level. In the epistemic readings of the conditionals as well as of the other expressions causality operates "between premise and conclusion in the speaker's mind" rather than between events in the external world. In general, Sweetser sees these epistemic-argumentative readings as metaphorical mappings of real-world causality onto the belief system of the speaker – a hypothesis that I have criticized elsewhere (Rocci 2005a). In the following sections I will present an alternative view of the meta-representational nature of epistemic modalities proposed by Papafragou (2000).

Example (29) is interesting also for another reason. It shows that it is not necessary to have an antecedent fully accepted in the common ground, like it was in (26), for a conditional to function as the enthymematic expression of an argument. In natural discourse conditional antecedent are always epistemically evaluated and this evaluation, expressed to some extent by their form, guarantees that they can function on occasion as weakly presumed premises, endowed with a variable degree of certainty.

In Chap. 6, the Kratzerian unified analysis of modals and conditionals will offer a sound basis to model the relationship between modal qualifiers and rebuttals which was investigated in Chap. 3, § 3.4.4. Rebuttals were reconstructed as negative non-asserted exceptive conditions, while the parallel phenomenon of positive presumed conditions, unaccounted by the Toulmin model, was also recognized. Both types of conditions seem to play a key role in presumptive reasoning according to some accounts (cf. Rescher 2006) and are constitutively tied to the modal qualification of the conclusion. The epistemic status of both types of conditions may vary according to context from simple possibility to a strong expectation.

4.2.9 Relative Modality, Anaphora and Discourse Relations

Having discussed the unified analysis of modality and conditionals proposed by Kratzer and having observed the similarities between conditional antecedents ad anaphoric antecedents in the interpretation of the modals, we are now in a better position to appreciate how the *discourse semantics* tradition – briefly introduced in § 4.2.7.3 – connected modal conversational backgrounds to the development of the common ground in discourse.

This tradition produced a substantial amount of work that examines the *anaphoric*[24] properties of the modals in discourse (cf. Roberts 1989; Frank 1996; Frank and Kamp 1997; Geurts 1999: 175–200). These works re-frame relative modality as part of a dynamic theory of discourse interpretation, where each incoming utterance in a discourse is seen as *updating* an evolving common ground with new propositional contents and, at the same time, as *presupposing* the satisfaction of certain propositional contents in the existing common ground.

In the Kampian discourse semantic tradition, following van der Sandt (1992), presuppositions are treated essentially as anaphorical, demanding the retrieval of an antecedent in the representation of the evolving common ground. In this climate it was natural to re-examine the context-dependent conversational backgrounds appearing in Katzer's theory of modals and conditionals as a kind of *presuppositional/ anaphoric* component of modal meanings that required satisfaction in the previously established discourse contexts.

This strand of research, however, did not focus examples such as the *bachelor* example in (22), discussed in § 4.2.8.2 and 4.2.8.3, where the proposition that appears in the antecedent simply *makes explicit* a component proposition of a strictly epistemic or, more generally, alethic conversational background without modifying the nature of the conversational background. Instead, discourse semanticists primarily looked into examples that represented discourse equivalents of conditional antecedents that do alter the composition of the conversational background by adding non-factual "premises" – even if not necessarily counterfactual ones. In other words, since Roberts (1989), work on modality and anaphora has almost exclusively focused on the phenomenon of *modal subordination*, where modals in discourse are interpreted as recovering their conversational background from another modalized discourse antecedent as if from a nonfactual conditional antecedent. The signature examples of this discourse semantic literature are the little *thief*-texts quoted in (30.a, b):

(30.a) A thief *might* break into the house. He *might* take the silver. (from Geurts 1999: 176)

(30.b) A thief *might* break into the house. He *would* take the silver. (from Roberts 1989: 697)

The basic idea behind the notion of modal subordination is a natural extension of Kratzer's views on the context dependency of modal expressions and on the modal nature of conditional constructions, combined with Stalnaker's view of a developing common ground. Here is, in informal terms, the gist of Craige Robert's (1989) analysis of example (30.b).

Two basic operations realized by utterances on the common ground are defined: utterances in the *factual mood* – or assertions – add the propositional content to the common ground, while utterances in the *nonfactual mood* – or hypothetical assumptions "add temporarily" their propositional content to the common ground "for the

[24] For some reason there is little trace of a discussion of cataphoric phenomena in the discourse semantic literature at large.

purposes of exploring the consequences" of the assumption (Roberts 1989: 687).[25] Adding assumptions to the common ground temporarily eliminates some worlds from the context set, and since we do not know whether the assumptions hold true the resulting reduced context set may or may not include the actual world. Roberts (1989: 687) assumes that for the kind of epistemic modals found in examples such as (30.b) the Kratzerian *conversational background* coincides, by default, with the Stalnakerian *common ground*. Thus the modal *might* in the first sentence of the *thief*-text is interpreted relative to a set of relevant factual propositions in the common ground, such as *Given that there has been a lot of theft in this neighborhood, Given that this house has poor security* (Roberts 1989: 698). On the other hand, the prejacent of *might* is non-factual: it is not introduced into the common ground, as it is merely compatible with it. Among the worlds of the context set there is at least one in which the prejacent is true.

The tricky part comes with the interpretation of *would*, which according to Roberts (1989: 699) "has the force of necessity". Clearly, its conversational background is not the same epistemic background of *might*: the second sentence is not saying that the taking of the silver is *necessary* in view of the common ground. Furthermore, the coreference between *a thief* and *he* is problematic: the hypothetical thief is merely a possible individual and we can refer to him only as long as we remain under the scope of the *might*.

In the case of (30.a), where we have a *might* followed by another *might*, extending the scope of the first modal to include also the prejacent of the second sense seems to do the trick. As Geurts (1999: 177) puts it "the second *might* is then construed as a flag which indicates that the hypothetical context introduced by the first *might* is being taken up again". According to this analysis, (30.a) means 'It is possible that a thief will break into the house *and* take the silver'.

This, however, does not work for (30.b), which is not merely saying that the breaking in and the taking of the silver are con-possible[26] instead saying something like "It is possible that a thief will break into the house, and *if he* does, he will undoubtedly take the silver". According to Roberts (1989: 699), the modal *would* functions "like the consequent of a conditional in need of an antecedent" and the required antecedent is provided by the "pragmatic accommodation of a contextually given hypothetical common ground" – that is of the prejacent of *might* – into the place of the antecedent.

Geurts (1999: 184), in revising Robert's initial proposal, contends that the recourse to the notion of accommodation is unnecessary and that the phenomenon of modal subordination can be explained by considering that modal predicates "pre-

[25] Note that the grammatical manifestations of the two "notional moods" are left admittedly vague and it is clear that they are intended as a mere signpost for a not-yet-existent Stalnakerian theory of speech acts in discourse. The interest of Robert's theory lies elsewhere and I will leave the matter at that. For those interested, subsequent works of Craige Roberts develop her Stalnakerian view of discourse pragmatics (Roberts 2012).

[26] By *con-possible*, I mean, in Kratzerian terms, compatible with each other as well as with the common ground. Note that two possibilities may be compatible with the same common ground without being compatible with each other, as in *Maybe a thief will break in. Maybe not.*

suppose their domains" and by explaining the modal subordination phenomena in terms of the anaphorical resolution of this presuppositional requirement.

Here I have limited myself to a very informal presentation of the literature on modal subordination, without giving any detail of Kamp's discourse semantic formalism in which these proposals are couched. At this level, it will be not possible to give an account of the differences between Robert's and Geurt's proposals – nor is it necessary for my present purposes. It will be useful, however, in view of the semantic analysis of modal expressions in Chap. 6 to pick up Geurt's presuppositional view of modal conversational backgrounds, which naturally translate in terms of the Congruity Theory framework I have adopted.

In congruity-theoretic terms a modal is a two-place relational predicate taking the prejacent and the conversational background as its arguments. As it is for any predicate in congruity theory, modal impose presuppostions of different kinds on both their argument places. The description of the presuppositional constraints associated with argument places is an essential component of the semantic analysis of any predicate, and, in the case of modal expressions, the constraints associated with the modal's conversational background are particularly delicate and crucial for explaining the functioning of modal expressions in discourse. Inferences can be triggered in discourse in the process of satisfying or accommodating the presuppositional constraints of any predicate in a discourse context. In the case of context-dependent expressions such as the modals, such a process can become more central as no complete syntactic saturation of the conversational background argument place is possible.

From Geurts (1999: 185) I pick up also a useful notational element. In order to describe the functioning of modals in discourse he defines two relational modal predicates □ and ◊ in terms that are equivalent to the Kratzerian notions of logical consequence and compatibility given above in Def. 10 and 11. So that, according to the Geurts-derived notation,

(Def. 19) p□q = q is a logical consequence of p, or, equivalently, q is true in all the worlds of the "universe" W in which q is true.

(Def. 20) p◊q = q is compatible with p or, equivalently, there is at least a world in W, at which p and q are true.

Note, also, that in Geurts' discursive framework the propositions p and q do not correspond necessary to "sentences" and may correspond to a complex discourse representation, for instance, one resulting from the update of the common ground with a series of successive utterances. The relations □ and ◊ will come in handy in Chaps. 5 and 6 in describing how modal expressions relate anaphorically to previous utterances in discourse.

To take stock of this brief review, the discourse semantic extensions of Kratzer's work have thus far mostly remained concerned with offering formally precise analyses of a limited set of constructed examples, where the conversational background of a possibility or necessity modal is recovered anaphorically from a non-factual assumption introduced by another modal. These analysis have little or no contact with the intricacies of authentic discourse data, a broader descriptive investigation of the 'phoric' properties of modal expressions "in the wild" remains to be done. Furthermore, the fact that they do not focus on the apparently more trivial case of modals pointing anaphorically to factual propositions in the common ground has possibly contributed to the disregard for the *discourse relations* that are created as a result of the anaphorical saturation of the conversational background. These discourse relations are the main reason for the interest in the anaphorical properties of the modals in the present investigation.

4.3 Argumentative Implications of Relative Modality

We are now in a position to draw some preliminary consequences of the Relative Modality analysis for the task of reconstructing the essential structure of arguments in naturally occurring discourse. Further implications of Relative Modality for the analysis of arguments will emerge in section 4, through the discussion of the conversational backgrounds corresponding to the different modal "flavors".

4.3.1 Reconstructing Modalities as Part of the Propositional Content of Standpoints

The most basic task of semantic analysis in argumentative reconstruction is facilitating the clarification of the *propositional commitments of the arguers* (cf. Chap. 2, § 2.3.2). Such a clarification is vital both for the content of the standpoints that discussants put forth and for the arguments that they muster in support of the standpoints. If the utterance of a standpoint contains context-dependent expressions, the task of reconstruction requires that the analyst provides a saturation for their contextual variables that is both pragmatically felicitous, maximally argumentative and charitable.

In this section I would like show that the Relative Modality analysis provides the means for clarifying the contribution of modals, typically non-epistemic modals, to the determination of the propositional commitments of arguers. As it will become clear in the following pages, specifying the exact contribution of context dependent modal meanings to propositional content can significantly impact further aspects of the reconstruction of an argumentative discussion. These further aspects notably include the semantic type of the propositions functioning as standpoints, the *loci*-relations supporting the inference from the premises to the standpoint and the macrostructure of the argument (see, again, Chap. 2, § 2.3.2).

Furthermore, the saturation of the conversational background will provide insights into the reconstruction and evaluation of the confrontation phase of argumentative discussions.

As it will turn out, the contextual specification of propositional modals is in many ways comparable to specifying the contribution of *quantifiers*. This is not surprising, given the well-known parallelism between modals and quantifiers, which Relative Modality further contributes to highlight. Interestingly, once their contribution to propositional commitment has been specified, quantifiers and propositional modals are similar *also* in the way they impact the reconstruction of arguments. It will be therefore useful to briefly discuss the indicative potential of quantifiers in Sect. 4.3.2.

The following pages represent a first reconnaissance of this rich indicative potential. A number of questions arising from it will receive further elucidation in section 4, from the discussion of individual flavors of modality and of their relationship with a semantic typology of standpoints.

4.3.2 Quantifiers as Argumentative Indicators: Revisiting Snoeck Henkemans' Analysis

According Snoeck Henkemans (1992: 117), a "careful interpretation of expressions referring to elements of the *propositional content* of the expressed opinion" impacts on the reconstruction of complex argumentation, as elements of the propositional content can provide "useful clues" to decide whether a *complex argumentation* has to be reconstructed as *multiple* or *coordinative*.[27] Here I revisit what Snoeck Henkemans' (1992) study on complex argumentation says about modals and quantifiers, to show how an analysis of propositional modals based on Relative Modality can enrich and complete it.

In looking at the modals and quantifiers, Snoeck Henkemans is concerned essentially with what I have called *indirect indicators*[28] of argumentation structure. The need of providing compound argumentation may depend "on the commitments that ensue for the arguer from the type of standpoint he is defending" (Snoeck Henkemans 1992: 103). To establish the kind and amount of support that the commitments of the protagonist require, the analyst has to clarify the meaning of the words used in the standpoint using "linguistic knowledge" and "pragmatic knowledge of how what is said should be interpreted in the context" (*Ibid.*).

Two aspects of the structure of the standpoint are investigated: its propositional *content* and the extra-propositional *force* of the standpoint. *Quantifying* expressions are investigated in relation to propositional content, while *modal* expressions are seen as indicators of the *force* (cf. Table 4.1).

[27] *Multiple argumentation* consists of independent arguments, each deemed sufficient to support the standpoint, while *coordinative argumentation* puts forward compound arguments jointly necessary to support the standpoint. Here I assume a full equivalence the *coordinative* argumentation of Pragma-Dialectics and *compound* argumentation in AMT, disregarding certain subtle differences in the approach (cf. Chap. 2, § 2.2.2).

[28] On the notions of *direct* and *indirect* argumentative indicators see the Introduction at § 1.3.

Table 4.1 Levels of meaning and indicators of argumentation structure in Snoeck Hekemans (1992)

Level of meaning	Indicators of argumentation structure
Extra-propositional force of the standpoint	Epistemic modality
Propositional content of the standpoint	Quantifiers

Table 4.2 Quantifiers and modals

	Domain	Force
Quantifiers	Domain of quantification	Universal – Existential
Modals	Conversational background	Necessity – Possibility

As observed in Chap. 3, according to Snoeck-Henkemans (1992: 109) modal words "can be used to indicate the extent to which the speaker is prepared to commit himself to the truth or acceptability of the propositional content of his standpoint". Like quantifiers, they are seen only as *indirect* indicators of argumentation structure: the expression of the degree of commitment "enables the analyst to determine what degree of justificatory or refutatory potential the argumentation should have, in order to lend sufficient support to the standpoint" (Snoeck Henkemans 1992: 113), which, in turn, can help the analyst decide whether each argument is independently capable of supporting the standpoint with the required modality (*multiple* argumentation structure), or the arguments need to be taken jointly in order to satisfy the force that the modal attributes to the standpoint.

Only epistemic modals are considered relevant for the indication of the extra-propositional "force" of the standpoint. Snoeck Henkemans (1992: 109, n. 6) briefly mentions *deontic modalities*, but only to observe that, while the epistemic modalities "do not belong to the proposition itself", the deontic ones "are part of the predicate of the proposition". Conversely, the discussion of the propositional content level only concerns quantifiers and does not involve modalities of any flavor, not even the deontic ones.

The specification of propositional modals can develop, in part, along the same lines that Snoeck Henkemans (1992) follows in examining the quantifiers. Snoeck Henkemans (1992) considers two basic aspects of quantifier meaning: their *domain* of quantification and their *strength*[29], which are indeed exactly parallel to the *conversational background* and *force* of the modals as analyzed in terms of Relative Modality (cf. Table 4.2).

[29] Actually, Snoeck Henkemans (1992: 106) talks about quantifier *scope* rather than *strength* but she uses the term *scope* in a very peculiar, idiosyncratic, way, which does not correspond to the well settled meaning of *scope* in logic. What Snoeck Henkemans (1992) means by *scope*, i.e. the meaning dimension varying along scales such as *all* > *almost all* > *most* > *some*, is more unequivocally captured by the term *quantifier strength*.

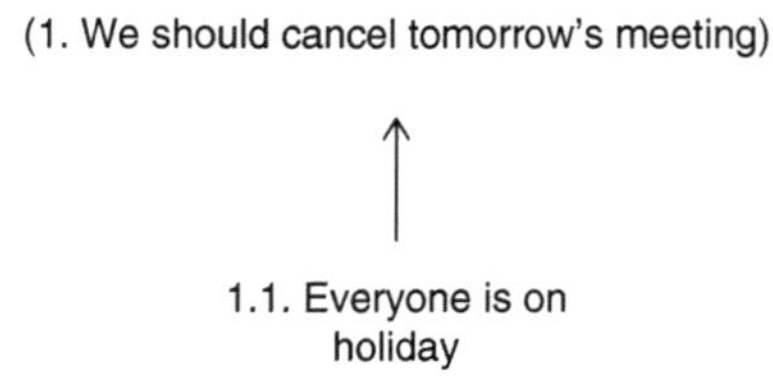

Fig. 4.5 Argumentation structure of example (31). Implicit parts are in parentheses

It is worth pausing on how Snoeck Henkemans Consider example (31), below:

(31) *Everyone* is on holiday (from Snoeck Henkemans 1992: 105)

Imagine, with Snoeck Henkemans, that the speaker has uttered (31) in the context of the scheduling of a meeting in a business context. The utterance has been put forth as an argument for cancelling a meeting that had to take place the following day (cf. Snoeck Henkemans 1992: 105), as shown in Fig. 4.5.

What would count as relevant criticism of the acceptability, relevance and sufficiency of this argument? What kind of further arguments would be needed to support it in the face of criticism? Snoeck Henkemans demonstrates how, in order to answer these questions, the analysts must first mobilize key aspects of the context[30] of utterance to flesh out the *domain* of the quantifier and establish its exact *strength* in context.

Providing the domain of quantification can be analyzed as the pragmatics-driven saturation of a context dependent variable. In fact, this operation closely parallels the saturation of the modal conversational background.[31] Snoeck Henkemans (1992: 105) observes that the physical setting of the utterance already provides indications that restrict the domain of *everyone*. Since the speaker and hearer are obviously at work they are not included in the domain of quantification, which makes the critical reaction in (32.a) inappropriate.

(32.a) But you and I are still here. (Snoeck Henkemans 1992: 105)

The "scheduling" *activity type* in which the arguers are engaged and, even more, the specific *issue*[32] they are discussing ("Should we confirm tomorrow's meeting?")

[30] Snoeck Henkemans' careful analysis covers three levels of context: the physical *setting*, the communicative *situation*, including the overall joint goal of the interaction, and the *institution* where the exchange takes place, which notably includes the social rules and regulations that apply in the communicative situation (Snoeck Henkemans 1992: 104). This three pronged analysis anticipates some aspects of later, and more developed, models of the context of an argumentative exchange, including, in particular Rigotti and Rocci (2006) and van Eemeren (2010). Referring to the former, which is adopted by the present study, the *situation* roughly corresponds to the *activity type*, while the *institution* closely matches the notion of *interaction field* (cf. Chap. 2, § 2.4).

[31] On the parallelism between quantifiers and modals see the discussion of *tripartite structures* in § 4.2.2.

[32] Here I have slightly adapted, I dare say updated, Snoeck Henkemans' (1992) analysis to the analytical categories employed in this book. The original analysis refers to the *situation* rather than to the *activity type* and does not mention the *issue*. In particular, the present research highlights the notion of *issue* as vital for understanding how argumentative discussions are rooted in the social and interactional context. See Chap. 2, § 2.4.

provide a decisive further restriction to the domain of quantification, so that critical reactions such as (32.b) become irrelevant as well.

(32.b) Well, there do seem to be a lot of people in town. (Snoeck Henkemans 1992: 106)

Thus, the domain over which *everyone* quantifies turns out to be an entirely *ad hoc* set composed by the people who are expected to take part in the meeting, minus the two interlocutors. It is remarkable observe how this contextual saturation of the quantifier domains is a necessary step to determine the actual propositional content of the argument and contributes to determine the relevance of critical reactions to it.

The contextual specification of the meaning of *everyone* in (31) is not limited to the saturation of the domain of quantification. Snoeck Henkemans also observes that the lexically determined *strength* of the quantifier is also reconfigured in context. She observes that in informal English usage universal quantifiers are often found used in a "loose sense" to mean *almost all.*

In fact, the interpretation of these loose uses is regulated pragmatically in view of the context of utterance, so that the loose universal quantifier get to mean a quantity that is close enough to the universal quantification to be equivalent to it *for the purposes of the present conversation*. Snoeck Henkemans (1992: 106–107) observes that for (31) the interpretation of the universal quantifier is guided by the knowledge that it is pointless to have a meeting if the majority of the expected participant is absent. So that *everyone* is understood as *almost everyone* or *the majority of the* members. More precise institutional or social knowledge, like the knowledge that "a quorum is required" or that the presence of certain people would be "of more importance than that of the other members" could further guide the loosening of *everyone* to mean *enough to invalidate the meeting*, or *all the important people,* etc.

Such a reconstruction, Snoeck Henkemans (1992: 107) observes, would rule out an objection to the acceptability of the argument such as (32.c):

(32.c) But Rob is still here

The presence of just one of the potential participants is normally not enough to make holding the meeting worthwhile and thus the reaction is (32.c) can be cast off as "over precise".

It is worth here to provide an integration to Snoeck Henkemans' analysis, which further confirms its relevance. It must observed that, contrary to what happens with quantifier *domain*, the pragmatic process at work here is not the contextual *saturation* of an incomplete, "gappy" meaning. While quantifier domains are inherently context dependent, the strength of the universal is semantically complete. The strength of the universal quantifier cannot be considered *semantically vague* either. Contrary to gradable adjectives such as *cool* and *warm*, *big* and *small*, to quantifiers such as *many* and probability expressions (*probable*, *probably*), universal quantifiers are semantically absolute terms as far as their strength is concerned: they do not indicate an indeterminate region of a scale, but its endpoint. The phenomenon at work with *everyone* has been analyzed as "pragmatic vagueness" by David Lewis (1979) and as "loosening" or "loose use" by Wilson and Sperber (2012). Lewis

(1979: 352) observes that the "standards of truthfulness" with which we use semantically absolute words are relative to the development of the conversation:

> When is a sentence true enough? [...] this itself is a vague matter. More important for our present purposes, it is something that depends on context. What is true enough on one occasion is not true enough on another. The standards of precision in force are different from one conversation to another, and may change in the course of a single conversation. Austin's "France is hexagonal" is a good example of a sentence that is true enough for many contexts but not true enough for many others. Under low standards of precision it is acceptable. Raise the standard of precision and it loses its acceptability.

Wilson and Sperber's analysis, instead of postulating context-dependent standards of precision, contend that the loose interpretations of semantically determined predicates arise from the search of a relevant interpretation, where relevance is understood in terms of the *inferences* that the interpretation enables in context. This view is particularly interesting for argumentative situations such as those considered by Snoeck Henkemans (1992). And indeed, as it often happens, the examples provided by Wilson and Sperber (2012) are argumentative in nature, even if the authors do not focus on argumentation in that work. Consider one of the examples they offer:

> Peter and Mary, who are both rather unfit, are discussing where to go on their next cycling holiday. Mary suggests Holland, adding "Holland is flat". From the strictly false proposition that Holland is flat [...] Peter can derive the true conclusion that cycling in Holland is would involve no mountain roads and would not be too demanding. (Wilson and Sperber 2012: 59)

According to Wilson and Sperber (2012: 60) "hearers have no objection to strictly false approximations as long as the conclusions they bother to derive from them are true". In my view, the example is more perspicuously understood if one explicitly addresses its argumentative nature: saying that *Holland is flat* is sufficiently precise for supporting the intermediate standpoint that cycling in Holland is not too demanding, which, in turn, is sufficient to support Mary's practical standpoint on the choice of holiday destination. From the perspective of argumentative analysis, a charitable reconstruction of Mary's propositional commitments will understand *flat* with just the degree of approximation that is sufficient for supporting the intermediate standpoint. This will rule out attacks to the acceptability of the premise that point out, for instance, that the country is not flat because there are small hills.

The strength of quantifiers in the standpoint can be exploited as an indirect indication of the multiple or coordinative structure of the supporting argumentation. When the propositional content of the standpoint is quantified *existentially* "giving *one* example is sufficient to defend the standpoint conclusively" (Snoeck Henkemans 1992: 118). In this case complex argumentation, consisting of several examples of the relevant kind, is more charitably construed as multiple argumentation. Multiple arguments do not add anything to the logical sufficiency of an argument that is conclusive in itself; they add instead to the dialectical robustness of the argumentation: "If one argument turns out to be unacceptable, the other argument still stands" (*Ibid.*: 119). Snoeck Henkeman's discussion of quantified standpoint contains insightful discussions of the quantificational structures hidden is a number of lexical predicates. I expand here on the discussion of one of her examples, concerning predicates like *complete* and *perfect* in order to better show the parallelism with modality. Consider, for instance, the following example adapted from Snoeck Henkemans (1992: 118):

(33) The reference list is not complete. Your article on relevance is missing. And Peter's paper is not on it either.

Absolute predicates such as *complete* and *perfect* contain a covert universal quantification over a certain relevant set of "components" or, respectively, "virtues". When they fall under the scope of external negation they create an existential quantification scoping over internal negation, which corresponds to the O corner of the traditional square of opposition (cf. *Introduction*, § 1.6.1) (Fig. 4.6).

Since the standpoint of (33) is existentially quantified, mentioning one missing item is sufficient to conclusively defend it. It is then charitable to reconstruct the supporting argumentation as multiple, as per Fig. 4.7.

It can be mentioned in passing that each of the arguments in (33) relies on the *Locus from opposition* (cf. Chap. 2, § 2.2.3.2), and, more specifically, on a maxim corresponding to the relation of opposition between the O and A corners of the square (cf. Introduction, § 1.6.1).

When we encounter standpoints whose propositional content includes entities or events that are quantified universally, converse considerations apply to the reconstruction of complex argumentation. Premises concerning individual entities or individual events will be more charitably reconstructed as jointly forming a coordinative argumentation. Even if the universal quantifier is interpreted loosely as meaning *almost all* or even *most* or *all the most important*, *most prominent* in the domain

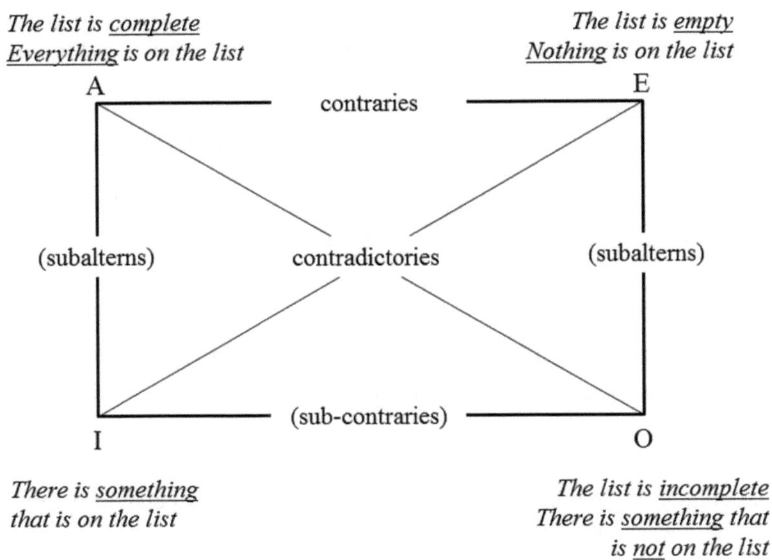

Fig. 4.6 Mapping of the predicates *complete* and *incomplete* onto the quantificational square of opposition

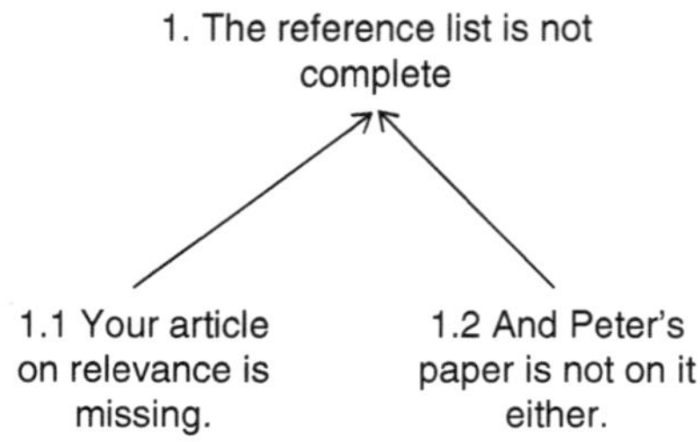

Fig. 4.7 Example (33) reconstructed as multiple argumentation

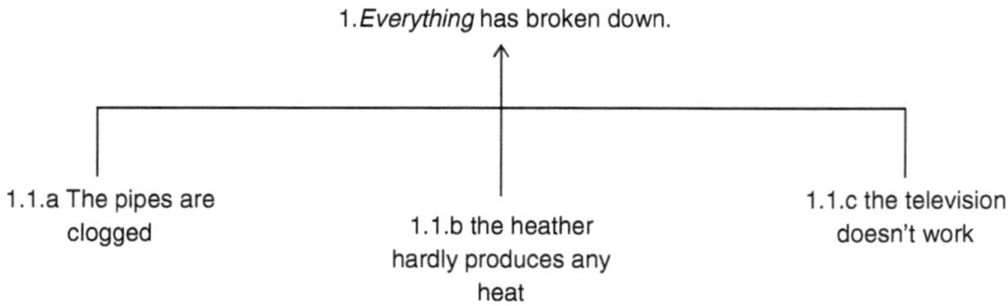

Fig. 4.8 Example (34) reconstructed as coordinative argumentation

presenting a single instance of the set would never amount to a conclusive defense. Consider, again, one of Snoeck Henkeman's (1992: 120) examples in (34):

(34) [The speaker rented cottage and is complaining to the landlord] Everything has broken down. The pipes are clogged, the heather hardly produces any heat, and the television doesn't work.

Even if we interpret *everything* loosely, "the arguer will have to give more than one example of things in the house that are not functioning properly" (*Ibid.*). The argument in (34) is then best reconstructed as a coordinative structure, as in Fig. 4.8.

In terms of the AMT model, the joint inferential structure of the coordinative argument is based on a variant of the locus *From Whole and Parts*, where 1.1a, b and c combine as multiple *data*, recognized as significant components of the rented house according to some appropriate *endoxon* about what equipment one should expect to find in a properly functioning rented cottage (cf. Chap. 2, § 2.2.3.2).

4.3.3 Propositional Modals as Argumentative Indicators of Argumentation Structure and Beyond

Much of what Snoeck Henkemans (1992) says about quantifiers as part of the propositional content of standpoints finds a close parallel in the behavior of modals. I will start here with considering primarily the *force* of non-epistemic, propositional, modals – i.e. the equivalent of quantifier strength – as an indirect indicator of argumentation structure, along the lines of Snoeck Henkeman's treatment of quantifiers. However, in the analysis of the examples of complex argumentation it will become progressively clear how the contextual saturation of the *conversational background*, which corresponds to the *domain* of quantification, contributes to determine what would count as relevant criticism of a standpoint and what kind of arguments would be needed to support it.

Let us start by considering possibility modals, which are the modal equivalent of existential quantification. It is widely acknowledged that the English possibility modal verb *can*, which appear in example (35), is incompatible with epistemic modality in its affirmative form. In fact, the modal is neither epistemic nor deontic. I will later argue that the conversational background in examples like (35) is *alethic* in nature. Portner (2009: 136) expresses differently a similar concept when he ranges such uses under the *dynamic* modals.

(35) Encounters with lions *can* be deadly. A seven-year-old boy and a woman, died in separate incidents after they were attacked by lions on Saturday, in Gir Forest, India. An illegal hunter was killed and four others barely escaped when they were attacked by two lions on a game farm earlier this week in Phalaborwa, South Africa. A few days ago, a safari guide has been mauled to death by a lion in Hwange national park Zimbabwe.

Here each of the separate incidents is sufficient to prove the point that death by lion is a real possibility, according to the Scholastic dictum *ab esse ad posse valet illatio.* As a consequence, the most charitable way to reconstruct the accounts of the separate episodes is by postulating a multiple argumentation, aimed at protecting the standpoint in case the acceptability or relevance of any of the stories is put into question (Fig. 4.9).

Note how in (35) we are clearly are not confronted with a protagonist who is weakly committed to the standpoint and qualifies it with the mere force of possibility (in contrast with "*Perhaps*, encounters with lions are deadly" or "If you walk in the bush, you *might* have a deadly encounter with a lion"). In (35) there is *no hint of doubt* of the arguer towards the propositional content of the standpoint. The standpoint is *epistemically unqualified* and states an objective possibility: the circumstances of encounters between people and lions are such that death by lion is not

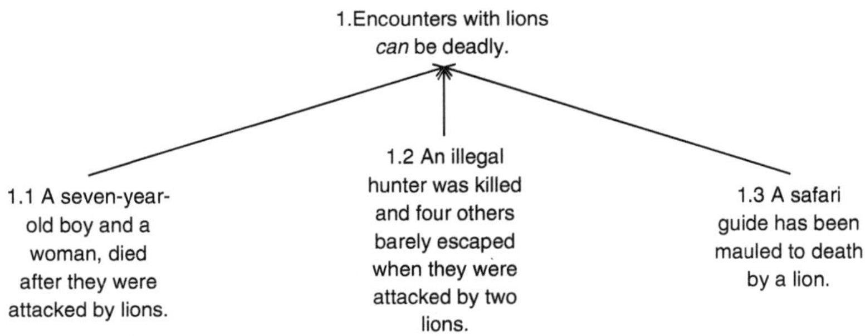

Fig. 4.9 Example (35) reconstructed as multiple argumentation

ruled out. One could have added as well a strong epistemic qualification to (35) without substantially changing its standpoint, as shown by (35.a, b).[33]

(35.a) *There is little doubt that* encounters with lions *can* be deadly.
(35.b) *It's a fact that* encounters with lions *can* be deadly.

As noted in Chap. 3, § 3.5.3, Snoeck Henkemans discusses the case of extra-propositional epistemic modals expressing a *weak commitment* of the arguer towards the propositional content of the standpoint and requiring only *weak evidence* in order to be defended. Thus, when arguers use "weak modalities" to qualify a standpoint supported by complex argumentation, it may be the case that "each of the arguments has sufficient justificatory potential to support the standpoint" (Snoeck Henkemans 1992: 114). As observed in Chap. 3, the overall rationale of this analysis is pragmatic and dialectical rather than semantic and logical: by advancing a standpoint *with a certain degree of confidence*, the arguer *ipso facto* commits himself to providing evidence supporting the standpoint with a *matching degree of force or weight*.

The case of (35) is genuinely different and hitherto unaccounted for. We do not have weak evidence matching a weak commitment to the standpoint, but, rather, the mention of *one situation* sufficiently matching the existential *quantification over situations* expressed by the force of the modal *can*. The rationale is here logico-semantic and closely comparable to what Snoeck Henkemans says of existential quantifiers.

As for the quantifiers discussed in the previous section, the relationship between the propositional possibility modal and the multiple argumentation structure is one of *indirect indication*. The signaling of multiple argumentation structure is not part of the meaning or function of *can*, not even of its post-pragmatic meaning resulting from the contextual saturation of its conversational background. In this respect, it is worth mentioning that there is no direct correspondence between the conversational background of the modal, the Toulminian *criteria*, and the arguments supporting the standpoint. In fact, example (35) closely resembles some of the examples that were presented in criticizing Toulmin's directly argumentative interpretation of modal criteria (cf. Chap. 3, § 3.2.3). Consider, for instance, the example reproduced here as (36)[34]:

(36) John *can* lift 100 kg single-handed. I've seen him doing the like often enough.

---

[33] I have already used this kind of test in Chap. 3, § Section 3.2.3, examples (10), (11) and (12) discussing Toulmin's pragmatic-argumentative analysis of the modals. The test that can be used to verify whether a modal is epistemic/ doxastic or not and consists in putting it into the scope of an epistemic modal expression. Non-epistemic modals, such as *can* in (35.a, b) are embedded without problems, while truly epistemic or doxastic modals produce three possible results: (a) polyphonic shifts in point-of-view as in *Mary believes that Max must be lonely*, (b) opaque, hardly interpretable results, as in *Maybe Max must be lonely*, or (c) *harmonic* interpretations, which do not amount to an embedding of one modal in the scope of the other: *Perhaps, Max may be lonely*.

[34] Example (36) corresponds to (9.c) in Chap. 3.

Both (35) and (36) are based on the locus *From Concomitances*, exploiting one version of the principle *ab esse ad posse valet illatio*: from the fact that certain happenings are concomitant with certain individuals or kinds of situations I infer that they are compatible with their intrinsic features. As suggested in Chap. 3, the conversational background selected of *can* in (36) corresponds to what is possible '*in view of* John's physical build', yet (36) is not a causal argument from John's physical build. Similarly, in example (35) *can* selects a conversational background corresponding to what is possible '*in view of* the dynamics of lion – human encounters', which, in turn, depend on the physical and behavioral features of lions and humans. As noted above, this causal reasoning does not correspond to the argument in (35). It is however not hard to come by in accounts of this kind of deadly incidents, as shown by (37.a) and (37.b).

(37.a) Dollar adds that the male lion was doing what it evolved to do. "Almost any organism around lions might be a potential prey item, and for people to think that they are an exception is folly," Dollar said in a previous interview following a fatal lion attack in South Africa in June. Dollar says danger arises when people allow themselves to be lulled into a false sense of security in the presence of lions or other carnivores. "We don't have claws or big canines or size as an advantage," he says. (news.nationalgeographic.com)

(37.b) While the death of the [...] guide [...] is extremely sad [...] no one should be surprised. Lions are the kings of the bush: the ferocious, 6ft-long carnivorous beast with teeth that can break bone, paws that can knock over a man and claws that can rip open tough hide and can bring down giraffe, buffalo – and in some instances elephant (www.telegraph.co.uk)

To sum up: the indicative potential of the modal *can* in (35) is not *direct* and, more specifically, is not directly tied to the relational structure of the modal captured by the conversational background – the relational structure which I explored in Chap. 3 following Toulmin's footsteps. The indirect indicative potential is rather connected with the nature of existential quantifiers of possibility modals.

The doubt may arise that the use of *can* in (35) is somewhat special. In fact, Portner (2009: 136, 213–220) characterizes these uses of *can* as *quantificational* dynamic modality. *Can,* in (35), says something close, but not exactly identical, to saying that *some* encounters with lions are deadly or that *sometimes* encounters with lions are deadly. According to Portner (2009: 219) quantifying modals, unlike other modals, do not quantify over *possible worlds*, which are *maximal*, and hence *alternative*, situations, but over what he calls "counting situations", which are discrete spatio-temporally connected situations, like the individual incidents in the supporting argument of (35). I will briefly come back to quantificational modalities in Chap. 5, discussing how the composition and internal structure of conversational backgrounds allow us to understand the different modal "flavors" and their relevance for argumentation. For now, it will be sufficient to show that the indirect indicative potential highlighted for *can* in (35) is not limited to the so-called *quantificational* modals. Consider the following example:

(38.a) You *don't have to* walk to the beach. There's a bus every hour. And Rudy has a car and will be happy to give you a lift.

The semantic functioning of *don't have to* in (38.a) is perfectly parallel to *not complete* in example (33). Falling under the scope of external negation the necessity modal *have to* becomes equivalent to a possibility modal scoping over an internal negation (*it's possible not to*), which, again, corresponds to the O corner of the square of opposition. Obviously, this time we have to do with the modal version of the square (cf. Introduction, § 1.6.1, Fig. 1.3).

Here the modality has to do with courses of action that are necessary or possible in view of the objective circumstances of the situation *and* the addressee's desire of going to the beach. We can see that the conversational background is not purely made of objective circumstances because (38.b) would not represent a valid alternative, unless an ironic interpretation is intended:

(38.b) You *don't have to* walk to the beach. You can always stay at the hotel and bore yourself to death.

Following Kronning (1996), I call *deontic-practical* these conversational backgrounds composed of objective circumstances and desires. They will be also discussed in greater detail in the upcoming sections. What is important here is that, in order to support a standpoint corresponding to the O corner, showing *one* alternative course of action not involving walking to the destination is sufficient. Since (38.a) mentions two alternative courses of action, the argumentation is best reconstructed as multiple, as in Fig. 4.10.

Here we have to do with truly alternative situations, alternative courses of action occurring in different possible worlds, rather than with the "counting situations" of the quantificational modal. Yet, the possibility modal still behaves like the existential quantifier as an (indirect) indicator of multiple argumentation structure.

The argument in (38.a) calls for a further observation concerning the correspondence between the propositions appearing in the two arguments and the propositions making up the conversational background of the modal. Clearly, the deontic-practical background of (38.a) concerns what courses of action are possible or required *in view* of the *goals* of the agent and the *means* at her disposal. And indeed the propositional contents of arguments 1.1 and 1.2 make explicit the *means* part of the deontic-practical conversational background. This correspondence is irrelevant to the issue of argumentation structure discussed above, but becomes interesting once the reconstruction of the argument schemes is taken into consideration. In terms of the AMT model of inferential configuration, the arguments in (38.a) are based on a form of practical reasoning drawing simultaneously from the *Locus from final cause* and the *Locus from instrumental cause.* This suggests a double matching (a) between the propositions in the conversational background and the premises of the argument, (b) between modal relation of necessity or possibility and the *locus*-relation. This clearly goes beyond the mere clarification of the propositional commitments of the arguers and I will return to these correspondences in due course.

Another interesting example of the indirect indicative power of modal quantification for argumentation structure is (39), a text circulating through the online social media and (spuriously) attributed to Pope Francis:

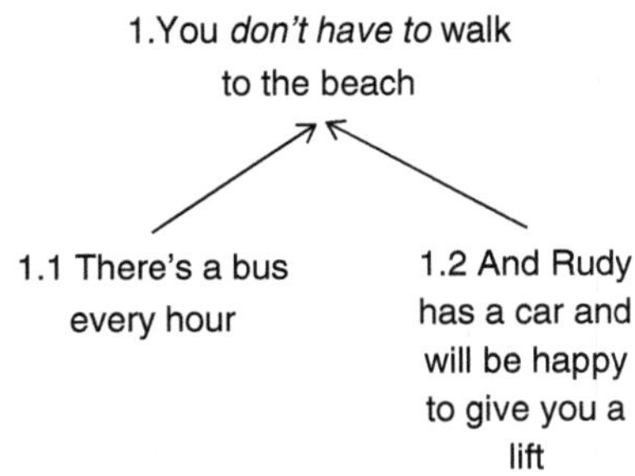

Fig. 4.10 Example (38.a), reconstructed as multiple argumentation

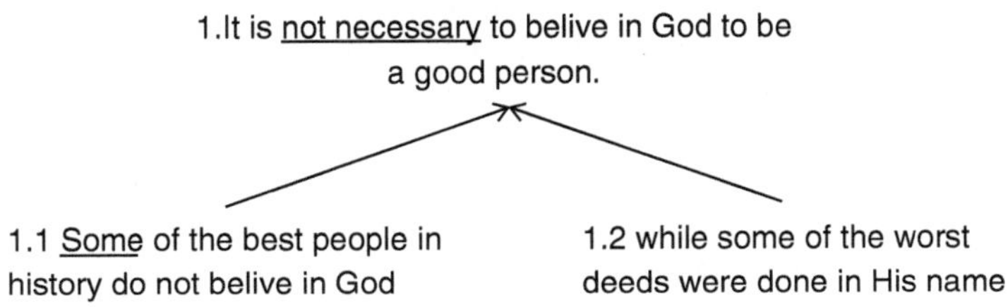

Fig. 4.11 Example (39), reconstructed as multiple argumentation

(39) *It is not necessary* to believe in God to be a good person. [...] *Some* of the best people in history do not believe in God, while some of the worst deeds were done in His name.

Like in the previous examples, the existential quantification that characterizes standpoints in the O corner only requires that *some* of the non-believers are good persons. No further arguments are needed. It is thus natural to reconstruct (39) as multiple argumentation (Fig. 4.11):

Interestingly, the second argument that is being offered (1.2) is irrelevant for supporting the expressed standpoint. The fact that bad deeds were committed by believers can be an argument for the standpoint that it is *not sufficient* to believe in God to be a good person, but says nothing about its *necessity* or lack thereof. The fact that the argument has been charitably reconstructed as multiple allows the first defense to stand on its own. Besides illustrating the dialectical impact of choices of reconstruction pertaining to the structure of complex argumentation, example (39) serves also to show how the concepts of *necessary condition* and *sufficient condition* need to be properly reconstructed as part of the propositional content of the standpoint in order to determine what counts as a relevant or sufficient defense of the standpoint. Both these converse concepts involve modal necessity and a conditional structure and can be clarified in terms of the integrated theory or Relative Modality, as shown later in this chapter.

Up to these point we have found that propositional modalities corresponding to the I and O corners of the modal square, respectively examples (35) and (38.a), (39), can be indirect indicators of multiple argumentation, much like the corresponding existential quantifiers. It remain to be seen if the modalities in the upper corners of the square can be taken as indirect signals of coordinative structures. Let us consider a canonical example used in Pragma-Dialectics to illustrate coordinative structures:

(40) We *must* go to eat out in a restaurant. The fridge is sadly empty. The supermarkets are closed.

As with example (38.a), the modal appearing in (40) is deontic-practical. This time, unlike (38.a), it's a deontic-practical necessity, i.e. what is necessary in view of the goals and factual circumstances. As it was sufficient to point to one alternative course of action to support (38.a) here it is necessary to mention the circumstances that rule out all alternative courses of action in order to make it practically necessary to proceed along the one selected. Each of the circumstances mentioned in the premises rules out *one* course of action and does not amount to a sufficient defense on its own. Thus, it is more charitable to reconstruct it as coordinative as illustrated in Fig. 4.12.

It can be observed that there is an exact correspondence between the factual, circumstantial, part of the conversational background and the arguments jointly supporting the standpoint. Note that the other fundamental part of the conversational background, the goals that, together with the circumstances, necessitates the course of action has been *explicitly* built into the reconstructed standpoint. Such a reconstruction, however, is appropriate only in certain contexts of discussion. The reconstruction in Fig. 4.12 is appropriate in discussions where the standpoint clearly arises as a viewpoint on the issue *what should we do in order to eat tonight?* There could be also situations where such a standpoint emerges in a less focused planning discussion where the issue is *what should we do tonight*? In this case also the goal part of the conversational background should be reconstructed as a (tacit) premise in the coordinative argumentation structure, as shown in Fig. 4.13.

If the argumentation is reconstructed as in Fig. 4.13, the match between the coordinate premises and the propositions that are contextually understood as part of the conversational background is perfect, with both components of the conversational background represented in the premises. Following the relational notation inspired by Geurts (1999), which I have introduced earlier in (Def. 19), the conversational background of *must* in (40) can be represented as in (41), or, by shorthand, as in

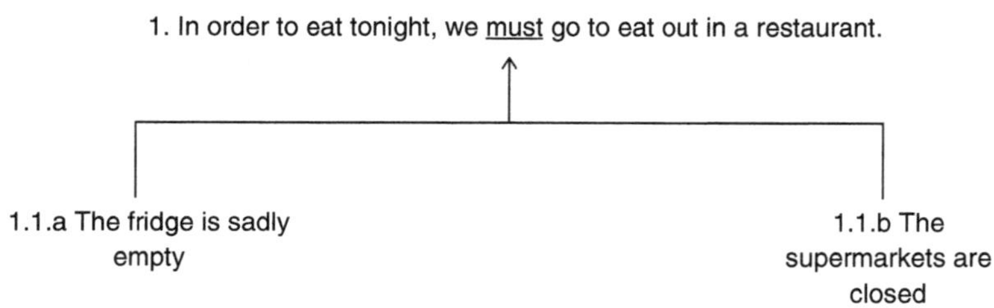

Fig. 4.12 Example (40), reconstructed as coordinative argumentation

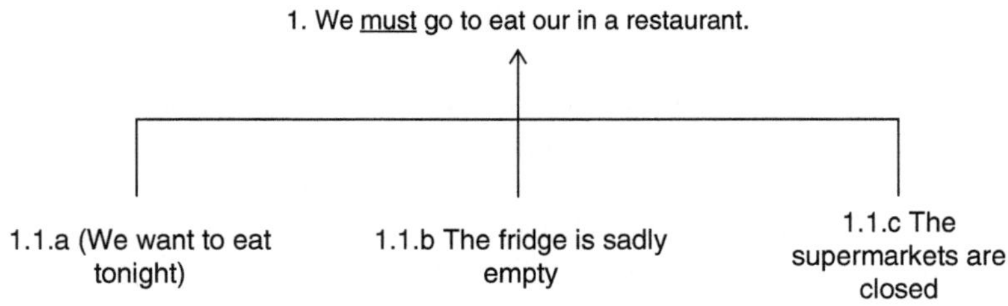

Fig. 4.13 Alternative reconstruction of example (40) as a coordinative argumentation with an implicit goal premise

(42). As customary, curly brackets enclose the elements of a set – in this case the set of propositions making up the conversational background.

(41) $B_{D\text{-}P} \Box$ 'We go to eat out in a restaurant'
$B_{D\text{-}P}$ = {'We want to eat tonight', 'The fridge is empty', 'The supermarkets are closed', …}

(42) {'We want to eat tonight', 'The fridge is empty', 'The supermarkets are closed', …} $_{BD\text{-}P} \Box$ 'We go to eat out in a restaurant'

This way of representing how the discourse context enters into the saturation of the conversational background will turn out to be particularly handy, especially in Chap. 6 where I will engage in fine grained analysis of the saturation of modal conversational backgrounds in discourse context.

Interestingly, an argument like (40) can occur also in context of discussion where the agreement on the goal of having dinner is not exactly presupposed as part of the common ground of the exchange. Rather the addressee is *invited* to accept it as a premise, which makes the whole argument conditional on this acceptance. This is equivalent to the argumentative functioning of conditional antecedents discussed above in § 4.2.8.4. The goal component of the conversational background is not an overt conditional antecedent, but, thanks to the integrated Relative Modality analysis can be shown to play the same logical role.

Example (43), below, presents a variant of the argument were this part of the conversational background is made explicit as a conditional antecedent. Such a move fits with a situation where the proposition is not firmly part of the conversational background, being only weakly presumed. The protagonist wants to prompt its acceptance as a premise, while, at the same time, safeguarding the possibility of easily retracting the standpoint in case the antecedent is challenged. The dialogue in (43) illustrates this possible development.

(43) A: If we want to eat tonight, we *must* go to eat out in a restaurant. The fridge is sadly empty. The supermarkets are already closed.
B: Actually, I think should skip dinner and go straight to bed. I'm not feeling great.
A: Cool. Obviously, I said *if* we want to eat.

In (43) the content of the prothasis *if we want to eat tonight* is a weakly presumed condition that functions like an affirmative counterpart of a Toulminian rebuttal (cf.

Chap. 3, § 3.4.4). As long as it is not challenged by the antagonist the protagonist can commit to the consequent as a standpoint. As soon as the acceptability of the antecedent is challenged, the protagonist can safely back-off from supporting the *consequent*, while maintaining support for the *entire conditional* as a standpoint. In particular when there is a partial manifestation of the conversational background through conditional sentences or other syntactic means, relative modality expressions can be ambiguous between a reading where the *truth* or *acceptability* of the propositions in the conversational background is presupposed and a purely conditional interpretation where there is no presumption of validity for the common ground. The different discursive dynamics arising from this double possibility for deontic and epistemic-doxastic conversational backgrounds would detailed consideration.

As we have seen in § 4.3.2, in addressing the proper reconstruction of the strength of universal quantifiers in arguments, Snoeck Henkemans (1992) pays considerable attention to the pragmatic "loosening" of universal quantifiers. Loosely used universal quantifiers are exempt from a number of attacks which would apply to strictly interpreted quantifiers. Looking at the argument in (40), it seems at first blush that something similar is at work with necessity modals. Consider, for instance, the more or less clearly marginal but strictly relevant attacks in (44.a, b, c):

(44)
(a) Well, we can still break into a closed supermarket and steal the food.
(b) Actually, we can start calling friends and relatives until we find hospitality.
(c) We can always borrow something from our neighbors.

Each of the attacks shows that *going to the restaurant* is not strictly speaking the only course of action capable of satisfying the goal of *eating tonight* in the current circumstances. How can the standpoint of (40) be immunized from (at least some) of these criticisms?

Is *must* used in a "loose sense" in (40)? If this is the case one should come up with an idea of the weakened modal force resulting from the loose use, as it was done with *everyone*, saying that in the context of (31) it means something like *almost all*. This is not easy.

Certainly, there are different shades of "weak necessity" modals such as *should* or *ought to*, whose analysis poses complex problems (cf. Portner 2009: 79–81, von Fintel and Iatridou 2008) but they seem quite different than the *must* in (40). In fact, according to the analysis proposed by Fintel and Iatridou (2008), weak necessity modals such as *should* or *ought to* do not really have a truly weaker force; they do not mean, so-to-say *almost necessary*. They instead add a stratification setting apart the *really good options* over which they quantify and a broader range of options, which are *not-so-good* but are still admissible in a marginal sense. If I say *we ought to eat out in a restaurant*, I roughly mean that this is the only truly good option we have, leaving open the door for other sub-optimal courses of action. With *must* this stratification is absent.

In fact, one can deal with the ruling out of the objections (44.a, b, c) without changing the force of the modal, simply by working on the composition of the conversational background. It is sufficient to assume that, beyond the focused goal of having dinner, other backgrounded goals, desires, preoccupations and values enter the conversational background such as *respecting the law*, which rules out (44.a), and *avoiding face-threatening* situations with relatives and neighbors (44.b, c). As long as these desires can be safely considered part of the conversational background, the objections can be ruled out.

In this respect, we can observe that while (44.a) will probably be irrelevant in most conversations among law-abiding sensible people, (44.b) and (44.c) require that we assume rules of politeness emphasizing strong protection of negative face, such as those of Western-type, urban, individualistic societies. It suffices that we move the imagined discussion to a less individualistic, more community-oriented culture, emphasizing positive politeness, or simply to a rural, tightly-knit, community, that (44.b) and (44.c) become perfectly acceptable suggestions, making the use of *must* in the standpoint of (40) unjustified.

4.3.4 Putting Order into the Conversational Background

In the previous section, building on Snoeck-Henkemans' (1992) remarks on the importance of paying attention to the *domain* and *strength* quantifiers in argumentative reconstruction, I have conducted a parallel examination of propositional, non-epistemic modals considering their *conversational background* and *force*. The *force* of propositional modals emerged as a relevant indirect indicator in determining the multiple or coordinative structure of complex argumentation, similarly to what Snoeck Henkemans (1992) had shown for quantifier strength.

4.3.4.1 A First Appreciation of the Argumentantive Importance of Conversational Backgrounds

While the indirect indication of argumentation structure is tied to the *force* of the modal rather than its *conversational background*, the discussion of examples (35), (38) and (40) has also revealed the important role of the latter. While the arguments supporting a modal proposition as standpoints and the propositions in the conversational background do not necessarily coincide, we can encounter cases when they clearly do. In these cases the nature of the conversational background seems to offer an indication of the *locus*-relation supporting the inferential step from the arguments to the standpoint.

Furthermore, determining which propositions are in the *conversational background* and whether they are part of the *common ground* of the discussion or not, is important to determine what would count as relevant criticism of a standpoint, and what kind of presumed conditions apply to the argument. As we have seen in Chap. 3, § 3.4.4, these conditions are functionally similar to Toulminian rebuttals. Indeed, the *presumptive*[35] nature of an argument seems to be tied more to the status of these propositions in the conversational background (conditions and rebuttals) than to a straight weaker *force* of the modal.

4.3.4.2 Developments of the Notion of Conversational Background: Modal Bases and Ordering Sources

It should be clear by now that the conversational background is a powerful construct. Relative Modality theory uses it to great effect to provide often elegant solutions to a variety of problems in the analysis of modality. As we have seen with the examples of practical reasoning discussed above in (38) and (40), in order to capture realistically the different flavors of modality it is necessary to postulate complex conversational backgrounds that combine different sets of propositions such as, for instance, *factual circumstances* and *desires*, or, similarly, factual circumstances and *moral values*.

A key issue in Kratzer's theory of modality is how precisely the components of complex conversational backgrounds should be combined. The simplest way of doing so would be via a set-theoretic union of the two sets of propositions:

$$B = B_{\mathrm{i}} \cup B_{ii}$$

The problem here is that we can end up with an inconsistent set of propositions. Consider the following example:

(45) You offended Mary. So, you have to apologize to her.

Commenting examples such as (45) Kratzer observes that the conversational background B resulting from the set theoretic union of B_i and B_{ii} would be inconsistent:

B_i (factual circumstances) = 'You offended Mary'
B_{ii} (values) = 'You don't offend people', 'You apologize to the people whom you have offended'

In other words, the necessity modal *have to* in (45) cannot indicate that the prejacent is true in all the worlds consistent with the factual circumstances and with *all* the moral values. This is because the moral values include not offending people and one has to apologize only when someone has been offended. This is a classic paradox of deontic logics.[36]

It is quite clear that an inconsistent B would completely kill the theory of Relative Modality, which bases possibility on the notion of logical compatibility and necessity on logical consequence. No proposition is compatible with an inconsistent set, but any proposition can follow from it. So, no prejacent would ever be *possible* with

[35] On presumptive argumentation see the Introduction, § 1.6.2.

[36] This is known as the Samaritan's Paradox, see Portner (2009: 58–63). Portner discusses different version of the paradox, all concerning obligations that only arise in imperfect situations.

Table 4.3 Modal base and ordering sources for example (40)

Conversational Background 2 (*Ordering Source*)	
Goals and desires:	We want to eat tonight
	We don't want to violate the law
	We don't want to embarrass anyone and create awkward situations
Conversational Background 1 (*Modal Base*)	
Factual circumstances:	The fridge is empty.
	The supermarkets are already closed.
	…

respect to an inconsistent B, and any prejacent would simultaneously be *necessary* and *impossible*.

The paradoxes that arise in the modeling of deontic and practical reasoning are one of the reasons that motivate Angelika Kratzer to go beyond the basic version of Relative Modality exposed here in § 4.2 and introduce, since Kratzer (1981), two distinct components of the conversational background, called the *modal base* and the *ordering source*, each corresponding to a set of propositions belonging to a distinct logical type, as illustrated in Table 4.3 above.

While the job of the *modal base* corresponds to what the conversational background does in the basic theory, the *ordering source* works differently. This set of proposition is used as a *set of standards*, a benchmark or point of reference, to sift through the alternatives compatible with the modal base and retain *only those that are best* as regards respecting the standards.

Ordering semantics involves additional formal machinery providing a ranking of the worlds at which the propositions of the *modal base* are true. The ranking is done on the basis of the propositions of the *ordering source*. In example (45) the deontic ordering source would tell us that while no world in which B_i is the case is entirely good – because in them 'You don't offend people' is always false. However, the worlds in which at least 'You apologize to the people whom you have offended" is true *are better* than those in which also this proposition is false.

Formally, an ordering source is simply a conversational background, that is a function *g* from worlds to sets of propositions. Like function *f* of the modal base, also *g* for any world of evaluation w, will give a set of propositions g(w). The difference between the sets f(w) and g(w) resides in their use. The set f(w) is used to provide a simple restriction of the set of worlds (e.g. we consider only the worlds at which 'The fridge is empty' and 'The supermarkets are already closed' are both true). The ordering source g(w) is used to provide an ordering of worlds $\geq_{g(w)}$. For instance, let us take as g(w) the set of goals in Table 4.3. If w_1 is a world where all the goals are realized and w_2 is a world at which the goal of abiding the law is not realized, then w_1 will be a better world than w_2 according to g(w). Consider now a world w_3 where I abide the law, but I embarrass myself by begging food from the neighbors. Both w_2 and w_3 will be *worse* than w_1 according to g(w), but nothing could be said on that basis about the ordering of w_2 with respect to w_3. Ordering sources provide a partial ordering of worlds.

In ordering semantics (cf. Kratzer 2012 [1981]: 40–41) the notion of *necessity* has to be redefined as truth at *all* the best worlds of the modal base according to the ordering source, those that "come losest to the ideal determined by the ordering source". *Possibility* becomes truth at *some* of the best worlds of the modal base according to the ordering source. Interestingly, while the notion of necessity is weakened by ordering semantics, the notion of possibility is strengthened: it is not sufficient for a proposition to be true in a "marginal" world to be a "real" possibility. I will come back to this idea when discussing the argumentative role of possibility modals.

Beside preserving the consistency of complex conversational backgrounds involving different kinds of propositions (e.g. facts and values, or facts and counterfactual propositions), ordering semantics is also aimed at deriving graded notions of modality, such as, for instance, comparative probability. Von Fintel and Iatridou (2008) use *multiple ordering sources* to render what I have called the "stratification" of weak necessity modals:

> Strong necessity modals say that the prejacent is true in all of the favored worlds, while weak necessity modals say that the prejacent is true in all of the very best (by some additional measure) among the favored worlds. While the standard Kratzer framework parametrizes the semantics of modals to two parameters (modal base and ordering source), we introduced a pair of ordering sources: (i) the primary one that is the only one that strong necessity modals are sensitive to and (ii) a secondary one which is the one that weak necessity modals use to refine the ranking of the worlds favored by the primary ordering source. (von Fintel and Iatridou 2008: 136–137)

According to their analysis, the conversational background of *ought to* in the context of (40) would look roughly as Table 4.4:

Table 4.4 Modal base and ordering sources for *we ought to eat out in a restaurant*, in the context of example (40)

Conversational Background 3 (*Secondary Ordering Source*)	
Goals and desires:	We don't want to violate the law
	We don't want to embarrass anyone and create awkward situations
Conversational Background 2 (*Primary Ordering Source*)	
Goals and desires:	We want to eat tonight
Conversational Background 1 (*Modal Base*)	
Factual circumstances:	The fridge is empty.
	The supermarkets are already closed.

4.3.4.3 The Argumentative Implausibility of Kratzer's Ordering Sources

The clear downside of the theory of Ordering Semantics with respect to the original simpler form of Relative Modality is that the neat connection of modality with logical consequence and compatibility is considerably blurred if not wholly lost. In particular, what is worrisome in view of an interface between modality and argumentation is that the beautiful isomorphism between the semantic structure of modalities and natural reasoning processes, which motivated the Lewis' nickname *premise semantics* for the theory, disappears.

In Ordering Semantics the "premise sets" are no longer used as premises are used in in ordinary reasoning. In our ordinary life we *do* sometimes compare alternatives with respect to a bundle of criteria. For instance we may want to do that when we buy a new smartphone: we compare the technical characteristics, the initial price to be paid, the contracts and consumption costs. In the end we may come up with the conviction that we *should* buy brand X because it fits best our criteria, or that we *can* buy either X or Y because they rank equally. We are usually not very good at doing that, especially when the criteria are many – and phone companies know that! If we are serious about that, we should sit down with paper and pencil lest something escapes us. This is a very specialized kind of reasoning and it is certainly *not* the kind that takes place when I decide that, since I offended Mary I should apologize to her. Here no comparison takes place: I simply pick up the normative proposition 'If you offend someone, you apologize to the offended' and I make it my B_{ii}, which I consider jointly with the facts in B_i. The proposition 'You don't offend people' is simply out of the picture as no longer relevant to determine my course of action, I don't pick up it when I create on-line the B_{ii} that the modal presupposes. I leave it out together with countless of other values that don't enter the picture in this particular case.

For this reason, in contrast with much of the current formal semantic literature on modality, this book does not adopt Kratzer's Ordering Semantics as a complement to the Relative Modality hypothesis. Here is the position I take in relation to the inconsistencies that might arise, for instance, between facts and values in deontic reasoning: I simply assume that no inconsistencies arise from the set-theoretic union of different sets of propositions to form a complex conversational background. In practice, the conversational background *is contextually created by invoking only the relevant propositions*. It is assumed that the pragmatic inference mechanisms that determine the saturation of B_i. and B_{ii} – to the extent to which this saturation is made precise – are powerful enough to rule out that inconsistencies of this type arise.

Admittedly, this amounts to setting up a pragmatic oracle in order to simplify the linguistic semantics – something that some of my semanticist readers are probably not going to find very exciting. Explaining how people manage to select the relevant set is certainly a complex issue, but, perhaps, this complexity is wholly pragmatic and should not be saddled on the semantics of modal expressions. See also the defense of a similar pragmatic view in Papafragou (2000).

It has also to be said that ordering sources, as implemented by Kratzer and others, do not provide a clear solution to the problems of teleological hierarchy that are encountered in moral and practical reasoning. The propositions in g(w) provide the ranking of worlds, but are not ranked themselves; they are a simple set. So, there is no *clear* way to say that abiding the law is a more important value than avoiding embarrassments, and worlds w_2 and w_3 turn out to be incomparable in the example $\geq$g(w) of Table 4.3.

4.3.5 Divergence, Vagueness and Shifts in Conversational Backgrounds

The pragmatic, contextually negotiated, nature of conversational backgrounds turns out to be crucial also in examining their contribution to the determination of the propositional content of standpoints in argumentative confrontations. Indeed, the contextual determination of the conversational background is necessary to determine the propositional content of the standpoint when it consists of a modal proposition. Consider example (46) below:

(46) You *must* give back the car to your brother.

The sentence in (46) can embody quite different standpoints according to the way in which the type of conversational background is narrowed down:

(46.a) <u>In view of the law</u>, you *must* give back the car to your brother.
(46.b) <u>In view of your previous commitments</u>, you *must* give back the car to your brother.
(46.c) <u>In view of the bonds of brotherhood</u>, you *must* give back the car to your brother.
(46.d) <u>In order to achieve your goals</u>, you *must* give back the car to your brother.

A protagonist can be committed to any one of the above separate standpoints without being committed to any of the others. Critical reactions pertaining to one of them may be utterly irrelevant for the others.

As observed by Lewis (1979: 354), the modal conversational background determines “the boundary between the relevant possibilities and the ignored ones”. Its specification is part of what Lewis calls the *conversational score*, i.e. of a dynamically developing common ground updated by each successive utterance. As such, the modal conversational background “may change in the course of conversation” (*Ibid.*).

4.3.5.1 Diverging Conversational Backgrounds and Defective Confrontations

Thus, narrowing down the type of conversational background is essential for properly reconstructing the *confrontation stage*[37] of the discussion (cf. van Eemeren and Grootendorst 2004: 60) and for detecting defective confrontations, where antagonists misrepresent the protagonist's standpoint, i.e. contradict or express doubt towards a proposition that has not been advanced as a standpoint.[38]

Furthermore, since in many cases the propositions making up the conversational background are taken to be part of the common ground of the arguers as *true facts*, *accepted norms* or *shared goals* the actual contextual saturation of the propositions making up the conversational background is desirable in view of establishing the *opening stage* of the discussion.

Improper confrontations can be the result of arguers intentionally and opportunistically committing the straw-man fallacy (cf. Chap. 2, § 2.1.1), but they could equally derive from a simple misunderstanding of the protagonist's standpoint. Careful consideration of the context dependency of modal meanings reveals a third possibility, where the confrontation is defective due to *unrecognized deep disagreement*.

An interesting case is represented by discussions involving deontic standpoints that take place between people that do not share the same set of values, norms, or behavioral rules, but might not know or might not be fully aware of the fact that they don't share them. This is because even if we narrow down correctly the contextual selection of the B, to a set of propositions such as *moral values* as opposed to the *positive by-laws*, this set of propositions might remain nevertheless a "black-box". The arguers might covertly disagree on *exactly* what propositions make up the B set.

This can happen, for instance, in situations of intercultural communication where people might tacitly refer to different traditional ethnic or religious values. For instance someone might utter

(46) You *must not* contradict John in that way!

The propositional content of the standpoint consists of a modal proposition where a necessity modal (*must*) scopes over an internal negation (*not*): the resulting modal force is *impossibility*, corresponding to the right upper corner (E) of the modal square of opposition. In terms of Relative Modality this is treated as *incompatibility* between the prejacent and the conversational background. Imagine that the speaker utters (46) intending a, supposedly shared, deontic conversational background B_1 that includes the value

---

[37] "In the *confrontation stage* of a critical discussion, it becomes clear that there is a standpoint that is not accepted because it runs up against doubt or contradiction, thereby establishing a [...] difference of opinion" (van Eemeren and Grootendorst 2004: 60).

[38] In Pragma-Dialectics this corresponds to the violation of the *standpoint rule*: "Attacks on standpoints may not bear on a standpoint that has not actually been put forward by the other party" (van Eemeren and Grootendorst 2004: 191).

(47) Openly contradicting people older than you is disrespectful

which is not made explicit. The antagonist might object to this standpoint, correctly singling out the standpoint as pertaining a social/ moral obligation, but having in mind a different set of cultural values B_2 which does not include the above commandment but rather the commandment

(48) Speak your mind openly and honestly.

One could even say that, since the saturation of the conversational background B is *covertly* different, the standpoint being *put forth* and the standpoint being *objected to* are not exactly the same:

(49) Standpoint put forth by the protagonist =
The action of you (= the respondent) *contradicting John is incompatible with the set of cultural norms* B_1.
Where B_1 = { 'Openly contradicting people older than you is disrespectful', ...}

(50) Standpoint objected to by the antagonist =
The action of me (= the respondent) *contradicting John is incompatible with the set of cultural norms* B_2.
Where B_2 = { 'Speak your mind openly and honestly', ...}

One thing is to argue *within* a certain set of norms – e.g. two lawyers arguing within the same legal system – another thing is to move to a higher level discussion about identified disagreements about the propositions making up the deontic conversational background and their hierarchy within the value system. If such a disagreement remains covert in the implicitness that the use of modals allows, such a discussion might never begin, to the detriment of intercultural understanding.

4.3.5.2 Vague Conversational Backgrounds in Argumentative Discussions

Whether they are exploited as sets of premises or not, conversational backgrounds of non-epistemic modals do enter in every case into the content of the standpoint and they have to be made explicit if we want to reconstruct the argumentation in order to evaluate its soundness.

We have already noticed that in order to use a modal is not necessary that the addressee should be able to recover precisely *every single proposition* in the B set. This level of specification of the conversational background is not necessary for the modals to function in communication. For instance, a sentence such as (51),

(51) Unfortunately, tonight Björk *cannot* sing.

can be uttered to hint that there are some factual circumstances that prevent Björk from singing tonight. Some sort of causal relation is inferred, but the nature of the presupposed *causes* that make the event impossible can remain completely unspecified. When is a modal interpretation too vague to be meaningful?

As Lycan (1994: 195) poignantly observes "when the context fails to supply any very specific cue" for the determination of the modal conversational background, "a modal assertion is often utterly pointless", as amusingly illustrated in the following quotation from a popular novel:

> "And the insurance?" Callaway asked. "When may the beneficiaries expect to have the claim approved?" Dora smiled sweetly. "As soon as *possible*," she said, and shook his hand. (L. Sanders, *The Seventh Commandment*, quoted in Lycan 1994).

Certainly, when a sentence such as (51) is used as a standpoint or as an argument in support of a standpoint (e.g. *Tonight's concert should be cancelled*) and a more precise saturation is not available in the common ground of the discussion, the antagonist is entitled to ask the speaker to provide it with a move such as (52):

(52) What exactly prevents her from singing?

The antagonist is entitled to ask the *wh-* question because the existence of causes preventing the singing is a factual presupposition associated with the conversational background of *cannot*. In the perspective of Congruity Theory, the presuppositions associated with the conversational background of a modal are not too different from those associated with argument places of ordinary predicates (cf. Chap. 2, § 2.2.3.1). In particular, the conversational background resembles those covert argument places which do not require syntactic manifestation and need to be saturated in context, such as those discussed earlier in this chapter at § 4.2.5.

From a dialectical viewpoint the speech-act performed by the antagonist with (52) is interesting. In pragma-dialectical terms (cf. van Eemeren and Grootendorst 2004), it can count as a *request of usage declarative*: by asking to saturate the conversational background more precisely, the antagonist is basically asking what exactly the protagonist means by *cannot.* At the same time, it cannot escape how this act can function also indirectly as a *request of argumentation.* By asking about the facts preventing the singing, the antagonist not only asks for arguments in support of the standpoint, but asks for arguments pertaining to one specific locus, the locus *From the efficient cause*, thereby steering the discussion towards the evaluation of the *acceptability* of the propositions composing the conversational background and their *relevance* as causes of the prejacent.

4.3.5.3 Shifts in Conversational Backgrounds in Argumentative Discussions

There are discussions where conversational backgrounds, instead of becoming more precise as an effect of antagonistic questioning, simply keep shifting, and shift in the direction of broader, less restrictive conversational backgrounds, admitting a further range of possibilities.

This phenomenon was hinted at by David Lewis in his famous article on *scorekeeping in a language game* (Lewis 1979). Lewis imagines a conversation where A – an elected official – discusses with B about the ways he might deal with an "embarassment":

(53) A: "You see, I *must* either destroy the evidence or else claim that I did it to stop Communism. What else *can* I do?"

B: "There is another *possibility*, you *can* put the public interest first for once!" (Adapted from Lewis 1979: 354)

Here the shift in the modal conversational background is quite sharp, and results in a broadening of the range of relevant possibilities as participant B proposes a course of action which is arguably not a possibility consistent with the quite restricted deontic-practical conversational background presupposed by participant A – something like *In view of my goal of being elected again.*

According to Lewis (1979: 354), once a conversational background is established it normally "remains in force until further notice to govern the interpretation of modal verbs in subsequent sentences". In (82), however, there is no advance notice of a change in the conversational background. As Lewis (1979) observes, here the hearer is asked to shift the conversational background by "accommodation".

In fact, the shift serves to open a new argumentative dialogue game embedded in the other: before deciding what to do, we need to discuss what kind of values motivate our actions.

In general, Lewis (1979: 355) observes that shifts in the conversational background of the modals generally operate in an *outward* direction, to bringing into consideration "possibilities hitherto ignored". On the other hand, "inward" shifts, asking to disregard possibilities that were previously considered, are seen as more difficult to realize. Lewis does not find any sound epistemological reason underlying this asymmetry. If this is the case, the shifting of conversational backgrounds may represent a powerful rhetorical device to constrain the development of argumentative discussions.

4.3.6 Modals as Indicators of Argumentative Discourse Relations

The preceding sections have explored the indirect indicative potential of propositional modals in the reconstruction of complex argumentation, noting, in passing, the importance of conversational backgrounds for reconstructing the *loci*-relations supporting the argument's inferential step and for determining the presumptive conditions attached on the argument. Finally, the dialectical consequences of the divergence, vagueness and shifts in the conversational background have been briefly explored.

It is now time to focus more tightly on what is possibly the simplest and more direct argumentative implication of the theory of Relative Modality: the possibility that modal expressions act as *direct* indicators of discourse relations between argument and standpoint. In other words, the presence of the modal helps the analyst to establish that a discourse unit realizes the advancement of a standpoint and other discourse units are functioning as arguments in support of this standpoint. One could say that modals act here *direct indicators* (cf. Introduction, § 1.3) of this relation as it is a direct consequence of the relational structure of the modal predicates.

In fact, things are somewhat less straightforward. A key distinction will have to be made between relations conveyed by the epistemic and doxastic modals and those conveyed by the other flavors of modality. As it will turn out, *neither* kind of relations directly coincides with the argumentative discourse relations at the pragmatic, illocutionary level, which Congruity Theory captures with the notion of *connective predicate* (cf. Chap. 2, § 2.2.3.2). They are connected, in different ways, to argumentative connective predicates, they guide hearers (and analysts) to reconstruct them, but neither kind of relation represents a truly performative straightforward expression of the illocutionary level of connective predicates. In this respect, the *directness-indirectness* of the indication is a matter of degree.

4.3.6.1 Establishing Discourse Relations Through the *Phoric Saturation of Conversational Backgrounds

Before going into the differences between the epistemic-doxastic modals and other flavors of modality, I shall briefly illustrate the basic discourse dynamics through which argumentative discourse relations arise from the relational structure of the modal predicates. These dynamics are common to both classes of modals.

Translating the insights of Relative Modality in the terms of Congruity Theory, we can analyze all linguistic expressions of modality (i.e. *morphemes*, *lexemes* or *constructions*) as (at least[39]) two-place relational predicates taking the *prejacent* and

[39] This leaves the door open to some modals having *more* than two arguments. This seems reasonable for at least some flavors of modality, as it will be shown in Chaps. 5 and 6. For epistemic modals, it makes sense to recognize an *epistemic subject* slot even if it has no regular syntactic mapping. For many epistemic modals, the epistemic subject by default maps deictically on the speaker of the utterance. These additional arguments I call *subsidiary arguments*, they are, in fact, subsidiary to the conversational background, in the sense that *their main job is precisely to determine the conversational* background: a *set of goals*, for instance, makes sense only relative to an agent or group of co-agents. They must be *someone's* goals (cf. Portner 2009: 187).

For instance, deontic-practical modals should include an argument place for the *agent* whose goals and actions are being considered. Similarly, among the alethic modalities, dynamic modals concerning what an agent can and cannot do should include an *agent* slot. Deontic modals corresponding to *positive law* should include two additional arguments: the *deontic subject*, i.e. the person – physical or juridical – who is under the obligation, and the *deontic source*, i.e. the juridical or physical person whose authority originates the positive obligation. Similarly, deontic modals corresponding to interpersonal *commitments* reasonably have slots for the person who is committed and the person with whom the commitment has been taken. For epistemic modals, it makes sense to recognize an *epistemic subject* slot even if it has no regular syntactic mapping. For many epistemic modals, the epistemic subject by default maps deictically on the speaker of the utterance. These arguments are subsidiary to the conversational background; in the sense that *their main job is precisely to determine the conversational* background: a *set of goals*, for instance, makes sense only relative to an agent or group of co-agents. They must be *someone's* goals (cf. Portner 2009: 187). When there is an overt syntactic manifestation of these arguments, it functions like the *in*

the *conversational background* as their arguments. Modals impose presuppostions of different kinds on both their argument places. As for any argument slot, the presuppositional constraints associated with the *conversational background* need either to be *satisfied* or *accommodated* in the common ground of the communicative interaction. In the case of context-dependent expressions such as the modals, these processes can acquire a special significance because no complete syntactic saturation of the conversational background argument place is possible. Hearers will seek to saturate the set of propositions required by the conversational background by recovering suitable propositions in the common ground of the exchange.

Within the common ground, previous utterances in an ongoing discourse constitute a privileged search space for the saturation of their context-dependent variable. In this sense we can say that modal expressions are *instructional*: they include "a virtual instruction to the addressee to examine the common ground of the conversation (along with other interpretive content of the sentence) to fill in some partially specified part of the intended interpretation" (Kay 2004: 676) When this *anaphoric* saturation is realized meaning relations are established between utterances in discourse as a simple byproduct of the saturation process. Consider the following trivial example from a newspaper headline:

(54) £3 for a sausage roll? It *must* have been gourmet. (www.telegraph.co.uk, collected via WebCorp)

A key proposition of the conversational background of epistemic *must* in (54) is easily recovered by anaphora from the preceding sentence, thereby establishing a relation between a standpoint and an argument, as per Fig. 4.14.

---

view of phrase in providing a restriction of the conversational background. A relative common view among linguists is that for modal verbs expressing deontic and dynamic modality the *agent/ deontic subject* argument is systematically mapped onto the syntactic subject of the modal. In generative syntactic parlance, they are seen as *control predicates* (cf. Portner 2009: 187), while epistemic modals, in which the syntactic subject is devoid of semantic function, are *raising predicates*. This is a mistaken view, at least for deontic and deontic-practical modals. One can easily find many examples where the agent or subject of obligation is not the syntactic subject. One of the first scholars to notice it was St. Anselm of Canterbury in the XI century (cf. Rocci 2005a, b: 141). Wurmbrand (1999) and Rocci (2005a), on the basis of examples of this kind, argue that *for all flavors* modal verbs should be regarded as raising predicates. Portner (2009: 188) proposes a nuanced position according to which "some deontic modals are control predicates and some are raising predicates". For the purposes of the present investigation it is not necessary to enter this debate. In the perspective of Congruity Theory arguments can be recognized on semantic grounds irrespectively of the systematicity of their syntactic manifestation (cf. the analysis of *foreigner* above in § 4.2.5), and subsidiary arguments remain semantically interesting even if they are only sporadically mapped on complements. Even for epistemic modals, which are, by general consensus, *raising predicates* it makes sense to recognize an *epistemic subject* slot even if it has no regular syntactic mapping. For many epistemic modals, the epistemic subject by default maps deictically on the speaker of the utterance. Yet, when the modal proposition is embedded in reported speech or in an attitude report (e.g. *Mary thinks John must be sick*) the epistemic subject is shifted to the subject of the attitude or speech verb (cf. Hacquard 2010: 80). Similar shifts can be obtained through free indirect speech or merely by clues of a shift in point-of-view. It is finally possible to have epistemic modals lacking a proper subject and referring back to a mere information base.

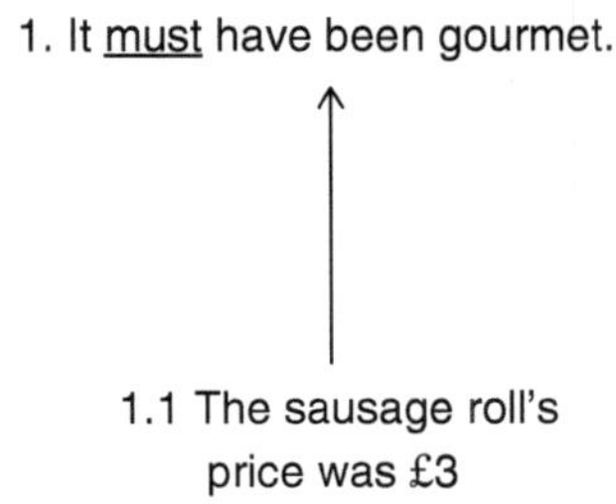

Fig. 4.14 Argumentation structure resulting from anaphoric saturation of the conversational background of *must* in (54)

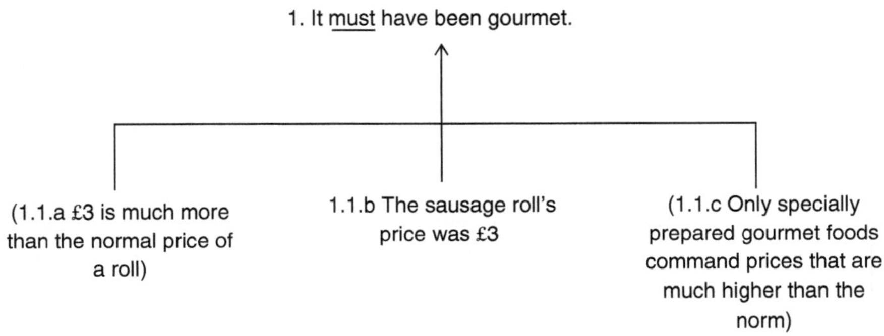

Fig. 4.15 Argumentation structure resulting from the justification of the conversational background of *must* in (54)

This process of *presupposition satisfaction* is typically complemented by a process *presupposition accommodation* through which the propositions of the conversational background are partly constructed on-line in order guarantee the congruity of the modal predicate in the context of utterance. The process of presupposition accommodation, however, should not be imagined as wholly separate from satisfaction. Accommodation does not create propositions *ex nihilo* when satisfaction fails. Kamp (2001: 58), speaking of presupposition accommodation in general, observes:

> [Accommodation] does not mean, however, that whenever direct verification fails, the unverified presuppositions get accommodated lock, stock and barrel. There are many instances where the context, while failing to verify the presupposition at issue, nevertheless contains much of what is needed for their verification; just a small bit of information is missing to make verification complete.

In the processing of argumentative discourse, justification of the conversational background leads to the construction of propositions that function as unexpressed premises in the argument. In the case of (54) this process of presupposition justification would plausibly at the set of implicit premises illustrated in Fig. 4.15, which justifies the force of the strong epistemic modal:

When analysts reconstruct unexpressed premises in arguments their reconstruction can find a justification if the premise naturally arises as a justification of the presuppositions of a modals conversational background. Consider also how the conversational background is reconstructed in example (55):

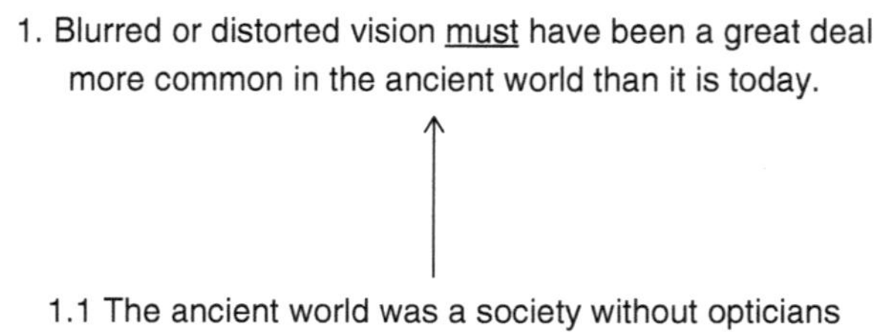

Fig. 4.16 Argumentation structure resulting from the justification of the conversational background of *must* in (55)

(55) As for blurred or distorted vision, this was disabling in the ancient world, as it is today, but it *must* have been a great deal more common in a society without opticians (*Los Angeles Times*, collected via WebCorp)

Here, a key proposition of the conversational background of *must* is recovered from a circumstantial complement which is part of the very clause in which the modal appears. This recovery leads to establish a local standpoint-argument relation (Fig. 4.16).

The establishment of a local standpoint-argument relation in (55) is an instance of a broader phenomenon that Jerry Hobbs (2010) *calls clause-internal discourse relations*. As it will be shown in Chap. 6, modal expressions often signal these short-range, local, argumentative relations.

In the article, the inferred proposition (i.e. local standpoint) is then used as an argument in support of the idea that Goliath's vision impairment (a consequence of his supposed acromegalic gigantism) was not noteworthy for his contemporaries. The argumentative step is again signaled by *must*, which, this time operates intesententially (56).

(56) Today, 75% of Americans use corrective lenses; how much less noteworthy *must* Goliath's vision impairment have been in his own time? (*Los Angeles Times*, collected via WebCorp)

In a combined rational reconstruction of examples (55) and (56), illustrated in Fig. 4.17, the first sentence of (56), reconstructed as 1.1.1.a, provides further support to the local standpoint of (55), now reconstructed as 1.1, and is only indirectly an argument for the standpoint 1 of (56). In should be observed, however, that from the perspective of a normal reader the process of saturation of the conversational background of the second *must* seems mostly tied to the establishment of the more *local* relation R2 between the two sentences of (56).

In other words, the instruction to saturate the conversational background of the modal *must* in (55) and (56) leads the reader to immediately establish discursive relations (R1 and R2) that are *consistent* with a rational reconstruction, but are *under-informative* with respect to it. A "reader-for-gist" (cf. O'Halloran 2003: 170–171), minimally competent from a pragma-linguistic point of view, knows less than a critical reader or analyst.[40]

[40] Kieran O'Halloran (2003) bridges insights from Critical Discourse Analysis and from psycholinguistics to build an interesting model of the inferences drawn by a "non-critical reader" of news texts. By "non-critical reader" O'Halloran means "a reader who (1) is largely unfamiliar with the

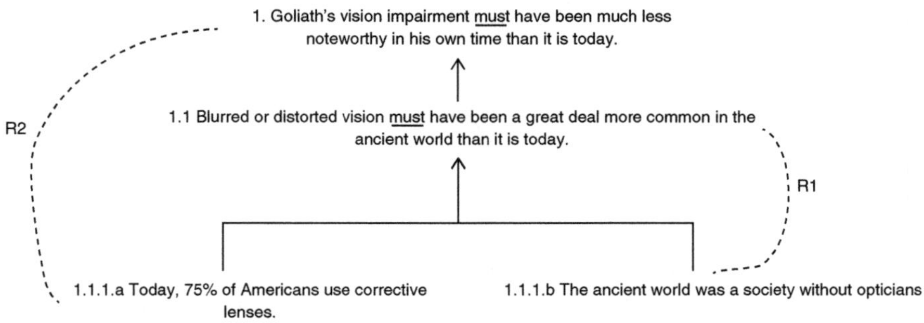

Fig. 4.17 Combined reconstruction of the co-textual examples (55) and (56). The dotted line marked as R1 represents the immediate cataphoric intraclausal relation established in the justification of the conversational background of *must* in example (55). The line R2 represents the immediate anaphoric relation resulting from the justification of the conversational background of *must* in example (56)

Thus, paying attention to linguistic signals guiding discourse processing such as the modals can help the analyst to single out those aspects of an argument that are obvious to any barely competent reader from those which require a reader aiming at a maximally argumentative reconstruction in the context of a critical discussion. This level of considerations may become relevant in analyzing the rhetorical strategy deployed by the arguer, the *strategic maneuvering* in pragma-dialectic terms (cf. van Eemeren 2010), and its potential fallaciousness for certain readers in a given context of interaction. As the present study remains mostly confined to matters of reconstruction, I will not be able to pursue further this line of inquiry.

There is one last, very basic, remark that should be made concerning examples (55) and (56). While in (56) the saturation of the conversational background resulting in the relation R2 is truly *anaphorical*, i.e. based on the recovery of an (abstract) referent in previous discourse segment, in (55) the relation R1 is not properly anaphorical. Not only the relation is *intra-clausal*, it is also *forward looking* as *must* precedes the segment *in a society without opticians*. Using a term from text linguistics, we can say that this is a case of *cataphora.*

In fact, also intersententially we can find both *anaphoric* and *cataphoric* saturations of the modal conversational background, as shown by the equally acceptable examples in (57.a, b):

(57.a) She never complains. She *must* be very brave. (anaphoric saturation)

(57.b) She *must* be very brave. She never complains. (cataphoric saturation)

Following Rigotti (1993, 2009) I use **phoric* as a generic term to refer to both types of discursive saturation of semantic relations. The term is used by Rigotti chiefly to refer to the way in which congruity theoretic connective predicates manifest their arguments, but it can be easily extended to the, closely related, discursive

events being referred to; (2) in reading for gist and so something akin to minimum effort, does not invest the effort to notice absences from a news text" (O'Halloran 2003: 169–170).

Fig. 4.18 Example of *Evidence* relation in Knott and Sanders (1998)

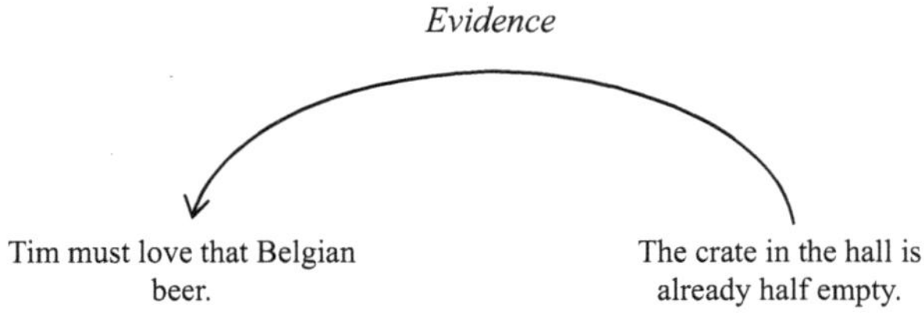

functioning of the modals illustrated above. In fact, in the reconstruction of the *connective predicate*, which the modals guide in a more or less indirect way, the anaphorically or cataphorically recovered B is made to correspond to the **phoric* argument $U_{+/-n}$ of the connective predicate (cf. Chap. 2, § 2.2.3.3).

The function of the modals in establishing argumentative discourse relations remains a severely under-investigated topic. As far as I can understand, this is not an intended application of the Relative Modality approach. As we have seen above in § 4.2.9, the discourse semanticists influenced by Relative Modality (cf. Roberts 1989; Frank 1996; Frank and Kamp 1997; Geurts 1999: 175–200) only discuss the anaphorical properties of the modals in relation to the phenomenon of modal subordination. This is even more remarkable in light of the fact that the same discourse semantic movement has given rise to very detailed models of discourse relations (cf. Asher and Lascarides 2003, in particular).

In parallel, there has been correspondingly little attention to modal expressions as the markers of discourse relations in the linguistic literature on discourse relations. For instance, in a well-known article on discourse relations and their markers, Knott and Sanders (1998) introduce the very notion of a *rhetorical relation* using the following example of the *Evidence* relation as defined by the typology of relations of Rhetorical Structure Theory (RST, cf. Mann and Thompson 1988: 251) (FIg. 4.18):

(58) Tim *must* love that Belgian beer. The crate in the hall is already half empty. (Knott and Sanders 1998: 136)

In (58) the clearest marking of the *Evidence* relation is provided by the cataphoric use of *must* in the conclusion. Yet, when Knott and Sanders (1998) move to their study of the linguistic markers of rhetorical relations in English and Dutch they do not devote specific attention to modal verbs like *must*, focusing instead on more "classic" instances of discourse connectives.

4.3.6.2 Modals and Anaphoric Connectives

The functioning of the modals illustrated above can indeed be compared and contrasted with that of the inferential *discourse connectives* with which they can alternate (59.a, b) or co-occur (59.c, d)

(59.a)	She never complains. *Therefore*, she is very brave.	The discourse connective marks the conclusion/ standpoint
(59.b)	She is very brave. *Because*, she never complains.	The discourse connective marks the premise/ argument
(59.c)	She never complains. *Therefore*, she must be very brave.	The discourse connective marks the conclusion/ standpoint The modal marks the conclusion, relying it anaphorically to the premise.
(59.d)	She must be very brave. *Because*, she never complains.	The discourse connective marks the premise/ argument. The modal marks the conclusion, relying it cataphorically to the premise.

In the descriptive pragmatic discourse representation provided by Congruity Theory examples (57.b), (59.b) and (59.d) correspond to the same succession of *connective predicates*, each expressing the argumentative relation quasi-redundantly from its specific vantage point (cf. Chap. 2, 3.3.3).

U_1: 'She [must] be very brave'	$\textit{Conclude\_f}_{U1}$ (Spk, Hr, U_1, U_X)	
U_2: '[because] she never complains'	$\textit{Argument}_{U2}$ (Spk, Hr, U_2, U_1)	$U_x = U_2$

In the doubly marked version of (58.d) this double realization of the argumentative relation is given full linguistic manifestation. As we have seen in the discussion of Congruity Theory in Chap. 2, in interpreting U_1 the addressee already recognizes, thanks to the presence of the modal *must*, that the utterance is performing an *act of conclusion* from some body of reasons corresponding to the unsaturated conversational background of the modal. This act is forward-looking (*cataphoric*) as it opens an argument$_S$ place to be saturated in the continuation of the discourse. The utterance U_2 is then understood as providing a reason, or argument, for U_1, thereby saturating the open argument$_s$ place.

Discourse connectives have been the object of numerous and diverse *instructional* analyses that can be instructively compared with the analysis of the context-dependent functioning of modal expressions that is being developed in the present study. Representative early research in this direction include Ducrot et al. (1980), Blakemore (1987), Moeschler (1989). A very explicit presentation of the generic structure of the instructional meaning of a connective is offered in Nølke *et al.* (2004: 23–24, 85–98). Connectives set up a relation between two or more *typed variables*, and *instructions* that constrain both structurally and logico-semantically the saturation of these variables. Typically, one variable is to be extracted from the utterance to which the connective is syntactically tied and at least one other variable is to be recovered from other parts of the text.

Recent computationally oriented work on discourse interpretation (cf. Webber et al. 2003, Forbes-Riley et al. 2005) has proposed to draw a distinction between two main types of connectives: *structural discourse connectives* and *anaphorical discourse connectives*. Both kinds of connectives express a relational predicate taking discourse contents as arguments. They differ, however, in the way in which the arguments are saturated.

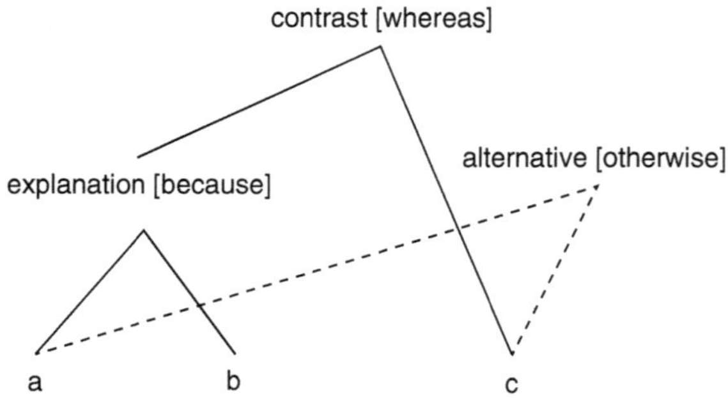

Fig. 4.19 In example (61), anaphoric relations (*dotted lines*) can cross dependencies with structural relations (Adapted from Webber et al. 2003: 551)

Structural connectives, like *but, on the one hand…on the other hand* continue to rely on structural principles at the level of discourse that mirror those of sentential syntax. They are predicates relating two propositional arguments deriving from clausal interpretations (or larger discourse units) to form a larger, macro-syntactic discourse unit.

Anaphoric connectives are predicates relating one argument deriving structurally from the discourse unit they modify and one anaphoric argument. Adverbials such as *then, otherwise, nevertheless* are free from the macro-syntactic constraints and simply recover their arguments anaphorically from information in the common ground of the discourse. For instance, as Forbes-Riley et al. (2005: 63) observe, anaphoric connectives "can take their prior argument intra-sententially" as illustrated by example (60), "where embedded *nevertheless* relates the interpretation of the matrix clause to the interpretation of the relative clause".

(60) Many people who have developed network software have *nevertheless* never gotten rich (from Forbes-Riley et al. 2005).

Being non-structural, anaphoric connectives, such as *otherwise*, are free to *cross dependencies* with structural connectives, as illustrated by Webber et al. (2003: 551) with the toy-discourse in example (61) and its analysis in Fig. 4.19.

(61)

(a) Because Fred is ill

(b) you will have to stay home

(c) whereas *otherwise* the two of you could have gone to the zoo.

Webber et al. (2003: 553) and Forbes-Riley et al. (2005: 63) also note that, contrary to structural connectives, anaphoric connectives can access "inferred entities as well as ones introduced explicitly" (*Ibid.*). To illustrate this property, they contrast *otherwise* with *or* in example (62).

(62)

(a) If the light is red, stop. *Otherwise* go straight on.

(b) If the light is red, stop. *Or* go straight on.

In (62.a) the anaphoric adverbial *otherwise* can access the inferred conditional antecedent *if the light is not red* and link it to the consequent *go straight on* in a conditional structure. In contrast, in (62.b), the structural connective *or* can only access structural units to its left: (a) the consequent clause (*stop*) or (b) the whole of the original conditional (*If the light is red, stop*).

It's worth noting that the pragmatic process through which *otherwise* constructs its implicit antecedent is analogous to the processes of presupposition accommodation/ justification examined in the previous section. A similar process of accommodation is at work in example (63), where the antecedent of *otherwise* is created from the negative answer to the preceding question.

(63) Do you want an apple? Otherwise you can have a pear (cf. Webber et al. 2003: 552)

It is perhaps interesting, in this context, to note the modal and conditional nature of the adverbial *otherwise*. In fact, in terms of Relative Modality, *otherwise* behaves like a necessity modal, which takes an appropriate conversational background – it's a deontic one in (62) and (63), but alethic Bs are also possible.[41] The accommodated negative antecedent is added to this conversational background according to Kratzer's analysis of conditionals (cf. § 4.2.8.2, above). The union of the initial conversational background and the conditional antecedent forms the conversational background that necessitates the scope of *otherwise*. Example (62.a) could be rendered roughly as (64), where the part in *italic* corresponds to the complex resulting conversational background:

(64) *In view of the street code and of the circumstances, if the light is not red*, you <u>have to</u> go straight on.

As a modal operator otherwise quantifies over all the alternatives where a previous modal or conditional does not hold. It exhaustively partitions a given set of worlds in two non-overlapping subsets, as highlighted by example (65):

(65) We must go out to eat out in a restaurant, *otherwise* we are not going to have dinner tonight.

In (65) the circumstances are such that in all the (circumstantially accessible) worlds where the goal of having dinner is realized, the participants eat out in a restaurant, and in all the worlds where the participants don't eat out in a restaurant they are also not realizing their goal of having dinner. I will come back later on certain implications of the use of *otherwise* in examples like (65).

Reasons of space preclude here the possibility of investigating further the extent the class of anaphoric connectives is composed by items with a modal semantics.[42] Irrespective of the precise semantics of the relations conveyed, anaphoric connec-

[41] As an alethic example of *otherwise* consider the following: *If the light is red, stop. Otherwise you'll get a ticket.* Here the scope of *otherwise* is a causal consequence *In view of the circumstances, if you don't stop at the red light, you will get a ticket.*

[42] The broader interactions between (relative) modality and issues of discourse-structural compositionality and discourse interpretation remain largely unexplored. For one, Webber et al. (2003: 567) actually admit that when they encounter issues of modality in their account of discourse structure they deal with them "fast and loose" in order not overburden their analysis.

tives and core modal expressions, such as modal verbs or adverbs, clearly exhibit a similar behavior in the establishment of discourse relations.

The key difference between the two classes seems to be that while the anaphoric connectives are strictly anaphoric, i.e. obligatorily backward-oriented towards a previous discourse segment or previous utterance event, modals (e.g. modal verbs, modal adverbs, modal inflectional morphemes) can be also cataphoric or purely contextual. Furthermore, modals can be content with a restriction of the conversational background to a set containing a certain type of propositions (e.g. *in view of your wellbeing*, *of the governor's plan*, *of the immigration law*, *of the way we do business here*, *of John's artistic preferences*, etc.) even if the actual propositions in the set are not part of the common ground. For proper anaphoric connectives, such minimal saturation of the relation is not possible.

In short, modals *can* function as anaphoric connectives, but the anaphoric resolution is just one possible saturation or justification of their conversational background. Proper anaphoric connectives always *require* some anchoring to previous discourse even if their actual antecedents do not necessarily coincide with formal discourse units and can be worked out through inferential *bridging* (cf. Webber et al. 2003: 555) processes.

4.3.6.3 Non-epistemic Modals and Argumentative Discourse Relations

In Sect. 4.3.6.1, all the examples of *phoric saturation of the conversational background leading to an argumentative discourse relation were based on the epistemic reading of the necessity modal verb *must*. However, if the context dependency of the conversational background is what drives the establishment of relations, these should be present also with other types modals, provided that their conversational background is indeed context-dependent, and with other "flavors" of modality. This seems to be the case in the following two simple examples:

(66) It's [= Portobello West] also a great place to find a distinctive piece for the home. With up to 100 designers on show, any fashionista *can* find something to suit their unique style. (WebCorp)

(67) We *must* also listen to young people, which we do not do enough. Young people come up with many wonderful scientific ideas (WebCorp)

In (66) the circumstantial prepositional phrase *with up to 100 designers on show* provides a fitting saturation for the conversational background of the "dynamic" modal can. The number of designers on show is what makes the objective attainable for "any fashionista". At the same time, the phrase is easily reconstructed as an argument presenting a *datum* in support of this possibility. In turn, the modal proposition functions as an argument in support of an evaluative standpoint about the Portobello West crafts market, as illustrated in Fig. 4.20, below.

In example (67), the conversational background of the deontic *must* is saturated cataphorically thanks to the evaluative proposition in the following sentence. This is also the reason why it is necessary to listen to young people. There is therefore again

Fig. 4.20 Argumentation structure of example (66)

1. Portobello West is a great place to find a distinctive piece for the home.

↑

1.1 In Portobello West any fashionista can find something to suit thier unique style

↑

1.1.In Portobello West are up to 100 designers on show.

a near-perfect correspondence between the recovery of the conversational background and the reconstruction of the argument.

This is exactly the kind of facts that prompted Toulmin to advance the bold hypothesis that *all* flavors of modality draw their force from the pragmatics of argumentative discussion and their criteria from the different fields of argument (cf. Chap. 3, § 3.2). Yet, the discussion in Chap. 3 pokes several holes in this view of non-epistemic modals, on which I will not return here (cf., in particular, § 4.2.3). I shall instead insist on one key difference between the epistemic-doxastic modals and the others.

If we compare the examples in (66) and (67) with the epistemic examples such as (54), (55), (56), (57) and (58) we notice a fundamental difference concerning the relationship between the modal and the propositional content of the discourse utterance in which it appears. In examples (54–58) the modal *must* does not seem to add anything to the propositional content of the standpoints or intermediate conclusions in which it appears. If I take the modal away, the propositions to which the protagonist is committed remain the same, as in the following examples:

(68)

(a) She is very brave.

(b) That roll was gourmet.

(c) Blurred or distorted vision was a great deal more common in the ancient world than it is today.

(d) Goliath's vision impairment was much less noteworthy in his own time than it would be today.

These are the propositions to which the arguer is committed and *must* does not add anything to them. The job of *must* in examples (54–58) seems to include: (a) signaling the relation between the premises and the conclusion, (b) specifying, as Freeman (1991) would put it, the strength of support that the premises lend to the conclusion, and (c) expressing the degree of commitment of the speaker towards the conclusion.

There are a few things about the working of epistemic *must* that remain still unclear. It is not clear, for instance, whether epistemic *must*, as it is used in ordinary discourse, truly is a "deductive modality" signaling, to use again the phrasing of Freeman's (1991: 112), "that the premises give complete support to the conclusion, that they necessitate it", or, as many linguists have contended, makes a weaker claim[43] expressing, in fact, a degree of confidence that is lower of that of the bare, unmodalized, assertion. More generally, the relationship between the strength of support (a) and the expression of commitment (b) should be better clarified. Finally, there might be further information that *must* conveys that is not captured by (a), (b) and (c). I will address these open issues – which are actually closely connected – in Chap. 6, discussing epistemic *must* in parallel with the epistemic uses of Italian necessity modal *dovere*, which are functionally very close to that of the English modal.

What is clear, though, is that all these clear and unclear aspects of the functioning of *must* pertain to the role of the proposition in the reasoning, belief formation and argumentation processes of the speaker; *not* to the content of what is being inferred, believed or argued for.

In contrast, if we turn to the non-epistemic examples (66) and (67) we find that the modal enters the propositional content of the standpoint or intermediate conclusion and enters it in a decisive way that affects the argumentatively relevant "logical type" (Toulmin 2003 [1958]: 13) or "type of statement" (Freeman 2005: 93–94) of the conclusion. Removing *can* and *must* from, respectively, (66) and (67) results in (68) and (69).

(68) Every fashionista finds something to suit their unique style.

(69) We also listen to young people.

Clearly, in (66) the content of the conclusion was not *a fact* about every fashionista, nor about any one in particular, but the dynamic *possibility of a course of action*. In turn, this possibility was used as a premise for supporting an *evaluative standpoint*. A specific maxim of the locus of *Instrumental Cause* presides over this inference: the goodness of a tool is established by the fact it makes a desirable outcome possible. There are cases where similar dynamic possibility statements are instead used to support *predictive standpoints*, lending them the force of epistemic possibility or, perhaps, probability. However, in (66), this is not the case.

There is little need to comment on the conspicuous difference between (67) and (69). The original conclusion is *prescriptive*, while the modified version is *descriptive*. It is trivial to say that what counts as evidence for a prescription does not necessarily counts as evidence for the corresponding description.

It is clear, at this point, that, in order to understand what non-epistemic modalities bring to argumentation we need to understand better the nature of the different flavors of modality and their relations to argumentatively relevant types of statements. This is what the next chapter sets out to do.

[43] See Karttunen (1972) for an early statement of the problem and von Fintel and Gillies (2010) for a comprehensive discussion arguing that, despite *prima facie* intuitions, the epistemic modal manifests the force of necessity.

References

Asher N, Lascarides A (2003) Logics of conversation. Cambridge University Press, Cambridge

Barwise J, Perry J (1983) Situation and attitudes. MIT Press, Cambridge

Beaver D (2001) Presupposition and assertion in dynamic semantics. CSLI Publications, Stanford

Bell VA, Johnson-Laird PN (1998) A model theory of modal reasoning. Cogn Sci 22(1):25–51. doi:10.1016/S0364-0213(99)80034-2

Blakemore D (1987) Semantic constraints on relevance. Blackwell, Oxford

Dancygier B, Sweetser E (2006) Mental spaces in grammar: conditional constructions. Cambridge studies in linguistics, vol 108. Cambridge University Press, Cambridge

Ducrot O et al (1980) Les Mots du discours. Les éditions de minuit, Paris

Fauconnier G (1994 [1985]) Mental spaces. Aspects of meaning construction in natural language. Cambridge University Press, Cambridge

Fauconnier G, Turner M (2006) Conceptual integration networks. In: Geeraerts D (ed) Cognitive linguistics: basic readings. Mouton De Gruyter, Berlin/New York, pp 303–371

Forbes-Riley K, Webber B, Joshi A (2005) Computing discourse semantics: the predicate-argument semantics of discourse connectives in D-LTAG. J Semant 23:55–106

Frank A (1996) Context dependence in modal constructions. Ph.D. thesis. Universität Stuttgart, Stuttgart

Frank A, Kamp H (1997) On context dependence in modal constructions. In: Lawson A (ed) Proceedings of SALT VII. Cornell University, Ithaca, pp 151–168

Freeman JB (1991) Dialectics and the macrostructure of arguments. A theory of argument structure. Foris Publications, Berlin

Freeman JB (2005) Acceptable premises: an epistemic approach to an informal logic problem. Cambridge University Press, Cambridge

Garson J (2014) Modal logic. The Stanford encyclopedia of philosophy (Summer 2014 Edition). Center for the Study of Language and Information (CSLI), Stanford. http://plato.stanford.edu/archives/sum2014/entries/logic-modal/. Accessed 6 Oct 2016

Geurts B (1999) Presuppositions and pronouns. Elsevier, Amsterdam

Girle R (2003) Possible worlds. Acumen, London

Groefsema M (1995a) Understood arguments: A semantic/pragmatic approach. Lingua 96(2-3):139–161. doi:10.1016/0024-3841(95)00002-H

Groefsema M (1995b) Can, may, must and should: a relevance theoretic account. J Linguist 31:53–79

Hacquard V (2010) On the event relativity of modal auxiliaries. Natural language semantics 18(1):79–114

Heim IR, Kratzer A (1998) Semantics in generative grammar. Blackwell textbooks in linguistics, vol 13. Blackwell, Oxford

Hobbs JR (2010) Clause-internal coherence. In: eds. Kühnlein P, Benz A, Sidner C (eds) Constraints in discourse 2. John Benjamins, Amsterdam/Philadelphia, pp 15–34

Johnson-Laird PN (1978) The meaning of modality. Cogn Sci 2(1):17–26. doi:10.1207/s15516709cog0201_2

Kamp H (2013 [1981]) A theory of truth and semantic representation. In: Heusinger K, ter Meulen A (eds) Meaning and the dynamics of interpretation. Selected papers of Hans Kamp. Brill, Leiden, pp 329–369

Kamp H (2001) Presupposition computation and presupposition justification: one aspect of the interpretation of multi-sentence discourse. In: eds. Bras M, Vieu L (eds) Semantic and pragmatic issues in discourse and dialogue. Elsevier, Amsterdam/New York, pp 57–84

Kaufmann S, Condoravdi C, Harizanov V (2006) Formal approaches to modality. In: Frawley W (ed) The expression of modality. Mouton de Gruyter, Berlin/New York, pp 71–106

Kay P (2004) Pragmatic aspects of grammatical constructions. In: Horn L, Ward G (eds) The handbook of pragmatics. Blackwell, Oxford, pp 675–700

Knott A, Sanders TJM (1998) The classification of coherence relations and their linguistic markers: an exploration of two languages. J Pragmat 30:135–175
Kratzer A (1977) What "must" and "can" must and can mean. Linguist Philos 1:337–355 (Republished in Kratzer 2012: 4–20)
Kratzer A (1978) Semantik der Rede: Kontexttheorie, Modalwörter, Konditionalsätze. Scriptor, Kronberg
Kratzer A (1981) The notional category of modality. In: Eikmeyer H-J, Rieser H (eds) Words, worlds, and contexts: new approaches in word semantics. Mouton de Gruyter, Berlin/New York, pp 38–75 (Republished in Kratzer 2012: 27–69)
Kratzer A (1991) Modality. In: von Stechow A, Wunderlich D (eds) Semantik/Semantics: An international handbook of contemporary research. Mouton de Gruyter, Berlin/New York, pp 639–650
Kratzer A (2012) Modals and conditionals. Oxford University Press, Oxford
Kripke SA (1963) Semantical analysis of modal logic I. Normal modal propositional calculi. Mathematical Logic Quarterly (Formerly: Zeitschrift Für Mathematische Logik Und Grundlagen Der Mathematik) 9(5–6):67–96
Kronning H (1996) Modalité, cognition et polysémie: Sémantique du verbe modal devoir. Upsala University, Upsala
Kronning H (2001a) Pour une tripartition des emplois du modal devoir. In: Dendale P, van der Auwera J (eds) Les Verbes modaux. Cahiers Chronos, 8. Rodopi, Amsterdam, pp 67–84
Kronning H (2001b) Au-delà du déontique et de l'épistémique. In: Prandi M, Ramat P (eds) Semiotica e Linguistica. Per ricordare Maria Elisabeth Conte. Franco Angeli, Milan, pp 97–118
Kronning H (2001c) Nécessité et hypothèse: « devoir » non déontique au conditionnel. In: Dendale P, Tasmowski L (eds) Le conditionnel en français. Centre d'études linguistiques des textes et des discours de l'Université de Metz, Metz, pp 251–276
Lewis D (1979) Scorekeeping in a language game. J Philos Logic 8(1):339–359
Lycan WG (1994) Modality and meaning. Kluwer, Dordrecht
Mann WC, Thompson SA (1988) Rhetorical structure theory: a functional theory of text organization. Text 8(3):243–281
McCawley JD (1993) Everything that linguists have always wanted to know about logic but were ashamed to ask. University of Chicago Press, Chicago
Moeschler J (1989) Modélisation du dialogue. Représentation de l'inférence argumentative. Hermes, Paris
Nølke H, Fløttum K, Norén C (2004) ScaPoLine. La théorie scandinave de la polyphonie linguistique. Paris, Kimé
Nunberg G (1993) Indexicality and deixis. Linguist Philos 16:1–43
O'Halloran K (2003) Critical discourse analysis and language cognition. Edinburgh University Press, Edinburgh
Papafragou A (2000) Modality: issues in the semantics-pragmatics interface. Elsevier, Amsterdam
Partee BH (1991a) Topic, focus and quantification. In: Moore SK, Wyner AZ (eds) SALT 1. Proceedings of of the 1st semantics and linguistic theory conference, April 19–21, 1991, Cornell University, pp 159–187. http://elanguage.net/journals/salt/issue/view/285. Accessed 6 Oct 2016
Partee BH (1991b) Adverbial quantification and event structures. In: BLSI 17. Proceedings of the seventeenth annual meeting of the Berkeley Linguistic Society: general session and parasession on the grammar of event structure, pp 439–456. http://elanguage.net/journals/bls/issue/view/322. Accessed 6 October 2016.
Peirce CS (1998 [1931]) Collected papers of Charles Sanders Peirce. In: Hartshorne C, Weiss P, Burks AW (eds). Harvard University Press, Cambridge, MA
Peregrin J (1993) Possible worlds: a critical analysis. Prague Bull Math Linguist 59-60:9–21
Portner P (2009) Modality. Oxford University Press, Oxford
Pustejovsky J (1995) The generative Lexicon. MIT Press, Cambridge, MA

Recanati F (1989) The pragmatics of what is said. Mind Lang 4(4):295–329. doi:10.1111/j.1468-0017.1989.tb00258.x
Recanati, François. 2002. Deixis and anaphora. In Semantics vs. Pragmatics, ed. Zoltan Szabo. Oxford: Oxford University Press. http://jeannicod.ccsd.cnrs.fr/ijn 00000293v2. Accessed 6 October 2016
Recanati F (2008) La philosophie du langage (et de l'esprit). Gallimard, Paris
Recanati F (2009) Compositionality, flexibility, and context-dependence. In: Werning M, Hinzen W, Machery E (eds) Oxford handbook of compositionality. Oxford University Press, Oxford, pp 1–29
Rescher N (2006) Presumption and the practices of tentative cognition. Cambridge University Press, Cambridge
Rigotti E (1993) La Sequenza testuale. Definizione e procedimenti di analisi con esemplificazioni in lingue diverse. L'analisi Linguistica e Letteraria 1(1):43–148
Rigotti E (2009) Whether and how classical topics can be revived in the contemporary theory of argumentation. In: van Eemeren FH, Garssen B (eds) Pondering on problems of argumentation. Springer, New York, pp 157–178
Rigotti E, Rocci A (2006) Le signe linguistique comme structure intermédiaire. In: de Saussure L (ed) Nouvelles perspectives sur Saussure. Mélanges offerts à René Amacker. Droz, Genève, pp 219–247
Roberts C (1989) Modal subordination and pronominal anaphora in discourse. Linguist Philos 6:683–721
Roberts C (2012) Information structure in discourse: towards an integrated formal theory of pragmatics. Semant Pragmat 5(6):1–69. doi:10.3765/sp.5.6
Rocci A (2005a) La Modalità epistemica tra semantica e argomentazione. I.S.U Università Cattolica, Milano
Rocci A (2005b) On the nature of the epistemic readings of the Italian modal verbs: the relationship between propositionality and inferential discourse relations. In: Hollebrandse B, van Hout A (eds) Crosslinguistic views on tense, aspect and modality. Rodopi, Amsterdam/New York, pp 229–246
Rocci A (2006) Pragmatic inference and argumentation in intercultural communication. Intercult Pragmat 3(4):409–442. doi:10.1515/IP.2006.026
Seuren PAM (1985) Discourse semantics. Blackwell, Cambridge, MA
Snoeck Henkemans F (1992) Analysing complex argumentation. SicSat, Amsterdam
Stalnaker R (1973) Presuppositions. J Philos Logic 2(4):447–457
Stalnaker R (1978) Assertion. In: Cole P (ed) Syntax and semantics, vol 9: Pragmatics. Academic, New York, pp 315–322
Stalnaker R (1991) Pragmatics. In: Davidson D, Harman G (eds) Semantics of Natural Language. Reidel, Dordrecht, pp 380–397
Sweetser E (1990) From etymology to pragmatics: metaphorical and cultural aspects of semantic structure. Cambridge University Press, Cambridge/New York
Toulmin SE (2003 [1958]) The uses of argument. Cambridge University Press, Cambridge
van der Auwera J (1999) On the semantic and pragmatic polyfunctionality of modal verbs. In: Turner K (ed) The semantics/pragmatics interface from different points of view. Elsevier, Amsterdam, pp 49–64
van der Sandt RA (1992) Presupposition projection as anaphora resolution. J Semant 9(4):333–377. doi:10.1093/jos/9.4.333
van Eemeren FH (2010) Strategic maneuvering in argumentative discourse. John Benjamins, Amsterdam/Philadelphia
van Eemeren FH, Grootendorst R (2004) A systematic theory of argumentation: the pragma-dialectical approach. Cambridge University Press, Cambridge
Veltman F (1986) Data semantics and the pragmatics of indicative conditionals. In: Closs-Traugott E, ter Meulen A, Reilly JS, Ferguson CA (eds) On Conditionals. Cambridge University Press, Cambridge, pp 147–167

von Fintel K, Iatridou S (2008) How to say ought in foreign: the composition of weak necessity modals. In: Guéron J, Lecarme J (eds) Time and modality. Springer, Berlin/New York, pp 115–141

Webber B, Stone M, Joshi A, Knott A (2003) Anaphora and discourse structure. Comput Linguist 29(4):545–587

Werth P (1999) Text worlds: representing conceptual space in discourse. Longman, London

Wilson D, Sperber D (2012) Linguistic form and relevance. In: Wilson D, Sperber D (eds) Meaning and relevance. Cambridge University Press, Cambridge, pp 149–168

Wurmbrand S (1999) Modal verbs must be raising verbs. In: Bird S, Carnie A, Haugen J, Norquest P (eds) Proceedings of the West Coast conference on formal linguistics 18. Cascadilla, Somerville, pp 599–612

Chapter 5
Types of Conversational Backgrounds and Arguments

Abstract The chapter reconstructs the most relevant "flavors" of modality (alethic, deontic, deontic-practical, anankastic, and epistemic) in terms of a *relational*, *discourse sensitive*, *context dependent*, *procedural* approach to the interpretation of modal constructions. The approach, while inspired by Relative Modality and the formal semantics tradition, integrates pragmatic insights and it is cast as a tool for the actual analysis of discourse and argumentation. This approach to the typology of modal "flavors" is needed because the full indicative potential of a modal emerges only when its conversational background has been specified. At this point, the modal will act as a guide in the clarification of argumentative confrontations, in the specification of the logical type of the standpoint, in the recovery of premises as well as in the determination of the inferential structure of the argument (argumentation scheme), where the flavor of modality can help by suggesting the *locus*-relation invoked by the argumentation scheme. The semantic category of *evidentiality*, which is closely related to epistemic modality, is also briefly examined in relation to argumentation, with a survey of the most basic categories of evidence sources encoded by languages (*perception*, *testimony*, *inference*).

5.1 Introduction

The context dependence of conversational backgrounds can ensure a highly situated and indefinitely nuanced definition of the "flavor" of modality involved in a discourse utterance. This specific, situated, flavor can be made explicit, to some extent, by various, non-equivalent, means.

Paying attention to these linguistic means of specifying the conversational background can be of considerable importance in the process of argumentative reconstruction. In fact, the full indicative potential of the modal emerges only when its conversational background has been specified to a reasonable level of precision. Only at that point the modal will act as a guide in the clarification of argumentative confrontations, in the specification of the logical type of the standpoint, in the recovery of premises, as well as in the determination of the inferential structure of the argument (argumentation scheme), where the flavor of modality can help chiefly by suggesting the *locus*-relation invoked by the argumentation scheme.

A. Rocci, *Modality in Argumentation*, Argumentation Library 29,
DOI 10.1007/978-94-024-1063-1_5

As amply illustrated by the examples used so far, some of the most common modal words, like the modal verb *must*, the adjective *possible* or Toulmin's *cannot* are compatible with a myriad conversational backgrounds covering several of the commonly recognized major flavors of modality. This seems to greatly reduce their usefulness as indicators. Three interconnected strategies can help to overcome these limitations:

(a) paying attention to the linguistic strategies of manifestation of the conversational background;
(b) relying on a cartography of broad classes of conversational backgrounds corresponding to commonly recognized modal "flavors" that have significant logico-ontological and linguistic properties;
(c) exploiting fine grained linguistic-semantic knowledge of modal expressions and the of the constructions in which they enter.

I will embark in strategy (c) in Chap. 6, where I consider in great detail a few modal constructions of the Italian language. It will emerge that even the seemingly more generic, all-purpose, modals impose subtle constraints on the construction of their conversational background arguments. These constraints often do not characterize lexical or morphological units as such, but rather more specific *constructions*[1], in which these units enter.

As for strategy (a), I shall proceed straight away in § 5.2.1 to list a bit more systematically the linguistic devices that contribute to make conversational backgrounds accessible, many of which we have already encountered. Strategy (b), the mapping of major modal flavors in terms of conversational backgrounds, makes the bulk of the present section, starting from § 5.2.2. A rationale for the classification of conversational background is presented in § 5.2.4. This approach integrates some recent proposals on the semantics of modality, gaining novel insights on the nature of different modal flavors as well as on the processes of meaning construction through which they emerge in discourse.

In Sect. 5.3, I will start reaping the fruits of the proposed cartography of modal conversational backgrounds examining how each of the major flavors contributes to the specification of the logical type of the standpoints, to the reconstruction of discourse relations at the propositional or illocutionary level, and, last but not least, to the recovery of the *locus*-relations and inferential structures.

---

[1] Here, as in the rest of the chapter, I use *construction* in a technical sense, albeit not far from the ordinary one, which comes from the theoretical trend of *construction grammars*, initiated by Charles J. Fillmore and Paul Kay (cf. Kay 2004, Kay and Michaelis 2008). In construction grammars a construction is "a form-meaning pair such that some aspect of its form or some aspect of its meaning is not strictly predictable from its component parts or from other constructions" (cf. Goldberg 1995: 11). Constructions can specify syntactic slots as well as a part of the lexical and morphological filling of said slots and conventionally associate them with semantic constraints at the level of the propositional content, illocutionary force and discourse function of the utterance. Specific construction inherit formal and semantic features from the lexical and morphological material they embed as well as from more abstract syntactic templates, but also feature further specifications that are conventionally associated to the whole.

With the non-epistemic flavors of modality (§§ 5.3.1, 5.3.2, and 5.3.3), the conditional-like relational structures of the modals, once contextually saturated, provide different kinds of semantico-ontological relations between propositional contents, which constrain the kind of *loci* (cf. Chap. 2, § 2.2.3) when either the modal proposition or the prejacent proposition appear as standpoint or as premise.

In the special case of the epistemic and doxastic flavors (§ 5.3.4), the context-dependent relational semantics will also contribute to clarify how these modals can function as *evidential* strategies. Evidential expressions specify the kind of evidence on which an assertion is based and can subtly guide the addressee to locate the required kind of evidence in the discourse context. In this way, epistemic-doxastic modal expressions that function as evidentials become argumentative indicators establishing argumentatively relevant discourse relations, and possibly providing a further layer of constraint on the *loci* exploited by the inferential structure.

5.2 Towards a Typology of Conversational Backgrounds

5.2.1 Types of Linguistic Manifestations of the Conversational Background

A number of different syntactic and discourse devices can provide a partial linguistic manifestation of the conversational background. These devices, taken jointly with the modal itself, can function as richly informative argumentative indicators.

1) **Prepositional adverbial phrases.** In her treatment of modality Angelika Kratzer elects adverbial prepositional phrases with the form '*in view of* NP' as standard syntactic manifestation of the conversational background. The choice seems felicitious, as in English these phrases can make explicit a wide variety of conversational backgrounds, including both epistemic (1) and non-epistemic ones (2).

(1) In view of what their tribal duties are, the Maori children must learn the names of their ancestors (Kratzer 2012 [1977]: 6)

(2) In view of what is known, the ancestors of the Maori must have arrived from Tahiti (*Ibid.*)

In her 2012 revision of the original Relative Modality studies she, however, observes that this phrase is not neutral and that "there are important differences between different adverbial phrases contributing conversational backgrounds for different types of modals" (Kratzer 2012: 21).

2) **Attribution to sources.** Kratzer (2012) contrasts phrases such as *in view of* NP or *Given* NP with phrases such as *according to* NP and their equivalents in other languages, illustrated by the German example in (3) and the Italian example in (4).

(3) Dem Gerücht nach, *soll* Roger zum Häuptling gewählt worden sein (Kratzer 2012: 35)
'According to the rumor, Roger was reportedly elected chief'

(4) Stando a quanto emerso ieri nella riunione del cda Bnl, Bnp *sarebbe* orientata a lanciare la prossima settimana l'Opa (*Il Sole 24 Ore*, 13/4/2006).
'According to what transpired from yesterday's meeting of the board of BNL, BNP *will reportedly* be inclined to launch next week the takeover bid'

The German necessity modal *sollen* in (3) and the Italian conditional mood in (4), which also diachronically originates from a necessity modal construction, function as *evidential strategies* manifesting that the justification of their prejacents consists in rumors or leaks – the so-called *hearsay* evidential. The prejacent is true in all the worlds consistent with the content of the hearsay. If the rumors or leaks turn out to be false, the sentences have still been uttered truly and felicitously. Kratzer contrasts the modals taking *according to* phrases with the use of *must* in (5):

(5) Given the rumor, Roger *must* have been elected chief (Kratzer 2012: 35)

Here the rumor is a fact not a set of propositions and *must* does not indicate what follows from the propositions composing the rumor, but what the speaker can infer from the fact that a certain rumor is circulating. If the rumor is unreliable and nothing can be really concluded from it, the assertion of (5) is infelicitous. In (5) the symptomatic reasoning of the protagonist is foregrounded, while (3) and (4) can be used to invoke an argument from authority while, at the same time, de-emphasizing the role of protagonist of the speaker.[2]

3) **Conditional antecedents.** Conditional sentences are a means of manifestation of the conversational background that has been extensively discussed in the previous chapter (cf. Chap. 4, § 4.2.8). A major distinction in this area is that between *content conditionals* and *epistemic conditionals*. The former truly add a new proposition to the conversational background thereby producing a qualitatively different, shifted, conversational background. The latter merely highlight a proposition that is already *de iure* part of the epistemic or doxastic conversational background.

(6) If you arrive late, you *must* wait outside. (content conditional)

(7) If Alfred is a bachelor, he *must* be unmarried (epistemic conditional)

4) ***Phoric relations.** *Anaphora* (and *cataphora*) have also been discussed thoroughly. With *phoric connections we also find a split parallel to that between content and epistemic conditionals. The so called modal subordination (8) is akin to content conditionals, while the anaphora of factual, asserted, material into an epistemic conversational background or an alethic circumstantial conversational

[2]As we will see in Chap. 6, this strategy is frequent in the corpus of Italian business and financial news reporting on which the analyses of Italian modals in Chap. 6 are based. The Italian example in (4) is, in fact, taken from that corpus.

background simply highlights what is already part of the conversational background, as in (9).

(8) A thief *might* break into the house. He *would* take the silver. (from Roberts 1989: 697)

(9) Alfred is a bachelor. He *must* be unmarried.

5) **Causes.** Another major group of devices manifesting conversational backgrounds includes causal subordinate clauses introduced by a causal connective (e.g. *because* S), illustrated by example (10), and adverbial prepositional phrases introducing a causal complement (e.g. *because of* NP). The scope of the latter can overlap with the more generic adverbial PPs (*in view of* NP).

(10) Padre Pio could not go to Foggia where at that time I [...] had prepared a triumphal reception for the traveling Madonna. Amongst other things he *could not* come because just when the statue arrived in Italy, Padre Pio was struck down by a bad attack of pleurisy, that made it impossible for him to even celebrate Mass from 5 May onwards. And the illness continued so that he was still prostrate when our Lady arrived in Puglia (WebCorp)

6) **Goals.** Somewhat complementary to causal clauses are *in order to* VP clauses (cf. Werner 2005) and subordinate clauses in general (e.g., *so that* S, *lest* S). *In order to* VP clauses are sometimes referred to as "rationale clauses", but, as Werner (2005: 305) observes, there are also instances where the proposition introduced by *in order to* cannot be a proper goal, rationale or end for an action. Consider the following examples:

(11) In order to pass his private accounts plan, Mr. Bush *must* first overcome approval ratings stuck near 50 percent, a united Democratic opposition, and Congressional Republicans for whom the President's proposals have little upside. (Werner 2005: 306)

(12) In order for me to sell those stocks, their price *must* first double (Werner 2005: 306)

(13) Green plants *must* absorb certain minerals through their roots to survive. (WebCorp)

While in (11) the clause introduces a goal of Bush, in (12) there is no goal to speak of, for the simple reason that the doubling of the stock price is not an action. As Werner (2005: 306) puts it "that is something which happens independently of anybody's intent", and only intentional actions can have a goal. Finally, in example (13) *green plants* cannot properly have the goal of surviving. The functioning of these *in order* clauses needs to be investigated further to clarify the nature of the conversational backgrounds.

7) **Disjunction.** A component of the conversational background of a modal can be expressed indirectly through a disjunction with another modal expression. Earlier (Chap. 4, § 4.3.6.2), I have commented on the modal functioning of the anaphoric connective *otherwise* in examples[3] where it provides a proposition

[3] See example (65) in Chap. 4.

that is added to the conversational background in a manner comparable to the addition of the antecedent of a conditional. The effect produced by the "disjunctive" modal expression *otherwise* can be obtained also through a disjunction between two modal sentences, as observed by Werner (2005: 303):

(14) John *must* call the plumber *or* the pipes *will* burst.

According to Werner (2005: 305), the second disjunct in (14) adds "the complement of the proposition embedded under *will* to the modal base" so that *must* in the first disjunct is interpreted as quantifying over "a background of worlds in which the pipes do not burst". Werner (2005) observes that not all disjunctions of modals are interpreted in this way. This strategy is typical of situations where "aversive consequences" are to be avoided. Yet, it does not require that the avoidance is the agent's goal. In this connection, Werner (2005: 304) comments example (14) as follows:

> The sentence allows that John is free to not call the plumber and let the pipes burst. But given the goal of preventing the pipes from bursting, the sentence says that John does not have a choice but to call the plumber. Notice that nothing is said about John's goals. Even if he wants the pipes to burst, the sentence can be true.

Clearly, disjunctions of this kind are closely related to the *in order to* VP clauses discussed above, of which they represent a negative version. In the following sections more details will emerge about the class of modal flavors associated with these peculiar linguistic means of manifesting the conversational background.

8) **Subject nominals.** As with Poe's purloined letter, linguistic clues of the conversational background can be often found hidden in plain sight in the most obvious place: the subject NP of the modal sentence. This happens not only with dynamic ability modals for which the agent's argument role coincides the syntactic subject, but also for cases of dynamic modality where the subject is not an agent:

(15) Small and defenseless lion cubs *can* fall prey to various predators, like hyenas, leopards, African wild dogs, jackals, eagles, and African rock pythons. (WebCorp, with adaptations)

(16) The weakening of the euro *could* help the crisis (WebCorp)

In (15) the subject corresponds to a *patient* role, while in (16) the subject NP is a nominalization referring to an event. As observed by Miecznikowski (2011), this use of abstract nominals to provide a propositional saturation of the conversational background and, eventually, also compactly express micro-argumentation, is frequent in the business and financial news corpus that provides the data for the case studies in Chap. 6.

5.2.2 Criteria for a Typology of Conversational Backgrounds

For every utterance of a modal proposition the conversational background is a contextually determined set of propositions, recovered with varying degrees of precision/ vagueness in the search space provided by the situational and discursive context, by the linguistic constraints conventionally attached on the B argument slot by the linguistic construction in question, and by linguistic structures, such as those examined above, which allow for a partial manifestation of the conversational background.

Nevertheless, the recognition of broad classes of conversational backgrounds[4], corresponding to distinct modal "flavors", is desirable in view of accurate linguistic description as well as in view of the task of argumentative reconstruction. There are several orders of reasons that make such a cartography of modal flavors desirable.

5.2.2.1 The Argumentative Relevance of Logico-ontological Considerations

The cartography of modal flavors is worth doing, first of all, in order to capture important generalizations about logical properties that characterize broad classes of conversational backgrounds.

Classes of conversational backgrounds can be characterized in terms of logical properties that mirror the properties of *accessibility relations* and the corresponding *axioms* of systems of modal logic (cf. Chap. 4, § 4.2.6). These abstract properties of conversational backgrounds are clearly argumentatively relevant properties in that they contribute to determine the inferences that can be drawn from modal propositions.

A case in point is *Axiom T* ($\Box\varphi \rightarrow \varphi$). Knowing that a certain class of conversational backgrounds is *realistic* (Kratzer 2012 [1981]: 32), i.e. that *Axiom T* holds for them, can have obvious and significant consequences for the analytical reconstruction and evaluation of arguments. It means that, for conversational backgrounds belonging to that class, once I have successfully defended a *necessity* or *impossibility* standpoint I have also successfully defended an affirmative or negative standpoint about the prejacent itself. For instance, once I have successfully defended (17.a) I have also successfully defended (17.b).

(17)

(a) Padre Pio *could not* go to Foggia <u>because he was struck down by a bad attack of pleurisy.</u>

(b) Padre Pio *did not* go to Foggia

In an argumentative perspective, also a further layer of somewhat more concrete properties is relevant for reconstructing the inferential configuration of arguments involving modal propositions. As we have seen, the kind of conversational

[4] In contrast, see Groefsema (1995) for an attempt at explaining the interpretations of the English modals with a sketchy context dependent semantics – broadly inspired by Kratzer – enriched by contextual inferences, without any recourses to broad classes of modal interpretation. In Rocci (2005a: 208–2011) I make some critical observation on the viability of this attempt. Partly similar observations can be found in Papafragou (2000).

background of a modal proposition can determine the *type of statement* (or *logical type*) of the standpoint and the *loci*-relations that *can be congruously applied* to it. Consider, again, the very first example offered in this book (cf. Introduction, Chap. 1, § 1.1):

(18)
(a) John *had to* resign.
(b) John *did* resign.
(c) I saw him signing the resignation letter

While (18.c) can offer adequate support to the standpoint in (18.b), it is not sufficient to support (18.a). This is because an argument based on the locus *From Authority* (*Testimony*) is capable of supporting *descriptive* statements, which are non-modal, extensional statements, but it's incapable of supporting statements of the *interpretation* class. Freeman (2005: 105) defines intepretations very broadly as contingent "intensional non-evaluative statements". Interpretations assert or presuppose a *nomic generalization*, a conditional holding for some set of possible worlds (thus for some conversational background, in terms of Relative Modality). If I say that John *had to* resign, I say that his resignation – the prejacent – is a necessary consequence of a presupposed conversational background. There is something that makes it necessary. The modal proposition is a covert interpretation. Proper discursive saturation of the conversational background can transform it in an overt interpretation, while, at the same time providing the adequate support:

(19) John *had to* resign. Otherwise his family would have suffered in the spotlight of the scandal-hungry media.

Here we understand that what makes resignation necessary is John's goal, need or desire of protecting his family from the "scandal-hungry media". Freeman (2005: 144) would most probably classify our modalized standpoint under the sub-type of *personal explanations*, which is composed by statements appealing to "the powers, beliefs, mental activities, desires, psychological dispositions, purposes or intentions of conscious personal agents". At the same time, the saturation of the conversational background provides an argument inferentially apt to support a *personal explanation* in the form of the *locus From the Final Cause*.

We can see how the information on the logical type of the standpoint and the *loci*-relations that *can be congruously applied* to it, once combined with the discourse relations emerging from *phoric saturation, conditional structures, adverbial clauses, etc., allows us to determine which *loci* are *actually* invoked in support of the standpoint (cf. Chap. 2, § 2.2.3).

In the following sections I will try to provide a first rough sketch of this functioning by coupling the typology of conversational backgrounds with the classification of statement types proposed by Freeman (2005).

5.2.2.2 Linguistic-Semantic Considerations

A second order of reasons for the utility of the cartography pertains to linguistic semantics. Broadly defined classes of conversational backgrounds can be associated with *linguistically encoded constraints* on the saturation of the B argument slot of the modal predicates. Certain modal expressions can only take certain kinds of conversational backgrounds. For instance, the adverb *perhaps* and the modal verb *might* are confined to epistemic-doxastic backgrounds. For other linguistic items the choice of a certain broad type of conversational background may entail a series of linguistically encoded restrictions on syntactic behavior and/or semantic context as well as additional constraints on the selection of the propositions forming the conversational background.

For instance, as it will be shown in Chap. 6, the epistemic readings of the Italian necessity modal *deve* and its near-equivalent English *must* are subject to restrictions concerning negation and future reference.

Furthermore, these epistemic interpretations acquire a further *evidential* specification, restricting the epistemic conversational background to propositions that are indirect evidence for inferring the prejacent. Furthermore, according to certain hypotheses (Hacquard 2010, 2011) very broad classes of conversational backgrounds may correspond to fundamentally different constraints on the saturation and anaphorical resolution of the conversational background that are enshrined in natural language syntax.

5.2.2.3 Pragmatic and Cognitive Considerations

Further reasons for distinguishing classes of conversational background pertain with their possible correspondence with properties of the pragmatic organization of the utterance or properties of the organization of human cognition. On the one hand, the selection of certain classes of conversational background may be associated with the pragmatic status of the modal predicate and of the modal proposition. For instance, only for certain classes of conversational backgrounds it makes sense to speak of *performative* uses of the modals.

On the other hand, different classes of conversational backgrounds might find correspondence in the empirical functioning of human cognition.[5] For instance, according to Papafragou (2000: 41), "propositions can be entertained and stored in memory in several different ways" including factual descriptions of the world, desires, interiorized systems of norms, etc. In this perspective broad classes of conversational backgrounds correspond to functionally different mental stores of "folders" in the human file-system. As Papafragou (2000: 42) puts it, they define "a sort of file-based organisation of our belief-desire system".

---

[5] I mean *empirical* as opposed to the *conceptually necessary* aspects of human cognitive functioning.

5.2.3 *Linguistic-Semantic Classifications of Modal Flavors*

Existing classifications of modal "flavors" differ as to the kind of kind of criteria that drive the selection of the chosen categories. Logico-philosophical proposals are centered on the fit between the formal properties of the semantics (e.g. the kind of modal frame) and philosophically important distinctions, which provide a philosophically relevant concrete interpretations of the abstract semantics. Classifications developed in linguistics often favor other criteria.

5.2.3.1 The Typological View

At the opposite end of the spectrum with respect to logico-philosophical classifications we find the proposals originating in linguistic typology and historical linguistics. "Semantic maps" of modality have been constructed by functional typological linguists on the basis of cross-linguistically recurrent avenues of semantic change and grammaticalization (cf. Bybee et al. 1994, van der Auwera and Plungian 1998, Nauze 2008). In linguistic typology the cartographic language of semantic maps is employed to reconstruct "the sum total of the semantic possibilities of the category under investigation" (de Haan 2006: 45), both in terms of synchronic polysemy and diachronic semantic change. Maps associate with a modal form a *polysemic network* of related but distinctly encoded meanings. Connections between the meanings in the semantic map correspond to reconstructed paths of historical semantic change. As an example, Fig. 5.1 reproduces a simplified version of the semantic map of possibility modals proposed by van der Auwera and Plungian (1998: 91).

To the left of the grey area premodal meanings provide the etymological sources of the modals. These include, for instance, *be strong* (cf. the etymologies of English *may* and Latin *posse*), *know* (cf. the origin of English *can*), which feed into

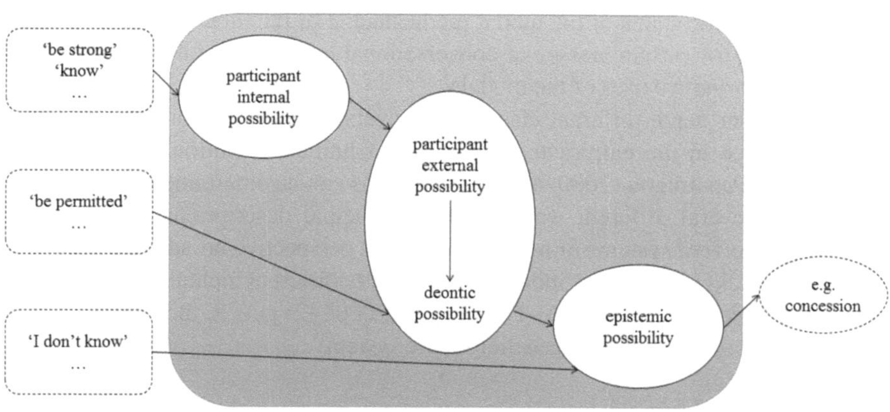

Fig. 5.1 Semantic map of possibility, simplified version of van der Auwera and Plungian (1998: 91)

meanings of *participant-internal possibility*, corresponding to capacity or potentiality in view of the characteristics of an event participant, typically mapping syntactically onto the subject of the modal. Other sources, like those indicating lack of knowledge, feed directly into epistemic possibility (cf. Spanish *quizá*). On the right of the grey area of modality one finds postmodal meanings: discursive meanings such as *concession* are one example (cf. *He may be a full professor, but sure he is dumb*). Within the modal meanings the most significant distinction from a typological viewpoint is that between *participant-internal* and *participant-external* modality, with *deontic* being just a special case of the latter.

Distinguishing conversational backgrounds prevalently centered on the internal characteristics of an agent from those comprising mostly external facts *does* have some significance also from the viewpoint of argumentation, but not to the point of overshadowing the distinction between alethic and deontic modalities, which has much broader logical consequences. While one has to acknowledge the interest of typological semantic maps in their own domain, it would be ill-advised to take them as primary source of insights for an investigation on the argumentative consequences of modal meanings.

5.2.3.2 Bipartite Views: Root vs. Epistemic Modality

In linguistics one minimal subdivision that has resisted through the decades is the one between *epistemic* interpretations and all other kinds of modality, which receive the, admittedly insignificant, signpost label of *root* modalities. The bipartition originated on (shaky) syntactic grounds in Generative Linguistics but continued to thrive in descriptive studies, cognitive linguistics and, in part, in formal semantics. In the process, the bipartition has been associated with a variety of linguistico-semantic differences in the behavior of modal expressions whose precise nature is difficult to pinpoint.

Originally, the *root-epistemic* bipartition was connected with the view that root modal verbs, expressing dynamic and deontic modality, have, respectively, an *agent* or *deontic subject* argument slot which is systematically mapped onto the syntactic subject of the modal (cf. Ross 1969). In generative syntactic parlance, they are seen as *control predicates* (cf. Portner 2009: 187), while epistemic modals, in which the syntactic subject is devoid of semantic function, are *raising predicates*.[6] This is a

[6] As it often happens for modern syntactic terminology, the terms *raising* and *control* are based on transformational metaphors rooted in early Generative Grammar. This makes them relatively opaque. In non-transformational terms (cf. Fried and Östman 2004: 63–66), raising and control structures are both constructions characterized by the co-instantiation of the subject of the non-finite verbal complement (e.g. the infinitival) and the subject of the main predicate (the modal verb in our case). The two constructions differs in that *control* main predicates (e.g. *try*) assign a semantic argument$_S$ slot to their subjects, while *raising* predicates (e.g. *seem*) do not. While filling the syntactic role of subject also in the main predicate, in raising structures the co-instantiated subject is semantically tied only to the infinitival complement.

mistaken view. For deontic and deontic-practical modals, one can easily find many examples where the agent or subject of obligation is not the syntactic subject. And dynamic uses include situations where an agent other than the subject is foregrounded. One of the first scholars to notice it was St. Anselm of Canterbury in the XI century (cf. Rocci 2005a: 141):

> Cum ergo dico, quia debeo amari a te, non ita dicatur, quasi ego aliquid debeam, sed quia tu debes amare me. Similiter cum dico, quia non debeo amari a te, non aliud intelligitur, quam quia tu non debes amare me. Qui modus loquendi est etiam in potentia et in impotentia. Ut cum dicatur: Hector potuit vinci ab Achille, et Achilles non potuit vinci ab Hectore. Non enim fuit potentia in illo, qui vinci potuit, sed in illo, qui vincere potuit; nec impotentia in illo, qui vinci non potuit, sed in illo, qui vincere non potuit. (St. Anselm, *De Veritate*, VIII, *De diversis intellectibus debere et non debere, posse et non posse*)

Wurmbrand (1999) and Rocci (2005a), on the basis of examples of this kind, argue that *for all flavors* modal verbs should be regarded as raising predicates.[7]

From a Cognitive Linguistics viewpoint, Eve Sweetser (1990) reframes the distinction between *root* and *epistemic* modals according to the well-known "force-dynamic" theory put forth by Talmy (1985) hypothesizing that the more abstract epistemic meanings are basically *metaphorical* extensions of the more experientially concrete force dynamic root meanings.[8] Compare the following two examples:

(20.a) I am starving! I *must* eat a sandwich, now!
(20.b) Bill had five sandwiches. He *must* have been starving.[9]

Sweetser proposes that the same relation of compulsion can hold in the "sociophysical world" between cause and effect or at the epistemic level "between premise and conclusion in the speaker's mind" (Sweetser 1990: 80):

> Here *must* is taken as indicating an *epistemic force* applied by some *body of premises* (the only thing can apply epistemic force) which compels the *speaker* (or people in general) to reach the *conclusion* embodied in the sentence. This epistemic force is the counterpart, in the epistemic domain of a forceful obligation in the sociophysical domain. (Sweetser 1990: 64)

A full discussion of Sweetser's proposal would exceed the modest scope of this brief review of the criteria used to classify modal flavors. The ambition of the force

[7]Portner (2009: 188) proposes a nuanced position according to which "some deontic modals are control predicates and some are raising predicates". For the purposes of the present investigation it is not necessary to enter this debate. Note that, in the perspective of Congruity Theory arguments can be recognized on semantic grounds irrespectively of the systematicity of their syntactic manifestation, and notions such as agent or deontic subject arguments remain important to understand certain flavors of modality even if they are only sporadically mapped ont complements.

[8]Note that force dynamic theory of Talmy (1985) and Sweetser (1990) has (had?) the ambition not only to offer a solution to the problem of the poly-functionality of modal expressions, but also, to offer a general semantic account of the underlying notional categories alternative to the logico-semantic treatments in terms of possible worlds. A discussion of the merits of the approach exceeds the concerns of this book, but see Rocci (2005a: 151–169) and Portner (2009: 110–115) for a critique.

[9]Examples adapted from Knott (2001: 129).

dynamic approach of offering a general semantic account of the underlying notional categories alternative to the logico-semantic treatments in terms of possible worlds has been criticized elsewhere (cf. Rocci 2005a: 151–169, Portner 2009: 110–115) and a critical discussion of the historical claims associated with the metaphorical hypothesis can be found in Traugott and Dasher (2002).

I shall instead be content with a basic remark closer to our argumentative preoccupations: Sweetser's analysis seeks to capture the idea that epistemic modals manifest the speaker's inference processes about the world rather than the ontological connections in the world. The link between the former and the latter is conceived in analogical terms as a metaphorical mapping between two domains. In other words, when we connect thoughts in a string of inferences it is *as if* facts were causally connected in the real world. However, a view of the relationship between inference and world ontology does not sit well with the view of argumentation and inference adopted by here.

According to the AMT view of the inferential structure of arguments (Chap. 2, § 2.2.3), the argument schemes used in ordinary discourse do not rely on the mere abstract logical form, but draw their force from the semantico-ontological makeup of arguments captured at a more concrete level. Thus, the inferential, epistemic, relations are not merely analogic or isomorphic with respect to content-level relations as they indeed draw their force from our understanding of the proper segment of the ontology.

Francisca Snoeck Henkemans (2001) takes a similar stance in her review of linguistic approaches to discourse relations conducted from the viewpoint of argumentation theory. She criticizes theories such as Mann and Thompson (1987) for positing that discourse relations hold *either* at the pragmatic/epistemic level *or* at the propositional level. According to Snoeck Henkemans (2001: 237) these theories do not recognize that "every illocutionary relation is by definition based on a subject matter (or 'propositional') relation. This recognition, she argues, is particularly important if we want to account for argumentation which holds at the speech act level, but inevitably rests on relations such as *causality*, *concomitance,* and *comparison* holding between the propositional contents of the standpoint and of the arguments.

If we go back to examples (20.a) and (20.b) we can see that the causal relation signalled in (20.a), for instance, can give rise either to a speech-act relation of *explanation* or to an *argumentative* relation, but at the same time it *still matters* to look at the content-level relation underlying (20.b) to understand what kind of argument it is. Argument criticism, for instance, either by analysts or by dialogue participants often targets the content-level relations rather than the act of inference as such.

Anna Papafragou (2000), who addresses modality from the viewpoint of the cognitive pragmatic framework of Relevance Theory, agrees with Sweetser (1990) that epistemic modalities in some way express relations – such as premise-conclusion relations – and processes – like inference – taking place in the speaker's mind rather than entities, events or even facts in the external world. The relationship between the mental operations and the external facts, however, is not a metaphorical mapping *seeing mental processes in terms of external world force dynamic processes*. Rather, epistemic modality involves the cognitive operation of *metarepresentation*, the speaker's own representation of the world is considered *qua* representation, and not simply as a set of facts. The propositional representations of facts are cognitively manipulated as representations of facts, rather than as facts. Here is how Papafragou summarizes the gist of her *metarepresentation* hypothesis:

> According to this hypothesis, epistemic expressions are markers of a logical relation between the complement of the modal and the set of the speaker's beliefs. Consequently, the appropriate use and comprehension of epistemics crucially presupposes the ability to conceive of evidential relations between propositions which form the content of beliefs, and is thus linked to the human capacity to metarepresent. (Papafragou 2000: 7)

The logical relation between the prejacent and the set of beliefs is conceived in terms of a minimal version of Relative Modality (cf. Papafragou 2000: 40), but the meta-representational nature of the propositions in the B set has important consequences, which Papafragou characterizes as follows:

> From the speaker's point of view, the employment of epistemic modality rests crucially on the ability to reflect on the content of one's own beliefs, to take into account the reliability of those beliefs (i.e. the relative strength with which they are entertained), and to perform deductive operations on them. The above processes jointly presuppose the ability to conceive of one's mental contents as representations distinct from reality which may bear a variable degree of correspondence to the actual world (Papafragou 2000: 70)

It is clear that Papafragou uses the term *epistemic* according to the prevalent usage in linguistics where it is synonymous of doxastic and not in the stricter and more etymological sense of logicians, where it refers to knowledge, which is true belief. So, epistemic possibility or necessity, here, is not what is possible or necessary *in view of the known facts*, but what is possible or necessary *in view of a set of beliefs*. Furthermore, when speakers use epistemic modal expressions they entertain their beliefs *qua beliefs*, separate from the world they purport to represent. This introduces the possibility of variable degree of correspondence of the representation to the actual world.

Clearly, if this notion of "variable degree of correspondence" were to be represented in the semantics it would amount to another epistemic modal, which hints to an undesirable *regressus ad infinitum*. This makes Papafragou's proposal rather unappealing from the viewpoint of a formal semantics viewpoint such as Kratzer's. However, Papafragou's cognitive-pragmatic stance is admittedly very distant (cf. Papafragou 2000: 10) from formal semantics and philosophical logic. In Relevance Theory, non-reflective, unconscious, inference is modeled as a deductive computation based on a subset of classical deductive logic and takes as input assumptions (propositions) that are salient in memory, or, to be more precise, in the *cognitive environment* of the reasoner. As a result of the computation, the cognitive environment is modified with the derivation of new assumptions, the reinforcement or weakening of existing assumptions and a new organization of memory whereby those assumptions that have been fruitfully used to derive new assumptions, or to strengthen or weaken existing assumptions have come on top of the salience hierarchy and are readily accessible to the reasoner. The strength or weakness of assumptions is not something that Relevance Theory deals with in terms of epistemic logic – or any other type of logic. In this perspective, I think one can conceive Papafragou's epistemic modality as the distinct conscious focalization of one step of an underlying process of reasoning, whose deeper parts will remain, so-to-say, inarticulate and inaccessible to the subject.

Angelika Kratzer (1991) provides her own version of the *root-epistemic* bipartition, based on a distinction between two types of basic conversational backgrounds, which highlights "two kinds of modal reasoning". Kratzer has recently abandoned this idea (cf. Kratzer 2012: 24), but it is still worth discussing it briefly. In Kratzer's early, view epistemic modals have as their *modal base* a conversational background composed of *all known facts*, additionally they can take other conversational backgrounds as their *ordering sources*. Root modals take a circumstantial modal base, which means a conversational background composed of *facts of a certain relevant kind*. Then, dynamic and deontic modals are set apart within root modality because deontic modals take a deontic conversational background[10] as their ordering source to single out, among the circumstantially compatible alternatives, those that best conform to the commands, laws or norm. Kratzer (1991: 646) argues for the significance of the *epistemic – circumstantial* split using the following example:

> Epistemic modality is the modality of curious people like historians, detectives, and futurologists. Circumstantial modality is the modality of rational agents like gardeners, architects and engineers. A historian asks what might have been the case, given all the available facts. An engineer asks what can be done given certain relevant facts.

Let us now consider two examples taken from Kratzer (1991), which are particularly clear thanks to the incompatibility of the English possibility modal *can* with purely epistemic conversational backgrounds (cf. Papafragou 1998) and the incompatibility of the possibility modal *may* with most purely alethic backgrounds:

(21.a) Hydrangeas *can* grow here.
= Growth is compatible with the conditions of soil and climate (circumstantial B)

(21.b) There may/might be hydrangeas growing here.
= 'They grow' is compatible with what I know to be the case (epistemic B)

In the first case we have an alethic modality where we assert the compatibility of the growth with some kind of real world conditions, most likely the conditions of soil and climate. In this case we can have an argument about *whether hydrangeas can grow here*, where we can take a standpoint like:

(21.c) I am absolutely certain that they can.
= I am absolutely certain that their growth is compatible with the conditions of the soil and the climate.

On the contrary, it would be bizarre to have an argument about whether they *may grow here*, because it is exactly the same thing as an argument about whether they *do grow here*. In fact, *There may / might be hydrangeas growing here* is not something *about which* one can have a discussion, it is rather an expression of uncertainty

[10] See Chap. 4, Def. 12 in § 4.2.7.1.

in a discussion about whether they *do* grow here. It means that my evidence – just *any* type of evidence I may happen to have – does not contradict the hypothesis, which is the interpretation of a possibility modal saturated with an epistemic conversational background. Someone could utter (21.b) to cast doubt on the negative standpoint that there aren't hydrangeas growing here, or, especially with *might*, to take a very weak positive standpoint on the issue. This is why an utterance like (21.d) would sound rather bizarre:

(21.d) *I am absolutely certain that they may/might.
= ? I am absolutely certain that the hypothesis that they grow here is compatible with all the evidence I have.

Such an utterance would involve making explicit the meta-cognitive operation implied by the doxastic qualification, which is usually not required in a discussion. Once we bring it on focus, epistemic modality seems to acquire a meta-representational, metacognitive nature.

The contrast established by Kratzer (1991) is between two ways of reasoning and two types of reasoners: "curious people" vs. "rational agents". From the example it seems that the contrast is basically between *cognitive argumentation,* oriented towards advancing the state of common knowledge and *practical argumentation*, oriented towards sound deliberation on the backdrop of shared goals. Tempting as it may be, the position does not correspond with what has surfaced from examples (21.a-c), where, rather than different *goals* of the reasoning, we have to do with different and possibly co-occurring *levels* of the reasoning: the propositional content and the cognitive and inferential operations of the speaker.

Kratzer (2012: 24) briefly comments on her reasons for abandoning this version of the bipartite subdivision of modal flavors, saying she now believes that it is dubious that the epistemic and circumstantial backgrounds really have distinctive semantic properties, as "both types of backgrounds are functions that map possible worlds to sets of factual premises" and it is difficult to establish which premises should be included in one type of background and not in the other. Note that, contrary to Papafragou (2000) above, Kratzer here takes *epistemic* in the strict logician's sense, which is not equivalent with doxastic.

Kratzer now believes that a better explanation of the *epistemic-root* divide is provided by Valentine Hacquard (2010, 2011) with her theory of the *event relativity* of modality. The theory hypothesizes that the *epistemic-root* split corresponds to a fundamental difference in the "very general recipes" (Kratzer 2012: 24) through which the (deep) syntax of the sentence "anchors" the search of the conversational background to different types of entities: circumstances of the main predicate of the prejacent for *root* modals, speech events or mental attitudes for *epistemic* ones.

In fact, the key elements of the *event relativity* analysis will be integrated in the typology proposed here, in combination with Papafragou's (2000) idea of the *meta-representational* nature of epistemics. I provide more details on *event relativity* as part of the integrated proposal in § 5.2.4.

Table 5.1, below, summarizes the conceptions of the *root* vs. *epistemic* split reviewed in this section:

Table 5.1 Conceptions of the *root* vs. *epistemic* distinction

Source	Root modality	Epistemic modality
Ross (1969)	control predicates	raising predicates
Sweetser (1990)	physical and social forces	metaphorical epistemic forces
Papafragou (2000)	representation	meta-representation
Kratzer (1991)	circumstantial base	epistemic base
Hacquard (2011)	anchored to the VP event	anchored to the speech event

5.2.4 A classification Balancing Linguistic and Logical Criteria

In this section I will propose an argumentatively relevant typology of conversational backgrounds allowing to capture distinctions between "flavors" of modality that matter from the viewpoint of argumentative reconstruction and evaluation. This classification combines linguistic-semantic criteria and logico-semantic criteria. The former concern the process through which the conversational background are constructed from linguistic and contextual information and the cognitive status of the propositions composing it. The latter concern the formal properties of the conversational backgrounds and the ontological status of the propositions composing them.

The proposal first introduces a distinction between *epistemic* and *root* modality. As it will emerge in this section, this distinction is not conceived here in logico-semantic terms. It makes reference to the instructional/ procedural semantics guiding the contextual saturation of the conversational background and to the resulting cognitive status of the propositions composing it.

- The conversational background of epistemic modalities is recovered in the context of a speech event or mental event and is composed of propositions entertained as metarepresentations (thoughts or utterances about states of affairs).
- The conversational background of root modalities is recovered in the context of the event of the prejacent proposition and is composed of propositions entertained as mere states of affairs.

As it will be shown in the following pages, This distinction takes inspiration from the works of Valentine Hacquard (2010, 2011) and of Anna Papafragou (2000) introduced above.

The second step of the classification operates according to a logico-semantic distinction within the category of root modals. The root modals, whose conversational backgrounds are composed by propositions entertained as mere states of affairs, can be subdivided in those for which the conversational background is entirely composed of factual states of affairs (*alethic modalities*) and those for which it also contains non-factual states of affairs such as desires, values, laws,

Table 5.2 Comparing classifications of modal flavors

<table>
<tr><td rowspan="4">Proposed classification</td><td colspan="3">Epistemic</td></tr>
<tr><td rowspan="3">Root</td><td colspan="2">Alethic root</td></tr>
<tr><td rowspan="2">Non-alethic root</td><td>Anankastic</td></tr>
<tr><td>deontic</td></tr>
<tr><td rowspan="3">Kronning (1996, 2001a, b)</td><td colspan="3">Epistemic</td></tr>
<tr><td colspan="3">Alethic (including anankastic)</td></tr>
<tr><td colspan="3">Deontic</td></tr>
<tr><td rowspan="6">Portner (2009)</td><td colspan="3">Epistemic</td></tr>
<tr><td rowspan="2">Dynamic</td><td colspan="2">Volitional</td></tr>
<tr><td colspan="2">Quantificational</td></tr>
<tr><td rowspan="3">Priority</td><td colspan="2">Deontic</td></tr>
<tr><td colspan="2">Bouletic</td></tr>
<tr><td colspan="2">Teleological</td></tr>
</table>

goals or just any kind of non-realized end-states singled out by the speaker.[11] Finally, these non-alethic root modalities are further subdivided into the properly *deontic* flavors and the *anankastic* modality, the modality of requirements and necessary conditions, a less commonly recognized kind of modality and yet an independent and indispensable one.

As a whole, if we look at the terminal categories, the resulting map of major flavors of modality is similar to other subdivisions of modality proposed on a balance of linguistic and logico-semantic considerations. In particular, Kronning (1996, 2001a, 2001b) and Portner (2009) propose roughly comparable categories (cf. Table 5.2).

To explain the rationale behind the proposed classification, I will first compare and contrast it with the proposals of Kronning (§ 5.2.4.1) and Portner (§ 5.2.4.2) highlighting the insights that the present proposal seeks to preserve, while showing, at the same time, how problems arising from these classification motivate the differences; I will then focus more closely on the role of Hacquard's theory of event relativity as a foundation for the distinction between root and epistemic modality (§ 5.2.4.3), proposing a procedural and discourse oriented reformulation of the theory (§ 5.2.4.4). Finally, I will turn to the distinction between alethic and non-alethic conversational backgrounds within root modality (§ 5.2.4.5).

[11] In principle, a similar, but not identical, distinction between alethic and non-alethic conversational backgrounds operates also within the epistemic conversational backgrounds separating the properly epistemic conversational backgrounds understood as selecting the set of known facts and the merely doxastic ones, comprising sets of beliefs which "may bear", as Papafragou (2000: 70) puts it, "a variable degree of correspondence to the actual world".

5.2.4.1 What Can Be Learned from Kronning's Treatment of *Alethic* and *Anankastic* Modalities

Kronning (1996, 2001a, b), proposes a tri-partition of modal meanings in *epistemic*, *alethic* – which is understood as including also the *anankastic* – and *deontic*. I am particularly sympathetic with Kronning's emphasis on the pervasiveness of *alethic* modalities in ordinary discourse, which was already observed in § 5.2.4.1, commenting Lycan's (1994) views on modal context dependency. Kronning, who focuses on the French necessity modal *devoir* 'must', includes in the alethic domain a diverse array of examples. A first kind of alethic modality is represented by sentences such as the following:

(22) Parmi les nombreux attributs de Dieu, il *doit*$_A$ exister, puisqu'il est la cause première. (WebCorp)
'Among the many attributes of God, he *necessarily* exists, given that he is the first cause'

The use of *devoir* in (22) corresponds to what Kronning (1996: 70, 115) calls *analytic alethic necessity* corresponding to sentences "that are true because of their meaning independently of any empirical verification".[12] This represents one extreme of the spectrum of alethic modality, as for Kronning (1996: 70) it is clear that also sentences involving a necessity deriving from "natural laws", such as (23), can be considered alethic.

(23) "Valar Morghulis", dans le langage de Game of Thrones veut dire "Tous les hommes *doivent* mourir" (WebCorp).
'"Valar Morghulis", in the language of Game of Thrones means "All men *must* die"'

Works in linguistics written in the English language seldom discuss alethic modality at length. They often quickly dismiss the category as irrelevant for the analysis of ordinary non-philosophical discourse and natural languages, declare it suspicious for its metaphysical flavor and hurry to conclude that it is linguistically indistinguishable from epistemic modality.[13] Palmer (1986: 11) is a particularly harsh dismissal:

> "[...] there is no distinction between [...] what is logically true and what the speaker believes, as a matter of fact to be true [...] there is no formal grammatical distinction in English, and, perhaps, in no other language either, between alethic and epistemic modality".

Views as the one expressed above are as philosophically naïve as they are linguistically unsupported and it beggars belief to find them so often repeated in the literature.

Now, semantically, there *is* a difference between saying that a certain prejacent proposition is a consequence of one's beliefs or knowledge, and saying that the truth of said proposition follows from the very concepts it puts together or that it follows

---

[12] Cf. Kronning (1996: 70) "qui sont vraies en vertu de leur sens indépendamment de toute verification émpirique".

[13] See for instance Nuyts (2006: 8–9) and the one footnote devoted to the issue by Portner (2009: 135).

from natural laws. For one, the two alethic modalities do not *entail* the epistemic. Certainly, *when I assert* that p follows from the laws of nature or from the semantics of its constituent concepts, I am also *committed* to it following from what I believe. But this is a pragmatic inference, not a semantic entailment. More generally, as we will see in the following sections, these semantic differences do matter in the reconstruction and evaluation of argumentation. To quip, they matter if we are interested in marginal and abstruse uses of language such as Cartesian philosophy or the *Game of Thrones* television series.

As for the lack of grammatical distinction of epistemic and alethic uses, one would be tempted to leave the defense to Kronning's (1996) painstaking analysis of the properties of the construction of *devoir* in the three major "flavors" based on fine grained observations of the behavior of the modal with respect to negation and focus. This would bring the discussion astray here, but in Chap. 6, I will discuss in some detail the grammatical properties that set apart the epistemic construction of Italian *dovere* ('must'). To provide just one example of linguistically observable semantic properties that set epistemic and alethic readings of the modals apart, one can observe the possibility of embedding alethic modals in nominals like *the fact that*, as shown in (24.a), while epistemic ones do not fare well in that position (24.b).

(24.a) *The fact that* all men must/ have to die is hard to accept.

(24.b) ??*The fact that* it must be raining outside is a bit depressing.

Another kind of alethic modal discussed by Kronning (1996: 63–66) is what he calls *alethic future devoir*, exemplified in (25).

(25) Dans les Vosges, lorsqu'il *doit*$_A$ pleuvoir, les abeilles se tiennent à la porte de leurs ruches (WebCorp)
'In the Vosges , when it *is going to* rain, the bees remain at the door of their hives'

The modal in (25) appears embedded in a temporal clause where epistemic *devoir* would be impossible and where English would rather employ the aspectual construction *be going to*. The generic context highlights that the modality expressed is *dispositional* (alethic)[14] rather than predictive (epistemic). While the modal semantics of future *devoir* in examples such as (25) is considerably bleached, the reference to atmospheric conditions in the discourse context can provide a partial manifestation of the conversational background and rekindle it.

A third class of alethic readings is represented by *anankastic* modality – a flavor of modality rarely discussed in linguistics, which Kronning (1996: 117–118, 2001b: 98–99) takes from Conte (2010 [1995]).

(26) Existence de l'objet au moment de l'exécution de la convention: Si l'objet *doit*$_A$ exister, il n'est pas *nécessaire*$_A$ qu'il présente ce caractère au jour de la formation du contrat. Il *suffit*$_A$ qu'il existe au plus tard, lors de son exécution Selon l'article 1130 alinéa 1 du code civil: « les choses futures *peuvent*$_A$ être l'objet d'une obligation ». (WebCorp)

[14] For a *dispositional* modal analysis of the French and Italian future tense see also Ludlow (1999) and Rocci (2000).

> 'Existence of the object at the time of execution of the agreement: While the object *must* exist, it is not *necessary* for it to have this character since the day of the formation of the contract. It is *sufficient* that at it exists at the latest when the contract is executed. According to Article 1130 , paragraph 1 of the Civil Code : "future things can_A be the object of an obligation"'

At first blush, given the legal context of example (26), the use of *devoir* and of the modal adjective *nécessaire* 'necessary' might appear deontic. They are not. The conditions of validity of a contract are not an obligation for anyone, they are institutional conditions for the contract to be binding. Clearly, it would not make sense for anyone to have *an obligation to exist*, least of all an obligation to exist not yet but later. And, generally speaking, the things that can be the object of a contract are not persons capable of having an obligation whatsoever. What is more, not even the signatories of the contract properly have an obligation to make this object exist. Simply, if it does not exist, the contract will be void.[15] Here the modals express the idea that existence of the object is a necessary condition of validity, it is necessary *in view of the contract's validity, in order for the contract to be valid.* Interestingly, the converse notion of sufficient condition also surfaces in the text.

Kronning (1996, 2001b) makes anankastic modality a sub-type of alethic. I do not agree. Alethic conversational backgrounds, at least in the view proposed here, are composed by propositions which are *facts* in the world of evaluation w. But the validity of the contract is not a *fact*. It looks more like a goal, but actually it is not exactly the goal of an action. The following sections will provide more detail on the nature and argumentative significance of this understudied category.

As an aside, I would like to observe that the French examples in (22), (23), (25) and (26) are particularly interesting also because they cover areas of modality, which *rarely* emerge in English language treatments of modality.[16] One partial exception is Nuyts (2006: 4), who briefly mentions *situational dynamic* modalities,

---

[15] This does not preclude that, by stipulating a null contract, the parties might have thereby violated some deontic obligation. For instance, if they knowingly did so in order to commit a fraud or if there was negligence on their part in ascertaining that the conditions of validity subsist.

[16] They are a good reminder of the fact that not only the peculiarities of each language – where central modal words may have restrictions that inhibit certain conversational backgrounds – but also the entrenched traditions of exemplification and grammatical description, can create blind spots in the most astute analysts. Ordinary language philosophers were sometimes criticized for being dilettante lexicographers, all too prone to see important philosophical distinctions under the patterns created by English idioms. I believe this criticism to be partly unwarranted. It may have merit as long as it concerns the linguistic dilettantism sometimes displayed by ordinary language philosophers, but not as regards the philosophical interest of linguistic analyses. The proper description of semantic distinctions enshrined by linguistic conventions regularly raises fundamental and difficult questions about the categories we use in reasoning. "Idioms" can be illuminating, provided that they are systematically analyzed with the right tools. That said, the "idioms" of other languages do not have a lesser status and should not be dismissed as something only descriptive linguists should care about. As we close towards the Italian language case studies of Chap. 6, I hope the reader will appreciate them for what they tell us about the nature of modality in connection with the essential structure of arguments.

which "characterize a potential or a necessity/ inevitability as inherent in the situation described in the clause as a whole". Ironically, at the same time, he seems to endorse Palmer's (1986) dismissal of *alethic* modality.

5.2.4.2 Portner and the Problem of the Argument Structure of Dynamic Modals

The tripartite classification of Portner (2009) cuts the space of modal flavors in *epistemic, dynamic* and *priority* modalities. Portner (2009: 135) coins the term *priority*, explaining that it conveys the idea that "such things as rules, desires and goals all serve to identify some possibility as better than, or as having higher priority than others".

As for the dynamic modals, he divides them into two quite heterogeneous subgroups: *volitional* and *quantificational* modals. Volitional modals relate to "the ways in which circumstances relate to the actions available to a volitional individual". According to Portner (2009: 201), they are *control predicates* where the syntactic subject is mapped onto an argument place that needs to be occupied by the volitional subject or agent. This subject is always an agent or cause of the process described by the main predicate and "at the very least is always volitional", i.e. "a sentient individual who is willfully involved" in the process described by the main predicate. In contrast, *quantificational* modals, which we have already encountered, "seem to produce existential or universal quantification over individuals" (*Ibid.*).[17] For Portner (2009: 197–201) the two subclasses are exhaustive, at least in English, and the cases of apparent dynamic modality that do not fit volitional modality have to be analyzed as quantificational. This a weak point in Portner's classification.

By focusing just on these two relatively small sub-classes Portner (2009) fails to embrace the full range of the modal meanings that linguists have labelled as *dynamic* or *circumstantial*, let alone to highlight that these *dynamic* or *circumstantial* modals are nothing else than everyday *alethic modalities*, as Lycan (1994) suggested (cf. Chap. 4, § 4.2.4.1), the most accessible of the modalities based on facts. Between them and the more abstract, more philosophically prestigious, alethic modalities there is a cline, not a difference in kind.

More technically, but relatedly, neither Portner's (2009) claim that all non-quantificational dynamic modals are volitional, nor his claim that they are control predicates seem tenable.

If the first claim is true, examples such as Kratzer's hydrangeas in (21.a) can only be interpreted as quantificational – i.e. as meaning something close to *Sometimes some hydrangeas grow here* – because they do not have "a volitional individual" in the subject role. Interestingly, this is not Kratzer's (1991) intended interpretation. She imagines a gardener surveying an unknown land and observing that the climate and terrain are indeed compatible with the growth of hydrangeas. Under this inter-

[17] On quantificational modals see also Chap. 4, § 4.3.3, which explores their indirect indicative power in the determination of argumentation structure.

pretation, the utterance of (21.a) would result in a true assertion even in contexts where hydrangeas have never lived in that place before. Thus, the idea that all dynamic modals are either quantificational or volitional as defined by Portner (2009) appears untenable.

Hydrangeas are not sentient and are not the only cause of the potential growth as it involves the climate, the terrain, and the actions that the gardener can perform. At this point, also the second claim appears to be questionable: if the modal *can* in (21.a) is a control predicate one should be able to assign to the subject *hydrangeas* a clear role in its semantics. Consider one of Toulmin's original *cannot* examples, slightly rephrased in (27)[18]:

(27) You *can't* get ten thousand people into Town Hall. There are simply not enough seats.

Here the conversational background is tied with the seating capacity of Town Hall and with the number of people, while the subject pronoun can be easily interpreted as a generic *you* and does not enjoy any special relation with the modal. Hacquard (2011: 1498) appeals to an example similar to (95) to conclude that in English, as in many other languages, "*all* modals are raising predicates", including *all* the dynamic modals.

The authentic occurrences of *cannot* in (28.a-c) offer an additional glimpse into the *variety of mappings* that occur between the syntactic structure of the modal sentence (including its subject, the complements of the prejacent, adverbial circumstantial complements and subordinate clauses), the discourse context, the semantic roles of *agent* and *cause* and the conversational background of propositions that lends the domain of quantification of the modal.

(28.a) Please note that the Limpopo River crossing at Platjan is on a 3m wide causeway. If the Limpopo River is flowing over the bridge, it *cannot* be crossed and the alternative then – if coming from South Africa – is to drive down to Martin's Drift which is approximately three hour's drive from Molema. (WebCorp)

(28.b) The bridge was destroyed so you *cannot* cross the river. And there are landmines around the trail. (WebCorp)

(28.c) But there is no bridge so you *cannot* cross the river here.

In (28.a), where the prejacent VP is an agentless passive, the subject (assuming *it = the bridge*) is neither agentive nor causal, and a causal relationship between events is recovered from the content conditional prothasis (*If the Limpopo River is flowing over the bridge*). This conditional antecedent is added to a conversational background of factual circumstances. In (28.b) and (28.c), which have an active prejacent VP, the subject maps on an agent role –which is not surprising given that the predicate of the prejacent can assign this role – but the key causal relation between events is recovered intersententially with the help of the coordinative conjunction *so* (a "structural connective" according to Webber et al. 2003). It is through this causal relation that the most prominent proposition of the factual conversational

[18] Cf. Chap. 3, § 3.2.3.

background is recovered. Contrary to the *if*-clause in (28.a), this proposition is factual and it is not truly *added* to the factual conversational background, of which the proposition is already part *de iure* (cf. Chap. 4, § 4.2.8.2 and 4.2.8.3).

5.2.4.3 How Hacquard's Notion of *Event Relativity* Fits into the Proposed Classification

Given the variety of syntactic and discursive structures that are used as scaffolding to saturate the conversational background of the modal via the reference to participants and events, it seems that focusing on the subject only and invoking a control analysis just for the dynamic modals is costly and provides little additional insight. Hacquard (2011: 1499) comes to the conclusion that a more flexible generalization about root modals needs to be captured:

> The purported thematic relation between a root modal and the subject instead seems to highlight the fact that root modals are centered around the circumstances of the event described by the main predicate, and especially, but not necessarily, those of its agent [...]

Her theory of the *event relativity* of the modals seeks to capture this more nuanced generalization by proposing a new way of looking at the *root* vs. *epistemic* distinction. This theory hypothesizes that the conversational backgrounds of *root* modals are anchored "to the circumstances of the event described by the main predicate", while the conversational backgrounds of epistemic modals relate to a speech or thought event, which is by default identified with the utterance event. As anticipated above, I adopt here a version of Hacquard's theory.

The version of event relativity offered here differs from the original in that it is formulated in procedural terms, as instructional constraints on the recovery of the B argument place, and in its explicit orientation towards the establishment of discourse relations. While Hacquard (2010, 2011) is mostly concerned with the ways in which the semantics of the modals is projected at different levels of syntactic structure, my main concern here is describing how root modals relate to the establishment of causal or final relations between events in discourse, while epistemic modals invite to recognize relations at an inferential and illocutionary level.

Innovative on many accounts, Hacquard's analysis uses the tools of formal semantics and generative grammar to rephrase and expand some ideas that have circulated for a long time in early XXth Century linguistics (cf. Jespersen 1924), and have been developed in quasi-formal accounts in various strands of functional and cognitive linguistics (cf., for instance, Hengeveld 1988; Kronning 1996; Nuyts 2001a; Pietrandrea 2005) about the *layered*, or *stratified*, structure of the utterance. In generative grammar, a syntactic version of this stratification has been developed, where different categories of meaning are associated with different positions in the syntactic tree (cf. Cinque 1999). By observing the scoping properties of different operators in natural languages (such as modalities of different flavors, tense, negation, etc.) different layers of semantic structure or, alternatively, "deep syntax" are recognized where operators take as their arguments different semantic entities (e.g. illocutionary act > proposition > fact > event, etc.) or scope over different syntactic units (e.g. taking scope over S or over VP).[19] For instance, epistemic modal expressions (cf. 21.c, above in § 5.2.3.2), including modal verbs (29), can routinely embed non-epistemic ones, while it is very difficult to construct examples where a root modal embeds an epistemic expression (cf. Haquard 2011: 1496).

(29) Stabilendo più o meno arbitrariamente il punto di partenza all'inizio dell'anno scorso – il gennaio 2005 – si *dovrebbe poter* dirimere la questione (*Il Sole 24Ore*, April 22, 2006)[20]

'By establishing more or less arbitrarily the starting point at the beginning of last year – in January 2005 – one *should be able* to sort out the matter.'

The outer layers of this stratification have been already discussed in Chap. 3, § 3.3.3, where we examined the evidence for the hypothesis that epistemic modals, or at least some of them, are un-embeddable, external to the propositional content, akin to performatives and (henceforth) subjective. The evidence is complex and nuanced. It suggests that certain epistemic markers are structurally non-propositional, others *can* be embedded under certain circumstances, even if they are subject to a variety of limitations regarding their embedded occurrence This shows a *proclivity* towards non-propositional status, which can be confirmed or disconfirmed in discourse context. As we have seen, non-propositionality is preferentially associated with m-performativity and with subjectivity, but intersubjective non-propositional interpretations are also possible.

Some functionalist hypotheses seek to identify the specificity of the *epistemic* flavors of context-dependent modals with *non-propositionality* and *m-performativity*. For instance, Kronning (1996: 52) concludes that what sets apart the epistemic interpretations of French necessity modal *devoir* 'must' from the others is that it belongs to that part of utterance meaning which is *shown* but not *asserted.* While based on fine grained observations of the behavior of the modal with respect to negation and focus, this analysis is not generalizable to the whole range of epistemic modals, and does not take into account the full range and the intricacies of the evidence.

The epistemic readings of French *devoir* 'must' and its analogue in Italian (*dovere*) are among the modal verbs for which the evidence of non-propositionality and m-performativity is stronger (cf. Kronning 1996; Rocci 2005a, b). The epistemic readings of these two modals scope *over* sentential negation, as shown in (30.a-b).

(30.a)	Pierre ne *doit*$_E$ pas être malade. = *I conclude* that Peter is *not* ill	(*doit*$_E$ > negation)
(30.b)	Pietro non *dev*$_E$'essere malato = *I conclude* that Peter is *not* ill	(*deve*$_E$ > negation)

[19] Here I remain agnostic as to the syntactic, semantic, pragmatic or simply cognitive nature of this stratification. Where these scope issues are dealt with is a highly theory-dependent issue exceeding the boundaries of the present work.

[20] This Italian example is taken from from the corpus of Italian business and financial news that I have exploited for the case studies expounded in Chap. 6.

The root modal readings of the two verbs, in contrast, are ambiguous between a wide scope reading ($doit_R$ > negation) and a narrow scope reading (negation>$doit_R$). As regards tense, the root occurrences of these two modals fall under the scope of past tense and are interpreted as holding true in the past. In contrast, their epistemic readings in (31.a-b), while formally exhibiting the morphology of simple/ remote past, *scope over* the past tense to be interpreted as inferences taking place at the moment of utterance.

(31.a) Elle m'a dit, comme elle **dut**$_E$–enfant – le dire à sa mère: – J'ai envie de rendre. (cf. Kronning 1996: 70)
'She said, as she **must**$_E$ have said it – as a child – to her mother:"I want to throw up"'.
dut = *devoir* ind. simple past 3rd sing.

(31.b) Negli anni della guerra, visse a Roma, città che amò ma che **dovette**$_E$ sembrarle un po' provinciale rispetto alle grandi capitali del mondo (WebCorp)
'During the war years, she lived in Rome, a city she loved, but which **must**$_E$ have seemed to her a bit provincial compared to the great capitals of the world
dovette = *dovere* ind. remote past 3rd sing.

This is an *unusually* neat situation. As Hacquard (2011: 1500) observes, several factors, among which the possibility/ necessity distinction, "seem to conspire to make the interaction of modals with negation a complex matter". I believe that similar remark could have been made for the interaction with tense.[21] For Italian, as soon as we move to the possibility modal *potere* the application of both tests of propositionality becomes more complicated and the results mixed, as shown in Rocci (2005a, 2005b).

It is therefore particularly significant that, also in the case of the "well-behaved" Italian and French necessity modals, the epistemic reading *can* be embedded by speech reporting and by some attitude reporting predicates, thereby ceasing to be the *m-performative*, i.e. ceasing to refer deictically to the beliefs of the speaker at the moment of utterance and referring instead to the beliefs of the attitude holder (cf. Hacquard 2011: 1499).

(32.a) Maria a detto che Giovanni *deve* aver perso il bus
Mary said that John *must* have missed his bus

(32.b) Maria pensa che Giovanni *debba* aver perso il bus
Mary thinks that John *must* have missed his bus

Interestingly, in these embedded examples it is the matrix clause that provides a specification of the conversational background of the modal. While the hypothesis that all epistemic modals are always non-propositional is clearly, untenable, it is also clear that the scoping properties of epistemic modals set them apart from other flavors.

[21] Actually, Hacquard (2011: 1508) believes that for tense the situation is more clear-cut: "epistemics cannot scope under tense". This may be correct for English, but the data about Italian modal verbs examined in Rocci (2005a: 256–259) suggest that the necessity modal *dovere* and the possibility modal *potere* behave differently with respect to tense and the interpretation of epistemic *potere* under the scope of tense, while impossible with most tenses, seems to be possible with the imperfect tense (cf. Rocci 2005: 258).

5.2.4.4 A Procedural, Discourse-Oriented, Version of Event Relativity

Hacquard's theory of *event relativity* has the merit of pulling together the various strings of suggestive but inconclusive evidence on the scoping properties and argument structure and weaving it into a credible and nuanced story based on Relative Modality. Here I will leave aside the original technical implementation of the proposal in terms of syntactic binding[22], and I will reformulate its core idea in procedural language, with a more explicit orientation towards discursive and contextual saturation. In doing so, I make use of a variant of the Reichenbachian coordinates that are commonly used in the procedural analysis of tense: *speech point* (s), *point of the event* (e) and, most importantly, *reference point* (r) (Reichenbach 1947; de Saussure 2003).

(Def. 1) *Generic event-relative procedure for the contextual saturation of conversational backgrounds:*

(a) A modal predicate manifests a relation M (B_r, p), where M is a relation of consequence or compatibility, p is the prejacent proposition of the modal, r is an event or state of affairs functioning as deictic or anaphoric reference point, and B is a set of propositions that makes up the conversational background of the modal.

(b) The set of propositions B is saturated indexically – deictically or anaphorically – in the context r.

(c) Retrieve r to saturate B.

(d) Consider whether $r = e$, where e is the prejacent VP-event. Retrieve the time of the tense t and the participants of the event i'-i^n. By default, B is a set of *circumstances* of e and of the participants i'-i^n at time t.

(e) Consider whether $r = s$, where s is the speech event. Retrieve the time of utterance t and the participants of the speech event i'-i^n, including the speaker and the addressees. By default, B is the content of representations of some participant i at time t.

(f) Consider whether $r = s'$, where s' is a reported speech or attitude event. Retrieve the time of the event t, the source or attitude bearer i. By default, B is the content of representations the source or attitude bearer i.

(g) If no salient r is recoverable as per (d)-(f), search for a saturation of B that makes the modal predicate congruous with the function of the utterance.

The procedure presented above is for an ideal generic modal expression, and thus not entirely adequate descriptively. Individual lexical, morphological or constructional modal expressions will place further restrictions on the recovery of the conversational background B. This could mean that certain clauses of the procedure have formal or semantic entry conditions or are simply inaccessible. For instance, for the English modal verb *can* the clauses (e) and (f) are not accessible at all, precluding truly epistemic readings (cf. Papafragou 2000: 43), while for the modal verb

[22] Leaving aside, here, does not imply a negative evaluation. Simply, the investigation syntax-semantics interface lays beyond the confines of the present study and a discussion of Hacquard's implementation of event relativity in term of syntactic binding of the modal's event variable will be out of place.

may clause (d) involves further semantic conditions, which results in a deontic interpretation. In general, even when all routes remain open, individual linguistic items will have a preference for one or the other kind of saturation of the conversational background.

Restrictions on the saturation of B are often associated with specific syntactic constructions rather than with the lexical item itself. The complementary interpretations of the two predicative constructions of the English adjective *possible* are a case in point: it seems that the '*possible for* NP *to* VP' is limited to route (d), while the '*possible that* S' construction selects routes (e) and (f) (cf. White 1975). Rocci and Musi (2014) have observed such a complementary distribution of interpretations for the equivalent constructions of Italian *possibile* in the corpus of business-financial news explored in Chap. 6. The fine grained semantic analyses of individual modal constructions in Chap. 6 will feature procedures which include *ad hoc* restrictions, while remaining along the lines of the generic procedure sketched in (Def. 1).

How does this procedural account compare with established accounts of the contextual determination of conversational backgrounds? Kratzer's classic conversational background is formalized as f (w), i.e. as a function *f* that takes any world w as its argument and gives a set of propositions as a value (cf. Chap. 4, § 4.2.7.1). Concrete examples could be *All the facts in* w, *What is commanded in* w, *What is normal in* w, etc. Clearly, the formalism is sketchy and world relativity is not enough to capture context dependency. For instance, given a whole world w it does not make much sense to speak of *one* set of propositions expressing what is commanded: commands are issued by different deontic sources to different deontic subjects at different times. Hacquard (2011) substitutes the f(w) function with a f(e), a function from events to sets of propositions. So that our concrete example would become *What is commanded in the context of* e. More precisely, since e is used to determine a time t and a participant i, the concrete function would become *What is commanded to* i *at* t. Needless to say, it all remains pretty abstract and sketchy, despite the precise mapping on the syntax. The extension of the B set is specified, but little is said of the eventual recovery of the actual propositions which compose the set.

In contrast, pragmatically oriented analyses such as Papafragou (2000), and even more radically Groefsema (1995), see the conversational background as a contextually built, highly *ad hoc*, set of propositions. In a sense, in these approaches the recovery from the cognitive environment of salient propositions comes before the "labeling" of the conversational background. These approaches, however, offer little scaffolding to the recovery of B, apart from invoking general principles of pragmatic processing.

The procedure in (Def. 1) sketches a third way, inspired by instructional accounts of context-dependent constructions (cf. Kay 2004) and by the procedural pragmatics of connectives and verbal tense (cf. Moeschler 1989, de Saussure 2003). It translates some key ideas of Hacquard (2011) in a procedural form, providing *a procedural analog of the* f(e) *function*, a rough and sketchy story of how you go from the event *e* to the set of propositions B. Importantly, the procedure suggests preferential paths for the saturation of B, but does not fully constrain it. If everything fails, peripheral interpretations could arise from the "escape clause" (g) of the procedure, provided that the context offers enough hints for the saturation of the conversational background.

One interesting feature that this procedure inherits from Hacquard (2010, 2011) is the contrast between *circumstances* and *content.* Hacquard (2011: 1505) proposes that "only modals relative to speech and attitude events have associated 'propositional content'[...] which provides an information state required by an epistemic modal base". In fact, in this way, event relativity provides a richer motivation to Papafragou's (2000: 7) idea that epistemic modality involves the human capacity to *metarepresent* in order to conceive of "evidential relations between propositions which form the content of beliefs". The metarepresentational nature of the epistemic conversational background emerges from its being indexically anchored to speech and thought events.

Where the present proposal differs markedly from Hacquard's original theory is its discourse orientation, which is expressed in clause (b). Hacquard is not really interested in the *saturation* of the B only in the determination of the general type of background via the syntactic binding of the event variable. Any further specification of the conversational background is the business of pragmatics and is not focused upon. While this choice of focus is perfectly legitimate, the close interaction between context, linguistic constructions and discursive relations remains outside the picture. Thus, the epistemic conversational background is simply identified with "the propositions that make up the attitude, such as a set of beliefs" (Hacquard 2011: 1505). In fact, the epistemic B often includes representations recovered from previous speech-acts, which involve the establishment of discourse relations that are at the same time illocutionary and inferential. For root modals, Hacquard insists in the recovery of the conversational background from the participants – agents first of all – of the VP-event. But there is a step beyond that: in many examples examined up to this point we have seen that causal processes involve relations between events that can be expressed interclausally and agents have goals that can be expressed by *in order to* clauses and other means. Also for root modals, then, interclausal and intersentential relations can be important. In this case the relations will be established, so-to-say at the *content level*, rather than at the illocutionary or inferential level.

In order to capture the distinctiveness of epistemic modals I was obliged to stray from purely logical considerations and delve deep in matters linguistic. This move was motivated by the unique role that these modals seem to have in signaling the essential structure of arguments, as Chap. 3 suggested. In fact, the complex machinery of event relativity not only draws the line between epistemic and root modality, but also casts light on the discourse relations that can result from the two kinds of conversational background saturation. Once the metarepresentational conversational backgrounds of epistemic modality have been set aside, the area of root modals still remains vast and amorphous.

5.2.4.5 Alethic and Non-alethic Root Modalities

The subdivisions of the vast region of root modality do not require us to espouse a particular linguistic theory of syntax, discourse or pragmatic processing. The criteria for obtaining the main classes of "flavors" within root modality can be drawn from relatively basic logical and ontological distinctions. The main subdivision within root modality is between *realistic* (cf. Kratzer 2012 [1981]: 32) conversational backgrounds and conversational backgrounds that are not entirely realistic. A realistic B is composed of propositions that are facts in the evaluation world *w*.

The realism of a conversational background is a formal property and corresponds to *reflexivity* of the accessibility relation in Kripkean frame semantics and to *Axiom T* in the syntactical characterization of a system of modal logic (cf. Kratzer 2012 [1981]: 32; Kaufmann et al. 2006: 82–86).

(Def. 2) *Realism*: $w \in \cap f(w)$ *Reflexivity*: wRw *Axiom T:* $(\Box\varphi \rightarrow \varphi)$

According to definition (2), a conversational background is realistic when there is no proposition in B which is false at the actual world *w*.[23] That is: B is a set of facts in *w*. It may not be the set of *all* facts in *w*, and may not even contain *all known facts* in *w*. But it does not contain propositions that are not facts in *w*.

For all root modals, the conversational background is preferentially reconstructed from the circumstances of the prejacent's event. Not all these circumstances, however, are facts in *w*. Agents' desires and goals, possible undesirable consequences of actions, the content of commitments and regulative rules, cultural norms and shared values are not facts; yet they are part of the circumstances affecting the realization of an action by an agent.

I call *alethic root modals* the root modals taking a realistic conversational background and *non-alethic root modals* those whose conversational background is not entirely factual. Terms like *realistic* or *ontological* could have been used instead of *alethic*. The choice fell on *alethic* because it is a common term in logic and is also the term used in Kronning (1996) and Lycan (1994) – two works from which the present views have taken inspiration.

---

[23] Note that in (Def. 2) this notion is expressed using the possible worlds semantics and ontology adopted by the theory of Relative Modality, saying that w belongs to the inter of the set of propositions selected by the conversational background function f for w. As observed in Chap. 4, § 4.2.7.1, since a proposition is a set of possible worlds (the set of worlds the proposition is true at), the set of propositions f(w) can immediately be translated in to a set of worlds $\cap f(w)$, corresponding the inter of all the propositions (i.e. sets of worlds) in f(w). Thus, saying that the inter of f(w) includes w is equivalent to saying that every proposition of f(w) is true at w. These idioms of possible worlds semantics may well seem counterintuitive to readers unfamiliar with the formal semantic tradition. Possible worlds provide a powerful framework for semantics, but there is a price to pay, as well as certain limitations, which will become more apparent in the fine grained linguistic semantic analyses of Chap. 6. To the reader perplexed by possible worlds semantics, it is perhaps worth reminding that, as observed by Kratzer herself, the key ideas of Relative Modality, are not tied to a possible worlds ontology and "can be implemented in any framework that provides suitable notions of logical consequence and compatibility" (Kratzer 2012 [1977]: 10, see Chap. 4, § 4.2.7.1 above).

Alethic root modality covers all the modalities traditionally classified as *dynamic* including not only the *volitional* and *quantificational* kinds considered by Portner (2009), but also dispositional future-oriented modalities, Nuyts' (2006) *situational dynamic* modalities, and a host of broadly alethic notions examined by Lycan (1994) and Kronning (1996). If the level of discourse makes them relevant, alethic root modalities can include possibility of necessity relative to more clearly defined sets of facts, such as *physical* possibility/ necessity.

5.2.4.6 Are the More Abstract Alethic Modalities Still "Root"?

The conversational backgrounds of root modals are constructed taking the VP as reference point. This typically results in selecting facts from the spatiotemporal circumstances of the VP event. Root alethic modals, thus, typically involve relations between events that are temporal and causal in nature. One can compare the range of causal root alethic possibilities to the Medieval notion of *potentia realis*, introduced by John Duns Scotus (see Vos 2006: 175). *Potentia realis*, which concerned possibilities that can be realized in time and thus does not admit synchronic alternatives, was contrasted with *possibilitas logica.* The latter, being conceived as the mere absence of incongruity between the terms of the prejacent proposition, included the abstract consideration of synchronic alternatives.[24]

The identification of root alethic modality with the range of temporal, real possibilities highlights an issue connected with the event-relativity approach to the distinction between root and epistemic conversational backgrounds. The problem is the following: as we proceed towards the most abstract strata of alethic modality, i.e. as we draw closer to Peirce's *essential possibility* (see Chap. 4, § 4.2.6) or to Duns Scotus' *possibilitas logica*, we abandon any notion of causality and of events and states occurring in time: notions of metaphysical, conceptual, semantic, mathematical and logical possibility or necessity are timeless and do not involve causality proper. Thus, their conversational background cannot be properly said to be composed of circumstances that have a bearing on the happening of the VP-event. It is thus legitimate to wonder how these conversational backgrounds are constructed and whether they fit at all with the dichotomy between root and epistemic modality as it is conceived by the event-relativity hypothesis. Two preliminary observations are very helpful in framing this problem in a way that makes it amenable to solution.

Firstly, these interpretations of the modals are rarer than the temporal ones and are mostly, but not exclusively, confined to specialized discourse. Certainly, this does not make them irrelevant and does not authorize the semanticist to discount them entirely from the explanatory burden of the theory. Yet, their comparatively lower salience, their more peripheral status is a fact that a theory of linguistic semantics should probably reflect.

---

[24] See the *Introduction* at § 1.6.1 and Chap. 3, § 3.4.2.3for more references about the Medieval discussion on modalities in which Scotus' distinction originates.

Secondly, event relativity, particularly in the procedural implementation provided here, does not directly provide conceptual or logico-semantics distinctions; it is a linguistic semantic hypothesis concerning how conversational backgrounds are constructed[25], rather than directly the nature of the propositions composing the conversational backgrounds. In other words, it models the paths of meaning construction, rather than directly the conceptual and logical structure of the product of meaning construction. Furthermore, in the procedural version, the saturation of the conversational backgrounds is presented as an open-ended contextual saturation guided along preferential paths, rather than fully constrained.

Taking into account the above remarks, it is not implausible to think that at least some of the abstract alethic meanings *do* arise procedurally from root modality, even if their conversational backgrounds do not concern the circumstances of an event. More generally, it seems reasonable to hypothesize that *for a wide variety of alethic modalities* the conversational background is recovered through the root branch of the procedure – (d) in (Def. 1) – even though they are far removed from causality between events. In Chap. 6, for instance, particular attention will be devoted to alethic conversational backgrounds composed by institutional facts and, more precisely, by economic facts. It will turn out that, despite the ontological specificities of the facts of the economy, possibility and necessity in the economic sphere are largely treated in terms of a notion of "economic causality" which draws closely on natural, physical causality. The event-relative procedure, prompting hearers to search for the participants and circumstances of the VP-event, will make certain interpretations of the modals more accessible and others more peripheral. Metaphorically, the more remote interpretations will often preserve a hint of the more central ones.

Consider the case of Euclidean geometric possibility and impossibility, illustrated by the following examples taken from user-generated contents of the *Yahoo Answers* website:

(33.a) How many obtuse angles *can* there be in a triangle? (answers.yahoo.com)

(33.b) Why *can't* a triangle have more than one obtuse angle? (answers.yahoo.com)

(33.c) Take a protractor, draw an obtuse angle. (For example: a 120 degree angle.) Next, make the next angle for your triangle another obtuse angle. Now, try to connect the opening without changing the angles of the group of lines you created. It's *impossible* for it to be a triangle. (answers.yahoo.com)

In the world of Euclidean plane geometry there are no events and there is no causality in time, only abstract objects on a two-dimensional space. Yet, the conversational background of *can* in (33.a, b) is plausibly identified following the route of root modality in the procedure. Starting from the scene, or frame, staged by the prejacent proposition the addressee is lead to focus on the properties of the participant enti-

[25] In Hacquard's (2010, 2011) original contribution, event relativity is presented chiefly as a hypothesis on the syntax-semantics interface, while the procedural variant presented here frames it as a hypothesis on the interaction between linguistic semantics and discourse interpretation, analogous to procedural models of tense. One can imagine that a comprehensive formulation integrating these two aspects is possible. In every case, it remains a linguistic-semantic, rather than logico-semantic, endeavor.

ties: the *triangle*, and the *angles* – the latter, as it turns out, being internal circumstances of the triangle. In fact, the conversational background, here, is composed by the properties that these abstract entities have in plane Euclidean geometry. To better account for these cases, one is tempted to substitute the term *VP-event* with *VP-frame* in the procedure; given that the latter is less ontologically charged and more clearly referring to linguistic-semantic or cognitive-semantic representations, multi-purpose gestalts through which we can apprehend ontologically diverse objects. At the same time, the answer given by a user in example (33.c) demonstrates the prominence of the experience of possibility and impossibility in action (*Now, try to connect... It's impossible*) in apprehending the more abstract level of ontology to which Euclidean plane geometry belongs. This is not limited to "naïve", informal, non-demonstrative arguments such as the user-generated answer in (33.c), as traditional textbook presentations of actual proofs in Euclidean geometry are also cast in the procedural and agentive language of construction.

Admittedly, there are alethic modalities that are even more abstract than those pertaining to the possibilities consistent with Euclidean geometry. Freeman (2005: 114) lists five categories of "alethic" or "broadly logical" necessities and impossibilities: formal, semantic, conceptual, mathematical and metaphysical. *Formal* alethic modalities concern what is possible or necessary in view of logical axioms only, while *semantic* alethic modalities also include in their conversational background "the meaning of certain non-logical expressions they contain" (cf. Freeman 2005: 115). The *conceptual* modalities, according to Freeman (2005: 116), depend "not on a semantic analysis of nonlogical concepts, but on other conceptual relations", which he calls *conceptual implications.* Furthermore, metaphysical modalities are associated with principles of meaningful "constructibility". Freeman presents (102.a) as an example of metaphysical impossibility:

(34) No object *can* have a property without existing (Freeman 2005: 117)

This sentence is as far as it gets from the prototype examples of non-deontic root modality, from Portner's (2009) "volitional" dynamic modality, and certainly does not involve any causal relation between events in time. Yet, under a strictly procedural interpretation of event-relativity, it is not difficult to see it arising from the root branch of the procedure rather than from an epistemic one. The "metaphysical impossibility" of (34) arises from the very terms of the prejacent proposition, the notions of 'having a property' and 'existing'. Thus, strikingly, the search for a viable conversational background for the impossibility does not have to go very far from the syntactic frame of the prejacent and its participants: it is not necessary to go back anaphorically to previously established textual premises, nor to refer to contextual assumptions and there is no need to examine the prejacent against a set of beliefs tied deictically to the act of utterance or connected with someone's beliefs or discourse.[26]

[26] The view of alethic modalities expressed here goes against the view aired by Papafragou (2000) and by others that the so-called "alethic modalities" are, in fact, a variant of epistemic interpretations.

In the end, the superposition of the procedural, linguistic semantic, distinction between root and epistemic and of the logico-ontological distinction between alethic and non-alethic leads to a surprisingly transparent classification of modalities.

The following sections refine the map of each major subdivision, starting from alethic root modalities in § 5.3.1, examining how the different kinds of conversational background relate to the logical types of the propositions functioning as standpoints and arguments and to the *locus*-relations holding between these propositional contexts and licensing the inferential step from premise to conclusion. The conceptual framework for this examination will be provided by Freeman's (2000, 2005) "typology of statements" and by the AMT theory of argumentative loci (Rigotti 2009; Rigotti and Greco Morasso 2010).[27]

5.3 Types of Conversational Backgrounds and Arguments

5.3.1 Alethic Conversational Backgrounds and Arguments

Alethic root modalities, as characterized above, can range from the *basic ontology* of the universe – logical, mathematical, semantic, metaphysical and physical – to very specific *sets circumstances*. These can be called *circumstantial alethic conversational backgrounds*. Among them we can distinguish the *external circumstances* and the *internal features* of an agent or entity imbued with causal power.

In contrast with the most abstract basic ontology modalities, which may abstract from time, circumstantial modalities are submitted to time, and are often causal in nature: they typically concern the possibility/necessity of a certain event to become the case in view of the causes. More precisely, when the prejacent denotes a non-stative eventuality B is made of a set of facts at t_0 that will *inevitably cause / not prevent* the event denoted by the prejacent to happen at $t_1>t_0$. When the prejacent is a state, or is habitual or generic, the causing circumstances in B sustain or allow the state or habit along its period of subsistence, as in (35).

(35) Lions **can** live in this savannah. There is sufficient water and plenty of prey all year round

In terms of Freeman's (2005) typology of statements, modal propositions featuring a root alethic modal will be classified either as *broadly logically determinate statements* – statements of formal, semantic, conceptual, mathematical, metaphysical possibility or necessity – or as *interpretations* (cf. Fig. 5.2 below). Interpretations have three defining characteristics: they are "broadly logically indeterminate or contingent statements" (Freeman 2005: 100), they are "intensional" and "nonevaluative" (*Ibid.*: 105). The truth of an interpretation does not simply depend on the state

[27] Both frameworks have been presented in Chap. 2, respectively in § 2.3.3 and § 2.2.3.

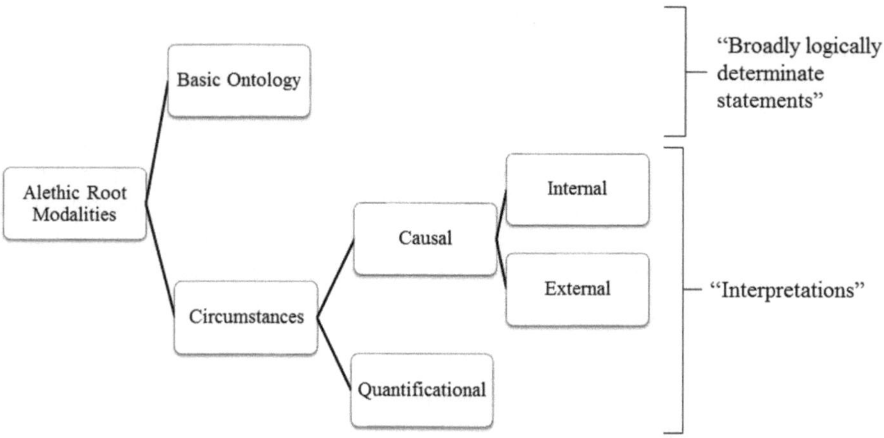

Fig. 5.2 Taxonomy of alethic root modalities and Freeman's (2005) types of statements

of the world: "for an interpretive concept to be truly predicable, some explanation needs to be true" (Freeman 2005: 107). By "explanation" here Freeman means what he calls a "nomic generalization" or "nomic regularity" which holds true, not in all possible worlds – like logical necessity – but "in all possible worlds of some sort" (*Ibid.*), such as physically possible worlds. Example (36), offers an illustration of this-less-than logical alethic necessity, discussing the Buddhist concept of *Sansara*:

(36) While we are content to live in and for Nature, in the Circle of Necessity, Sansara, we doom ourselves to perpetual change. That which is born *must* die, and that which dies *must* be reborn. It is change evermore, a ceaseless series of transformations (WebCorp)

The nomic necessity of perpetual change that forms the conversational background of the two occurrences of *must* in (36) does not hold absolutely, but only as long as "we are content to live in and for Nature".

Interpretive statements either assert or presuppose nomic regularities. Comparable notions of nomic regularity were introduced by the philosopher Roger Wertheimer (1972) and by the linguist Hans Kronning (1996) in their treatments of modality: both see modal propositions as presupposing nomic regularities of some sort.

Nomic generalizatons can be distinguished by mere empirical, observational, descriptive generalizations because they involve what Freeman calls "a subjunctive conditional", that is, a kind of intensional conditional rather than a mere material entailment. Let us contrast the following two examples, inspired by Freeman (2005: 105):

(37.a) All the subjects in the experiment displayed symptoms of nausea and dizziness.
(37.b) All human beings have to die.

The descriptive generalization in (37.a) is a mere relation of set inclusion holding in the world of evaluation. If I know that John was a participant in the experiment I can conclude that he displayed symptoms of nausea and dizziness. Yet, no causal connection is established between participating in the experiment and manifesting these symptoms and my generalization does not provide a deductive argument for a counterfactual claim such as *If John participated in the experiment he would display symptoms of nausea and dizziness*. The claim in (37.b) is different: rather than making an observational generalization, the sentence states a nomic generalization, holding for a certain set of worlds, where certain physical and biological conditions hold. For all worlds belonging to this set, *human* (x) is sufficient for *to die* (x) to take place (at a certain moment in time). In other words, we have an entailment that is necessary according to some specified sense of nomic necessity:

(Def. 3) $\forall x \, \Box_n$ (*human* (x) → *die* (x))

Contrary to (37.a), this holds also for the future generations and for merely hypothetical human beings provided that we remain within the bounds of worlds for which certain physical and biological laws hold. Freeman (2005: 154) introduces the connective of "nomically sufficient condition" ($\Box_n\rightarrow$), for which he provides the following semantics by means of a Kripkean accessibility relation R and a truth-value assignment function α:

(Def. 4) A $\Box_n\rightarrow$ B is true on α at w if and only if for all w' such that wRw' if A is true on α at w' then B is true on α at w'. (Freeman 2005: 154)

He also introduces a notion of *nomic necessity* for states of affairs are true whenever the causal laws and other relevant nomic regularities hold (Freeman 2005: 170) and *nomic possibility* for states of affairs whose negation is not a nomic necessity. I will not discuss in detail here, Freeman's style of formalization of these notions. It will suffice to say that what he models in terms of Kripkean accessibility relations could also be modelled in terms of Kratzerian conversational backgrounds consisting of the nomic regularities themselves. This would also have the advantage of making the relations between nomically sufficient condition and nomic necessity more transparent.

I will not pursue this formal translation in terms of Relative Modality here, preferring to focus instead on the argumentative consequences of the recognition that everyday root alethic modal propositions are *interpretations*. Compare the following examples involving Björk, the famous Icelandic singer:

(38.a) Björk *could not* sing that night. Her amazing voice was gone for good. (permanent internal physical circumstances)
(38.b) Björk *could not* sing that night. She had a sore throat. (temporary internal physical circumstances)
(38.c) Björk *could not* sing that night. Her boat remained ice-trapped in a bay. (external physical circumstances)
(38.d) Björk *could not* sing that night. The concert organizers were outraged at her political statements and rescinded the contract. (external social circumstances)

The first sentence of each example already goes beyond the pure description of facts: it represents an interpretation, in the sense that it presupposes some sort of causal explanation. Yet, without a specific saturation of the conversational background it remains a most generic an uninformative interpretation: *something caused*

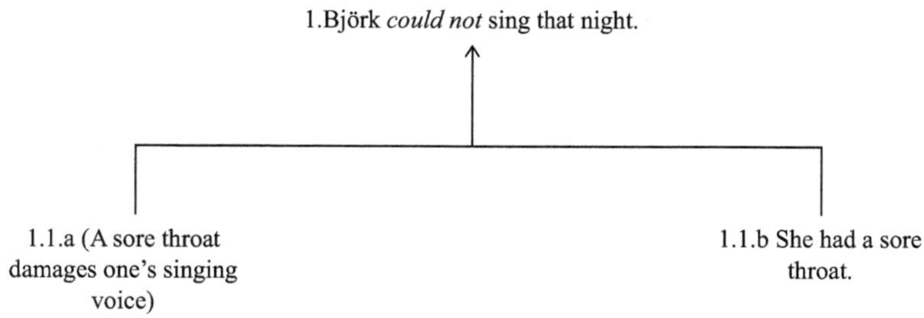

Fig. 5.3 Argumentation structure of (104.b) with an interpretive standpoint

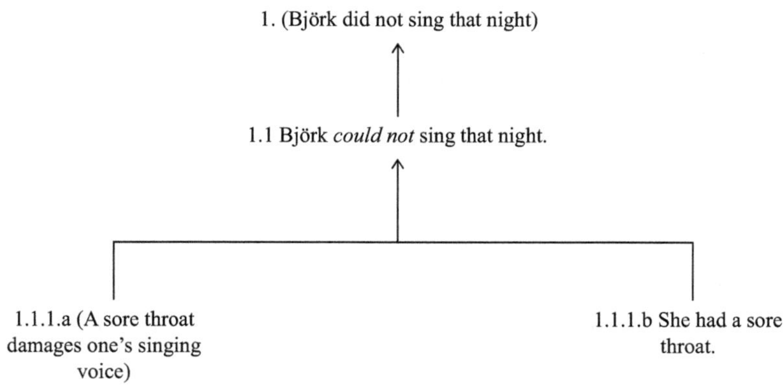

Fig. 5.4 Argumentation structure of (38.b) with an implicit descriptive standpoint

Björk not to sing that night. The second sentence of examples (104.a-d) provides the *something* specifying the kind of alethic conversational background that the modal takes. At the same time, each of the second sentences can be reconstructed as presenting an argument.

What kind of standpoint does this argument support? According to the most direct interpretation, the standpoint being supported is an *interpretive* standpoint corresponding to the *entire modal proposition*. The reconstructed argument should be roughly the one presented in Fig. 5.3.

However, utterances of (38.a-d) are also often used indirectly to support an implicit negative standpoint towards a proposition such as 'Björk did not sing that night'. This second standpoint is *derived from the prejacent* only. In terms of Freeman's typology of statements this a *descriptive*, non-modal, purely extensional, standpoint, its truth-conditions depending solely on the world of evaluation. It is as if the alethic root modality expelled from the propositional content of the standpoint reappeared outside of it as an epistemic qualification, or, to use another metaphor, as if the alethic modality was spent to buy an epistemic one.

In this case the interpretive statement represented by the modal proposition becomes an intermediate step between the premises and the descriptive standpoint, as in Fig. 5.4.

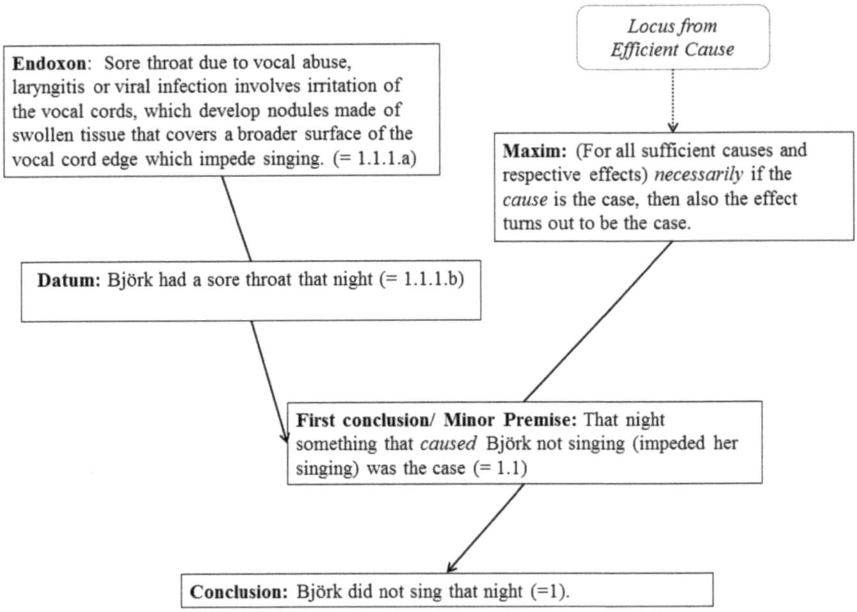

Fig. 5.5 Inferential configuration (Y-structure) of (38.b) with an implicit descriptive standpoint

The inferential step from the interpretive proposition (1.1) to the descriptive proposition (1.) is a very direct one. If we analyze this kind of argument with the tools of the AMT, it seems reasonable to assume that this step does not correspond to an autonomous Y-structure, but to an internal sub-step within the same Y-structure and, namely, to the step between the first conclusion and the final conclusion, realized through the application of the maxim alone. In other words, the inference between the causal interpretive and the descriptive statement does not involve the invocation of specific culturally shared beliefs as endoxical premises. This can be represented by the Y-structure in Fig. 5.5.

The *endoxon* corresponds to a cultural configuration that established a relation of causation between sore throat and not singing. The argument treats this relation of impeding causation as sufficient. This is obviously debatable in a myriad of ways – as it is always the case with real world circumstantial causation: maybe not all sore throats are equally serious and only the most serious make singing impossible, maybe they do not make singing *completely* impossible, but just painful, dangerous for your health and horrible to listen to, maybe we don't know exactly how wretched a sound must be to cease to be an instance of *singing*, and so on. All this would be relevant in the critical evaluation of the quality of the argument in (38.b). Thanks to the application of the datum to this endoxon we can conclude that a (presumedly sufficient) cause of not singing is the case. Finally, the maxim derives a consequence from the nomic necessity inherent in the concept of sufficient cause and obtains the descriptive standpoint as final conclusion.

An AMT analysis of the inferential configuration such as the one in Fig. 5.5 already provides a sketchy informal framework for examining how the structure of a relative modal maps onto the components of the inferential structure and for clarifying the indicative potential of the modal for argumentative reconstruction.[28] There are several orders of considerations that can be made from this comparison.

First of all, it can be observed how the conversational background, which has been conceived up to this point abstractly as a "set of propositions" or sometimes "set of premises" (Kratzer (2012 [1977]: 10), maps onto the more concrete articulation of premises comprising the *endoxon* and the *datum*. Interestingly, while the *datum* is typically an extensional premise – a fact in w, for alethic modalities – the endoxon turns out to be intensional, an interpretation itself, corresponding to the statement of a nomic regularity. Similar observations are made by Hans Kronning (1996: 33–34) when he hypothesizes an underlying enthymematic structures associated with the different uses of the French necessity modal *devoir*. The underlying enthymemes reconstructed by Kronning, include both factual minor premises – typically expressed in the discourse or recoverable from discursive "traces" and major premises that are themselves nomic regularities, rather than mere descriptive generalizations.

Thus, the Y-structure reveals a modal stratification of the conversational background, which remains invisible in the abstract perspective of the "set of premises". It emerges how very specific circumstantial necessities, possibilities and impossibilities – sometimes concerning *singular* prejacents, like in the Björk examples in (38) – arise from major premises that are themselves alethic modal propositions via the composition with descriptive minor premises.

Secondly, we can observe how the event of the 'sore throat', which represents the *terminus medius* between the endoxon and the datum, is discharged – as it should be – in the first conclusion. As a consequence, the first conclusion simply states that 'a sufficient cause of not singing is the case', without specifying what this cause is. This is not a special feature of this example and represents the normal and intended functioning of the AMT model: the material component needs to be discharged so that the maxim can work with higher level concepts (*intentiones secundae*), such as 'part', 'cause', 'genus', to derive the conclusion. What is interesting, however, is that this discharge of the concrete cause – which one might be tempted to consider an artefact of AMT – corresponds very well to what the modal statement *Björk could not sing* conveys by itself without considering the cataphoric saturation of the conversational background in the second sentence of (104.b): *Something prevented her from singing*. First conclusions had been introduced into the AMT theory to get the inferential job done in the Y-structure, and weren't supposed to receive linguistic

[28] Admittedly, putting side by side Relative Modality – a very abstract but also formally precise semantic theory – with AMT – a framework for the study of argumentative inferences adopting a richer ontology, but still lacking proper formalization and a semantics– may seem an exercise in futility. It is important to stress, though, that what is mapped onto AMT here is not the possible world implementation of Relative Modality, but just the key idea of a premise semantics based on notions of logical consequence and compatibility.

manifestation. Thus, it is striking how close the actual root alethic modal proposition is to the content of the first conclusion.

Thirdly, it is interesting to see that in Fig. 5.5, the maxim does the inferential job of axiom T ($\Box\varphi \rightarrow \varphi$). Generally speaking, maxims are supposed to simply bring out the logical properties of the locus-relations. Here, we have a root alethic modality that is carried down from the endoxon to the first conclusion. More concretely, this alethic root modality is understood as a kind of efficient causation. At this point, in the place of the maxim we could have used Axiom T to extract the prejacent and derive the conclusion. The AMT reconstruction appeals instead to the maxim, a less general inference warrant, holding for sufficient efficient causation, rather than for any kind of alethic necessity or impossibility.

Concerning the indicative potential of root alethic modalities for argumentative reconstruction it seems reasonable to hypothesize, as anticipated in Chap. 2, § 2.2.3.2, that the interpretive root alethic modalities can function as indicators of the causal loci, and, in particular, of the *locus from efficient cause*, illustrated by the examples in (38) and of the *locus from instrumental cause*, illustrated by (39), reproduced from Chap. 2.

(39) We *cannot* expand our business because we do not have enough capital (Palmieri 2014: 32)

In relation to example (39) it is interesting to observe how arguments based on these causal loci are not limited to physical causality. It is not uncommon in ordinary arguments to treat causality between pieces of social reality much in the same way as physical reality. The language of economics and finance is a case in point. In Chap. 6, we will several instances where seemingly alethic modals are used to reason about economic causality in terms of efficient causality, as exemplified by (40):

(40) The price of crude oil can only increase in the present market situation.

In (39), expanding one's business and having capital at one's disposal are not purely physical states of affairs: they are social facts grounded in institutional states of affairs. The same is true, in (40) for the price of crude oil, which depends on a cascade of conventions, including the existence of money and of markets for the public trading of commodities. One marked difference between the "business" perspective of (39) and the "economics" perspective of (40) is the apparent effacement of human agency in the efficient cause scenario of (40) as opposed to the centrality of human agency in the instrumental cause scenario of (39).

This contrast between agentive causality and an agent-less view of economics as quasi-physics will be discussed in more detail in Chap. 6; here I would like to concentrate on what these examples have in common. What (39) and (40) have in common is that, ultimately, the institutional states of affairs that appear in them are made of *commitments* (cf. Searle 2005). In terms Freeman's typology of statements they are *institutional interpretations* as opposed to *physical* ones (2005: 193).

Given that these states of affairs arise from commitments and conventional rules it is natural to ask whether the modalities that take them as prejacents should be regarded as more akin to deontic rather than to alethic modality. The answer is no, as long as we are speaking of *constitutive* rather than *regulative* rules. According to Searle (1969: 33), regulative rules regulate "antecedently or independently existing forms of behavior", while constitutive rules "do not merely regulate, they create or define new forms of behavior". In other words, constitutive rules create new institutional states of affairs. As a consequence, modalities taking a conversational background of constitutive rules are alethic in their own peculiar way. They are alethic in that Axiom-T holds for them, contrary to conversational backgrounds composed by regulative rules, which are genuinely deontic, for which Axiom-T does not hold. People can violate regulative rules by behaving against them, that's why deontic necessity based on a conversational background of regulative rules does not entail the truth of the prejacent proposition. In contrast, you cannot properly violate constitutive rules.

Consider the following examples:

(41.a) You cannot exceed the speed limit of 30 mph

(41.b) No, Arnold Schwarzenegger Can't Be President (Headline, thinkprogress.org)

The action of driving a car at, say, 50 mph would make the deontically impossible prejacent true (107.a) without making the deontic proposition false. In other words: $(\Box\neg p) \wedge p$ is the case, which makes Axiom T false. In (41.b) there is no way for Schwarzenegger to be U.S. president (p) *in violation* of the constitutional (and constitutive!) rule stating that "no person except a natural born citizen shall be eligible to the office of President" ($\Box\neg p$).

And that's because being president is an institutional rather than a natural state of affairs. Even supposing the unlikely event that somehow the Austrian-born former governor of California manages to campaign and be voted to office, his election will be *null* (or *void*) and thus the prejacent would continue to be false: he would *not really be* the president. Suppose, on the contrary, that constitutional law is changed: then Schwarzenegger would be really president (p), but in that case the constitutional restriction in the common ground is no longer in force and $\Box\neg p$ ceases to be true. The same would happen if he installed himself in the presidency via a military coup: constitutional law would no longer be in force and new institutional states of affairs would become possible. To some extent, we can compare these changes to changes in in natural circumstances affecting the conversational background of previously holding circumstantial alethic modalities so that these possibilities or necessities cease to hold (e.g. a moment ago the cheetah *could* still catch the gazelle, but now the herbivore is too far away and the feline *can't* any more).[29]

[29] The comparison cannot be drawn beyond a certain point, though. Institutional and natural alethic modalities are alike in their respect of Axiom T. There are, however, crucial differences connected with the ontological specificities of institutional facts In brief, while changes in physical circumstantial alethic modalities originate in physical causality, changes in institutional alethic modalities originate from a renegotiation of a collective assignment of *status function* (cf. Searle 2005: 7–10), in our example the status function that "Such and such a set of procedures counts as the election of a president of the United States" (Searle 2005: 7).

Everyday circumstantial alethic uses of the modals often pack together physical and institutional circumstances in a rather opaque manner. This kind of opaqueness or vagueness may become relevant in the context of argument reconstruction, and, even more significantly, argument evaluation. When example (104.d) states that Björk could not sing because the concert organizers rescinded the contract a causal relation is established between an institutional event (termination of the contract) and the event of *singing*. Now, if the predicate *to sing* is understood purely as a physical action the argument *from the efficient cause* does not hold water: the institutional event does not have direct causal power over the physical one. In contrast, if the *singing* is reconstructed – more charitably – as a complex social event involving physical and institutional aspects, corresponding to the whole frame of the concert (performing in front of an audience at a proper venue, obtaining permits, getting paid, etc.), then the termination of the contract arguably makes this whole event impossible by striking down its institutional aspects. At the same time, such a composite event makes an ideal target for *dissociation* tactics in argument (*Well, she could have sung for free in the town square, if it really mattered to her!*).

When the interpretive propositions in examples (38.a-d) are understood as an intermediate step in support of a descriptive standpoint (as per Fig. 5.3) the alethic modal indirectly provides an epistemic qualification for the implicit negative descriptive standpoint. This epistemic implication stems from an interaction between the flavor of the conversational background and the modal force of quantification.

Several examples discussed in this chapter feature the negative form of the English modal *can*; and the reader will remember that examples based on *cannot* were used by Toulmin to argue for the argumentative nature of the field-invariant force of the modals (cf. Chap. 3, § 3.2.3). The choice of *cannot* was not by chance, given the epistemic implication of its alethic uses. Let us see why.

Circumstantial alethic modals take a conversational background of facts of a certain, more or less clearly defined, kind. They may be internal circumstances of an agent or external ones, physical or institutional facts, structured by the relevant underlying nomic regularities. Such a conversational background does not include *all* known facts, as epistemic background would, but, when it comes to impossibility, the incompatibility of the prejacent with any set of facts entails also the incompatibility with the set of *all* facts. Going back to (104.a-d), if Björk's sore throat rules out her singing, it won't help observing that her boat arrived on time, or that she is extremely talented, that her contract is fine, etc. If the prejacent proposition is ruled out by a specific set of facts (e.g. Björk's sore throat), no *additional* fact that I might know (e.g. that she arrived on time) is going to put the proposition in play again. The proposition will continue (monotonically) to be impossible no matter what additional kinds of facts we consider. In other words, any kind of alethic impossibility entails Peircean *substantial* impossibility (cf. Chap. 4, § 4.2.6), that is impossibility with respect to "supposed information of the present in the present, including among the objects known all existing laws as well as special facts" (Peirce 1998: 44). Substantial impossibility is, however, not sufficient to entail epistemic impossibility for any given epistemic subject, because a subject might ignore the facts and nomic regularities that rule out a proposition. Consider the following examples:

(42.a) Paul thinks John *might* be the murderer, but this is *impossible* because that day John was abroad.

(42.b) Paul thinks there *might* be "triangles" with two obtuse angles, but they are geometrically *impossible*.

(42.c) ?? John *might* be the murderer, even though this is *impossible* because that day John was abroad.

(42.d) ?? There *might* be triangles with two obtuse angles, even though they are geometrically *impossible*.

In (42.a) it suffices for Paul to ingnore John's alibi to make the practically impossible an epistemic possibility. Even in the case of (42.b), supposing that John's grasp of the concept of triangle is shaky or incomplete (hence the use of quotes in the example), triangles with two obtuse angles can be an epistemic possibility for him. The implication from alethic impossibility to epistemic impossibility is pragmatic: as shown in (42.c) and (42.d), when the epistemic subject coincides with the speaker at the moment of utterance the assertion of epistemic possibility and alethic impossibility of the same prejacent result in pragmatic incongruity. This is another variant of Moore's paradox, which has been already discussed in Chap. 3, § 3.3.2. The same holds for asserted alethic necessities such as the one in (43):

(43) All living beings **have to** die

If the biological nomic regularities indeed necessitate eventual death, there is no other fact I might know that can rule out these propositions. No *increment* of the conversational background would rule out a necessary proposition. Only a *revision* of the conversational background could.

Alethic possibility does not implicate epistemic possibility in the same way. What is alethically possible with respect to certain kinds of facts (e.g. the singer's physique) might still happen to be made impossible by other circumstances (e.g. the boat being ice-trapped in a northern bay). Alethic possibilities do not entail Peircean substantial possibility. Yet, pragmatically, the use of an alethic possibility modal often implies that the prejacent is also consistent with what is known.

This can be explained as an implicature based on the Maxim of Quantity (Grice 1989): if I assert that Björk *can* sing, because her throat got well, when I *know* that *she won't* because the organizer rescinded the contract, or the theater did not have the safety certificate, I am being logical but quite uncooperative, as it would be *more informative* to say, counterfactually, that she *could* sing, or *would be able* to do so as far as her voice is concerned. The latter expressions can convey a counterfactual nomic possibility: the prejacent is consistent with nomic regularities and facts of a certain kind, but not with all nomic regularities and facts true at w. So, also *can* ends up as frequently having *epistemic implicatures* in causal argumentation and support descriptive or predictive standpoints derived from the prejacent only.

Up to this point the interpretations corresponding to alethic modal propositions have been examined as arguments in support of a – typically implicit – descriptive standpoint, where the alethic modality of the underlying interpretation translates into an epistemic modality in the standpoint. This is not necessarily the case. Sometimes the interpretation itself – i.e. the alethic possibility, necessity or impossibility – represents the propositional content of the standpoint. It is also possible that the point of inferring an alethic modality is to support an *evaluative*

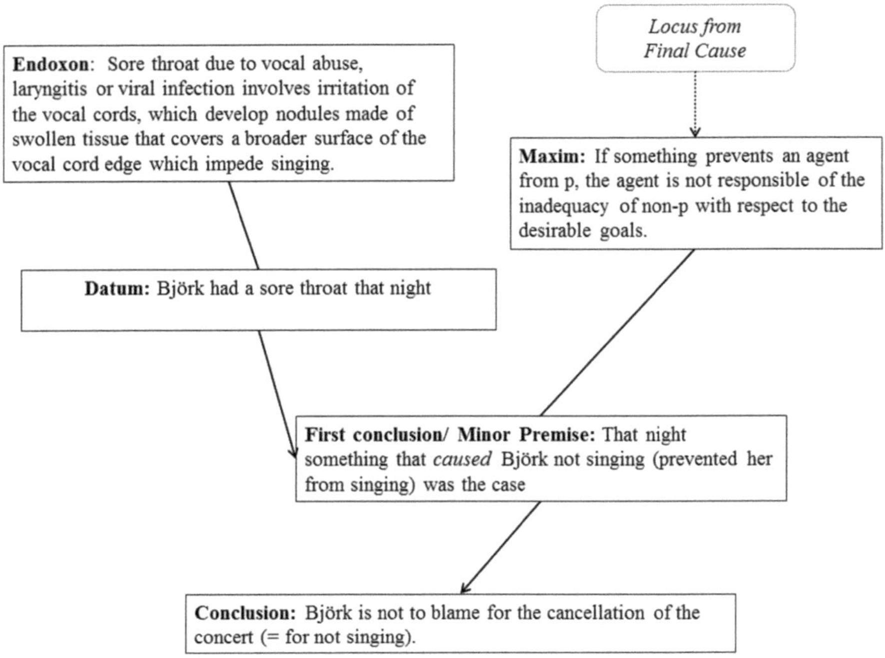

Fig. 5.6 Inferential configuration (Y-structure) of (38.b) with an implicit evaluative standpoint

standpoint. This can happen in different ways, through the mediation of concepts such as opportunity, competence or responsibility.

Here I will limit the exemplification to the case of *responsibility*. Freeman (2005: 170) observes that "a very significant use of the concept of nomic possibility occurs in the explication of statements attributing responsibility". A person can be held responsible of a certain behavior or a certain result only if it is nomically possible, i.e. "consistent with the laws of this world" (*Ibid.*) that the person refrains from the behavior or avoids the result. Thus, the impossibilities in (38.a-c) could be used in support of the standpoint that *it's not Björk's fault* if the concert has been cancelled. She is not responsible. One could also argue for such a standpoint using (38.d), but to such a move might have to face criticism stating that, in fact, she is indirectly responsible of the cancellation, having infuriated the organizers with her political statements. I propose to analyze these arguments as based on a particular maxim of the locus *from the final cause*, as illustrated in Fig. 5.6. Clearly, only modalities based on agent external.

Clearly, the maxim invoked by the enthymematic inference in Fig. 5.6 also involves the concept of efficient cause. Yet, responsibility and blame cannot be understood without recurring to finalistic, evaluative, concepts such as goals, desires or values. More generally, as Rigotti (2008: 565–566) observes, "the locus from the final cause belongs to the ontological area of action" and within this ontology it focuses "on the relation connecting the end (goal, purposes) of an action with the action itself". Clearly, other causal loci such as efficient cause, material cause, and, especially, instrumental cause can participate to the same ontology focusing on

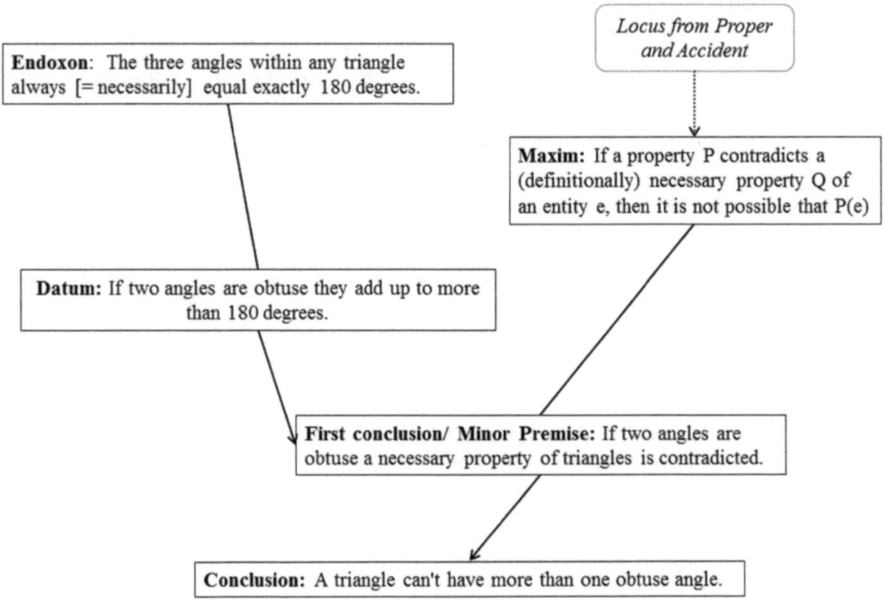

Fig. 5.7 Inferential configuration (Y-structure) of (44)

other nodes of the ontological network of human action. Individual maxims, such as the one in Fig. 5.6 often draw from more than one of these interrelated loci.

Given the apparent affinity of circumstantial alethic modalities with the causal loci, one could wonder whether the alethic modalities corresponding to Freeman's broadly logically determinate statements also manifest an affinity with a group of loci. Definitional loci are a very plausible candidate for this role. Definitional arguments obviously mirror the structure of the conversational background of semantic and conceptual necessity and seem to apply equally well to some kinds of mathematical necessity, as illustrated by example (44), where the definitional locus *From Proper and Accident* is activated as a necessary property of triangles is exploited, as the rough reconstruction in Fig. 5.7 tries to capture.

(44) Why can't a triangle have more than one obtuse angle? […]The three angles within any triangle always equal exactly 180 degrees. […] If two angles are obtuse — even if they are both only 91 degrees — they add up to more than 180 degrees. (www.ask.com)

At this point of the exploration of the relationship between different types of conversational backgrounds and arguments it is useful to pause to look at the emerging pattern. In the first part of this chapter Relative Modality has been introduced with a special attention to the notion of conversational backgrounds as functions determining "sets of premises" and on the theory's ambition of being a "premise semantics" relying the meaning of the modals to "principles of rational inquiry that apply whenever we reason from a set of premises" (Kratzer 2012 [1977]: 10). Of these sets of premises, the possible worlds implementation of Relative Modality offers a rigorous but inherently static and abstract view. The attention devoted to the *phoric saturation of conversational backgrounds and the sketch of a procedural view of their specification allowed to flesh out, in part, how premises are actually located in discourse processing. Now, with the tools offered by Freeman's typology of statements and by Rigotti and Greco's AMT model, the relationship between conversational backgrounds and the fine-grained inferential configuration of arguments is explored.

Conversational backgrounds reveal an internal modal stratification as alethic circumstantial backgrounds combine contingent facts with nomic regularities, which embody more fundamental kinds of alethic modality. They also appear as structured sets of premises as nomic regularities map on *endoxa* and contingent facts with *data* in the inferential configuration. Finally, the logical type of the modal proposition – captured in terms of Freeman's typology – illuminates the kind of loci with which it can combine as a standpoint and as a premise.

In a way, the approach adopted here compares to the formal semantic method of exploring conversational backgrounds, as trekking the wilderness compares to satellite photography. As trekkers we get a richer, more detailed, picture, observed at human scale, but also an incomplete one.

5.3.2 Deontic Conversational Backgrounds and Arguments

The conversational backgrounds of alethic root modals are composed *entirely* of propositions that are true in the world of evaluation w. Non-alethic root modals have conversational backgrounds that include also propositions that are not a fact in w. Two main sub-types of non-alethic root modals can be distinguished: the *deontic* and the *anankastic*. While for these two types of conversational backgrounds realism does not hold, they still have to be *consistent* sets of propositions, a property that corresponds to *seriality* ($\exists w'$: wRw') and to *Axiom D* ($\Box\varphi \rightarrow \Diamond\varphi$) in modal logic (cf. Kaufmann et al. 2006: 86).

Broadly defined *deontic* conversational backgrounds are those composed of propositions corresponding to some sort of *norm, ideal, desire* or *goal* – states of affairs that are "good" with respect to some normative system or set of preferences. It can include values, laws and regulations, contracts, commitments as well as the simple desires, preferences and goals of an agent. Deontic modality, thus conceived can be straightforwardly connected with one of the types of statement defined by Freeman's typology, namely *evaluations*:

> An evaluation characterizes something as either good or bad, better or worse, preferable or avoidable, an act as either right or wrong, obligatory, permissible or forbidden; a person or character as praiseworthy or blameworthy, an action as morally good or morally bad (Freeman 2005: 101)

It is not possible to delve here in the semantic structure of the variety of value judgments of people, things or acts which is evoked in the above passage and subsequently explored from an epistemological viewpoint in the book (see Freeman 2005: 218–280). To characterize the structure of deontic modalities it will be sufficient to say that they relate a human act – or any state of affairs that can result from human acts – with a conversational background composed of two distinct subsets of propositions: a set of states of affairs to which value is ascribed – good, desirable, states of affairs – and a set of factual circumstances. According to the force of the modality, the act is seen as necessarily entailed, compatible or incompatible with the values in the present circumstances. Thus, deontic modalities depend (also) on a set of evaluations and are themselves a kind of evaluation.

More precisely, in a possible worlds implementation of Relative Modality, the conversational background of deontic modals is not formed by a set of value judgements (e.g. *It is right to punish wrongdoers*), but by the set of states of affairs on which the attribution of value is supervenient (e.g. *Wrongdoers are punished*). On *supervenience* see Freeman (2005: 220 ff.).

Technically, in the standard version of Kratzer's theory the factual circumstances of w act as modal base f(w), while the states of affairs forming the set of values in w function as an ordering source g(w) (cf. also Portner 2009: 185). In the simpler procedural and contextual account adopted here, the deontic conversational background B_D corresponds to the simple union of the two subsets B_V and B_C, with a pragmatic oracle ensuring consistency. The semantics for deontic modalities is given here in (Def. 5.a, b), drawing on the notation introduced in Chap. 4 (Def. 19, 20). Following the style of analysis of Congruity Theory[30], the conditions on the prejacent and on the conversational background are treated as *presuppositions* and the proper force of the modal is treated as its semantic entailment or update condition (see also Seuren 1985: 401).

(Def. 5.a) *Deontic Necessity*$_w$ (e, a, B_D, p):

Presuppositions:

- *e* is the VP-event of the prejacent *p*
- *a* is the deontic subject, agent of *e*
- $B_D = B_V \cup B_C$
- B_V is a set of value judgements in w relevant for the agent *a* of *e*
- B_C is a set of circumstances of *e* that are the case in w

Semantic entailment:

- $B_D \Box p$

(Def. 5.b) *Deontic Possibility*$_w$ (e, a, B_D, p):

Presuppositions:

- *e* is the VP-event of the prejacent *p*
- *a* is the deontic subject, agent of *e*
- $B_D = B_V \cup B_C$
- B_V is a set of value judgements in w relevant for the agent *a* of *e*
- B_C is a set of circumstances of *e* that are the case in w

Semantic entailment:

- $B_D \diamond p$

---

[30] The format for semantic representation of Congruity Theory and its presuppositional analysis of selectional restrictions that predicates impose on their argument places have been introduced in Chap. 2, § 2.3.4.

While cast in terms of Freeman's typology of statements, the present account of deontics and evaluations departs in many ways from Freeman (2005). First of all, Freeman does not devote special attention to the semantic analysis of evaluations cast in modal form. Secondly, he uses the term *deontic* in a stricter sense to refer to "judgements of moral obligation" (Freeman 2005: 218), which represent only a subset of the category of deontic modality outlined here. Finally, Freeman's account is precisely and systematically cast in terms of the "ethical intuitionism" of William David Ross, which is not taken into consideration here. Nevertheless, I will still benefit, on occasion, from Freeman's precious insights.

Semanticist Paul Portner (2009: 185) also adopts a stricter view of deontic modality, bemoaning as "misleading" the broad use of the term among linguists. As reported earlier (cf. Table 6, above) he calls the broader class *priority modals*, characterizing it in terms of an ordering source "representing reasons for preferring, or assigning pfriority to, one possibility over another". The subcategories of priority modality are *deontic* modality, which is strictly understood as having to do with "obligations, right and wrong, and other normative notions", *bouletic* modality, which relates "to someone's desires" and *teleological* modality based "on someone's goal". Portner admits that "it is not always easy" put an actual sentence under one or the other sub-category. Suppose that young person comes home from a long journey abroad and utters (45):

(45) I *must* immediately go home and kiss my mother!

Is kissing the mother a *goal*, a *desire*, a *moral obligation* or the three of them? Or is it simply *good*? Separating law-like moral deontics from goals and desires reflects what might be called an ethics of the *right*, centered on obedience to laws determining the right actions, as opposed to an Aristotelian ethics of the *good*, where the knowledge of virtues emerges in the pursuit of happiness. In an Aristotelian perspective happiness has the status of a *final cause*, which determines moral action as an ultimate goal rather than as a model (*formal cause*) or rule (cf. Yarza 2000: 37). As it will become clear also from the proposed typology of deontic modality (cf. Fig. 5.8, below), I adopt here a broadly *teleological* view of moral norms (cf. von Wright 1963b), without attempting to offer a fully-fledged justification – which would obviously exceed the modest limits of this book.[31]

The present proposal instead of immediately operating a separation between the closely related concepts of desire, goal and moral value, gives priority to two other distinctions that affect the semantic functioning of deontic modals in discourse and reasoning. The first is the distinction between *intrinsic value* and *instrumental*

[31] Nor I will discuss finer, but philosophically important, distinctions among different types of teleological ethics such as eudaimonism vs. utilitarianism. In his Guildford Lectures von Wright (1963b) contrasts the *teleological* views of moral norms with *prescriptivist* views (moral norms are akin to commands) and d*eontological* views (moral norms are entirely sui generis and do not need to be brought back to a source or justification).

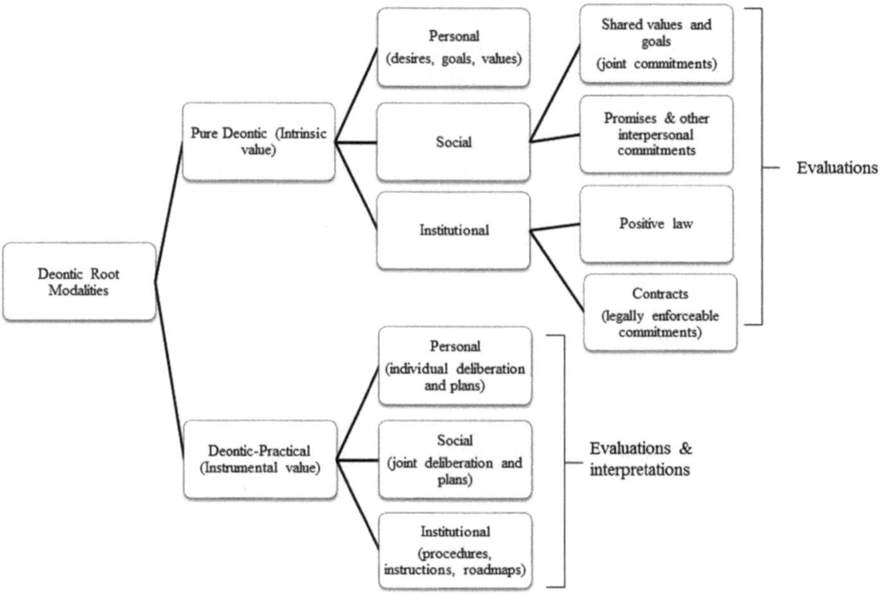

Fig. 5.8 Map of deontic modality

value. The second is the distinction between *personal*, *social* and *institutional* deontics. These two sets of distinctions provide more useful and clear-cut categories than those singled out by Portner. For instance, Portner's *teleological* modality will cover both prejacents that are ultimate goals – *per se* desirable – and prejacents that are goals inasmuch they are causally necessary steps to achieve another goal. Portner's *deontic modality* will bundle together personal convictions and moral values, interpersonal commitments and legally enforceable contractual obligations. The map of broadly conceived deontic modality resulting from the proposed taxonomical principles is illustrated in Fig. 5.8.

The category of deontic-practical modality is introduced by Kronning (1996: 112) in his book on the semantics of French necessity modal *devoir* ('must'), in connection with von Wright's notion of *practical inference* (von Wright 1963a). According to Kronning, the underlying "inferential composition" of deontic practical necessity modals corresponds to von Wright's "practical syllogism":

(Def. 6) (a) One wants to attain *x*.
(b) Unless *y* is done, *x* will not be attained.
(c) Therefore *y* must be done. (von Wright 1963a: 161; cf. Kronning 1996: 112)

For Kronning this syllogism differs from the ones underlying "theoretical" deontic modals in that premise is "teleological" rather than "nomic", drawing a distinction not unlike Portner's distinction between *teleological* and *deontic* priority modals.

The classification proposed here adopts Kronning's label of *deontic-practical* modality. As in Kronning's account, the structure of the conversational background of a deontic-practical modal can be compared to the premises of von Wright's practical syllogism. However, unlike Kronning's account, the demarcating trait between pure deontic and deontic-practical is not represented by the teleological nature of premise (a), but rather by the fact that premise (b), as von Wright (1963: 160) observes, rests "upon a causal relationship", suggested by words such as *done* and *attained*. It is interesting to observe that von Wright's discussion of practical inferences is not limited to inferences that feature goals or wants in premise (a). Instead of a teleological premise we can have a normative one, as in (Def. 7), below.

(Def. 7) (a) It is my duty to do *x*.
(b) Unless I do *y*, I cannot do *x*.
(c) Therefore I <u>must</u> do *y*.
(von Wright 1963a: 176)

Von Wright (1963: 175) insists that "the 'must' here is the 'must' of practical necessity". He calls the conclusion (c) in (Def. 7) a *derived obligation*, a notion he defines as a practical necessity of action relative to a primary obligation, or "something which must be done for the sake of doing one's duty". In the next section I will further elaborate on derived obligations as a sub-type of deontic-practical modality.

In the end, the distinction between deontic and deontic-practical, as formulated here, corresponds to the classic Aristotelian distinction between *intrinsic value* – state of affairs desirable in themselves – and instrumental value – what is desirable only because it is instrumental to attaining an intrinsic value. Freeman (2005: 219) draws this line very clearly:

> We distinguish instrumental from intrinsic value. As Ross points out, judgments of instrumental value have causal components. Such judgments involve both interpretations – the causal component – and evaluations – the state of affairs caused is intrinsically good or bad. Clearly, there will be a presumption for a judgment of instrumental value only to the extent that there is a presumption for its component judgments.

A different kind of discontinuity in the range of broadly deontic conversational backgrounds is introduced when we move from *personal* values, goals or desires, to *socially* shared values and commitments and then to *institutional* ones.

Theorists of social ontology, such as Margaret Gilbert (2013), have argued that shared values are not purely summative – values that happen to be held by several people – and neither are purely based on common knowledge: they require a *joint commitment* to the values. This commitment component makes shared values more alike to the deontic modals based on more familiar interpersonal commitments created by *commissive* and *directive* speech acts (cf. Searle 1969). Institutional deontic modalities represent a step further: they take a conversational background containing, legally enforceable commitments, either law-like or contract-like. Institutional deontic modalities involve a sanction or penalty for the deontic subjects when they incur in a violation. A parallel distinction in three levels can be hypothesized also for the deontic-practical modals, as illustrated in Fig. 5.8.

Being root modalities, deontic interpretations are derived by setting the reference point r for the search of the conversational background to $r = e$, where e is the frame of the prejacent VP-event. Certain key participants to the frame of *e* need to be recovered. For instance, deontic-practical modals should include an argument place for the agent whose goals and actions are being considered. I call these participants subsidiary arguments of the modal. They are, in fact, subsidiary to the conversational background, in the sense that their main job is precisely to determine the conversational background: a *set of goals*, for instance, makes sense only relative to an agent or group of co-agents. They must be *someone's* goals (cf. Portner 2009: 187). Co-agents presuppose joint commitments and bring us into social deontic modality.

Deontic modals corresponding to *positive law* should include two additional arguments: the *deontic subject*, i.e. the person – physical or juridical – who is under the obligation, and the *deontic source*, i.e. the juridical or physical person whose authority originates the positive obligation. Similarly, deontic modals corresponding to interpersonal or institutional *commitments* reasonably have slots for the person who is committed and the person with whom the commitment has been taken.

5.3.2.1 Pure Deontic Conversational Backgrounds and the Inferential Configuration of Arguments

Let us now briefly examine how the conversational background of a pure deontic modal might map onto the inferential configuration of an argument. Kronning (1996: 33) observes how occurrences of deontic modals can amount to normative enthymemes where the modal indexes implicit normative premises – which he sees as a type of nomic regularity,[32] Kronning presents a very clear biblical example, which I reproduce here in a modern English version:

(46.a) 7 The Jewish leaders insisted, "We have a law, and according to that law he **must** die, because he claimed to be the Son of God." (John 19:7, NIV)

Kronning recovers from biblical sources all the premises of the underlying enthymeme:

(46.b) 16 anyone who blasphemes the name of the Lord **is to** be put to death. The entire assembly **must** stone them. Whether foreigner or native-born, when they blaspheme the Name they **are to** be put to death. (Leviticus 24: 16, NIV)

(46.c) Whoever calls himself son of God claims to be God (from John 5:18)

(46.d) Claiming to be God is blasphemy (from John 10:33)

(46.e) Jesus called himself son of God (= called God his father) (again from John 5:18)

Once these inter-textual premises are recovered, the argumentation structure (or macrostructure) of this complex inferential chain, as well as the invoked loci, can be reconstructed as per Fig. 5.9.

The example is particularly neat because the law underlying the deontic can be precisely recovered from biblical intertextuality. The example shows also another

[32] In this respect Kronning's view on nomic regularities differs from Freeman's. Freeman (2005) does not use the notion of nomic regularity in his treatment of evaluations.

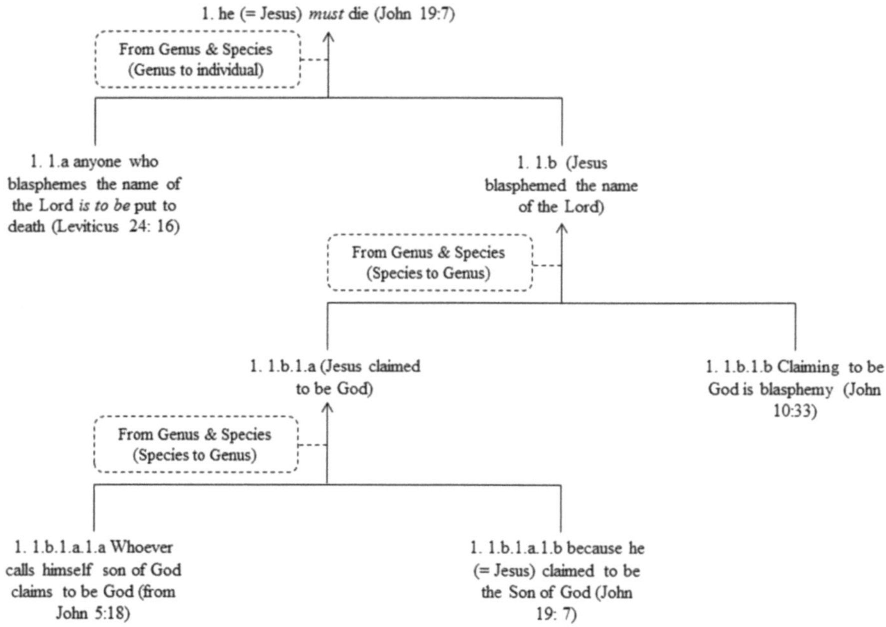

Fig. 5.9 Implicit argumentation structure and loci of John, 19.7 (46.a)

important thing about pure deontics: in order to necessitate particular deontic conclusions, one typically needs both universal deontic nomic regularities – such as 1.1 (46.b) – and particular facts – such as 1.1.b.1.a.1.b (46.e). One can also observe that there is no causal reasoning going on: only definitional loci are applied and, more precisely, the taxonomic locus *From genus and species*. In two steps, moving from species to genus, the act of Jesus is brought back to the genus ("blasphemy") for which the law applies, the deontic implications for the genus are then transferred to the specific case in a way that is typical of the so-called "legal syllogism". Figure 5.10, below, zooms-in to focus on the inferential configuration of this last step:

Kronning, for whom necessity modals are always signaling "the product of an inferential composition" worries about the modal presentation of the universal nomic regularities. For instance, the nomic regularity appearing as endoxon in Fig. 5.10, i.e. example (46.b), contains the necessity modal construction *is to be*. If nomic regularities "are themselves the product of an inferential composition"[33] (Kronning 1996: 36) there seems to be a *regressus in infinitum* in the semantics of modality. Kronning observes that the use of necessity and obligation expressions is not the only, nor the prevalent way of presenting norms in laws, statutes, and constitutions as well as in religious texts. He lists several French examples where the present tense, the future, the subjunctive or the imperative are used. Similar

[33] I translate from the French original: "ces régularités nomiques sont donc, elles-mêmes, le produit d'une composition inférentielle" (Kronning 1996: 36).

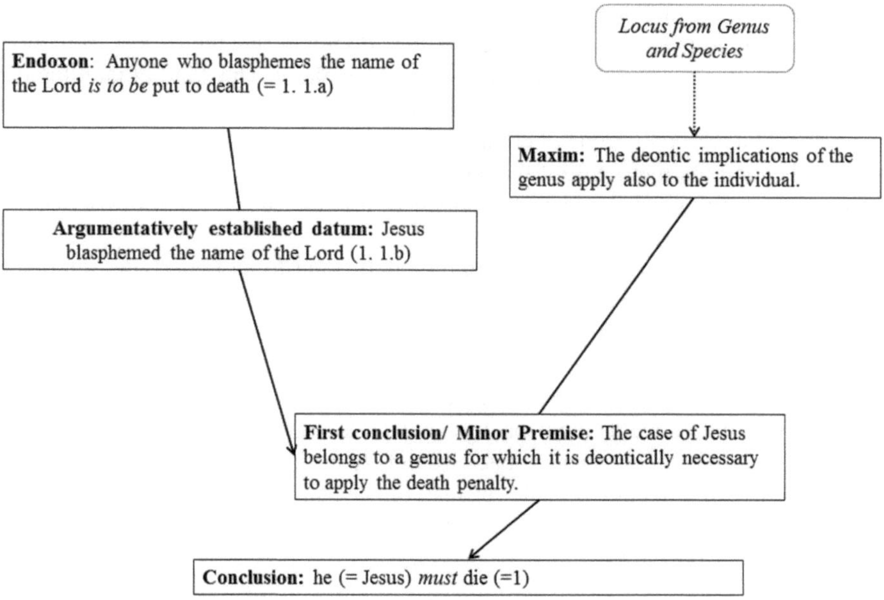

Fig. 5.10 Partial inferential configuration (Y-structure) of John, 19.7 (46.a)

examples could be found for English. The choice of a necessity modal verb like French *devoir*, he concludes, is meant to explicitly present the nomic regularity as the result of an inference (Kronning 1996: 37).

Kronning's remarks are interesting but do not solve the problem for those laws that *are* expressed as modal propositions. In fact, if we consider the original utterance of positive law, we see that its deontic force *qua law* does not derive from anything else: it is the result of the *thetic* speech act of the legislator. From the outside, one can debate the expediency or the morality of a legislation or question the legitimacy of the lawmaker's thetic speech act, but all that requires readjusting the score to a different, broader, conversational background. From within the conversational background of a certain body of laws, the deontically necessary status of a legal norm does not derive inferentially from anything else.

Technically, Relative Modality does not incur in Kronning's problem. This is because it defines necessity in terms of logical consequence from the conversational background rather than in terms of inference and because the conversational background will not include the modal proposition of the norm, but only its prejacent. As observed above, the deontic conversational backgrounds include the states of affairs to which the evaluation is supervenient. Thus, a sentence such as *Those who blaspheme the name of the Lord* ***must*** *be put to death* indicates that its prejacent 'Those who blaspheme the name of the Lord are put to death' follows necessarily from a normative-deontic conversational background containing the proposition 'Those who blaspheme the name of the Lord are put to death'.

There is thus a tautological flavor to universal deontically modalized statement corresponding to the positive law itself as they do not derive their force from anything but the thetic act of the (human or divine) legislator. This kind of tautological emptiness can affect also singular deontically modalized statements when the conversational background is one of command and there is nothing more general to fall back to beyond the mere will of the issuer of the command. Clearly, in these limit cases the tautological logical consequence relation does not correspond to Kronning's "inferential composition".

While one cannot say that *all* deontic modals point to an underlying inference involving deeper values or norms, Kronning's view still seems to correspond to their *preferential* reading. Kronning's observation that in French modal statements of deontic necessity are not the preferred mode of expression of the statement of positive law may have to do with the fact that the event-relative contextual procedure of saturation of the conversational background preferentially prompts the hearer to reconstruct an *ulterior* normative background. Also, in the case of singular commands the use of a deontic modal – instead of, say, an imperative – seems to point to an unsaid background of norms behind the issuing of the command. Compare the two utterances in (47.a, b):

(47) a. Your library book is overdue. You *must* return it to the library.
b. Your library book is overdue. Return it to the library.

The use of *must* in (47.a) immediately activates an assumed background of norms and establishes a strong anaphoric link with the violation described in the previous sentence. In contrast, the imperative utterance in (47.b) has a weaker link with the previous sentence and highlights the issuing of a singular command whereby the speaker's will and personal authority are put on the line rather than relying on the application of a general norm.

The possibility of (47.a) being a mere descriptive "deontic report" rather than the performance of a directive speech-act is consistent with this contrast. The constructed example in (47.a,b) echoes the findings of a corpus based analysis of Dutch modal verbs: examining the contrast between the use of the modals and the imperative mood, Nuyts *et al.* (2010) found that modal utterances are often inextricably ambiguous between a performative and a descriptive meaning, while, obviously, imperatives "can only be used performatively, not descriptively" (p. 29). They observe that modals are used descriptively "to refer to prescriptions or instructions contained in or issued by laws and regulations, people in charge of things" (p. 30) and even when the speaker is indeed responsible for issuing the directive speech act the recourse to a modal can still "raise the suggestion that the speaker is rendering a wider opinion and is issuing the directive also on behalf of others or by virtue of general rules or principles" (p. 31). Clearly, these observations fit well an analysis of the modals as procedural expressions guiding the hearer in the contextual saturation of a conversational background of propositions.

The findings Nuyts *et al.* (2010) suggest that it is *only marginally* possible to have purely performative deontic necessity modals where the relation of logical consequence is trivially tautological as the content of the B_v coincides with the prejacent. These correspond, for instance, to singular directives that do not have other justification than the will and authority of the utterer or to the general statements of positive law originating directly in the thetic act of the legislator. Following Carmello (2009: 216–216), I use the term *thetic* to indicate a duty that comes into existence thanks to the very utterance imposing it.[34] The marginality of thetic uses of core context-dependent modal expressions such as modal verbs (e.g. *should, must, can,* etc.) and modal adjectives (e.g. *possible, necessary*) has a special significance in view of the goals of the present investigation and the relational analysis of the modals it adopts: it tell us that understanding necessity in terms of logical consequence is not a mere and sterile rewriting of universal quantification over a set of accessible possible worlds, for a real, non-trivial, inferential pattern typically underlies the logical consequence and propositions in the conversational background do function as premises, making *premise semantics* worthy of its name. Even when premises are not really accessible to the addressee (or even to either dialogue participant) a potential inferential path, accessible to the speaker or to a third party, is typically hinted at.

5.3.2.2 *And If America Is to Be a Great Nation, This Must Become True*: A Case Study of Deontic Modality in Argument

I will now turn to a historical example where modal expressions refer to what states of affairs are necessarily entailed (or compatible, or in-compatible) with respect to a special kind evaluation: the legitimacy of an institution. We are here in the area of the *social deontic modalities* (cf. Fig. 5.8) where the conversational background consists of shared values to which members of a community are jointly committed, and certainly we are not yet in the area of fully institutional modalities, which simply originate from a thetic act of the legislator. The legitimacy of an institution is, in fact, external to any positive norms emanating from such an institution, which *presuppose* its legitimacy.

As we will see, simple reference to shared values is not sufficient to capture the kind of deontic modality at work here, which refers to a historically traceable commitment of the founders of the institution to certain values as constituting the *raison d'être* (van Eemeren 2010: 129) of the institution itself.The example is taken from Martin Luther King's famous *I have dream* speech. This address was delivered on August 28, 1963, at the Lincoln Memorial, Washington D.C., before a crowd of more than 200,000 civil rights demonstrators, as part of the March on Washington "For Jobs and Freedom". This momentous event in the struggle for civil rights and racial equality in the United States– in King's words "the greatest demonstration for freedom in the history of our nation"— had the immediate objective of supporting

[34] While I adopt here the term *thetic*, which Carmello (2009) derives from philosophers of law, such as Amedeo G. Conte, the view of the landscape of deontic modality espoused by Carmello's chapter could not be farther from the one defended here. For Carmello theticity is the very essence of deontic modality, which covers just performative thetic statements and their reports. On the (presumable) thetic origin of moral norms nothing is said.

the promulgation of the Civil Rights Act, which was stalled in Congress at the time. The modal proposition on which I focus my analysis is the following:

(48) And if America is to be a great nation, this *must* become true.

The utterance in (48) appears at a strategic junction of King's speech: it appears shortly after the end of the lyrical evocation of King's "dream" through the repetition of the formula *I have a dream that one day* ... and immediately precedes the very final part of the speech where, in keeping with the rhetorical template of the Black Sermon genre, the speech transcends the exhortation and becomes a direct prayer to the Lord. I believe this utterance and this *must* have a pivotal role in the argumentative structure of King's speech, as if it served to summarize King's key argument in support of his standpoint from the viewpoint of a civic and institutional logic as he was about to move more decidedly into the religious territory of messianic promise. In fact, the *must* in (48) can have both a religious reading, which is alethic in nature ('Given that America is destined to be a great nation, this will not fail to become true') and *requires* the audience to embrace a very specific religious perspective on the nature and destiny of their national community, and a *deontic* reading, which stands also in a purely secular perspective. The two readings are not in themselves contradictory, and it is not far-fetched to think that King wanted to communicate *both*.

Reconstructing the tripartite structure of *must* in (48) requires extensive recourse to its discourse context, prompted by the anaphorical functioning of the pronoun *this* and of the modal *must* itself. This process is not limited to the immediately adjacent or otherwise directly structurally accessible[35] discourse sequences, but involves also more indirect and inference based operations of resumption of key themes repeatedly made accessible to the audience in King's speech.

In order to reconstruct the prejacent of *must* in (48) we need to provide an antecedent to the anaphoric pronoun *this*. In doing this we are guided by converging semantic constraints imposed by the deontic modal *must* and by the main predicate of the prejacent: *become true* (x_1). In order to be congruous with *become true*, the antecedent of *this* has to be a non-factual state of affairs, which is alethically possible in the future. In order to form a congruous prejacent for a deontic *must* this state of affairs has to be further specified as something *doable*, something that can

[35] According to a widespread hypothesis in formal linguistic studies of discourse, attachment relations between discourse sequences are subject to the *right frontier constraint* (Asher and Lascarides 2003): a discourse sequence can "attach" only to sequences that are either adjacent to it or to sequences that dominate hierarchically over those adjacent to it. These sequences are said to be *on the right frontier*. Attachment to sequences deeply buried down and left in the discourse is presumed to be impossible. While the right frontier constraint has a certain predictive power in calculating some attachment phenomena – e.g. the functioning of *structural discourse connectives* (cf. Chap. 4, § 4.3.6.2, and see Forbes-Riley et al. 2005)– there are also more indirect and inference-driven forms of anaphoric recovery that can circumvent it, and pick-up virtually any discourse meaning, making it the object of reprise. Anaphoric connectives in the sense of Forbes-Riley et al. (2005), and modals even more so can function in this manner. To these phenomena, I refer generically with the term *resumption*.

become true as the result of human action. We thus need a propositional content introduced in the discourse in an appropriate, non-factual, intensional context. The most obvious choice – and, I believe, ultimately the correct one – is identifying the antecedent of *this* with the propositional content of King's "dream" as manifested by the series utterances introduced by the pattern by the pattern *I have a dream that...will*, exemplified in (49).

(49) I have a dream that one day this nation will rise up and live out the true meaning of its creed: "We hold these truths to be self-evident, that all men are created equal."
I have a dream that one day on the red hills of Georgia, the sons of former slaves and the sons of former slave owners will be able to sit down together at the table of brotherhood.
I have a dream that …

I have analyzed elsewhere (Rocci 2009: 27–32) the rich semantics and "vague resonances" of the attitudinal predicate manifested by the pattern *I have a dream that... will,* showing that its prejacent is not a counterfactual dream but a state of affairs (a) accessible as a possible future from the world of evaluation w and (b) the object of desire. In fact, as observed in Rocci (2009), in the speech this same propositional content, let us call it *d*, becomes in turn the prejacent of a series of intensional predicates. With some slight simplifications, this is the series of predicates we encounter as the speech progresses, including the modal *must*:

(50) *Deeply rooted in* (d, 'American Dream') […] *I have a dream that one day...will* (d) […] *is our hope* (d) […] *is the faith that I go back to the South with* (s) […] *must* ('America is to be a great nation, d)

The observant reader will have noticed that in (50) I have already analyzed the modal *must* as a two-place predicate, according to a Relative Modality analysis which takes the conditional antecedent *If America is to be a great nation* as a plausible if partial linguistic manifestation (cf. § 5.2.1) of the conversational background of the deontic modal. If we accept this identification[36], we already have a minimal reconstruction of the conversational background and of the prejacent of *must*, presented below in (Def. 8).

[36] I note in passing that accepting that the conditional antecedent *if America is to be a great nation* contributes to the evaluative – i.e. properly deontic – part of the conversational background would not be obvious from the viewpoint of Kratzer's original version of Relative Modality, where *if*-clauses contribute propositions that are added to the modal base f(w). The problem is that in the original theory (cf. Kratzer 1991: 649, Portner 2009: 72) *deontic* conversational backgrounds are supposed to only contribute *ordering sources* g(w). The semantics of a deontic reading of a modal like *must* is obtained through a modal base f(w) of *factual circumstances* selecting a set of worlds which is then ordered thanks to the norms in the deontic ordering source. As a consequence, the theory fits well *if*-clauses that modify the circumstantial base of the deontic (e.g. *If you arrive late, you must wait outside*) but does not allow for *if*-clauses contributing to the properly normative, bouletic or teleological part of the conversational background – as (114) seems to do – because conditional antecedents are not added to ordering sources.

(Def. 8) *d* logically follows from B_D
$B_D = B_V \cup B_C$
d is the propositional content of King's dream of freedom and equality for black citizens
B_V is the shared goal/obligation/duty that America be a "great nation"
B_C is the present political and social circumstances of the USA on August 28, 1963

While (Def. 8) already provides a minimal congruent reconstruction of the prejacent and conversational background of *must* in (48) it remains a singularly hollow one, quite far from the pragmatic optimum: what does 'becoming a great nation' really mean? In fact, all along King's speech, from its exordium to the sentences that immediately precede (48) we find both open and subtler reminders of the common ground of values that King and his audience share, not only *de facto*, but also *de iure* because of their status as citizens of the institutional interaction field of the United States of America.

Abraham Lincoln looms large even before the beginning of the speech thanks to the silent presence of the memorial's colossal statue and is immediately mentioned with a stylistically coded reference to his *Gettysburg Address* (1863) and the explicit mention of the centennial of the *Emancipation Proclamation* (1863) he signed. The stylistic echo of the exordium of the *Gettysburg Address* is as easy to spot, as it is significant in building King's demand for civil rights for African American citizen's on the foundations not only of Lincoln's *Proclamation*, but first of all, on the *Declaration of* Independence:

(51) (a) <u>Five score years ago</u>, a great American, in whose symbolic shadow we stand today, signed the Emancipation Proclamation (Martin Luther King Jr, *I Have a Dream*)

(b) <u>Four score and seven years ago</u> our fathers brought forth, upon this continent, a new nation, *conceived in Liberty, and dedicated to the proposition that all men are created equal* (Abraham Lincoln, *Gettysburg Address*)

(c) We hold these truths to be self-evident, that all men are created equal, that they are endowed by their Creator with certain unalienable Rights, that among these are Life, Liberty and the pursuit of Happiness (*Declaration of Independence*).

It is not necessary here to track all the explicit references and less overt allusions found in King's speech to the foundational texts and to the values from which the United States draw their raison d'être as an institution and their legitimacy.

King also says that his dream is *deeply rooted in* the American Dream. As suggested by Rocci (2009: 29), we can conceive the American Dream simultaneously as a grand narrative and as a set of values including freedom and equality of opportunity, noting that "these propositions are not stored in the common ground as facts" but "describe an ideal world" so that "it would be natural to preface them with a deontic modal word". As with any culturally shared master narrative, the American Dream is extremely plastic, but we also know – from King's sermons (cf. Rocci 2009: 31). – that he identified the *Declaration of Independence* as the source of the American Dream.

It is tempting, at this point to compare the vague intensional relation suggested by *deeply rooted in* with the relationship between the prejacent of a deontic proposition and its conversational background (cf. Rocci 2009: 30). In fact, King's dream is nothing else that the authentic realization of the values that constitute the Nation's *creed* expressed in the *Declaration*:

(52) I have a dream that one day this nation will rise up and live out the true meaning of its *creed*: "We hold these truths to be self-evident, that all men are created equal. (Martin Luther King Jr, *I Have a Dream*)

As observed in Rocci (2009: 30), given the institutional and foundational value of the *Declaration of Independence*, the deontic force of the *creed* transcends the status of a mere set of culturally shared values (the indication of what is good in a given culture) assuming the nature of "a commitment to which the United States are bound as an institution". The status of binding commitments of the principles of the *Declaration of Independence*, and the fact that they had not been honored in the present state of affairs is conveyed in a particularly forceful way when King uses the ironic analogy of a financial obligation that the debtor failed to honor, providing the "citizens of color" with a "bad check":

(53) In a sense we've come to our nation's capital to cash a check. When the architects of our republic wrote the magnificent words of the *Constitution* and the *Declaration of Independence*, they were signing a promissory note to which every American was to fall heir. This note was a promise that all men, yes, black men as well as white men, would be guaranteed the "unalienable Rights" of "Life, Liberty and the pursuit of Happiness." It is obvious today that America has defaulted on this promissory note, insofar as her citizens of color are concerned. Instead of honoring this sacred obligation, America has given the Negro people a bad check, a check which has come back marked "insufficient funds. (Martin Luther King Jr, *I Have a Dream*)

Given this richly stratified, but easily accessible discourse context both the evaluative and the alethic-circumstantial parts of a deontic the conversational background can be given a deeper interpretation as per (Def. 9).

(Def. 9) *d* logically follows from B_D

$B_D = B_V \cup B_C$

d is the realization of a radical change in the life of the nation so that for "citizens of color" the unalienable rights of life, liberty and the pursuit of happiness are truly guaranteed with equal opportunities with respect to white citizens.

B_V is the constitutive commitment of the United States in the *Declaration of Independence* to the proposition that that all men are created equal with unalienable rights of life, liberty and the pursuit of happiness, solemnly reaffirmed by Lincoln and formally implemented for black citizens in his *Proclamation of Emancipation.*

B_C is the present political and social circumstances of the USA on August 28, 1963 characterized by substantial inequality (nationwide) and lack of formal and substantial recognition of basic civil rights to black citizens (in southern states)

But even this richer reconstruction does not yet account for what the conditional antecedent *If America is to be a great nation* suggests. After all, the B_V of *must* is

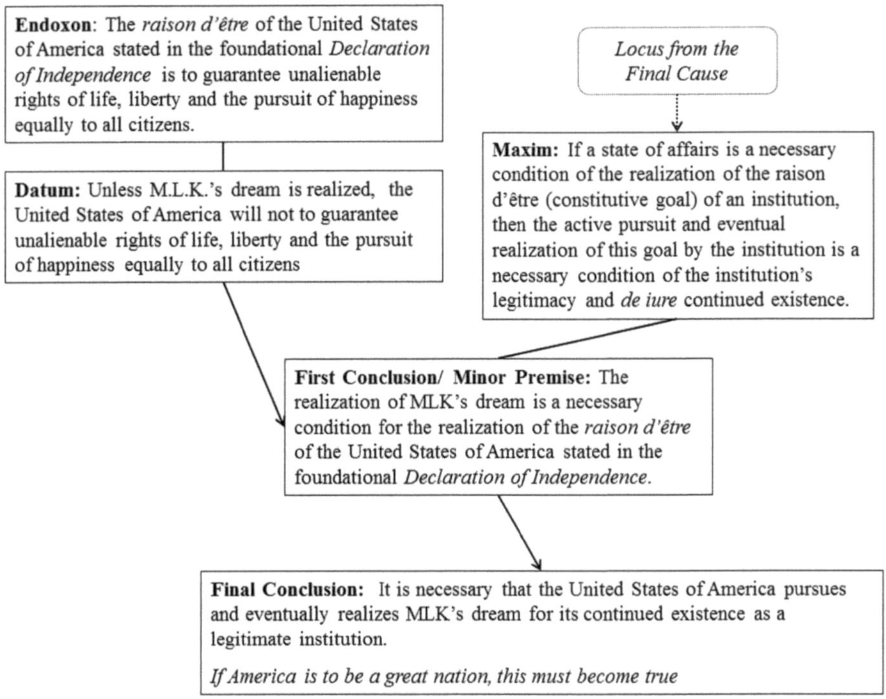

Fig. 5.11 Inferential configuration (Y-structure) of one argumentative line of the *I have a dream* speech evoked by (48)

prima facie represented by the ideal of a "great nation" and not directly by the commitments of the *Declaration of Independence*. And King gave us a hint at what makes a nation great: living out the true meaning of its *creed*, fulfilling the commitments towards the fundamental goods that that justify the nation's continued existence as an institution. The alternative, to keep the financial metaphor of the passage reproduced in (119), is "defaulting" on these constitutive obligations and becoming a (morally) "bankrupt" institution. The possibility for the United States of being a "great nation" fully realizing its *raison d'être* and the risk of becoming a morally bankrupt, delegitimized institution are a good and an evil in themselves, pertaining the sphere of civic life, *supervening* upon the primary moral and social good and evil deriving from the realization or betrayal of the values of the *Declaration of Independence.* It seems to me that the idea of the legitimacy or moral bankruptcy of the United States of America as an institution is pivotal in one line of argumentation underlying the whole *I Have a Dream* speech. This line of argumentation is tentatively reconstructed in Fig. 5.11:

5.3.2.3 Deontic-Practical Modalities

The map of deontic modalities proposed in Fig. 5.8, § 5.3.2 separates between *pure* deontic conversational backgrounds and *deontic-practical* ones. The latter term is taken from Hans Kronning (1996), who coined it with reference to the concept of *practical inference* in von Wright (1963a). Here I have taken von Wright's idea that practical inferences involve a causal component relating means to ends as criterial and I have identified the distinction with the contrast between *intrinsic* and *instrumental* value.

For the sake of clarity, it is important to set apart the definition of practical inference adopted here for from Aristotle's view of arguments oriented towards action and the modern views of pragmatic, practical or prescriptive argumentation directly dependent on it. Von Wright (1963a) observes from the outset that his definition of *practical inference* does not coincide with Aristotle's notion of *practical syllogism* in the *Ethica Nicomachea* (VII, 3). Aristotle sets apart the practical syllogism from its theoretical counterpart as follows: while in theoretical reasoning the mind of the reasoner is compelled to *affirm* the proposition resulting as a conclusion from the premises, in a practical syllogism the reasoner is compelled to take straight away – if not prevented – the course of action resulting from the premises. As noted by von Wright, although the relationship between goals and means and the distinction between intrinsic and instrumental value are prominent in Aristotelian thought, they do not enter his definition of practical syllogism.

For instance, Aristotle's notion of practical syllogism could be plausibly applied to both our examples of inferences with pure deontic modalities, whose Y-structures are reconstructed in Figs. 5.10 and 5.11. As long as the addressee of the argument coincides with the deontic subject[37], for a reasonable critic who is also a reasonable agent, the acceptance of the inferential path proposed does not simply result in the acceptance of the deontic proposition but in the very performance of the action. This step from reasons to action, as recognized also by von Wright (1963a: 162, 165), does not concern reasoning "in the third person", but only the subjective reasoning of the deontic subject "in a first person pattern of inference". Aristotle's distinction between inferences that compel to asserting a conclusion and inferences where reasons compel to action a rational agent is an important one for argumentation theory. It is clear, though, that it does not coincide with the distinction between pure deontic and deontic-practical modality drawn here. Instead it partially overlap both categories when the subjective inferences of an agent are concerned. Let us set aside, for the moment, the Aristotelian practical syllogism and examine the argumentative relevance of distinguishing the two types of deontic modality.

Von Wright's basic scheme for logically valid practical inference in (Def. 6) can be easily mapped onto the conversational background of the necessity modal appearing in the conclusion of the inference – typically *must*, in von Wright's

[37] By *deontic subject* I mean the person – physical or juridical – who is under the obligation to act or has the faculty to act.

examples. This conversational background is a special case of the deontic conversational background template given in (Def. 5). Consider the following example:

(54) The Socceroos *must* keep cool heads in the heat of the world spotlight, however, for despite their high hopes they are the underdogs of their group, with a world ranking of 42. They must face Asian champion Japan (No.18), defending champion and five times World Cup winner Brazil (No.1) and the powerful 1998 semifinalists Croatia (No.23). (from an Australian football news website just before the 2006 World Cup)[38]

Despite the implicitness and vagueness of the argument of the sport columnist, it is fairly straightforward to see von Wright's scheme as licensing the inferential step immediately supporting the standpoint:

(55) (a) The Socceroos want to advance in the 2006 football World Cup (B_V)
(b) Unless the Socceroos keep cool heads in the heat of the world spotlight, they won't be able to advance in the 2006 football World Cup. (B_C)
(c) Therefore, the Socceroos *must* keep cool heads in the heat of the world spotlight.

Both premise (a) and premise (b) are implicit. The reconstruction of the goal premise (a) is however completely obvious, given the context of the competition. As for (b), all the explicit arguments presented by the commentator can be seen as supporting it: an excess of confidence deriving from the good results of the team can prove fatal in a situation where they have to compete as the underdog against stronger and more experienced teams. At the same time, things can go wrong also if they are too scared of their adversaries. That's why without "cool heads" they cannot advance in the tournament. This inferential configuration is easily translated into an AMT structure:

In terms of the deontic conversational background, the goal premise (a) can be seen as a special type of the evaluative conversational background B_V, while the causal premise (c) and the factual statements supporting it are a subtype of the circumstantial conversational background B_C. The semantic of *must* in (120) can be captured by (Def. 10), using the notation inspired by Geurts (1999) which was introduced in Chap. 4, (Def. 19).

(Def. 10) {'The Socceroos want to advance in the 2006 football World Cup'}$_{BV}$ ∪ {'Unless the Socceroos keep cool heads in the heat of the world spotlight, they won't be able to advance in the 2006 football World Cup'}$_{BC}$ ❑ 'Socceroos keep cool heads in the heat of the world spotlight'

What sets the deontic-practical conversational background apart is that B_C contains a *causal connection* between a human action and a goal.

Consider the difference between the deontic-practical modality in Fig. 5.12 and the pure deontic modality in Fig. 5.11. Realizing King's dream may be a necessary condition for being a "great nation" but it is not a means, a tool for *causing* it to happen. Actually, recognizing black citizens' "unalienable rights" is not a means at all: it is part and parcel of the goal itself. If my goal is to be a just person, acting

[38] *The World Game*, June, 9, 2006. Retrieved August 15, 2006 from http://www6.sbs.com.au/socceroos/index.php?pid=st&cid=71987

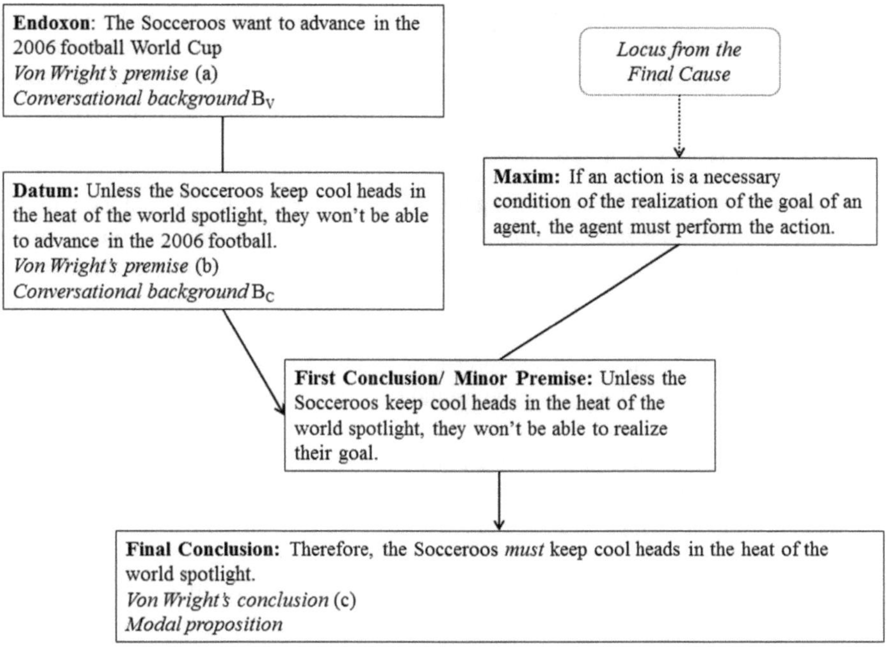

Fig. 5.12 Inferential configuration (Y-structure) of (54) according to von Wright's scheme of practical inference in (55)

fairly in my role of teacher is surely a necessary condition of the realization of the goal, but it is not a cause nor a means. It seems better to say that it is necessarily entailed by the realization of the goal as a *part* is entailed by the *whole*. It clear that both in Fig. 5.11 and in Fig. 5.12 the *datum* presents a necessary condition, but while in Fig. 5.12 the this necessary condition derives from a relation of *efficient cause*, in Fig. 5.11 the underlying relationship is *part-whole*. Even if the reconstruction of these two inferential patterns in terms of the Y-structure does not make the difference immediately apparent (both invoke the *locus of the final cause*) the AMT typology of *loci*-relations turns out be useful, so-to-say, at a deeper level of stratification, to distinguish how the datum of each is constituted.

As anticipated in § 5.3.2, not all deontic practical modalities arise from a conversational background B_V of goals. Deontic-practical modality can also correspond to a *derived obligation*: the practical necessity of "something which must be done for the sake of doing one's duty" (von Wright 1963a: 175). That derived obligation arise is an everyday occurrence with positive law as the law "leaves it to the individual citizen to consider the various measures and steps which he must take in order to fulfil his legal obligations", von Wright (1963: 176), uses the obligation to make an income tax return as an example – adapted in (56):

(56) A: In order to file my income tax return, I *must* get the proper form, fill it and then mail it to the Tax Office.

B: Actually, now you *can* also fill and submit the income tax form online.

In *Norm and Action*, von Wright (1963b) use the term *anankastic statement* to indicate statement of a necessary condition, such as those underlying practical reasoning from goals and practical reasoning arising from obligations. He maintains that an anankastic statement such as (57) is not *normative*, but purely *descriptive*.

(57) If the house is to be made habitable, it ought to be heated

In terms of Freeman's typology of statement we should say that it is not a combination of *evaluation* and *interpretation*, as deontic-practical modalities, but a purely *interpretive* statement. Von Wright defends the non-evaluative (non-normative) of anankastic statements saying that their truth or falsity is independent of whether "anyone wants to make the house habitable and aims at this as an end". In contrast, deontic-practical utterances always presuppose an actual goal or an actual duty of an agent in the world. This marks the boundaries of deontic modality and of evaluative statements.

Deontic-practical utterances have a closer or more distant relationship with their anankastic counterparts depending on their *objective* (so-called, "third person") or *subjective* ("first person") status. For von Wright the conclusion of an objective practical reasoning amounts to the simple statement that the action is a necessary condition of the realization of a goal (or duty) of the agent. In other words, the conclusion is a special kind of anankastic statement. For an objective practical inference to be valid it is not required that the agent knows the anankastic premise (b) nor that the conclusion is recognized by the agent. A sports columnist can utter (54) also in a context in which the football players are so engrossed by their past wins that they do not recognize the anankastic premise in (55.b) and do not pay heed to the advice. As noted by von Wright, from an objective practical inference it does not follow that the agent will perform the action. It is not the conclusion of a practical syllogism in the Aristotelian sense. In contrast, when a rational agent performs the practical inference in the first person she will necessarily believe the anankastic premise and, consequently, will be rationally bound to adopt the conclusion as goal and to act accordingly. This is a practical conclusion also in the Aristotelian sense (cf. von Wright 1963a: 165–166).

In his essay on practical inference von Wright subscribes to a version of the Kantian principle that whoever wants to attain an end is bound also to want to use the necessary means towards its attainment. This is to be understood as an analytical principle of the will. What happens then that it turns out that I am unwilling to use the necessary means to attain a goal of mine? According to von Wright, the practical inference "collapses" as my supposed goal is no longer something I really want to make happen. The goal is demoted to a *wish*, something that I would cherish if the circumstances were different (von Wright 1963a: 170).

This type of inconsistency between desires occupies the center stage in Kratzer's comments on practical reasoning (Kratzer 1981: 65—67, 2012: 62–64). These comments are interesting because they are probably the most explicit discussion of the interaction of conversational backgrounds and arguments found in the original Relative Modality theory.

In these passages Kratzer also argues that augmenting Relative Modality semantics with *ordering sources* (cf. Chap. 4, § 4.3.4) allows to better model practical reasoning in situations where agents are unwilling to adopt the necessary means to attain a goal, predicting which conclusions can be correctly drawn in these situations. In this book (Chap. 4, § 4.3.4.3), I have set aside ordering sources because this powerful formal tool blurs the correspondence between modal conversational backgrounds and sets of premise and the correspondence between the modal force (e.g. logical consequence) and inferential moves in a naturalistic inference process. It is therefore interesting to assess whether significant insights and predictive power are lost in the area of practical inference by renouncing to this formal device.

Kratzer introduces the Relative Modality analysis of practical reasoning with a simple example, reproduced in (58). There are small but significant differences in the examples between the original paper (Kratzer 1981) and its 2012 revision. I will need to keep track of them.

(58) (a) I want to become mayor
(b) I will become mayor only if I go to the pub regularly
(c) Therefore I *must* [*should* 2012] go to the pub regularly.
(c') Therefore, considering the relevant circumstances and what I want I *should* [it is *necessary* that I 1981] go to the pub regularly.

The modal in (58.c) is interpreted relative to a circumstantial modal base f(w) – a function mapping the evaluation world w "to the set of propositions that correspond to the relevant circumstances in w" and an ordering source g(w) mapping from w "to the set of propositions that correspond to what I want in w". Kratzer starts by "assuming" that f(w) only contains the proposition that 'I will become mayor only if I go to the pub regularly' and g(w) only contains the proposition 'I will become mayor'. In this simple case the two conversational backgrounds f(w) and g(w) can be placed on an equal footing and their set-theoretic union $f(w) \cup g(w)$ can provide the conversational background from which the prejacent 'I go to the pub regularly' necessarily follows.

This looks very much like the simple account I have proposed for example (54). There are, however, subtle but interesting differences. Kratzer nowhere claims that the premises of the reasoning (a) and (b) provide the conversational background. As observed earlier (cf. Chap. 4, § 4.2.7.3), anaphora and discourse dynamics do not enter semantics in the original theory. Instead, the "hidden assumptions" of (58.c) are "spelled out" to construct a more explicit version of the modal sentence (58.c') and the phrases *considering the relevant circumstances* and *considering what I want* are said to contribute, respectively, f(w) and g(w). This is, in fact, an expository expedient: contextual assumptions provide a ghost sentence, whose phrases provide the required conversational backgrounds. In turn, the sets of propositions provided by f(w) and g(w) is only world-relative, which, for a set of "relevant circumstances", seems quite a simplification. Of course, expository expedients and simplifications are more than legitimate when endeavoring the arduous task of formalization. But it is equally legitimate to note them, when tacking stock of the insights gained through the formalization. Here it is important to note that it is only by assumption that f(w) and g(w) contain only one proposition each. The original version of Relative Modality tends to have a rather "holistic" view of conversational backgrounds, as observed earlier (cf. Chap. 4, § 4.2.7.1). Kratzer wants to show that the simple analysis based on the union of the two conversational backgrounds collapses once we examine a scenario where an agent entertains two wants that turn out to be inconsistent in view of the factual circumstances. This scenario is described in (59):

(59) All I want is two things, namely avoid going to the pub and become mayor. The relevant circumstances are such that I will become mayor only if I don't avoid going to the pub (Kratzer 2012: 63)

Here the modal base f(w) rules out worlds where both desires are satisfied. According to Kratzer, the use of the set g(w) as an ordering source to rank the remaining worlds by closeness to the wants correctly predicts which practical conclusions are sound given (59). Now, there are three sets of worlds:

A. Worlds where I go to the pub and I become mayor
B. Worlds where I avoid going to the pub and I fail to become mayor
C. Worlds where I go to the pub and still fail to become mayor

The worlds in C rank lower than either B or A, while it is not possible to rank A and B among themselves. Kratzer's notion of necessity and possibility with respect to a modal base and an ordering source predict that conclusions of practical necessity such as (59.a) and (59.b) are false in this scenario:

(59.a) I *should* go to the pub (2012) / It is *necessary* that I go to the pub regularly (1981)
(59.b) I *should* avoid going to the pub (2012) / It is *necessary* that I do not go to the pub regularly (1981)

Neither course of action fits *all the best worlds*, as (59.a) is false in B worlds and (59.b) is false in A worlds. So, there is nothing that I have to do or that I should do. But, is there anything I *can* or *could* do? Yes, according to the predictions of ordering semantics. Conclusions (59.c) and (59.d) are predicted to be true:

(59.c) I *could* go to the pub (2012) / It is *possible* that I go to the pub regularly (1981)
(59.d) I *could* avoid going to the pub (2012) / It is *possible* that I do not go to the pub regularly (1981)

This is because their prejacents are true in *some of the best worlds*: the prejacent of (59.c) is true in A worlds, while the prejacent of (59.d) is the case in B worlds. Kratzer (2012: 64) concludes that "separating modal bases and ordering sources allows us to make the correct predictions in practical inferences". But, is this the case? Are (59.c) and (59.d) utterances that someone would naturally utter as conclusions following from (59)? I do not think so.

Obviously, the conclusions are (trivially) correct if we assume that *could* and *possible* have the ordering semantics that Kratzer assigns to them, but I am not sure that this semantics is descriptively correct either for *possible* or for *could* in (59.c,d) *and* that this semantics captures realistically the notion of deontic-practical possibility. Let us see why. The version with the adjective *possible* immediately strikes as most unnatural for introducing such a conclusion in an actual argumentative passage. This is quite apparent in (60.a):

(60) (a) All I want is two things, namely avoid going to the pub and become mayor. But I know that I will become mayor only if I go to the pub regularly. So, it is *possible* that I go to the pub regularly, but it is also *possible* that I do not go.

(b) All I want is two things, namely avoid going to the pub and become mayor. But I know that I will become mayor only if I don't avoid going to the pub. So, I *could* go to the pub, but I *could* also avoid going to the pub.

The *possible that* construction in (60.a) tends to have an epistemic flavor and it seems hard to interpret it as deontic-practical possibility. This possibly the reason why the example was amended in the revised paper. *Could* in (60.b) sounds a bit less strange, but, given the artificial nature of the example and the lack of real context of utterance it is hard to say what it could mean and whether it really conveys a deontic-practical possibility. Let us observe an uncontroversial example of deontic-practical possibility:

(61) In order to reach the summit of Monte Brè you *can* take the old funicular railway. Of course, you *can* also hike along an enchanting woodland path.

In (61) the prejacent of *can* indicates a course of action that is among those compatible[39] with reaching the goal stated in the *in order to* phrase; this is consistent with situations where the prejacent is not a necessary condition for reaching the goal. This type of *can* fares very badly in a situation such as the one described in (54), *especially* if the set of goals g(w) is made explicit in a *in order to* phrase, as shown by the utterly nonsensical (62.b).

(62) (a) In view of the current circumstances, I will become mayor only if I go to the pub regularly. I want to avoid going to the pub and I want to become mayor. ?? So, I *can* either go to the pub or avoid going to the pub.

(b) In view of the current circumstances, I will become mayor only if I go to the pub regularly. *So, in order to become mayor and avoid going to the pub, I *can* either go to the pub or avoid going to the pub.

In view of Kratzer's analysis it is not clear why *can* should not able to express the kind of possibility postulated by ordering semantics. In contrast, in the

[39] It has to be observed, though, that a view of deontic-practical possibility as mere compatibility with reaching the goal is not adequate. Examples such as *To reach the summit of Monte Brè you can eat tomato salad* or *To prepare a good coffee you can wear pink underwear* strike us as nonsensical even if eating tomato salad does not prevent us from reaching the summit and the color of underwear is not negatively affecting our ability to prepare coffee. Should we conclude that deontic-practical possibility modals indicate *sufficient conditions* for attaining the goal? Actually, a sufficiency presupposition would be too strong. In ordinary causal interpretations, as observed by Freeman (2005: 144), causes are hardly sufficient conditions. When I say *The match lit because it was struck* I presuppose (or "presume" as Freeman says) a host of out of focus necessary conditions (a force of sufficient magnitude, the presence of oxygen, the match being dry). A fully fledged account of deontic-practical *can* would require this predicate to have a presupposition of sufficiency given a conversational background of *normal conditions*. Such a presumptive sufficiency operator would be a special kind of *necessity* operator.

circumstances of (59), a statement of deontic-practical *impossibility* such as (63) seems perfectly warranted:

(63) I will become mayor only if I go to the pub regularly. So, there is *nothing* that I *can* do to become mayor *and* avoid going to the pub.

In my examples (62.b) and (63) the modal clearly takes the set of goals as an indivisible solidarity. Kratzer's approach divides the propositions and ranks the worlds according to their closeness to the full set. The *could* sentences in (59.c, d) are considered possibilities because they occur in some of the best alternatives of a situation where the full set of my desires cannot be realized.

I do not think that this the way in which we rank our alternatives in less than ideal situations. Faced with the fact that the goals of avoiding the pub and the goal of becoming mayor cannot not be both realized, agents will have to revise their set of goals to avoid the "collapse" of practical reasoning and decide which goal is more important and which one can be set aside and demoted to a wish. This seems to require another kind of priority ordering in the conversational background, not among worlds but among the propositions. Normally it would mean very little that the options in (59.c) and (59.d) are each compatible with one desired state of affairs and incompatible with the other: the options would be equivalent only in the limit case where we do not prioritize a goal over the other. But, normally, we do rank our goals. This is something that ordering semantics does not seem to provide, at least not directly and not in a cognitively realistic way.

What to make, then, of the sentences with *could* in (59.c) and (59.d)? Should we accept that *could* is more acceptable than *can* or *possible* in a context like (59), the explanation may hinge on the counterfactual component of a "distal" or "past-tense" modal like *could*. Imagine that someone replies to (59) by uttering (64):

(64) Well, as things stand, if you really want to become mayor so badly, you *could* just go to the pub and put up with the annoyance. But, if you really cannot stand the pub, you *could* avoid going and just forget about becoming mayor.

One strategy for analyzing *could* in (64) is to treat it indicating each time that a course of action is compatible with a restructured set of preferences, should the agent give priority to one of the conflicting goals, demoting the other to a mere wish. The conversational background is not the actual set of desires, but what the set would be provided the contrasting desires are amended in one way or the other.

I should bring here to a close this brief discussion of practical inference from the viewpoint of modal semantics, having barely scratched the surface of the key problems in practical reasoning. Introducing practical inference, an argumentation scholar like Douglas Walton insists that it is "neither deductive nor inductive in nature" but rather presumptive (Walton 2006: 300), while von Wright (1963a) successfully defends the validity of the basic pattern of practical inference deducing the deontic-practical necessity of a means that is an anankastic necessary condition of an agent's goal. The apparent contrast between these views of practical inference, though, is largely a matter of perspective.

Von Wright is successful in defending the validity of practical inference under strong assumptions, which notably includes "Kantian" firmly set wants or goals. In practice, we rarely firmly commit to a goal. What we normally have is *presumptive goals* based on incomplete information on the circumstances (and limited ability to process the information). As new information is acquired a presumptive goal can be demoted when it turns out that an unwanted action is necessary to achieve it, or simply that the mere realization of the presumed goal has unforeseen negative consequences. Considering real-world conditions, it is hard not to agree with what Walton says about the presumptive nature of practical inference:

> Practical reasoning is normally evaluated or weighed on a balance of considerations. The importance of a goal has to be weighed against the negative or positive consequences of carrying out the means to the goal, to the extent that these consequences can reasonably be anticipated. However, because the future is never known with certainty, and most typically not even known with any precise degree of probability, practical reasoning characteristically is tentative and presumptive in nature. A plausible practical conclusion, even one that is strongly indicated, should generally be seen as open to revision should new, relevant information come in. (Walton 2006: 302)

Additionally, our actions are rarely *necessary conditions* for bringing about a goal so that we have to evaluate the pros and cons of different courses of action that are compatible with the attaining the goal (i.e. are deontic-practical possibilities). When this is the case, not even the presence of a firm goal is sufficient to determine our course of action, as additional preferences would enter into consideration to weigh an option against the other.[40] Nor our actions are usually *sufficient conditions* for bringing about what we want. As observed by Freeman (2005: 144) they are, at best, sufficient assuming a (conversational) background of normal conditions.

Eddo Rigotti provides an effective characterization of the complexities of reasoning about human action, which nicely complements Walton's remarks:

> Often, many different, not strictly constitutive factors become relevant: given the situated nature of decision making, different competitive desires and different costs of the causal chain may induce the agent to abandon or substantially change the purpose; the degree of adequacy of the causal chain may show to be insufficient and transform the action in an unhappy attempt; the possible positive or negative side effects, including the informative and relational implications of action, the possible presence in the causal chain of subservient instrumental actions and the quality of their ends and of their possible side effects turn the action into a complex and hardly manageable process, in which the human subject intensively "negotiates" the realization of its purposes with the surrounding context. (Rigotti 2008: 565)

As the *Introduction* (§ 1.6.2) admitted, the treatment of non-demonstrative inference exceeds the limits of this book. Thus, we will leave these complexities for another day.

---

[40] If these preferences fail to make a difference we enter a "Buridan's ass" type of situation.

5.3.2.4 Deontic Modality and the Meta-argumentative Uses of the Modals

Before leaving the area of deontic modalities, it worth pausing on a very special case of deontic conversational background, which deserves consideration in a discussion of modality and argumentation. One encounters it in examples such as (65) and (66):

(65) No matter how high the correlation between two variables is you **cannot conclude** that one of them causes the other based on a correlational study alone. (Google via WebCorp)

(66) It **can be argued** that the conclusions that Professor Zimbardo and others have drawn from the Stanford Prison Experiment are not valid. Professor Zimbardo acknowleges that he was not merely an observer in the experiment but an active participant and in some cases it is clear he was influencing the direction the experiment went.

One way of looking at these examples of *can* and *cannot* is considering them *meta-argumentative deontics*. Like other deontic conversational backgrounds the meta-argumentative deontic conversational background is composed of a set of ideals B_V and a set of factual circumstances B_C. Normative-pragmatic models like Pragma-Dialectics (van Eemeren and Grootendorst 1984, 2004, van Eemeren 2010) can help in figuring out what these conversational backgrounds could be.

For instance, a meta-argumentative possibility can be conceived as an argumentative speech-act whose performance is *consistent* with the *set of dialectical rules* governing the discussion[41] (B_V), with the *material starting point* of the discussion and with the *store recoding the preceding moves* of the discussants (B_C). Given that dialectical rules are typically conditional to the stage of the discussion, a number of derived commitments can arise from the union of B_V and B_C. The nature of the discussion is such that B_C is always part of the common ground, while B_V may be partially ignored by ignorant, dialectically inept discussants.

Sometimes people use meta-argumentative modalities in a way that suggests that the set of dialectical rules is to be conceived as an ideal model – like the Pragma-Dialectical model of the critical discussion. In other contexts, however, it is clear that people refer to discussion in a precise institutional context (e.g. criminal justice in a given country). In these cases the B_V is a set of institutionalized argumentatively relevant conventions (cf. van Eemeren 2010, Chap. 6) that hold in the relevant argumentative activity type, like, for instance, in an appellate court in (67):

[41] This set of dialectical rules can be conceived either as an ideal model – like the Pragma-Dialectical model of the critical discussion – or as a set of institutionalized argumentatively relevant conventions that hold in a given argumentative activity type, like, for instance, an adjudication in court (Cf. van Eemeren 2010, Chap. 6). This gives rise to two distinct flavors of meta-argumentative possibility: something may well be at the same time a *possible standpoint* to defend in a critical discussion and an *impossible case* to bring to court in view of the procedural constraints of the relevant judicial system.

(67) No matter how little you dislike the fact that Johnny X was found to be dishonest, if it states this in the facts of the case, and it is not disputed in the points of appeal, you *cannot* argue against it. The appellate court does not have access to the evidence which was in front of the original court, and so *cannot* dispute the findings it made on the basis of that evidence (WebCorp).

This gives rise to two distinct flavors of meta-argumentative possibility: something may well be at the same time a *possible standpoint* to defend in a critical discussion and an *impossible case* to bring to court in view of the procedural constraints of the relevant judicial system.

Examining the consistency of the performance of the argumentative speech act with respect to the above mentioned sets of information means, at the same time, characterizing it with respect to the development of the discussion. At every moment in the development of an argumentative discussion, a range of developments consistent with the dialectical rules and with the state of the discussion remains open. Argumentation theorists have sometimes envisaged dynamically this set of argumentative possibilities with the aid of *dialogue profiles* (Walton and Krabbe 1995) or *dialectical profiles* (van Eemeren 2010). Saying, meta-argumentatively, that one '*must* conclude that p' means characterizing the speech-act of 'concluding that p' as the only alternative left that is consistent with the rules and the state of the discussion.

In the previous sections it has been argued that non-epistemic modals become argumentatively relevant when they refer to the logico-ontological loci-relationships underlying the inferential configuration of an argument. Meta-argumentative deontic modals are relevant in a different way: their prejacents contain predicates describing argumentative speech acts. These predicates are either used descriptively or form complex performative constructions together with the modal. In sum, meta-argumentative deontics are either used to describe and evaluate argumentative speech acts, or to perform them.

The obvious alternative to considering the meta-argumentative modals deontic is to consider them alethic. This can work in two ways. One way is to say that they are analytic and pertain to semantic or conceptual necessity and possibility (cf. §§ 5.2.4.6 and 5.3.1). This would require us to consider that an unreasonable conclusion is *not really a conclusion* and that an unacceptable argument is *not really an argument*, which is, admittedly, a perspective that people adopt on occasion when confronted with presumed fallacy (e.g. *But, that's not an argument!* as a reaction to an allegedly unreasonable argument). Another route would be to consider them circumstantial alethic taking a background of institutional social circumstances. This position has some merit with respect to institutional rules in institutionalized argumentative activity type: these rules are often very close to *constitutive* rules in Searle's sense, while deontic modalities are associated with *regulative* rules. A non-institutional human act, say *singing aloud*, can occur against regulative rules. If it does it constitutes a *violation* of the rules (e.g. of the regulations against loud behavior during the night hours). An institutional act, in contrast, cannot occur against constitutive rules: an attempt at performing it would be void and will not bear any result for the institutional record. Imagine that during a volley match someone kicks the ball into the net and shouts *Goal!* It would be reasonable to say that:

(68) No matter how hard you try, you *cannot* score a goal in volley.

This is a special kind of *alethic* impossibility: the impossibility for institutional acts of occurring against constitutive rules. What about a lawyer attempting to argue about facts in an appellate court? I believe this falls in between the goal and the singing. The argument will be there as a verbal and rational act, but it will be irrelevant and devoid of consequences from an institutional point of view. In fact, the institutional irrelevance would entail also a kind of defectiveness on a rational level if the argument was uttered under the pretense of full institutional relevance. So, quite probably, is not a good argument. And we are back to the semantic question whether defective arguments still deserve the name of arguments.

In the next section I will continue to deal with constitutive rules in connection with *anankastic* uses of the modals.

5.3.3 The Anankastic Modalities: Necessary Conditions and Telos

Anankastic modality (cf. Gr. ἀνάγκη, 'force', 'constraint', 'necessity') is a distinctive kind of non-alethic modality, rarely acknowledged by linguistic treatments of modality. Anankastic modals are routinely confused either with deontic or with circumstantial modalities to which they superficially resemble. In fact, they are set apart from alethic root modalities by the non-realistic nature of their conversational background: contrary to alethic modalities, anankastic conversational backgrounds are not entirely composed of propositions that are facts in the evaluation world w. Like deontic modalities they have composite conversational background, including a set of factual circumstances (B_C) but also another set of propositions that are non-factual in nature. However, contrary to deontic modalities, this set of propositions is not made of evaluations.

Faute-de-mieux, I will use the term *telos* (cf. Gr. τέλος 'end', 'goal'), which I borrow from Tom Werner (2005: 306), to characterize these conversational backgrounds, which contain propositions describing a potential, non-factual, end-state in view of which the prejacent is considered. The telos of anankastic modalities, however, is certainly an *end*, but only figuratively a *goal*. If it were a proper goal, we would have already transgressed into deontic-practical territory. Additionally, in many instances of anankastic modality there are no actions to speak of, and thus no goals. It is in this very limited acceptation that I will speak of the *telic conversational background* (B_T) of anankastic modalities.

We have seen that von Wright (1963b) uses the term *anankastic statement* to indicate a statement, like (57) in § 5.3.2.3, to the effect that an action is a necessary condition of a certain end-state. Von Wright maintains that an anankastic statement such is not normative, but purely descriptive as its truth or falsity does not depend

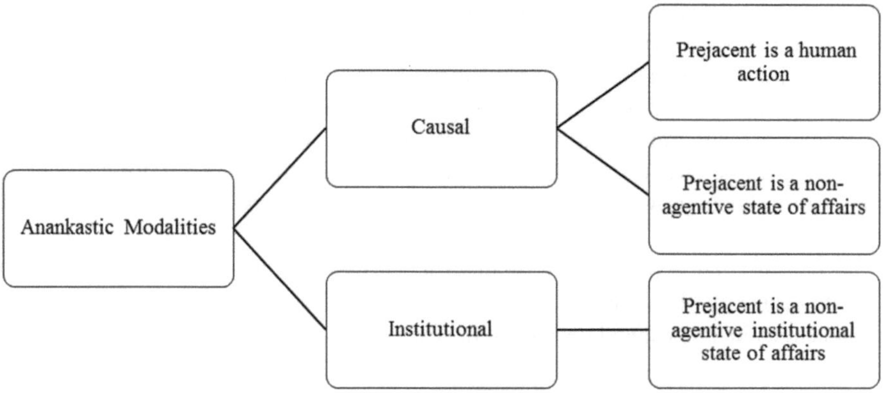

Fig. 5.13 Map of anankastic modalities

on anyone actually wanting to realize the telos, but on the prejacent being a necessary condition of the telos. The notion of anankastic modality presented here is a generalization of von Wright's.

It is also clear from the above that, according to the typology of statements proposed by Freeman (2000, 2005), the modal propositions created by anankastic modalities are *interpretations*. In fact, they are either *causal interpretations* or *institutional interpretations* (cf. Freeman 2005: 145). Figure 5.13 above offers a map of anankastic modalities distinguishing these two main subtypes:

The following are examples of anankastic necessity:

(69.a) In order for me to sell those stocks, their price *must* first double.

(69.b) There *must* be water on a planet for life to develop.

(69.c) A river must keep flowing, otherwise it begins to stagnate, to suffocate and strangle its very own life force (WebCorp).

(69.d) To be elected in the Italian Senate, you *must* be at least 35 years old.

Anankastic necessity modals tell us that the prejacent is a necessary condition for the realization of the telos, given the relevant circumstances. Formally, the prejacent is a logical consequence of a set of propositions corresponding to the set-theoretic union of B_C and B_T (Def. 11.a):

(Def. 11.a) *Anankastic Necessity*$_w$ (e, B_{AK}, p):

Presuppositions:

e is the VP-event of the prejacent *p*

$B_{AK} = B_T \cup B_C$

B_T is a set of propositions which are not facts in w, describing an end-state *t* of *e*, in view of which *e* is considered

B_C is a set of circumstances of *e* that are the case in w

Semantic entailment:

$B_{AK} \, \square \, p$

It has been already observed, in an earlier discussion of ordering semantics (Chap. 4, § 4.3.4.2), that multiple conversational backgrounds weaken the notion of necessity, and indeed an anankastic necessity is a weaker notion than the corresponding circumstantial alethic necessity. For instance, in (69.a) the circumstances do not necessarily imply that the stock price doubles, yet the realization of the telos of selling *in* the circumstances does. The natural circumstances do not necessitate that a river keeps flowing (69.c), but the circumstances *and* the realization of the telos do.

It can also be observed that anankastic necessity is always equivalent to the alethic necessity of a conditional having the original telos as its antecedent and the prejacent as its consequent, as shown in (Def. 12):

(Def. 12) *Anankastic Necessity*$_W$ $(e, B_{AK}, p) \Leftrightarrow$ *Circumstantial Necessity*$_W$ $(e, B_C, (t \rightarrow p))$
where
Anankastic Necessity$_W$ (e, B_{AK}, p) is defined as per (Def. 11.a)
t is an end-state of e, as per (Def. 11)

German-Italian linguist Maria Elisabeth Conte (2010 [1995]) was among the first to consider anankastic modality in linguistics. She came into contact with this category through the philosophy of law and focused her attention on institutional anankastics such as (69.d). Contrary to positive law deontics, the anankastic modality expressed by (69.d) does not concern actions, and cannot be translated into an imperative, as demonstrated by the absurdity of (69.e):

(69.e) *Be 35 years old, if you want to be a Senator!

If it turns out that someone was elected and was not 35, the election is null. As observed above, nullity is different from violation. As it has been observed in the discussion of alethic modality, this difference is related to Searle's distinction between *regulative* and *constitutive* rules (Searle 1969). In fact, Conte's institutional anankastics relate to institutional circumstantial alethics just like Wright's causal anankastics relate to physical circumstantial alethics. Let us take up again the Schwarzenegger example discussed in § 5.3.1. to observe how the institutional anankastic statement in (70.a) relates to the institutional circumstantial alethic statement in (70.c).

(70) (a) The President of the United States *must* be "a natural born citizen"
(b) Arnold Schwarzenegger was born in Austria
(c) So, Arnold Schwarzenegger *cannot* be President

It is clear that the anankastic statement in (70.a) can function as an endoxical premise to conclude the circumstantial alethic (70.c) from the datum in (70.b), using a definitional locus. Indeed the anankastic statement is one and the same thing as the underlying nomic regularity of the circumstantial modal, as it results from (Def. 12).

It is worth pausing briefly on the differences between the two types of anankastic. As Freeman (2005: 145) says of institutional interpretations "they are not matters ultimately of empirical discovery but are rather given or derivable from some body of statute law, an institution or a facet of an institution". Institutional anankastic ultimately depend, as Carmello (2009: 222) observes, on a thetic act of the legislator arbitrarily establishing a norm as necessary condition of an institutional state of affairs. This thetic act corresponds to what Searle (2005: 7) calls "status function assignment".

It therefore not too surprising that we know the necessary conditions of institutional states of affairs much better than we know those of causal ones. Indeed we – as a society – did *decide* ourselves what are the necessary conditions for an institutional state of affairs to be the case.

Having spoken of anankastic necessity it remains to discuss *anankastic possibility*, first of all, to determine whether there is such a thing. I have not found examples of anankastic statements with possibility modals in Conte (2010 [1995]) or Kronning (1996). And, as a term, anankastic possibility, sounds vaguely contradictory. Despite this, I believe that the following examples are at least credible candidates for an analysis in terms of anankastic possibility:

(71.a) The stock price *can* go up 100%. He will not sell.

(71.b) The stock price *can* go up 100% without him selling.

(71.c) Gerard Depardieu says he *can* drink up to 14 bottles of wine a day "But I'm never totally drunk." (WebCorp)

(71.d) Boris Johnson has boasted that he *can* "drink an awful lot at lunch" and still work well [...] (WebCorp)

Treating the examples in (70.a-d) as anankastic means considering their conversational background as composed of a factual B_C and of a non-factual, non-evaluative B_T. This interpretation is captured by (Def. 11.b), parallel to the definition of necessity in (Def. 11.a):

(Def. 11.b) *Anankastic Possibility*$_w$ (e, B_{AK}, p):

Presuppositions:

e is the VP-event of the prejacent p

$B_{AK} = B_T \cup B_C$

B_T is a set of propositions which are not facts in w, describing an end-state t of e, in view of which e is considered

B_C is a set of circumstances of e that are the case in w

Semantic entailment:

$B_{AK} \diamond p$

This definition works with the examples in (71.a-d) provided that we recognize the B_T in *He will not sell* in (71.a), *without him selling* in (71.b), *But I'm never totally drunk* in (71.c), *and still work well* in (71.d). Interestingly, contrary to what happen with necessity, anankastic possibilities are *stronger* than purely alethic possibilities. An anakastic possibility always entails the corresponding alethic possibility but not vice versa: if for Boris Johnson drinking "an awful lot at lunch" is compatible with the circumstances and with the telos of "working well", then the action is also circumstantially possible *simpliciter*. The reverse is not true: I may be able to actually drink "an awful lot", but it does not follow from it that I will be also fine.

Admittedly, when the telos is syntactically tied to the prejacent as in (71.b) and (71.c) a simple analysis in terms of circumstantial alethic possibility might seem more economic. This is not surprising as also for anankastic possibility we can formulate a rule of equivalence with the circumstantial modality, as above with necessity in (Def. 12).

(Def. 13) *Anankastic Possibility*$_W$ (e, B_{AK}, p) ⇔ *Circumstantial Possibility*$_W$ (e, B_C, $(t \wedge p)$)
where
Anankastic Possibility$_W$ (e, B_{AK}, p) is defined as per (Def. 11.b)
t is an end-state of e, as per (Def. 11.a)

What (Def. 13) says is that the anankastic possibility of a given prejacent is equivalent to the circumstantial possibility of the conjunction of the original prejacent and the telos. Even if fully reducible to circumstantial possibility in cases such as (71.d), the notion of anankastic possibility seems interesting for unifying the treatment of utterances where the telos is syntactically expressed and those where it is expressed by discursive, anaphorical means – as in (71.a) and (71.c). The conversational background seems the right tool to capture this dynamic.

An interesting aspect of anankastic possibility is the way its use is suggestive of a dialogical, and potentially dialectical, configuration. When the telos is positive, as in (71.d), anankastic possibility can become relevant in the refutation of a presumed *necessary condition* of the telos (e.g. presumably, one has to be sober to work well). When the telos is negated, as in (71.a), the possibility modal can become relevant as a refutation of a presumed *sufficient condition* of the underlying positive telos (e.g. presumably, if the price goes up 100% he will sell to reap the profits).

This phenomenon is possibly part of a broader tendency of possibility modals suggest a dialogical configuration or, to use a term commonly used in French linguistic studies of argumentation, function as markers *polyphony*. It seems plausible to trace back this tendency to the weak informative content of the core semantics of possibility: mere logical compatibility of p with the conversational background B. It is clear that one situation where such a weakly informative utterance can be communicatively relevant is whenever p has been previously ruled out (in view of p), or, more generally, a negative presumption against p has been established. Hence, the polyphonic vocation of possibility modals.

This is, however, only part of the story. Despite their weak basic semantics possibility modals become stronger as their conversational background is further restricted by additional sets of propositions, just like necessity modals become weaker as further restrictions are added to their B, as observed by Kratzer in discussing the effect of ordering sources on possibility and necessity modals (cf. Kratzer 2012 [1981]: 40–41).

5.3.4 *Epistemic Conversational Backgrounds in Arguments*

Consistently with the procedure in (Def.1) in § 5.2.4.4, epistemic[42] conversational backgrounds can be characterized as per (Def. 14):

(Def. 14) Epistemic conversational backgrounds are composed by propositions corresponding to representations of the world w by an attitude bearer or information source relative to a certain speech or thought event, which, by default, corresponds to the deictic *origo*, the speech event.

According to this definition, epistemic modals involve the cognitive operation of *metarepresentation* in the sense of Papafragou (2000): the speaker's own representation of the world is considered *qua* representation, and not simply as a set of facts.

These representations of w, need not be actually true in w. Thus, from a logico-semantic viewpoint, my *epistemic* conversational backgrounds are actually closer to *doxastic* conversational backgrounds, than to the logician's "epistemic". Even the notion of *doxastic*, though, is too narrow.

While all *epistemic* backgrounds are composed of propositions that are representations of states of affairs by a given "source" or "store" of information not all of them are *anyone's beliefs*. Von Fintel and Gillies (2011: 113) noted that "stores of information – ship logs, interview notes, and computers – can *ceteris paribus* count as agents" for anchoring the B of epistemic modals. They craft an example where two people are playing the popular Mastermind codebreaking board-game. The game has gone through a few rounds of wrong guesses on the secret sequence of colored pegs (each guess generating a feedback from the codemaking player according to the game's rules). At some point the codebreaking players ask whether there might be two reds in the secret sequence. The codemaker answers with (72):

(72) That's right. There *might* be. (von Fintel and Gillies 2011: 121)

Obviously, it would not make sense to interpret (72) as referring to the codemaker's own beliefs as this player knows very whether there are two reds or not. In fact, here the epistemic conversational background is anchored deictically to the information stored in a peculiar semiotic artifact: the tiny red and white pegs used to communicate the results of guesses each round. As von Fintel and Gillies observe, the coder can utter (72) even if she knows there aren't two reds as long as the guess is compatible with the information that is stored in the board.

Such broad definition of epistemicity in terms of representations of the world w by an attitude bearer or information source relative to a certain speech or thought event can be more advantageous in dealing with issues of evidentiality, subjectivity vs. intersubjectivity, polyphony and point of view that arise with epistemic modals.

---

[42]The nature of epistemic modality in ordinary language and the path through which epistemic interpretations of context-dependent modals are obtained have already been discussed at some length in this chapter (§ 5.2.4), as well as in Chap. 3 (§ 3.3.3, in particular). Here, I take stock of what has emerged in the previous s and prepare the terrain for Chap. 6, which will delve deeper into the argumentative properties of epistemic readings of specific modal constructions in Italian.

As far as their logical properties are concerned, epistemic conversational backgrounds are *consistent* sets of propositions (Def. 15). A consistent conversational background ensures that the modal does not quantify over an empty set of worlds.

(Def. 15) *Consistency*: $\cap f(w) \neq \varnothing$ *Seriality*: $\exists w': wRw'$ *Axiom D*: $\Box\varphi \rightarrow \Diamond\varphi$

5.3.4.1 M-Performativity and (Inter-)Subjectivity

When their conversational backgrounds are anchored to the speech event epistemic modal expressions are *m-performative* (Faller 2002: 211) as they involve the speaker's commitment towards the epistemic evaluation at the moment of utterance. As observed in Chap. 3, § 3.3.3, this notion of performativity originates in Nuyts (2001a: 40–41) and indicates that a linguistic element expresses "a mental act of evaluation of a state of affairs" that is being "performed here and now". According to Nuyts, this notion of performativity differs from the speech-act theoretic notion because acts of communication towards a hearer only exist "by virtue of the utterance", while in the case of epistemic expressions "the verbal act only expresses the mental act without being constitutive of it", as a mental act could happen silently, without outward expression.

The attitude bearer or information source of an epistemic conversational background can be an individual, an institutional subject, or an epistemic community, which may or may not include the speaker and hearer, or even a document or semiotic artifact carrying some sort of information. An important distinction, again derived by Nuyts (2001b: 399), is that between *subjective* and *intersubjective* epistemic conversational backgrounds: if an epistemic expression is *subjective* its conversational background is composed by evidence that is only available to the speaker, while *intersubjective* epistemic modalities are based on bodies of knowledge that are "more widely known (including to the hearer)". This is important from an argumentation perspective as conversational backgrounds can be made to correspond to bodies of recoverable premises only if they are part of the common ground between the speaker and the hearer.

Note that an epistemic modal can be at the same time *m-performative* and *intersubjective* as the propositions in the conversational background can be at the same time those at work in the here and now of the speaker's reasoning and be part of a publicly available body of knowledge. In fact, being simultaneously m-performative and intersubjective seems to be a necessary condition for a modal to become a direct argumentative indicator.

5.3.4.2 Epistemic Modality and (Non-)Propositionality Revisited

Generally speaking, epistemic modals can take as their prejacent any kind of statement: be it descriptions, interpretations, evaluations or even logically determinate statements. At the same time, for the purposes of argumentative confrontation and argumentative support and refutation, epistemic modals do not alter the nature of the underlying statement according to Freeman's (2000, 2005) typology. In contrast, with other modalities, they are, so to say, transparent or immaterial, as regards the type of statement of their prejacent: an epistemically modalized description will continue to be a description for the purposes of argumentative support, an epistemically modalized interpretation or evaluation will continue to be an interpretation or evaluation.

Consider the following examples of epistemically modalized *description* (73.a), *interpretation* (73.b) and *evaluation* (73.c):

(73.a) *Maybe* he is sleeping (WebCorp)

(73.b) Andy Burnham says it *may* be "*impossible*" for Labour to reach an agreed position on Trident. Mr Burnham, who backs renewal of the UK's nuclear weapons, said a review by shadow defence secretary Emily Thornberry would be "very difficult" (WebCorp)

(73.c) It *must* be that I'm *unworthy* of my good office (WebCorp)

The arguments that can be reasonably advanced in support of (73.a) do not differ in kind from those that can be advanced in support of its extensional prejacent 'He is sleeping' – only the burden of proof is reduced. And the same happens with (73.b) and (73.c) where the nature of the intensional prejacent remains untouched by the epistemic modal as far as being the target of argumentation is concerned. This clearly runs counter the idea that modals are intensional propositional operators, working as functors from propositions to propositions, taking the prejacent as their argument and yielding a modal proposition as value (cf. Chap. 2, §2.3.1). If that were the case, epistemic modals should be considered as functors taking any type of statement and turning it into a very special kind of *interpretation*, an intensional non-evaluative statement about the cognitive functioning of the epistemic source, i.e. the speaker in the m-performative (73.a) and (73.c), or Andy Burnham in the embedded (73.b).

In a sense, this is precisely what epistemic modals are: intensional operators lending interpretations about the relationship of entailment or compatibility of a propositions with respect to the content of the belief system of an individual or group.

Yet, this modal proposition is rarely the target of support or criticism. In the case of m-performative occurrences, one can easily understand that it is not normal for subjects to inquire and even less to argue about what they believe at the moment of utterance. But even in the case of (73.b), where the conversational background of *may* is anchored to a reported speech event (cf. § 5.2.4.4) arguments are much more likely to concern *whether it is practically impossible for Labour MPs to agree on the Trident nuclear defense system* rather than *whether this impossibility is really compatible with what Andy Burnham believed at the moment of the reported utterance*. At most, the reported speech-act could become an issue: *whether Andy Burnham really said that*, etc. In sum, arguments can strike, so-to say, *below* the epistemic modal and refer to the prejacent alethic modal proposition, or *above* it and refer to the embedding speech-act, but they have a hard time targeting the epistemic modal proposition itself.

It is clear at this point, that we have come back to the ever controversial issue of the presumed non-propositionality of epistemic modals, of which we have perhaps uncovered a new angle, concerned with the behavior of epistemically modalized utterances as standpoints in an argument. Interestingly, it seems plausible that this particular kind of non-propositionality is not strictly tied to unembedded, m-performative, occurrences of the epistemic modals, but concerns, to some extent, also occurrences anchored to a reported speech or thought event.

In this argumentative sense the non-propositionality of epistemic modalities means that the modal propositions virtually formed by the modal and its prejacent are usually not the object of argumentative confrontations and are not the target of argumentatively relevant relations of support or rebuttal, which typically target the prejacent only. The tendency to disregard epistemic modals in argumentative confrontations had already emerged in Chap. 3 (§ 3.3.2) in discussing the behavior of epistemics in dialogical contradiction:

(74.a) A: This *must* be John's car entering the driveway.
B: No. John's Trabant does not sound like a V12 Ferrari!

(74.b) A: It *might* rain tomorrow.
B: No. There is no way it is going to happen this season.
(cf. 29.a, b, Chap. 3)

As noted in Chap. 3, the critical reactions (*No*) of the antagonist B in (74.a, b) and the subsequent rebuttals (e.g. *John's Trabant does not sound like a V12 Ferrari!*) cannot target the state of A's subjective beliefs at the moment of utterance and concern the sole prejacent instead.

Robert Pinto (2001) addresses the difficulty of reducing standpoints involving what he calls "secondary doxastic attitudes" – our epistemic modalities – to "second-order meta-cognitions about first-order propositions", i.e. to interpretations about the relationship of entailment or compatibility of the prejacent with respect to the content of the belief system:

> Might one then try to construe the cases involving the secondary doxastic attitudes as concerned essentially with second order meta-cognitions about first-order propositions (e.g. one presents reasons for straightforwardly believing the proposition that the evidence for a first order proposition is counterbalanced by the evidence against it)? At least some cases can and probably should be interpreted along these lines. But even in these cases the *point* of getting someone to accept the second-order proposition is typically to modify his or her attitude toward a first order proposition. Moreover, it is far from clear that *all* attempts to modify attitudes toward first-order propositions attempt to do so by arguing for the straightforward acceptance of meta-cognitional statements (Pinto 2001: 13).

In Chap. 3, § 3.3.3, having examined the inconclusive linguistic evidence concerning the presumed non-propositionality of epistemic modals I suggested to distinguish between (a) linguistic constraints preventing individual lexical items from occurring in embedded contexts, and (b) the non-propositionality of an epistemic modality as a discourse phenomenon. The former is a linguistic semantic phenomenon, which can take the form of rigid semantico-syntactic constraints preventing embedding or softer, more probabilistic, proclivities associated with an expression, the latter is a discourse-pragmatic phenomenon, which ultimately depends, to use Pinto's word, on the *point* of the utterance. Constraints and proclivities can guide interpretation, but the propositional or non-propositional status of an epistemic expression will ultimately depend on its function in discourse and, for what concerns us here, in the dialectical and inferential structure of argument.

As anticipated in Chap. 3, I use Congruity Theory to capture the function of non-propositional and performative epistemic modalities, which are seen as contributing to the presuppositions and pragmatic effects of the *pragmatic predicate* of the discourse sequence at issue. In the argumentative contexts that concern us here, the context dependent modals in their epistemic non-propositional m-performative readings are seen as contributing directly to the expression of acts of *concluding* the prejacent from premises recoverable through the saturation of the conversational background.

As observed in Chap. 2, *pragmatic predicates* offer a relatively rich *descriptive* analysis of the relational structure of discourse at the speech-act level. They do not substitute normative models such as Pragma-Dialectic's *critical discussion*. Instead, they can be rather straightforwardly *mapped onto* the model of the critical discussion, an operation which can be considered as part of the "transformations" involved in analytical reconstruction (cf. van Eemeren and Grootendorst 2004: . Mapped onto the ideal model of the critical discussion, these acts of concluding *count as* commitments towards the advancement of the prejacent as a standpoint. It can be the main standpoint of the arguer in that particular argument, or a sub-standpoint that is meant to provide an argument for the main standpoint. Furthermore, depending on the organization of the actual discourse the act of conclusion may be the only outward sign of the advancement of the standpoint or a reiteration of its expression. Finally, depending again on the discourse and on the context of interaction, the performance of an act of concluding can be seen as evidence of the realization of the *concluding stage* of a critical discussion.

5.3.4.3 Epistemic Modalities from m-Performativity to Performative Signals of Argumentative Moves

It is now time to put the pieces together and examine in detail how epistemic modals, analyzed along the lines outlined above can become indicators of argumentatively relevant speech acts in a more direct fashion than their non-epistemic counterparts. Here I will concentrate on the functioning of an m-performative, intersubjective, context dependent necessity modal as an indicator of conclusion.

The paradigm case here is represented by the English modal verb *must*. For now, I will keep the exemplification to this single verb, without contrasting it with other modal expressions of epistemic necessity. In Chap. 6, I present an extended case study of the epistemic readings of two morphologically distinct expressions of necessity in Italian – the verb forms *deve* and *dovrebbe*. This study will allow to recover, by example, a sense of how subtle differences in the semantics of modal expressions can have far reaching implications for their role as argumentative indicators. The study concerns Italian expressions, but the perceptive reader will find that it gives a good starting point for understanding similar contrasts in English. By focusing on *must* and *Conclusion*, I also forfeit to deal with the more troublesome behavior of markers of epistemic possibility in arguments. To these I will devote special attention in the review of related research I carry out in the final sections of this book.

The various shades of alethic, deontic and anankastic modality examined in previous sections express relations between the prejacent and sets of facts, values, norms, goals or non-factual end states (τέλη). These semantic relations are part of the *content level* of argumentation and may or may not be exploited inferentially through the inferential configurations activated by the different *loci.* It is thus clear that non-epistemic modals are *indirect indicators* both with respect to the argumentatively relevant semantic types of the standpoint and, once the conversational background is contextually and/or anaphorically saturated, also with respect to the selection of *loci* and the explicit and implicit premises entering the epistemic configuration.

That said, as observed already in Chap. 3, § 3.2.3, nothing ensures that the relation between the prejacent and the conversational background is exploited by the Y-structure of the argument as *locus*-relations. This is demonstrated by examples such as (75.a-c), which are reproduced from Chap. 3:

(75.a) John *can* lift 100 kg single-handed. He has an exceptionally powerful physique. (9.b, Chap. 3)

(75.b) John *can* lift 100 kg single-handed. I've seen him doing the like often enough. (9.c, Chap. 3)

(75.c) John *can* lift 100 kg single-handed. The trainer told me so. (9.d, Chap. 3)

In example (75.a) the inference relies on the intrinsic *Locus from Instrumental Cause* and the causing circumstances are, at the same time, the conversational background of the circumstantial alethic modal. In contrast, (75.b) rests on the extrinsic *Locus From Analogy* projecting inductively past occurrences onto comparable future situations. Yet, the B of the modal is not made of these past occurrences – *can* does not acquire a quantificational reading here – but remains composed of the causing circumstances of John's physique. Even more clearly, in (75.c), which is based on the complex *Locus from authority – expert opinion* subtype, does not feature a B composed of the expert's utterances, which would make it akin to a meta-representational epistemic conversational background. Also in this case the B of the modal remains causal and does not coincide with the activated *locus*-relation. Thus the indication of the *locus* by the non-epistemic modals remains a fairly indirect one.

Epistemic modals are different. As observed above, the do not affect the type of statement, as they do not really combine with the prejacent to form a modal proposition. Their indicative power concerns more directly argumentative discourse relations at the pragmatic, illocutionary level captured by Congruity Theory in terms of *connective predicate*. Compare the following examples already examined in Chap. 4, § 4.3.6.1:

(76.a) £3 for a sausage roll? It *must* have been gourmet. (cf. 54 in Chap. 4, www.telegraph.co.uk, collected via WebCorp)

(76.b) As for blurred or distorted vision, this was disabling in the ancient world, as it is today, but it *must* have been a great deal more common in a society without opticians (cf. 55 in Chap. 4, *Los Angeles Times*, collected via WebCorp)

Both (76.a) and (76.b) are based on a causal locus relation (efficient cause), but the direction of causation exploited by the inference differs: in (76.a) a likely cause is inferred from an observable effect – only gourmet quality could justify such a high price for a sausage roll – while in (76.b) it is the lack of opticians (the cause) which justifies the illation that blurred or distorted vision was more common in Biblical times than today. Example (76.a) is an inference from effect to cause, (76.b) is an inference from cause to effect. Interestingly, contrary to what happens with alethic modals, the B of *must* is not tied either to the causes or to the effects. It is tied to the *premises* of the inference. This is why we can say that they are more *direct* indicators of argumentative discourse relations.

Yet, epistemic meanings have been characterized in terms of m-performativity, rather than true performativity and it remains to be seen how one moves from the mental act of inferring of the speaker to the communication act of arguing.

Looking closely into the interactions between the semantico-ontological level, the level of mental acts and the properly pragmatic level is interesting also because it provides a testing ground for linguistic approaches to discourse maintaining that the relations between text spans that account for the coherence of an extended text *hold at different levels*. As observed by Knott (2001: 128), "one of the fundamental dimensions to have been proposed [to account for the semantics of coherence relations] distinguishes between the content of the text spans they link and those that hold between the utterances of the text spans themselves, or the belief which underlie them". To reduce the inventory of discourse relations, and to provide a monosemous semantics for polyfunctional discourse connectives it has also been proposed that the same relation, say *consequence*, can hold at the content level between cause and effect occurring in the world or at the epistemic level "between premise and conclusion in the speaker's mind" (Sweetser 1990: 80) conferring to the relation "an argumentative or rhetorical force" (Knott 2001: 128).

Yet, a serviceable account of inference in ordinary discourse cannot be content of conceiving epistemic relations as *alternative* to content-level relations, as Sweetser (1990) does, holding in a world of belief objects whose relationship with real world causality is simply "metaphorical". Theories of argument schemes based on underlying semantic relations, such as the AMT approach adopted here, by zooming into the ontological connections that instantiate the logical form of the inference, make it clear that classic *either-or* approaches to the level of discourse relations are incomplete. Argumentative relations hold *simultaneously* at different levels: the pragmatic, the inferential and the semantico-ontological.

In Congruity Theory (Rigotti 2005: 85–86) we can implement this stratified conception of the argumentative discourse relation by considering the propositional level relation of the *locus* as a *presupposition* of the Connective Predicate. The stratified structure of an argumentative connective of corresponding to the act of *Conclusion* in examples such as (76.a) or (76.b) can be therefore represented as follow:

(Def. 16) *Conclude* (Spk, Hr, X, U_{-n}, U_0):
Presuppositions:
(a) $p_1 \ldots p_n$ being the propositional contents of $U_{-1} \ldots U_{-n}$ (and $X_1 \ldots X_n$), $p_1 \ldots p_n \in CG_{Spk,Hr}$ – that is belong to the common ground between Spk and Hr;
(b) q being the propositional content of U_0, $q \notin CG_{Spk,Hr}$; $\neg q \notin CG_{Spk,Hr}$
(c) There is some ontological-semantic relation L ($\{p_1 \ldots p_n\}$, q) instantiates $\{p_1 \ldots p_n\} \Box_n\!\rightarrow q$
Pragmatic effects:
With U_0, Spk presents q as *reasonably acceptable* by Hr thanks to the premises $p_1 \ldots p_n$ presented in $U_{-1} \ldots U_{-n}$ (and $X_1 \ldots X_n$).

Presuppositions (a) and (c) in the structure of the connective predicate ensure that the material ensure that the conclusion p can be inferred: (a) deals with the *material* aspect by anchoring the propositional contents of the premises to the current common ground of the interaction, (c) regards the *procedural* aspect, where a given locus-relation L instantiates a nomic connection between the premises and the conclusion. Presupposition (b) serves to ensure that the act of conclusion is not trivial nor contradictory as neither p nor its negation already belong to the conversational background CG. As the Aristotle of the *Rhetoric* would put it (I, 1357a), people do not argue about established facts nor on what has already turned out to be false.

Presupposition (b) of the connective predicate *Conclude* clearly fits the epistemic interpretation of the modal verb *must*. Already Lauri Karttunen (1972: 12) compares utterances with the epistemic necessity modal (77.a) and naked assertions (77.b),

(77.a) John *must* have left.
(77.b) John has left.

observing that "in general one would use […] the epistemic *must* only in circumstances where it is not yet an established fact that John has left". Karttunen combines this observation with two other remarks, namely, that the utterer of (77.a) "has no first hand evidence about John's departure nor has it been reported to him by trustworthy sources" and that in (77.a) "the truth of *John has left* in some way logically follows from other facts the speaker knows and some reasonable assumptions he is willing to entertain". Karttunen then provides the following informal semantic analysis of (77.a):

(Def. 17) From the things I either know or regard as very probable, it logically follows that John has left (although I cannot report this as an established fact) (Karttunen 1972: 12)

As it will be clear from the semantic analyses proposed in Chap. 6, I believe that the old analysis proposed by Karttunen (1972) is fundamentally correct both for English *must* and for the Italian epistemic *deve* construction, which behaves very similarly to its English counterpart. In order to better understand how this kind of semantics relates to the argumentative relation of *Conclusion* in (Def. 16) it is useful to note that Karttunen incorporates in (Def. 17) three conceptually independent claims:

I. The prejacent of epistemic *must* is not "an established fact".
II. The speaker's knowledge of the prejacent of epistemic *must* does not derive from *direct observation* nor from *report*.
III. The prejacent of epistemic *must* logically follows from an *epistemic-doxastic* B including both facts known to the speaker and assumptions the speaker entertains with a variable degree of confidence.

An interpretation of (I), is offered by presupposition (c) in (Def. 16): when I draw a conclusion, the content p of the conclusion is not an established fact in the sense that neither p nor $\neg p$ are part of the common ground of the speaker and the hearer.

Another interesting "reading" of (I) is found in the analysis of epistemic *must* offered by Pieter A.M. Seuren (1985). Seuren's semantic analysis is interesting because of the closeness of its style of semantic analysis to that of Congruity Theory. For Seuren (1985: 248) the extension σ of a predicate is expressed by a set of n-uples of term extensions characterized by the *predicate conditions*. Among predicate conditions, Seuren distinguishes between *preconditions* (presuppositions) and *satisfaction conditions*, in a manner not unlike Congruity Theory's. In the notation, the two sets of predicate conditions are separated by a vertical line (|). The predicate conditions of *must* are given as follows:

(Def. 18) σ (*must*) = { ⟨p, **T**⟩ : **T** is known to be true; **T** gives rise to the question of the truth-value of p; there is no **T'** ⊃ **T** known to be true and containing either p or ¬ p | **T** |= p} (Seuren 1985: 401)

The variable p ranges over propositions, while Seuren's **T** is a "set of statements", not unlike the conversational background B. The analysis if far from complete and fully formalized and Seuren renounces to provide an account of what it means for a set of statements "to give rise to a question". It maintains, though, that T not including either p or ¬ p is a "natural condition" for it to give rise to the question of the truth-value of p. In short, for Seuren *must* presupposes that the prejacent is unsettled in relation to a maximal set of relevant statements known to be true, and asserts that the prejacent logically follows from that same set of statements known to be true. This analysis is meant to explain why it is unacceptable "to make a modal statement about something that is already known to be true or false" (Seuren 1985: 402), as demonstrated by the unacceptability of examples such as (78.a, b).

(78.a) *Murphy is guilty of embezzlement, and he must be guilty of it (Seuren 1985: 402),.

(78.b) *Murphy is guilty of embezzlement. Therefore, he must be guilty of it (Seuren 1985: 402),.

Broadly speaking, what Seuren proposes is a variant of Relative Modality. There are, however, some key differences with respect to the possible worlds machinery underlying Kratzer's classic treatment. In classic Relative Modality there is no proviso about the proposition p not being part of B and, given the equivalence of

Kratzer's "premise semantics" with regular possible-worlds semantics of modality, such a proviso would have to deal with the *problem of logical omniscience*[43], i.e. if the conversational background corresponds to what is known and p logically follows from it, then p would be also part of the conversational background.

It seems that Seuren's idea of a "set of statements" requires a representational view of the body of knowledge and the inclusion of p or lack thereof is to be interpreted in representational terms as the explicit inclusion of a mental representation. In this sense, it is possible that p – as a representation – is not included in T and, at the same time logically follows from it. The same representational view of propositions as mental objects underlies Papafragou's (2000) view of epistemic utterances as meta-representational.

Seuren (1985) does not specify exactly *whose* knowledge **T** represents. It seems, however, that his interpretation of this body of knowledge is *intersubjective* as he speaks of it in terms of the "availability" of truth-values for the statements of the set and observes that a *must* statement can be canceled by dialogical replies stating that there is insufficient knowledge and, conversely, that the denial of a *must*-statement can be "rectified by revealing new information", as in example (79).

(79) A: Alex needn't be guilty. We have no convincing proof yet.
B: I've just heard that he lied about his alibi
A: Ah, well, if that is correct, he *must* be guilty. (Seuren 1985: 402)

Thus, Seuren's intersubjective view of the conversational background of epistemic *must* is very close to the idea of the conversational background one finds in the analysis of *Conclusion* in (Def.16).

Certainly, an act of conclusion from arguments requires that the propositional content not be part of the common ground yet. It has the status of a *standpoint*, endorsed by the arguer and doubted by the antagonist. It becomes part of the common ground if the speech-act of concluding realizes its perlocutionary goal. But, let us zoom more closely on *how exactly* the act of *Conclusion* seeks to realize this perlocutionary goal. This will be important to understand properly the relation between the meaning of an epistemic *must* and the performance of the argumentative act of *Conclusion*. In particular, it will be important to understand *whose inference* is envisaged by the argumentative speech act.

We know that in argumentation reasons as put forth in order to effect *reasonable acceptance* of a standpoint by the critic, or antagonist (cf. van Eemeren and Grootendorst 2004). It is an attempt at persuading of the standpoint standpoint *because* of the support offered by the premises. We can see arguing as trying to lead somebody to assent to a standpoint, *making it follow inferentially* from arguments that are part of the common ground. This is the view proposed by Pinto (1996), when he treats arguments as "invitations to inference" and emphasizes that the spe-

[43] The *problem of logical omniscience* affecting epistemic and doxastic logics interpreted via a possible worlds semantics arises because under these logics knowers know all the logical consequences of their knowledge. Whether this obviously unrealistic feature truly is a problem or an acceptable idealization depends very much on the intended applications of the logic.

cific goal of argumentation is "to effect an *inference* in the person to whom it's addressed" and not simply "to effect acceptance of its conclusion" (Pinto 1996: 168). What counts, in the end, is the inference of the reasonable critic, that is of the addressee. Thus, we can rewrite more perspicuously the illocutionary pragmatic effect of the *Conclude* pragmatic predicate as follows:

(Def. 19) *Conclude* (Spk, Hr, X, U_{-n}, U_0):
Presuppositions:
As per (Def. 16)
Pragmatic effects:
With U_0, Spk invites Hr to infer q making it logically follow from the premises $p_1 \ldots p_n$ presented in $U_{-1} \ldots U_{-n}$ (and $X_1 \ldots X_n$).

It is now clear where linguists get it wrong when they reduce relations in argumentative discourse to relations between premise and conclusion *in the speaker's mind* (cf. Sweetser1990: 80). It is also clear what is the gap between the *m-performativity* of the epistemic modal and the true *performativity* of the connective predicate *Conclude.* A mental act of inference by the speaker, albeit simultaneous with the utterance, is, in principle, independent from it. In contrast, the act of presenting a proposition as the conclusion of an argument, inviting the addressee to infer it from the premises, is a communication act that can be realized only *through* the utterance.

If we go back to back to claim (II) of Karttunen's analysis of *must* we can see that it primarily concerns the inferences *of the speaker*, not of the hearer. The prejacent of must has not been acquired through direct observation nor through the discourse of a source, it is not known *to the speaker* independently from the m-performative act of inference. In linguistic semantics there is a standard term for meaning of constraints of the kind associated with epistemic must: *evidentiality* – *evidential* being the term designating morphological or lexical expressions systematically encoding the source of the speaker's knowledge of the propositional content of the utterance. The use of *must* excludes *direct* and *reportative* evidentiality and conveying *inferential* evidentiality, as shown by the unacceptability of the sequences in (80.a, b):

(80.a) Murphy *must* be guilty of the murder. *I've seen it with my own eyes.

(80.b) Murphy *must* be guilty of the murder. *His sister told me so.

This has interesting consequences on the role of *must* as an argumentative indicator. It means that *must* is, in some ways *less* informative than a straightforward linguistic manifestation of the *Conclude* connective predicate, and, in some other ways, *more* informative. Let us see why.

Commenting on Sweetser's view of argumentative-inferential uses of causal connectives Knott (2001: 138) remarks:

> Analyzing an argumentative relation as providing information about why the *speaker* believes something does not by itself how it can have this effect [=causing the hearer to believe a proposition], at least not directly. It requires in addition the stipulation that when the hearer learns about the causal processes amongst the speaker's own beliefs, this will prompt similar processes amongst his own beliefs.

The path of inference followed by the speaker in obtaining the prejacent, once manifested, can be followed also by the addressee: arguments work in a similar way in convincing myself and in convincing the others. The "power" of arguments presented through an interpersonal dialogue is not different, in principle, from the power of reasons one evaluates in her own internal forum in an intra-personal decision process. One of the reasons why a speaker would open a window on her own inference processes is to propose them also to the addressee. In this case the *m-performative* act of inference becomes a true *performative* act of argumentation, inviting the addressee to conclude from the reasons made available.

But this is not necessarily so. Inferential evidentiality does not always correspond to interpersonal argumentation. Consider, for instance, the following example from Rigotti (2005: 89). John was sleeping on the grass in the garden. Mary wakes him up and John says:

(81) The sun is setting. I *must* have slept several hours.

Here we obviously have the presentation of an inference, but we do not have an argumentation, at least not in the interpersonal sense: he is not trying to persuade her, rather he shows what has persuaded him by presenting his internal inference process to her. Since in our constructed example Mary is in a position to know, Sean's utterance also counts as an indirect request of confirmation of inference. We might argue, perhaps, that apart from interpersonal argumentation one should recognize also the existence of a monologic or soliloquial kind of argumentation. It is clear, though, that in (81) *must* does not translate into the connective predicate *Conclude*, as defined in (Def. 19), falling short of being an invitation to inference for the addressee. This is why, considered as an argumentative indicator, *must* is *less informative* than a straightforward sign of the act of conclusion.

At the same time, *must* carries more specific information that a mere sign of argumentative conclusion would not. If it is true that not all indications of inferential evidentiality correspond to interpersonal arguments, it is also true that not all interpersonal arguments require inferential evidentiality. We can very well have a situation where the speaker is not directly presenting her own inferences but nevertheless she is certainly arguing. The alibi is the classic case: in a trial a suspect may provide evidence that she was in London at the time of the murder (credit card receipts of payments for hotels and restaurants, witnesses, etc.) in order to prove that she is innocent. Of course, this is not the way she *learned* that she was in London and she was innocent; as the source of this knowledge is obviously her direct experience and not the argument. What the speaker proposes to the hearer is still an inferential path (and in order to propose it the speaker has to simulate this path in her own mind). Consider also the evidential context where the use of *must* proves to be impossible:

(82.a) **1.** Murphy is guilty of the murder.
  1.1 I've seen it with my own eyes.

(82.b) **1.** Murphy is guilty of the murder.
  1.1 His sister told me so.

The lack of inferential evidentiality does not make these sequences less argumentative. They can be both analyzed as instances of single argumentation where the standpoint is supported by the argument via a complex locus from authority (cf. Chap. 2, § 2.2.3.2). In fact, what we have here is two variants of the *testimony* subtype, the testimony of the speaker in (82.a) and the testimony of a third party in (82.b). Being incompatible with direct reportative evidentiality means that *must*, as an argumentative indicator rules out acts of *Conclusion* relying on most types of locus from authority.[44] Thus, it turns out that this specific epistemic modal expression offers also an indication concerning the allowable *loci*. This indication originates because of the *evidential* constraints associated with the modal expression.

This is why an epistemic *must*, as an argumentative indicator is in certain respect *less* informative than a direct sign of the argumentative act *Conclusion* and in other respects more informative. On the one hand, the m-performative act of inference is compatible also with moves other than the presentation of a *Conclusion* in an interpersonal argument. On the other hand, when *must* does signal a conclusion it restricts the range of possible loci.

5.3.4.4 Epistemic Modality and Evidentiality

In Chap. 6, I will discuss the role of two constructions based on the Italian modal verb *dovere* as indicators of argumentative relations of *Conclusion*. I will argue that an accurate analysis of their context-dependent semantics can explain the different, finely grained, constraints that they impose to the interpretation of their immediate discourse context. Looking closely at the differences between the two modal constructions will lead us to clarify how semantic constraints concerning underlying content-level relations – in particular *causal* relations – and semantic constraints concerning the type of evidentiality conspire to make each construction a functionally very specific argumentative indicator, associated with certain types of standpoint, a certain range of loci, and certain typical discourse contexts.

In the concluding section of this book, I summarily report further research on the verbal and adverbial expressions of epistemic modality and evidentiality in Italian and English. This ongoing research which is being carried out in collaboration with students and colleagues uncovers for each expression fine grained constraints on the kinds of inferential relations they can support, and, more broadly on their use in argumentative discourse.

These findings are in stark contrast with the rather negative assessments by argumentation scholars that I have reported at the end of Chap. 3. Freeman (2011: 120) believes that expressions such as *obviously*, *evidently*, *surely* and *apparently* modals "all have force but, at least in some cases, their cognitive meaning is sufficiently vague as to render their force the only aspect of their meaning worthy of interest". Similarly, Tseronis (2009: 54) suggests that "there is no difference in the discourse

---

[44] The expertise variant of the locus from authority remains marginally possible.

effect" of using a *modal* or an *evidential* expression when qualifying a standpoint. Tseronis adds that these expressions are to be treated equivalently because the question he is interested in is not the "lexical meaning" of the expressions but "the language user's choice to mark a high or low degree of probability/ certainty in argumentative discourse".

In fact, many lexical (and grammatical) expressions of epistemic modality in European languages such as English, French and Italian are, at the same time, carriers of subtle evidential meanings. Indeed these evidential aspects of their "lexical meanings" – jointly with the relational structure on which the past two chapters have insisted – play a big role in shaping their role in arguments and in making them more than mere indicators of a degree of force on a scale degree on a scale (e.g. force of commitment, degrees of support offered by the evidence, discrete steps in a scale of doxastic attitudes).

Before proceeding with the semantic analyses and the contextual studies of next chapter it is worth providing a thumbnail sketch of the notional-semantic category of *evidentiality*.

Customarily defined as the morphological or lexical indication of the speaker's source of knowledge (evidence) of the propositional content of the utterance (cf. Givón 1982; Dendale 1994; Squartini 2004, 2008), the category of *evidentiality* originated fairly recently from a field of inquiry far removed from the logical and philosophical analysis of language where discussions of modality have been carried out for centuries. The idea first emerged in the grammatical description of American Indian languages where the grammatical marking of the information source – typically distinguishing *direct perception, hearsay* and *inference* – is obligatory to produce a grammatically acceptable sentence. The notion and term were then used in typological linguistics to refer to the (many) languages of the world where such a grammatical system exists.

It is only recently that indicators of the source of evidence have become the object of focused attention in semantics, also in languages (such as English or Italian) were their presence is not obligatory and that their close connection with the semantic of epistemic modality has begun to be explored. Surprisingly, the role of markers of evidentiality in argumentative discourse has received relatively little attention in linguistics and discourse analysis and has been, so far, virtually ignored in the community of argumentation scholars.

Let us review the main evidential distinctions that are typically drawn by grammaticalized systems of evidentiality (Fig. 5.14).

Evidential systems (cf. Willett 1988, Aikhenvald 2007) typically distinguish between *direct*, sensory evidence and indirect evidence, which includes both *inference,* and *reports*. Finer distinctions are often drawn within each of these domains. Some evidential systems discriminate sight from other sensory evidence within *direct evidence* and, within the domain of reports, set apart *quotative* evidentiality with an overt reference of the quoted source from *hearsay* with no reference to the source. Within the realm of inference there seems to be a cross-linguistically significant distinction between *inferences based on results*, or more generally, direct observable circumstances that function as a sign of non-observable states of affairs

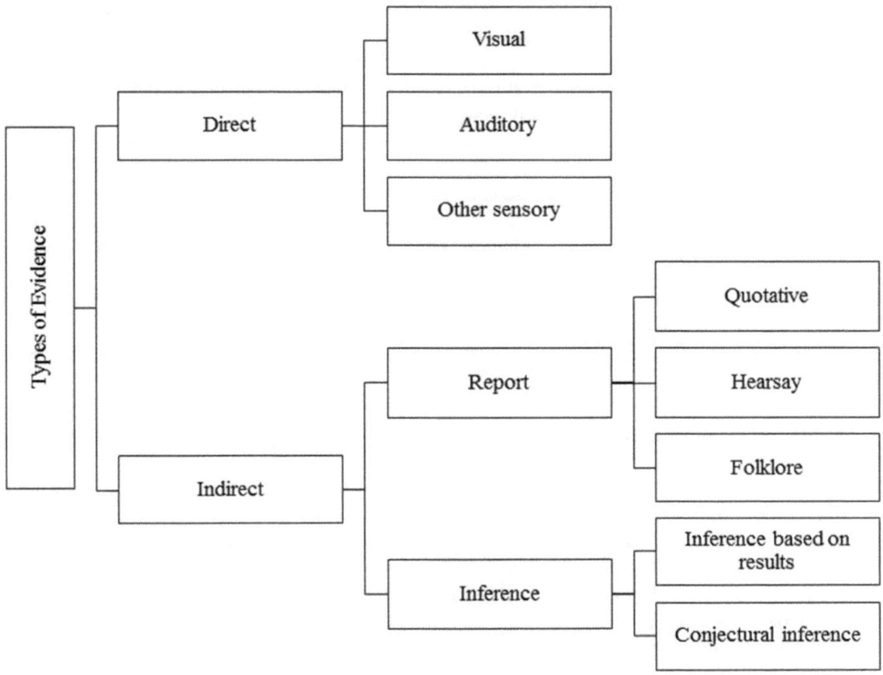

Fig. 5.14 Types of evidence relevant for grammatical systems of evidentiality (adapted from Willett 1988: 57)

and *conjectural inferences*, which are variously characterized as based on "logical reasoning", "general knowledge", or simply "non-perceptual knowledge" (cf. Willett 1988, Aikhenvald 2007, Squartini 2008, Faller 2002 and 2011).

Within a Relative Modality framework, the evidential component of modal verbs such as English *must* can be accounted for in a semantic analysis in terms of finer presuppositional restrictions on the type of propositions that make up an epistemic conversational background, by formulating restrictions that exclude direct evidence and reports. Conversely, recent work on languages with fully grammaticalized systems of evidentiality (cf. Faller 2011on Cuzco Quechua) has set out to analyze morphemes expressing direct, inferential or reportative evidentiality as different kinds of modals which impose different presuppositions concerning the evidential source on their conversational backgrounds.

References

Aikhenvald A (2007) Information source and evidentiality: what can we conclude? Rivista di Linguistica (Special Issue on Evidentiality between Lexicon and Grammar) 19(1):207–227

Asher N, Lascarides A (2003) Logics of conversation. Cambridge University Press, Cambridge

Bybee J, Perkins R, Pagliuca W (1994) The evolution of grammar: tense, aspect and modality in the languages of the world. University of Chicago Press, Chicago
Carmello M (2009) Considerazioni semantiche e pragmatiche sull'anankastico. A partire Da Maria-Elisabeth Conte. In: Venier F (ed) Tra pragmatica e linguistica testuale. Ricordando Maria-Elisabeth Conte. Edizioni dell'Orso, Alessandria, pp 205–235
Cinque G (1999) Adverbs and functional heads: a cross-linguistic perspective. Oxford University Press, Oxford
Conte M-E (2010[1995]) Epistemico, deontico, anankastico. In Venier F, Proietti D (eds) Vettori del testo. Carocci, Roma, pp 151–159 (Originally published in From pragmatics to syntax. Modality in second language acquisition, eds. Anna Giacalone Ramat and Grazia Crocco Galèas, 3–9. Tübingen: Gunter Narr)
de Haan F (2006) Typological approaches to modality. In: Frawley W (ed) The expression of modality. Mouton de Gruyter, Berlin/New York, pp 27–69
Dendale P (1994) Devoir épistémique, marqueur modal ou évidentiel? Langue française 102(1):24–40
de Saussure L (2003) Temps et pertinence. De Boeck – Duculot, Bruxelles
Faller MT (2002) Semantics and pragmatics of evidentials in Cuzco Quechua. Stanford University, Stanford
Faller MT (2011) A possible worlds semantics for Cuzco Quechua Evidentials. In: Li N, Lutz D (eds) Proceedings of SALT 20. CLC Publications, Ithaca, pp 660–683
Forbes-Riley K, Webber B, Joshi A (2005) Computing discourse semantics: the predicate-argument semantics of discourse connectives in D-LTAG. J Semant 23:55–106
Freeman JB (2000) What type of statements are there? Argumentation 14(2):135–157
Freeman JB (2005) Acceptable premises: an epistemic approach to an informal logic problem. Cambridge University Press, Cambridge
Freeman JB (2011) Argument structure: representation and theory. Springer, Amsterdam
Fried M, Östman J-O (2004) Construction grammar. a thumbnail sketch. In: Fried M, Ostman J-O (eds) Construction Grammar in a cross-language perspective. John Benjamins, Amsterdam/ Philadelphia, pp 11–87
Geurts B (1999) Presuppositions and pronouns. Elsevier, Amsterdam
Gilbert M (2013) Joint commitment. How we make the social world. Oxford University Press, Oxford/New York
Givón T (1982) Evidentiality and epistemic space. Stud Lang 6(1):23–45
Goldberg AE (1995) Constructions: a construction grammar approach to argument structure. University of Chicago Press, Chicago
Grice HP (1989) Studies in the ways of words. Harvard University Press, Cambridge, MA
Groefsema M (1995) Can, may, must and should: a relevance theoretic account. J Linguist 31:53–79
Hacquard V (2010) On the event relativity of modal auxiliaries. Nat Lang Semant 18(1):79–114
Hacquard V (2011) Modality. In: Von Heusinger M, Portner P (eds) Semantics (HSK 33.2). Mouton de Gruyter, Berlin/New York, pp 1484–1515
Hengeveld K (1988) Illocution, mood and modality in a functional grammar of Spanish. J Semant 6:227–269
Jespersen O (1924) The philosophy of grammar. Allen & Unwin, London
Karttunen L (1972) Possible and must. In: Kimball J (ed) Syntax and semantics, vol 1. Academic, New York, pp 1–20
Kaufmann S, Condoravdi C, Harizanov V (2006) Formal approaches to modality. In: Frawley W (ed) The expression of modality. Mouton de Gruyter, Berlin/New York, pp 71–105
Kay P (2004) Pragmatic aspects of grammatical constructions. In: Horn L, Ward G (eds) The handbook of pragmatics. Blackwell, Oxford, pp 675–700
Kay P, Michaelis LA (2008) Constructional meaning and compositionality. In: Maienborn C, von Heusinger K, Portner P (eds) Semantics: an international handbook of natural language meaning. Mouton de Gruyter, Berlin/New York, pp 2271–2297

Knott A (2001) Semantic and pragmatic relations and their intended effects. In: Sanders T, Schilperoord J, Spooren W (eds) Text representation: linguistic and psycholinguistic aspects. Benjamins, Amsterdam, pp 27–151

Kratzer A (1977) What "must" and "can" must and can mean. Linguist Philos 1:337–355, republished in Kratzer 2012: 4–20

Kratzer A (1981) The notional category of modality. In: Eikmeyer H-J, Rieser H (eds) Words, worlds, and contexts: new approaches in word semantics. Mouton de Gruyter, Berlin/New York (Republished in Kratzer 2012: 27–69), pp 38–75

Kratzer A (1991) Modality. In: von Stechow A, Wunderlich D (eds) Semantik/Semantics: an international handbook of contemporary research. Mouton de Gruyter, Berlin/New York, pp 639–650

Kratzer A (2012) Modals and conditionals. Oxford University Press, Oxford

Kronning H (1996) Modalité, cognition et polysémie: sémantique du verbe modal devoir. Uppsala University Press, Uppsala

Kronning H (2001a) Pour une tripartition des emplois du modal devoir. In: Dendale P, van der Auwera J (eds) Les verbes modaux. Cahiers Chronos, vol 8. Rodopi, Amsterdam, pp 67–84

Kronning H (2001b) Au-delà du déontique et de l'épistémique. In: Prandi M, Ramat P (eds) Semiotica e linguistica. Per ricordare Maria Elisabeth Conte. Franco Angeli, Milan, pp 97–118

Ludlow P (1999) Semantics, tense and time. An essay in the metaphysics of natural language. The MIT Press, Cambridge

Lycan WG (1994) Modality and meaning. Kluwer, Dordrecht

Mann WC, Thompson SA (1987) Rhetorical structure theory: a theory of text organization. Technical report. USC Information Sciences Institute

Miecznikowski J (2011) Construction types and argumentative functions of possibility modals: evidence from Italian. In: Eemeren FH, Garssen B, Godden D, Mitchell G (eds) Proceedings of the 7th conference of the International Society for the Study of Argumentation. Sic Sat, Amsterdam, pp 1284–1297

Moeschler J (1989) Modélisation du dialogue. Représentation de l'inférence argumentative. Paris, Hermes

Nauze FD (2008) Modality in typological perspective. Institute for Logic, Language and Computation, Amsterdam

Nuyts J (2001a) Epistemic modality, language, and conceptualization: a cognitive-pragmatic perspective. John Benjamins, Amsterdam/Philadelphia

Nuyts J (2001b) Subjectivity as an evidential dimension in epistemic modal expressions. Journal of Pragmatics 33(3):383–400

Nuyts J (2006) Modality: overview and linguistic issues. In: Frawley W (ed) The Expression of Modality. Mouton de Gruyter, Berlin/New York, pp 1–26

Nuyts J, Byloo P, Diepeveen J (2010) On deontic modality, directivity, and mood: the case of Dutch *mogen* and *moeten*. J Pragmat 42(1):16–34. doi:10.1016/j.pragma.2009.05.012

Palmer F (1986) Mood and modality. Cambridge University Press, Cambridge

Palmieri R (2014) Corporate argumentation in takeover bids. John Benjamins, Amsterdam/Philadelphia

Papafragou A (1998) Inference and word meaning: The case of modal auxiliaries. Lingua 105:1–47

Papafragou A (2000) Modality: issues in the semantics-pragmatics interface. Elsevier, Amsterdam

Peirce CS (1998 [1931]) Collected papers of Charles Sanders Peirce. In: Hartshorne C, Weiss P, Burks AW (eds). Harvard University Press, Cambridge, MA

Pietrandrea P (2005) Epistemic modality. John Benjamins, Amsterdam/Philadelphia

Pinto RC (1996) The relation of argument to inference. In: van Benthem J, van Eemeren FH, Grootendorst R, Veltman F (eds) Logic and argumentation. North-Holland, Oxford/New York/Tokyo

Pinto RC (2001) Generalizing the notion of argument. In: Pinto R (ed) Argument, inference and dialectic. Collected papers on informal logic. Kluwer, Dordrecht, pp 10–20

Portner P (2009) Modality. Oxford University Press, Oxford

Reichenbach H (1947) Elements of symbolic logic. Free Press, New York

Rigotti E (2005) Congruity theory and argumentation. Stud Commun Sci (Special Issue: Argumentation in Dialogic Interaction):75–96

Rigotti E (2008) Locus a causa finali. L'analisi Linguistica e Letteraria XVI(Special Issue: Word meaning in argumentative dialogue):559–576

Rigotti E (2009) Whether and how classical topics can be revived in the contemporary theory of argumentation. In: van Eemeren FH, Garssen B (eds) Pondering on problems of argumentation. Springer, New York, pp 157–178

Rigotti E, Greco Morasso S (2010) Comparing the argumentum model of topics to other contemporary approaches to argument schemes: the procedural and material components. Argumentation 24(4):489–512

Roberts C (1989) Modal subordination and pronominal anaphora in discourse. Linguist Philos 6:683–721

Rocci A (2000) L'interprétation épistémique du futur en italien et en français: une analyse procédurale. Cahiers de Linguistique Française 22:241–266

Rocci A (2005a) La Modalità epistemica tra semantica e argomentazione. I.S.U Università Cattolica, Milano

Rocci A (2005b) On the nature of the epistemic readings of the Italian modal verbs: the relationship between propositionality and inferential discourse relations. In: Hollebrandse B, van Hout A (eds) Crosslinguistic views on tense, aspect and modality. Rodopi, Amsterdam/New York, pp 229–246

Rocci A (2009) Doing discourse analysis with possible worlds. In: Renkema J (ed) Discourse, of course. An overview of research in discourse studiesi. Benjamins, Amsterdam/Philadelphia, pp 15–35

Rocci, Andrea, and Elena Musi. 2014. From modal semantics to discourse: exploring the functioning of the adjective "possibile" in a corpus of Italian financial news. In: Blochowiak J, Grisot C, Laenzlinger C, Durrlemann-Tame S (eds) Papers dedicated to Jacques Moeschler, Geneva, pp 1–21

Ross JR (1969) Auxiliaries as main verbs. In: Todd W (ed) Studies in philosophical linguistics (Series 1). Great Expectations Press, Evanston, pp 77–102

Searle JR (2005) What is an institution? J Inst Econ 1:1–22

Searle JR (1969) Speech acts. An essay in the philosophy of language. Cambridge University Press, Cambridge

Seuren PAM (1985) Discourse semantics. Blackwell, Cambridge, MA

Snoeck Henkemans F (2001) Argumentation, explanation and causality: an exploration of current linguistic approaches to textual relations. In: Sanders T, Schilperoord J, Spooren W (eds) Text representation: linguistic and psycholinguistic aspects. John Benjamins, Amsterdam/Philadelphia, pp 231–246

Squartini M (2004) Disentangling evidentiality and epistemic modality in Romance. Lingua 114(7):873–895

Squartini M (2008) Lexical vs. grammatical evidentiality in French and Italian. Linguistics 46(5):917–947

Sweetser E (1990) From etymology to pragmatics: metaphorical and cultural aspects of semantic structure. Cambridge University Press, Cambridge/New York

Talmy L (1985) Force dynamics in language and thought. In: Eilfort WH, Kroeber PD, Peterson KL (eds) Papers from the Parasession on Causatives and Agentivity. Chicago Linguistic Society, Chicago, pp 293–337

Traugott EC, Dasher RB (2002) Regularities in semantic change. Cambridge University Press, Cambridge

Tseronis A (2009) Qualifying standpoints: stance adverbs as a presentational device for managing the burden of proof. LOT, Utrecht

van der Auwera J, Plungian V (1998) Modality's semantic map. Linguist Typol 2(1):79–124 van Eemeren FH (2010) Strategic maneuvering in argumentative discourse. John Benjamins, Amsterdam/Philadelphia

van Eemeren FH, Grootendorst R (1984) Speech acts in argumentative discussions. Foris Publications, Dordrecht

van Eemeren FH, Grootendorst R (2004) A systematic theory of argumentation: the pragma-dialectical approach. Cambridge University Press, Cambridge

von Fintel K, Gillies AS (2011) Might made right. In: Egan A, Weatherson B (eds) Epistemic modality. Oxford University Press, Oxford, pp 108–130

von Wright GH (1963b) Norm and action. A logical enquiry. Routledge & Kegan Paul, London

von Wright GH (1963a) Practical inference. Philos Rev 72(2):159. doi:10.2307/2183102

Vos A (2006) The philosophy of John Duns Scotus. Edinburgh University Press, Edinburgh

Walton D (2006) Foundamentals of critical argumentation. Cambridge University Press, Cambridge

Walton D, Krabbe E (1995) Commitment in dialogue: basic concepts of interpersonal Reasoning. State University of New York Press, New York

Webber B, Stone M, Joshi A, Knott A (2003) Anaphora and discourse structure. Comput Linguist 29(4):545–587

Werner T (2005) You do what you gotta do, or why "must" implies "will." In: Georgala E, Howel J (eds) Proceedings of SALT XV, pp 294–308. http://elanguage.net/journals/index.php/salt/article/download/15.294/1766. Accessed 13 Oct 2016

Wertheimer R (1972) The significance of sense: meaning, modality, and morality. Contemporary philosophy. Cornell University Press, Ithaca/London

White AR (1975) Modal thinking. B. Blackwell, Oxford

Willett T (1988) A cross-linguistic survey of the grammaticization of evidentiality. Stud Lang 12(2):51–97

Wurmbrand S (1999) Modal verbs must be raising verbs. In: Bird S, Carnie A, Haugen J, Norquest P (eds) Proceedings of the West Coast conference on formal linguistics 18. Cascadilla, Somerville, pp 599–612

Yarza I (2000) L'attualità dell'etica di Aristotele. In: Brock SL (ed) L'attualità Di Aristotele. Armando Editore, Rome, pp 33–52

Chapter 6
Case Studies of Italian Modal Constructions in Context

Abstract The chapter examines how specific modal constructions of the Italian language in a concrete context of argumentative discussion can function as indicators for the reconstruction and evaluation of arguments put forth in ordinary discourse. The chapter examines a segment of the Italian modal system comprising different constructions with verbs *potere* 'may, can' and, especially, *dovere* 'must', which have direct or indirect epistemic interpretations. A special focus is represented the constructions DEVE_E and DOVREBBE_E based on the verb *dovere.* In these constructions the basic context-dependent relational semantics is enriched by subtle constraints on the evidential source. As a result, the two constructions end up guiding the establishing of argumentative relations of very specific kinds, differing not only in the expressed strength of support or degree of commitment towards the conclusion, but also along a series of parameters including the nature and epistemic status of the premises, the presence of rebuttals, the semantic type of the standpoint, and the argumentative *locus* invoked. At the same time, the studies in this chapter aim at understanding how the social and material ontology of the contexts in which argumentative discussions take place is reflected in the semantic structure of argumentative discourse in the area of modality. To do so, the chapter unravels how argumentation and modality intertwine in the genre of business-financial newspaper articles by focusing on acts of *prediction*, which represents a key speech-act in view of the whole functioning of the interaction field of financial communication.

6.1 Introduction to the Case Studies

The case studies in the present chapter are meant to apply the broad framework introduced in Chaps. 4 and 5 to the examination of specific modal constructions of the Italian language in a concrete context of argumentative discussion. The studies are based on earlier published analyses[1], which they expand and update, making the underlying theoretical framework more transparent in the light of the previous chapters.

[1] In particular, the analyses of the Italian modal verbs proposed herein are expanded and revised versions of those published in Rocci (2012, 2013) and reproduce passages from those publications.

A. Rocci, *Modality in Argumentation*, Argumentation Library 29,
DOI 10.1007/978-94-024-1063-1_6

Any suchlike investigation is bound to provide only very partial coverage of the phenomena of potential relevance for argumentation scholars dealing with modality in argumentative reconstruction and linguists interested in the indexing of argumentation by linguistic constructions. Additionally, here I have made a three-fold conscious choice for depth of insight instead of breadth of coverage. Let us see how.

6.1.1 Taking Advantage of Native-Speaker Competence

I have restricted myself to the Italian language, where I can better trust my native speaker intuitions in judging the acceptability of constructed examples and in appreciating the nuances of authentic ones. All the examples analyzed are duly accompanied by their English translation. Thus, readers unfamiliar with Italian will be able to follow the analyses through the English translation and glosses provided.

As I will mention from time to time, most of the phenomena I consider here do have parallels in English, often very close ones. Attentive readers will be very soon able to surmise from the translations which English modal constructions are characterized by similar semantic constraints on their functioning in arguments. While comparisons with English phenomena will be used in the discussion of certain issues, I will not venture into a systematic account of English constructions or into a contrastive analysis of the two modal systems. I am confident that this cautious policy will add to the book's rigor without detracting to its usefulness for readers who have no interest in Italian linguistic semantics or the study of discourse structure in the Italian language or in other Romance languages.

In fact, as stated in the *Introduction* (§ 6.5), the book wants to remain primarily a semantic contribution to the core of argumentation theory and the use of examples in a language other than English does not detract from that goal: it is not that Italian modal verbs and verbal morphology are *inherently more parochial* than the English adverb *probably*, which Toulmin (and others) took as paradigm example for studying modal qualifiers in arguments.

Quite the opposite. The lingering feeling of *strangeness* for the non-Italian-speaking readers will act as a healthy reminder that whatever attempt at providing an account of argumentative discourse in its semantic and pragmatic facets will inevitably rely on a language-specific sample. There is no corpus of language independent arguments showcasing only generalizable phenomena. Pretending that the English data used by the vast majority of argumentation studies offer such a corpus surely has a cost.

6.1.2 A Corpus of Italian Financial News Articles

To explore the case studies, I have chosen a homogeneous corpus of texts: journalistic texts reporting financial news and related business news. These texts are representative both of one discourse genre and of one field of interaction. As anticipated in Chap. 2, § 2.4, the discourse genre is the specialized journalistic genre of business-financial news, and the field of interaction focused on is represented by the financial domain in a broad sense, that is by the field of human activity concerned with investment decisions.

The corpus was collected as part of a broader research project entitled *Modality in argumentation. A semantic-argumentative study of predictions in Italian economic-financial newspapers*[2]. Ideally, the present book constitutes the main report of the theoretical findings of the project. The project combined the general theoretical aim of better understanding the role of modalities in arguments with the context-related goal of uncovering the characteristic hitherto largely overlooked argumentative functioning of ordinary business-financial news. The two goals were combined under the assumption that modality was key to understand how the arguments presented in financial news texts deal with the uncertainty surrounding investment activities, which are constitutively future-oriented, as well as with other physical, cognitive and normative constraints of the field. Conversely, it was hypothesized that the genre of business-financial news offered an ideal corpus for investigating modality in argumentation precisely because of these constraints.

The main results in the direction of a better understanding of the argumentative nature of business-financial news, which largely focus on the centrality of *predictions* in this genre, are published in Miecznikowski et al. (2012), Zlatkova (2012) and, in a more mature version, in Palmieri and Miecznikowski (2016). Some of these results are summarized in this chapter, in section § 6.2.

6.1.2.1 The Rationale for Focusing on *Predictions* as the Main Nexus of the Interaction of Argumentation and Modality in Business-Financial News

The main advantage of focusing on one particular genre and interaction field is that it illuminates the tight fit that can be observed between what I have called in Chap. 2, § 2.4 the *constraints* and *affordances* of the *sociopragmatics* of a certain field of interaction and the semantic structures involved in argumentation.

As stated in Chap. 2, the centrality of prediction (Merlini 1983; Bloor and Pindi 1990; Walsh 2004, 2006; Donohue 2006) emerges from the basic concern of the financial markets as a field of interaction – which is taking investment decisions –

[2] The project was supported by the Swiss National Science Foundation (Grant: 100012-120740/1) from 2008 to 2011.

and from the basic ontology of the physical world – as investment decisions concern the possibility of future returns. Speech acts about future events are so central to investment activities that the field of finance created its own indigenous speech-act label to deal with them metalinguistically: "forward looking statements" (cf. Mc Laren – Hankin 2008).

In the financial markets, the uncertainty inherent in making statements about future, and hence "unsettled", events (cf. Kaufmann et al. 2006: 99)[3] is combined with the uncertainty deriving from an incomplete knowledge of the present situation, which typically takes the form of the so-called *information asymmetry* between corporate insiders and outsiders (Barone-Adesi 2002).

Given these fundamental features of the social ontology of investment activities, the work of financial journalists is characterized by epistemic constraints of its own, as observed by (cf. Cooper and Ebeling 2007; Rocci and Luciani 2016). Financial journalists heavily rely on written corporate sources (disclosures) that are tightly regulated but also carefully designed to persuade investors and often do not have the possibility of recourse to alternative sources to check the accuracy of the corporate disclosures. The often highly technical nature of the subject matter – which combines financial technicalities with industry-specific ones – and the tight time constraints in which newsrooms operate conspire to severely limit the journalists' ability to critically scrutinize their sources. Yet, both Cooper and Ebeling (2007) and Rocci and Luciani (2016) found ethnographic evidence that business-financial journalists do try to adopt a critical stance towards their sources, despite the severe limitations of time and knowledge resources to which they are subjected.

The kind of modalities that become relevant in arguments in financial news articles depends on the predictive nature of the standpoints and, more generally on the physical and social ontology of the activity. As the case studies will show, the kinds conversational backgrounds of the modals that appear or cannot appear in these predictive contexts are tightly connected with the kind of arguments that can or cannot support a predictive standpoint – more precisely, with the kind of *topoi* that can or cannot be invoked to license a predictive inference.

Other aspects of the interaction field such as, for instance, the expertise gap between journalists and financial analysts and the reliance on corporate "forward looking" disclosures (cf. Mc Laren Hankin 2008) and guidance emanating from company insiders as well as on rumors concerning plans and objectives also have a deep impact both on the kind of modalities that become relevant and the kind of *topoi* invoked. As the case studies will show a prominent role is played by conversational backgrounds that have an *informational* component (cf. Kratzer 2012: 33) pertaining to *reportative* evidentiality or a *deontic* component pertaining to verbal commitments and formulated plans of individual or corporate subjects.

[3] See also Peirce's notion of "substantial possibility" as possibility with respect to the set of present facts and relevant laws, discussed in Chap. 4, § 4.2.6.

6.1.2.2 Corpus Design and Annotation

The corpus was designed and partially annotated by the original project team, which included Johanna Miecznikowski, Gergana Zlatkova and myself. The entire sample consists of one full month (April 2006) of three Italian business newspapers: *Il Sole 24 Ore*, *Italia Oggi* and *MF/Milano Finanza*, corresponding to roughly 4 million words.

Since these business dailies do not report only about business and finance, but include sections on largely unrelated topics such as the foreign affairs or Sunday's literary supplement, we used the running section titles used by each newspapers to exclude sections that would likely contain mostly noise for a researcher interested in exploring the journalistic genre of business and financial news.

At the same time, we did not want to restrict the sample to the point that it risked looking very much like our preconceived ideas about the business news genre. Such a corpus would not represent a good testing ground for hypotheses about genre structure, the presence of argumentation or the role of modality in that genre. For instance, if we had selected only news items dealing the stock markets and other strictly financial topics we would not be in a position to show that the broader genre of business news is dominated to a large extent by financial concerns and that business news writing tends to adopt the perspective of an investor reader– even if oftentimes the real reader is at best a potential investor, sometimes just a virtual or imaginary one. Thus a conservative approach was adopted including not only sections *exclusively* devoted to the financial markets but all sections in which articles about finance, the economy or business can appear.

The resulting broad selection still included the front pages and the top stories, where business and financial stories can be found alongside politics and other general interest news. It also included features on very specific industries (e.g. food industry, furniture) or family finances advice columns, neither of which conforms to the prototype of business-financial news. The resulting business-financial news corpus included 6512 texts and 3,087,056 running words. From this corpus we extracted a balanced sample of 200 texts (101,974 running words) from these financial sections was used to create a manually annotated sub-corpus. The composition of the 200 text sub-corpus is presented in Table 6.1, below.

The different sections of the three newspaper were represented proportionally, except for the overrepresentation of small sections, for which at least a text was selected. To offer a more vivid snapshot of the thematic composition of the sub-corpus, Table 6.2 presents the section by section detail of the 116 texts from the

Table 6.1 Composition of the 200 text annotated sub-corpus

Newspaper	N. Texts	%
Il Sole 24 Ore	116	58%
Milano Finanza - MF	46	23%
Italia Oggi	38	19%

Table 6.2 Distribution of the texts from *Il Sole 24 Ore* (N = 116) across sections

Section(*)	N. Articles	Section (*)	N. Articles
Finance & Markets	24	Top Stories	12
World & Markets	2	Furniture Industry Report	1
Finance	1	Saving & Family	1
Italian Economy	11	Annual Reports Special	7
Comments and Inquiries	6	IPO Special	1
Agricultural Industry	1	Italian Style	1
Home & Real Estate	3		
Dossier	3		
Finance and Markets Plus	14		
Job 24 Hours	1		
Jobs & Career	1		
Managers and Corporate	1		
World	10		
Nova 24	6		
Page One	9		

(*) Section titles have been translated from Italian

newspaper *Il Sole 24 Ore* – Italy's main financial daily – which were included in the 200 text annotated sub-corpus.

The annotation of the 200 texts was carried out using *UAM Corpus Tool* (O'Donnell 2008)[4], a freely available open source environment for the annotation of text corpora. The multi-layer annotation scheme used there includes tags for basic formal categories (notably including the formal structure of articles) as well as for functional discourse-semantic units, with a special reported speech, speech acts of prediction[5], and future states of affairs. Later on, for a random sample of 50 texts, Rudi Palmieri and Johanna Miecznikowski carried out an annotation of the *loci* invoked in arguments supporting predictive standpoints.

[4] The software is freely available from Mick O'Donnell (http://www.wagsoft.com/CorpusTool/).

[5] The unit of annotation chosen for predictive speech acts is the *discourse act* (cf. Béguelin 2002), or, using the terminology of Congruity Theory *minimal discourse sequence* (cf. Rigotti 1993). Theoretically, it corresponds to the minimal semiotic gesture characterized by a relatively autonomous meaning, or, equivalently, being assigned a *pragmatic predicate / connective predicate*. For purposes of delimitation it corresponds (a) to the sentence, that is to the maximal domain of binding relations or (b) to all parts of a sentence that the journalist singles out as a relatively autonomous unit of meaning via the use of punctuation. We have annotated as *predictive* those sequences that contain at least a reference to the future construed as outside the control of the speaker or of the hearer (cf. Miecznikowski et al. 2012: 75).

6.1.3 *An in-Depth Look at Specific Necessity Constructions in Italian*

In this chapter I present in-depth semantic analyses of two epistemic constructions based on the Italian necessity modal verb *dovere*. These two constructions are contrasted in relation to their ability to occur in predictive standpoints, in relation to the kind of *loci* that can support the conclusion, as well as with regard to the epistemic and pragmatic status of the premises on which the inference is based. Thus, the center stage in the chapter is occupied by lexical-constructional units whose basic quantificational force is necessity.

Possibility expressions also receive some attention. In fact, the chapter also includes a closely related investigation concerning the epistemic implications and argumentative relevance of non-epistemic readings of possibility and necessity modal verbs in the context of predictions.

Clearly, this choice leaves out linguistics units directly expressing probability and appears to eschew the thorny problem of the proper analysis of probability qualifications, which plays an important role in Toulmin's (2003) account of real "working" argument as opposed to idealized logic, and, in various guises, occupies the center stage in later treatments of modal qualification in arguments, from Freeman (1991) to Bermejo-Luque (2011). In this choice, however, there is more than avoiding to directly tackle the issue of probability. The analyses presented in this chapter will show that different lexical units whose basic quantificational force is necessity or possibility, once they enter a specific construction and have their conversational backgrounds contextually saturated end up as expressing an overall qualification corresponding to the "median probability" values[6]. They express less than necessity and more than mere possibility.

Interestingly, the two constructions based on the necessity verb *dovere* examined in this chapter can be both said to sit in the median region together with *probabilmente* ('probably'). However, they greatly differ in the precise way in which this

---

[6] It is worth noting that the practice of characterizing modalities primarily by placing them on a scale of strength is widespread well beyond the confines of argumentation theory. In applied linguistics and discourse analysis, for instance, a particularly influential proposal in this sense is the one originating from M.A.K. Halliday and Systemic Functional Linguistics, which involves a scale encompassing high, median and low modalities. According to Halliday (1994: 356): "modality is the area of meaning that lies between yes and no – the intermediate ground between negative and positive polarity". A recent elaboration of Halliday's ideas on modal qualification in discourse is found in Martin and White (2005: 13-17), who include a treatment of modal qualifying expressions in terms of probability scales in the descriptive apparatus of their Appraisal Theory. This work has become increasingly influential in discourse analysis as well as in the design of sentiment analysis software applications. Here probability is one of several scalar dimensions along which a modal expression's contribution is plotted. Three "regions of meaning" are distinguished along the probability continuum: *low*, *median* and *high*. We are not, however, speaking of probability values in any mathematical sense. For Martin and White (2005: 54–55) the probability scales correspond to an intuitively characterized judgment or "social sanction" of *veracity*.

derived median value is arrived at starting from the basic quantificational force and from the contextual saturation of the conversational background. This different path of interpretation is reflected in their divergent functioning as argumentative indicators at multiple levels (types of standpoints, range of *loci*, epistemic and pragmatic status of premises).

Thus, it turns out that the indicating function of the two constructions goes well beyond the translation of their quantificational force into an indication of the *degree of certainty* with which the prejacent proposition is put forth as a standpoint or into an indication of the *degree of strength* that the premises confer to the conclusion.

6.1.3.1 Two Epistemic Construction of the Italian Modal Verb *dovere* ('must', 'have to')

Different constructions of the of the Italian modal verbs *dovere* and *potere* can function as signals of argumentative discourse relations by recovering anaphorically co-textually available premises and relating them as arguments for the conclusion represented by the prejacent proposition. This role is most easily recognized when modals acquire an epistemic reading.

Since for these verbs the possibility of acquiring an epistemic reading is subject to conventional morphological, syntactic and semantic constraints, I will speak of *constructions* to denote the conventional configurations which the lexical unit can enter to license an epistemic saturation of the conversational background. One such form-meaning pairing, which can be realized by the modal verb *dovere* in the *present*, *imperfect* and *remote past* tenses of the *indicative* mood is exemplified by (1):

(1) Ratan Tata, presidente del gruppo indiano Tata dal 1991, è noto per aver svecchiato l'impresa di famiglia e per l'affermazione 'Niente mi stimola di più di una sfida'. **Deve** essere stato questo che ha fatto scattare un feeling istintivo tra lo stesso Tata e Sergio Marchionne, a.d. del gruppo torinese, durante le trattative che hanno portato ad un accordo commerciale tra i due gruppi in India. (*Il Sole 24 Ore*, April 19, 2006)

'Ratan Tata, chairman of the Indian group Tata since 1991 is known for having renewed this family company and for having declared: "Nothing stimulates me more than a challenge". It **must** have been this that triggered an instinctive feeling between Tata and Sergio Marchionne, CEO of the Turin based group, during the negotiations that lead to a commercial agreement between the two groups in India'.

Following the custom of linguists working on modal verbs in Romance languages I will call this form-meaning pairing DEVE_E[7]. In (1) the role of DEVE_E is not limited to expressing – in fact rather vaguely!– a certain degree of confidence lower than that of a bare, non modalized, assertion. It also contributes to establishing an

[7] The naming follows the more or less established convention in studies on Romance modal verbs of using the third person singular form of the present indicative (*deve*) adding a subscript E, which stands for *epistemic* (see, for instance, Kronning 1996; Dendale 1994).

argumentative discourse relation of *Conclusion* between two discourse sequences. Note that if we eliminate the modal the level of discourse coherence is affected:

(2) Ratan Tata, presidente del gruppo indiano Tata dal 1991, è noto per aver svecchiato l'impresa di famiglia e per l'affermazione 'Niente mi stimola di più di una sfida'. È stato questo che ha fatto scattare un feeling istintivo tra lo stesso Tata e Sergio Marchionne, a.d. del gruppo torinese, durante le trattative che hanno portato ad un accordo commerciale trai due gruppi in India. (*Il Sole 24 Ore*, April 19, 2006)
'Ratan Tata, chairman of the Indian group Tata since 1991 is known for having renewed this family company and for having declared: "Nothing stimulates me more than a challenge". It was this that triggered an instinctive feeling between Tata and Sergio Marchionne, CEO of the Turin based group, during the negotiations that lead to a commercial agreement between the two groups in India'.

In (2) it is no longer clear that the first utterance functions as a premise supporting an inferred conclusion in a partially manifested argument. Note, also, that the original version (1), DEVE_{E} truly functions as an inferential evidential, signaling that the writer is inferring that the reason why Marchionne and Tata get well together *must* be that they both like challenges.

In terms of the typology of statements introduced by Freeman (2005) the standpoint introduced by the prejacent of DEVE_{E} is an *interpretation*. More precisely, it is a *personal interpretation* explaining the occurrence of an event by appealing to "the powers, beliefs, mental activities, desires, psychological dispositions, purposes or intentions of conscious, personal agents" (Freeman 2005: 145). What is inferred is the plausible explanation of the getting along of the two managers in terms of their stable psychological dispositions. This is a highly implicit inference path resting on a number of unstated premises, which notably include:

(a) the belief that Mr. Marchionne likes challenges, which can be recovered as an inference from his record as a manager and, in particular, for the fact that he accepted the challenge of restructuring FIAT, which is highly salient in the common ground of the Italian readers of *Il Sole 24 Ore* and made more accessible by the remark on Tata restructuring his family company;
(b) some major premise supporting the inference from similarity to getting together well, which, in Aristotelian terms, is an *endoxon*[8] in the cultural common ground: *Chi si somiglia si piglia* 'Birds of a feather flock together'.

A sketchy reconstruction of the largely implicit argumentation structure of (1) is offered in Fig. 6.1, below.

[8] See Chap. 2, § 2.2.3.1. According to Aristotle "those opinions are reputable (*endoxa*) which are accepted by everyone or by the majority or by the wise—i.e. by all, or by the majority, or by the most notable and reputable of them." (*Topics* I 100b 21-23, transl. by W.A. Pickard-Cambridge, Revised Oxford Translation in Aristotle 2014).

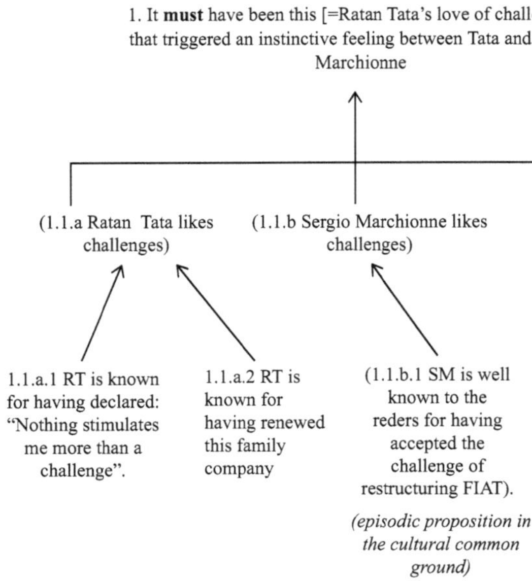

Fig. 6.1 Argumentation structure of example (1)

The inferential procedure putting together the four premises in 1.1.a-d is based on the locus *from efficient cause*. In the underlying Y-structure 1.1.a, 1.1.b and 1.1.c appear as multiple data. The endoxon slot corresponds to 1.1.d. The maxim licensing the inference has to be something like 'If x is the case and y is also the case, and y is an *obvious* efficient cause of x, then, *ceteris paribus*, y did cause x'.

Clearly, this is a weak maxim which does not have the force of *ontological necessity*, but rather corresponds to a kind of abductive reasoning which admits exceptions[9]. If the framework presented in the previous chapter is correct, though, we can maintain that there is no direct connection between the force of the maxim and the force of the modal: *deve* here does not express an alethic necessity – which would have to be immediately reflected in the nomic connection of the maxim – but an epistemic necessity in the weak sense discussed in Chap. 5 (§ 5.3.4.3): the conclusion, to paraphrase Karttunen (1972) follows from "things I either know or regard as very probable".

In the modified version in (2) the writer seems to be just *reporting* that the reason why Marchionne and Tata get well together is their common liking for challenges. It seems, therefore, that a modal like *deve*, like English *must*, can function as an indicator that the prejacent is *inferred* by the communicator from evidence available in the context, and not *independently known* from *direct experience* or *hearsay*. In the case of (1) the premises supporting the inferential operation signaled by the modal are partly identified anaphorically with the content of the preceding utter-

[9] On the variable force of maxims see Chap. 2, § 2.2.3.3.

ances in the text, partly supplied from the cultural common ground, and partly constructed on-line through wholesale presupposition accommodation of a factive construction (in the case of premise 1.1.c).

This kind of observations lead linguists (cf. Dendale 1994, Squartini 2004, Rocci 2005a, b, von Fintel and Gillies 2011) to the conclusion that certain modals verb constructions like $DEVE_E$ and its analogues in several Romance and Germanic languages function as lexical markers of *inferential* evidentiality, much like the grammaticalized evidential morphemes and particles specifying the source of knowledge of the propositional content of the utterance, which are found in many languages of the world.

In keeping with this line of thought and with the procedural analyses of modal context-dependency sketched in Chaps. 4 and 5, I hypothesize that $DEVE_E$ not only signals that the prejacent is the result of an inferential process, but also constrains the interpretation of the immediate co-text allowing the addressee to establish an argumentative relation – which can be analyzed in terms of the *Conclude* connective predicate[10] – between the utterance in which it appears and co-textually recoverable evidence. The modal functions as an evidential both *conceptually* (by restricting the source of evidence) and *procedurally* (by prompting the addressee to the recovery of evidence of the required kind in the context)[11].

As observed in Chap. 5, the hypothesis that $DEVE_E$ – and Eglish epistemic *must* – incorporate an element of *inferential* evidentiality can be corroborated by showing that that they are incompatible either with direct perceptual evidence (3) or with *reportative* evidence (4).

(3) * Giovanni **dev'**essere andato. L'ho visto che usciva.
*John **must** have gone. I've seen him leaving.

(4) ?? Giovanni **dev'**essere andato a casa. Me l'ha detto Laura.
?? John **must** have gone home. Laura told me so.

The interpretability of (4) can be saved assuming that the speaker does not trust Laura completely as a source of information and some further inference is needed to conclude that John has left. Alternatively, the first sentence in (4) can be read as a free indirect discourse, where Laura is the subject performing the inference. In both cases the possibility that the prejacent is accepted on the basis of a simple testimony is excluded.

The hypothesis $DEVE_E$ functions *procedurally* as a pointer for the addressee to recover the evidence from the situational or discourse context can be shown by

---

[10] See the analysis of this discourse relation in terms of Congruity Theory in Chap. 5, § 5.3.4.3, Deff. 16 and 19. In Congruity Theory, *connective predicates* are constructs allowing the analysis of illocutionary-level discourse relations in terms of the (presuppositional) constraints they imposes on their conjuncts as well as on discourse participants and in terms of the illocutionary effects that the interpretation of the discourse sequence achieves.

[11] On the distinction between conceptual and procedural encoding in language see the classic paper of Wilson and Sperber (recently republished as Wilson and Sperber 2012).

comparing $DEVE_E$ or epistemic *must* with other epistemic expressions, such as belief predicates (*I think, I believe, I'm sure that*), which are devoid of this procedural element:

(5.a) E' andato a casa presto. **Doveva** essere stanco.
He went home early. He **MUST-ind-imperf-3rd –sing** be tired.
He went home early. He **must** have been tired.

(5.b) E' andato a casa presto. **Sono sicuro che** era stanco.
He went home early. **I'm sure** he was tired.

While the modal in (5.a) unequivocally points anaphorically to the preceding utterance as the evidence supporting the conclusion, the belief predicate in (5.b) is more fuzzy: it expresses a subjective state of certainty of the speaker, which might derive from a variety of sources, and which may or may not include what is observed in the preceding utterance.

Apart from $DEVE_E$, there is a second construction with *dovere* that has been associated with an epistemic *probability* or *plausibility* meaning and with *inferential evidentiality*. It corresponds to the epistemic reading that the verb can acquire in the *present* of the *conditional* mood (cf. Pietrandrea 2005; Squartini 2004). A closely parallel construction exists in French (Kronning 1996, 2001; Dendale 1994, 1999, 2000, Tasmowski and Dendale 1994). Epistemic *should* in English also bears a looser, but still interesting, resemblance to it (cf. Stone 1994; Salkie 1996). Henceforth, I will call this construction $DOVREBBE_E$.

It is widely observed that, despite the obvious similarity, the meanings of $DEVE_E$ and $DOVREBBE_E$ differ in subtle ways, which are hard to pinpoint. Their distribution is largely complementary and substitution of one form for the other can result in incongruity or in sharp changes of interpretation. Consider the following examples, cited in Rocci (2011: 66)[12]:

(6.a) Ciò detto, ci dobbiamo preparare ad agire in un mercato negativamente influenzato dall' andamento fiacco del Pil, un mercato che nel 2002 **dovrebbe** calare dell'8% in Italia a 2,2 milioni di vetture [...] (Corpus IL SOLE 24 ORE).

'That said, we must be prepared to act in a market negatively influenced by the weak GDP figures, a market that in 2002 **is expected to** decrease by 8% in Italy to 2.2 million cars'

(6.b) Poco dopo la seconda guerra mondiale, ha ricordato, l'Italia ha vissuto il famoso miracolo economico, diventando in poco tempo uno dei Paesi più ricchi del mondo. **Dev'** essere scattata una molla particolare per innescare un processo di sviluppo così rapido, ha sottolineato l'economista. (Corpus IL SOLE 24 ORE)

'Shortly after WWII – he reminded – Italy witnessed the famous Economic Miracle, becoming in a short time one of the richest countries in the world. Some very particular trigger **must** have gone off that sparkled such a rapid economic development process – the economist emphasized'.

[12] These examples are not part of the main corpus used for the present Chap.. They are part of a chronologically earlier sample of the same Italian newspapers collected for Rocci (2011).

In (6.a) the substitution of $DOVREBBE_E$ with the form *deve* would create a bizarre effect of absolutely certain inevitability – which is not at all the usual meaning of $DEVE_E$: with *deve calare* the "8% decrease" of the automotive market in Italy would be somehow presented as an inevitable necessity. On the other hand, in (6.b) changing $DEVE_E$ to *dovrebbe* destroys discourse coherence, making it impossible to consider the event of the 'Economic Miracle' as evidence that 'some special trigger went off'. In other words, it becomes impossible to establish an argumentative discourse relation where the prejacent proposition is interpreted as a conclusion, and the fact mentioned in the previous utterance as a premise.

I will show that an analysis based on an instructional and constructional version of relative modality semantics can capture insightfully the subtle semantic differences behind these data, and, more importantly perhaps, can demonstrate that these differences play a crucial role in defining the functioning of the two constructions as argumentative indicators.

It will turn out that an important phenomenon to observe in order to understand the difference between $DEVE_E$ and $DOVREBBE_E$ is the impossibility of obtaining the $DEVE_E$ construction in predictive contexts such as (6.a), while $DOVREBBE_E$ not only is at ease in predictive contexts, but also occurs in predictive speech acts with significant frequency in the corpus (see § 6.5.2, below).

6.1.3.2 Broadening the View: Evidential and Argumentative Implications of Non-epistemic Modals in Predictions

It is this observation that suggested the opportunity of broadening the view by surveying the kinds of interpretation of the modals *dovere* and *potere* that are found in predictions in the corpus and examining their evidential and argumentative functioning. This reconnaissance reveals that different readings of the modals *potere* and *dovere* can function as indicators of argumentative discourse relations with respect to predictive standpoints. In contrast with $DEVE_E$, many of these uses cannot be considered *epistemic*, except in a very indirect sense. Typically, they give rise to an epistemic evaluation only as a context dependent implication on the basis of other kinds of modal meanings, such as, for instance, alethic or deontic meanings. Similarly, some of these uses can be considered to convey an *evidential* specification only as a contextual effect in discourse. Some of them do not function as evidential at all, and yet they are deeply intertwined with argumentation.

More generally, it will be argued that the different constructions examined impose different constraints on the saturation of their conversational backgrounds B and, as a result, they differ with respect to several dimensions:

(a) the level of the relation they directly signal and which, in turn, allows to infer an argumentative relation;
(b) the possible types of reasoning, argumentation schemes, *topoi, loci*, on which the argument is based;
(c) the epistemic evaluation and evidential source of the premises making up B.

Concerning (a), it is worth remembering that, according to the event-relative procedure for the contextual saturation of conversational backgrounds presented in Chap. 5, § 5.2.4.4, the relation between the conversational background B and the prejacent can correspond either to a relation between the VP-event of the prejacent and a set of circumstances of the event or to a relation between a propositional representation and a set of representations anchored to the speaker or to another source.

6.1.3.3 A Procedural and Constructional Approach to the Linguistic Semantics of Modality.

The present chapter is at the same time more contextual and more linguistic than the preceding ones. As we move from the general discussion of Relative Modality of Chap. 4 and the exploration of different kinds of conversational backgrounds in Chap. 5 to the fine grained analysis of the semantics of lexical and constructional modal expressions in Italian, it is worth delving a little deeper into the linguistic side of the semantics of highly polyfunctional, context-dependent, units like Italian modal verbs.

Since a Relative Modality approach offers a core semantics for possibility and necessity modals and a mechanism for explaining how this basic semantics combines with contextual information it becomes tempting to treat highly polyfunctional words like modal verbs as fundamentally monosemous units. According to this view, only the core semantics assigning the quantificational force is conventionally encoded by the linguistic system and the whole range of modal meanings is realized contextually through the saturation of conversational backgrounds. It is not entirely clear whether the original papers by Kratzer subscribed to this view[13], but it is explicitly espoused, among others, by Groefsema (1995), and, in a more moderate fashion, Papafragou (2000).

The approach presented in Chap. 5 differs from basically monosemic treatments because it maintains that, for many modal expressions, certain broad classes of conversational backgrounds are encoded as semiotic structure of the language, conventionally associated with the presence of certain formal or semantic constraints, while the fine grained composition of conversational backgrounds remains a contextual affair.

The approach can be qualified both as *procedural* and *constructional.* The term *constructional* comes from the theoretical trend of *construction grammars*, initiated by Charles J. Fillmore and Paul Kay (cf. Kay 2004; Kay and Michaelis 2008). In construction grammars a construction is "a form-meaning pair such that some aspect of its form or some aspect of its meaning is not strictly predictable from its

[13] It is clear, however, that Kratzer (2012) does not. In the book Kratzer endorses the idea that root and epistemic modals "differ syntactically" (Kratzer 2012: 50). Translated into the constructional terminology adopted here this means that natural languages encode them as separate.

component parts or from other constructions" (cf. Goldberg 1995: 11). Constructions can specify syntactic slots as well as a part of the lexical and morphological filling of said slots and conventionally associate them with semantic constraints at the level of the propositional content, illocutionary force and discourse function of the utterance. Specific construction inherit formal and semantic features from the lexical and morphological material they embed as well as from more abstract syntactic templates, but also feature further specifications that are conventionally associated to the whole.

There is a natural fit between constructional approaches and the procedural treatment of context dependency. Constructions can also feature contextual slots and associate them with constraints guiding the recovery of their content as in procedural approaches. As Kay (2004: 676) puts it, constructions can behave like "a virtual instruction to the addressee to examine the common ground of the conversation (along with other interpretive content of the sentence) to fill in some partially specified part of the intended interpretation". Kay and Michaelis (2008: 12) associate this behaviour with the satisfaction of presuppositions in the common ground:

> One of the areas in which constructional approaches have contributed to semantics is that of conventional implicature or pragmatic presupposition. It seems appropriate to allow the notion of compositionality to comprise these 'pragmatic' instructions embedded in the grammar that provide the addressee with a certain semantic structure and instruct him or her to find content in the context that satisfies that structure.

This is exactly the role of the B variable in the semantics of context dependent modal expressions. The procedural aspect of modal meanings is not limited to simply setting up the variable B to saturate in context, but includes *preferential paths that constrain the saturation*, such as those given in the general event-relative procedure presented in Chap. 5, § 5.2.4.4. These paths of saturations can be conventionally associated with certain formal or semantic restrictions encoded in the language.

In the spirit of construction grammars (cf. Boogart 2009), the conventionalization of preferential and marked paths of saturation for a given modal expression can be conceived *declaratively* as a set of more specific constructions inheriting the basic context-dependent semantics of the modal and adding more specific, but still very abstract, presuppositions constraining the saturation of B.

Such an approach to the meanings of modals, which I adopt here for the verb *dovere*, was largely anticipated by Hans Kronning's (1996) Cognitive Linguistics inspired "schematic network" for the French modal verb *devoir* 'must'. Kronning recognizes the utility of postulating three highly generic linguistically encoded schemata: *alethic, deontic* and *epistemic*. These schemata or templates have the semiotic status of linguistic signs, they correspond to what current construction grammar approaches would call a *construction*.

Among these schemata, the *deontic* one enjoys a sort of primacy, being, at the same time, historically and ontogenetically prior, statistically more frequent, and cognitively prototypical for French *devoir* (cf. Kronning 1996: 134- 143). Interpreted in instructional terms, this primacy would translate in the indication of a *preferential path of saturation* of the B variable[14].

The three schemata, on the one hand, correspond to salient broad classes of interpretations, which are then further specified in a number of more concrete "contextual and situational" interpretations, whose "cognitive salience and semiotic status" is hard to determine (Kronning 1996: 109). They may or may not have autonomous constructional status.

On the other hand, the three schemata are subsumed under a very abstract superschema corresponding to what he calls *apodictic necessity.* While operating in a Cognitive Linguistic framework that is very distant from Kratzer's formal semantic tradition, Kronning (1996: 30) adopts a hypothesis that is actually very close to Kratzerian "premise semantics". In fact, he defines *apodictic necessity* as necessity resulting from an "inferential composition" resulting from different types of premises. These include data which are often recoverable in the context of utterance of the verb (Kronning 1996: 33), but also systems of *nomic regularities* of different types: laws, religious norms, customs and traditions, rules of games, rules of grammar, laws of logic and mathematics, stereotypical scenarios and frames, commonplaces, topoi, etc (*Ibid.*: 34–35).

For Kronning (1996: 107) the superschema subsuming the alethic, epistemic and the deontic meanings of *devoir* is characterized by "configurational incompleteness" which needs to be saturated. This means that it correspond to a 'gappy' semantic representation rather than to a very generic or vague meaning. According to Kronning (1996), pure apodictic necessity is not something that one can mean or think organically, not even in generic or vague terms, but just a gappy template, a partial recipe for constructing meanings. In constructional terms, the three basic schemata *inherit* the features of the incomplete superschema, but each of them fills it in different, incompatible, ways.

Here I maintain that the modal *dovere* can take two main classes of conversational backgrounds: root and epistemic, defined according to the procedure provided in Chap. 5. At a lower level of conventionalization sits the distinction between the alethic and non-alethic root interpretation. Generally speaking, the most salient interpretations of *dovere* are *root* and *non-alethic*; those statistically more frequent are the deontic ones. But the accessibility of the various interpretations can change dramatically as the lexeme *dovere* combines with verbal tense and mood morphology. Individual constructions, as we will see, can be characterized by very specific instructions for the saturation of the conversational background.

[14] This does not mean that the deontic interpretation is flatly the *default* interpretation of the modal. The context still has to satisfy certain fundamental semantic requirements for a deontic interpretation to be available. See, for instance, the discussion of deontic vs. anankastic readings in Chap. 5.

6.2 Argumentation and the Genre of Financial News Articles

> The customer wants the economist to be an expert forecaster, telling that simplest and most charming of economic stories: Once upon a time there was a newspaper reader who was poor; then she read a column by a wise economist, who for some reason was giving his valuable advice to her and two million other readers; and now as a result she is rich (Mc Closkey 1990: 3)

6.2.1 The Financial Markets: Interaction Field and Discourse System

Recent work by myself and others (cf. Miecznikowski et al. 2012, Zlatkova 2012) on financial newspapers articles as a genre and as a venue of argumentative interaction showed that one can very profitably start the investigation from the interaction field of the financial markets that financial journalists seek to mediate for their readers.

Interaction fields are institutional realities that can be characterized in terms of their chief concern or legitimizing *raison d'être*, and activity types belonging to a given field serve the purposes relevant to realizing this "institutional point or rationale"(van Eemeren 2010: 129). The *raison d'être* has the status of a joint goal, broad as it may be, to which all field participants are *de iure* committed in their different capacities.

Following (Palmieri and Palmieri 2012), I start from the hypothesis that the *joint goal* of the interaction field is the creation of value through investments. There are two basic kinds of participants in the interaction, who contribute to this joint goal in very different ways: *investors* have capital at their disposal, but lack the skill, time or knowledge to increase its value, while *entrepreneurs* have potentially profitable business projects, but lack the capital to realize them. This sets up specific *participatory goals* for the two roles: provide capital and develop a business project, respectively. An investment decision is the negotiation of a commitment to this joint goal, with the participatory goals that it entails.

Obviously, joint and participatory goals are not all that there is in an interaction: participants commit to a joint goal such as the creation of value because they have their own *individual goals*. Prominent among these individual goals is the appropriation (of a part of) the value that is expected to be created. Thus, the contracts being negotiated between the participants define different kinds of commitments as regards the appropriation of value. For instance, a *bond* is a (legally enforceable) promise to repay a certain sum of money in the future.

The negotiation of contracts between investors and enterprises can be pursued directly and privately, as it happens for private companies (e.g. start-up companies that negotiate privately with a venture capitalist). This is called *direct private finance* and represents the conceptually simplest case. Alternatively, two other kinds of participants could intervene in mediating the transaction, making the interaction field more complex.

The first is represented by institutions such as *stock exchanges*, which provide – for a price – a public marketplace where the two basic participants can meet. This is called *public direct finance*. By bringing the negotiation to the public, the stock exchanges also provide an efficient mechanism for the formation of the price. Prices have an informative value: they react to news and come to embed the information available to investors at a given moment, as well as their expectations. Yet, the investment decision does not become trivial because of that. Considerable uncertainty continues to linger on investment opportunities.

As reminded by Barone Adesi (2002), this uncertainty is due both to the intrinsic incompleteness of information concerning the occurrence of future events, and to private, undisclosed, information available only to insiders (e.g. firms' management). This uncertainty creates a demand for information from investors to which the financial communication of enterprises tries to respond. Being oriented towards the decision making of investors, this communication is, for the most part, overtly or covertly argumentative.

The second kind is represented by *financial intermediaries*, such as *banks*, which, in various guises, offer their expertise and operational capacity – for a price – to investors and to enterprises to help them matching their needs either publicly or privately. In this case each of the two basic roles negotiates the investment separately with the intermediary. The actual counterpart of the investment decision could be far removed and even be made largely "invisible" by the intermediary. Consider, as an example, the very remote relationship between an investor buying a "structured product" from a bank and the company making up the so-called *underlying* of the product.

The picture of the interaction field would not be complete, however, without the inclusion of *information intermediaries*. These include *financial analysts* – both those working for institutional investors (buy-side) as well as those working for banks (sell-side), but also the specialized business and financial *news media*. Analysts intervene with their recommendations and reports to offer a (supposedly unbiased) third party evaluation for supporting investors' decisions. The texts they produce are often explicitly argumentative – they do contain arguments supporting the recommendation – and can function in themselves as *argumenta ex auctoritate* in view of investors' decisions. The media, which count analysts among their sources, have a function of information intermediation which is somewhat similar, which we examine more closely below.

The resulting picture of the interaction field can be summarized through the map in Fig. 6.2.

The map in Fig. 6.2, which draws on those in Palmieri and Palmieri (2012) and Miecznikowski et al. (2012), represents economic transaction flow with continuous lines and communication flow with dotted lines.

The next section moves the genre of business-financial news articles to show how the discourse pragmatic organisation of this genre is crucially shaped by the

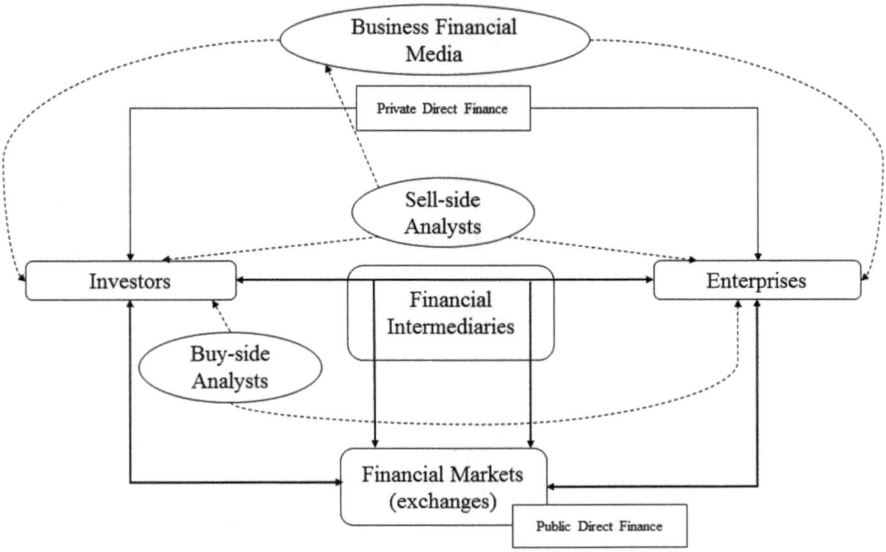

Fig. 6.2 A basic map of the interaction field of finance

basic nature of the interaction field, as it reflects the very nature of the investment decision.

6.2.2 *Prediction and the Discourse Organisation of Financial News*

Consider the headline and highlight of an article from the corpus, published on *Il Sole 24 Ore*, the main Italian business-financial daily after the publication of the quarterly financial results of British Petroleum:

(7) **Headline**:
Bp, utili in frenata ma il futuro è rosa: L'a.d. Browne: 'Il prezzo del greggio resterà alto'
'BP, earnings are slowing down [lit. braking] but the future is bright [lit. rosy]: the CEO Browne: 'The price of crude oil will remain high'
Highlight:
I risultati trimestrali sono in calo del 4% ma al di sopra delle attese del mercato.
'Quarterly results are down 4% but above the expectation of the market'
(*Il Sole 24 Ore*, April 26, 2006. Doc. 22)

This short example showcases many of the defining features of that "relatively stable" type of construction of the whole utterance (cf. Bakhtin 1986: 60) which is the genre of business-financial news.

The first striking feature is that the headline and highlight are mostly about future events rather than about something that has happened: this is not only signaled explicitly by the use of the word *futuro* and by the use of the future tense (*resterà* 'will remain'), but is also implied by the progressive aspect constructions *in frenata* and *in calo*.

It should be noted that both the Italian future tense and the progressive aspectual constructions can, and probably should, be analyzed as containing some kind of restricted alethic modality (Portner 2005: 152). The modal nature of the progressive emerges clearly in sentences like (8):

(8) Max was crossing the street, when he was hit by a bus.

Portner observes that the first clause of (8) remains true even if Max did not, in fact, cross the street. This is because "he did cross it in some non-actual possible worlds", that is in worlds where he was not interrupted by the bus. According to this analysis progressive expressions include a necessity modal operator quantifying over a set of *inertia worlds*, that is "worlds in which what was going on in a given world at a given time continues to its normal completion" (Portner 2005: 152). The prejacent of the progressive sentence is represented by a non-progressive proposition, let us follow Portner rendering it with *Max cross the road*. The progressive sentence is true just in case its non-progressive prejacent is true in all inertia worlds.

To translate this possible worlds analysis terms of Relative Modality and event-relativity – something that Portner (2005) does not do – one could envisage a double conversational background. A first conversational background has to be root and alethic and correspond to the actual circumstances B of the event participants at the time of tense *r* – past in (8) – while the second conversational background would correspond to a set B' of inertia assumptions. This second conversational background is not alethic and, as a result, neither is the combination of the two backgrounds. As it will turn out, conversational backgrounds of a similar kind will be necessary to account for the modal verb construction that constitute the main focus of this chapter. For now, suffice to say that the progressive Italian locutions *in frenata* and *in calo* project the events into the future – for an indefinite span of time – but also modalize the proposition.

In (7) the headline contains two distinct acts of prediction. The second is attributed to BP's CEO, while the first remains unattributed, so that we can attribute it, by default, to the journalist's voice. The highlight contains a reference to past acts prediction (the expectation of the market), which will be made clearer in the body of the text.

6.2.2.1 The Speech Act of Prediction as a Connective Predicate

Predicting, according to Searle and Vanderveken (1985: 186) corresponds to asserting "with the propositional content condition that the propositional content is *future* with respect to the time of utterance and the additional preparatory condition that the speaker has evidence in support of the proposition. Evidence is a special kind of reason."

According to the Austin inspired typology of speech acts proposed by Sbisà (1989), predictions are *verdictive* acts concerning the future. It presupposes a body of knowledge of the speaker, which makes him/her competent to accomplish the act and justifies his/her assertive commitment (cf. Sbisà 1989: 123–124). Contrary to the commissive and directive speech-acts concerning future events (e.g. promises and commands), in prediction the realization of the concerned event does not depend either on the speaker or the hearer.

In Congruity Theory, the minimal constraints of a prediction speech-act can be captured by the following connective predicate template:

(Def. 1) *Predict* (Spk, Hr, B, U_0):

Presuppositions:

(a) p being the propositional content of U_0, p $\notin CG_{Spk,Hr}$; $\neg$p $\notin CG_{Spk,Hr}$ – that is: neither p nor $\neg$p belong to the common ground between Spk and Hr;

(b) e_0 being the event of the utterance U_0, p denotes an eventuality e_p such that $e_p > e_0$

(c) The realization of e_p does not depend on either Spk or Hr

(d) B is some epistemic conversational background of propositions known to Spk so that B

Pragmatic effects:

With U_0, Spk commits to the truth of p on the basis of its inferrability from B

Predictions are not necessarily standpoints, or *Conclusions* in the technical sense of Chap. 5, § 5.3.4.3. When they do become standpoints the connective predicate template is enriched so that Spk commits to making the evidence in B accessible to Hr so that Hr can infer p from it. At this point the Speaker's evidence of presuppositional clause (d) becomes part of the common ground: $CG_{Spk,Hr} \supseteq B$. When predictions are standpoints in an argumentative discourse the connective predicate has the potential of establishing argumentative discourse relations with anaphorically recovered utterances (U_{-n}) and/or cataphorically anticipated ones (U_{+n}) which are the arguments that make accessible the evidence B to Hr. If this is the case U_{-n}/ U_{+n} become part of the enriched argument frame of the resulting *Conclude* connective predicate.

Since Aristotle's (*De Interpretatione*, 9) the issue of truth of future statements has puzzled philosophers. According to a commonsense view, there is an asymmetry between the past and the future. This asymmetry has both an *ontological* facet and an *epistemic* one.

From an *ontological* point of view, one could say that the occurrence of future events is indeterminate, unsettled. This non-deterministic position has been sometimes translated into a modal analysis of future statements. According to this view, future statements undergo an analysis not unlike the one discussed above for progressive aspect. This has surprising consequences on Freeman's (2005) typology of statements: if we accept a modal analysis of the future, the propositional content p of an act of *Prediction* cannot be a mere description. All predictive claims turn out to be, overtly or covertly, claims about interpretations. This ontological side of the past-future asymmetry will be further discussed in § 6.4.1.

From an *epistemic* viewpoint, people are limited in the sources of evidence they can rely on when trying to predict the future. There are thus interesting constraints on the evidence that can constitute the B in an act of prediction and, consequently, on the types of arguments that can support a prediction when it is put forth as a standpoint or conclusion.

6.2.2.2 Prediction and Types of Evidence

In order to examine these restrictions let us briefly go back to the main evidential distinctions that are typically drawn by grammaticalized systems of evidentiality (cf. Chap. 5, § 5.3.4.4) (Fig. 6.3).

With respect to this typology of sources of evidence, predictions exhibit a series of constraints, the most obvious one being the fact that future events are *not directly observable*. In terms of the *loci* available to the arguer this translates into the impossibility of recurring to one's own *testimony* (a subtype of the locus *from authority*).

Less obvious is the restriction that predictions impose on evidence from reports: it is possible to base one's prediction on what other people say about the occurrence of future events, but in this case the words of others, cannot be taken as a form of *testimony* as the impossibility of direct evidence holds also for the other subjects. Reports can merely have the value of *expert opinion*, presupposing or embedding an inference of the source. Alternatively, reports may refer to the *verbalized intentions and plans* of an agent, again embedding the agent's reasoning concerning the feasibility of the planned course of action and the agent's *commitment* to the course of action. With the latter we shift from the locus *from authority* to a variant of the locus *from promise/ warning* within the area of complex loci (cf. Chap. 2, § 2.2.3.2).

If the evidential source is inference further restrictions apply to the inferential procedures that can be invoked. Linguists working on grammaticalized evidentiality include *inference from results* as one linguistically relevant category within inferential evidentiality. It is clear that one cannot base predictions on inferences from results, because there are no observable results of future events. This has, again, a clear translation in terms of the AMT model. For certain causal loci, only maxims moving *from the cause to the effect* can be applied. This applies, to the *efficient cause* as well as to the *instrumental cause*. For instance, I can predict that the concert will not take place because the singer has a sore throat. In contrast, I cannot use the subsequent non occurrence of the concert (effect) to predict a future sore throat (cause) as the effects of future events are not observable. Similar considerations apply to the *material cause* and the *formal cause*. The absence of a raw material can be used to predict the lack of a product. It is however not possible to predict the

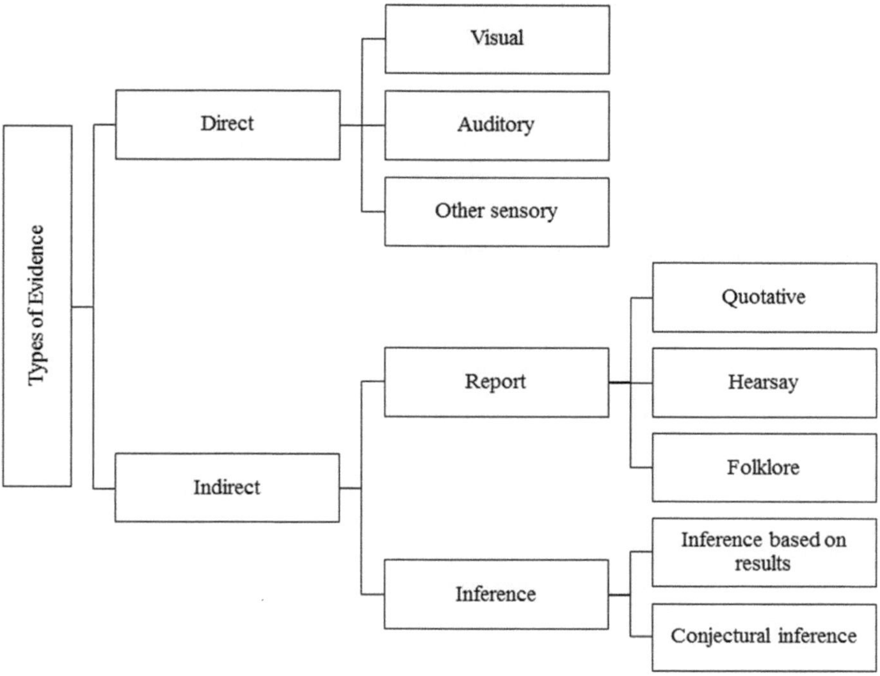

Fig. 6.3 Types of evidence relevant for grammatical systems of evidentiality (Adapted from Willett 1988: 57)

future absence of a raw material from the future lack of the product[15]. Similarly, I can move from a *project* (formal cause) to make predictions from a future *realization*, but not the other way round. The key exception to these restriction on the direction of causal maxims is represented by the locus *from final cause*: it is quite common to infer the desired goal of an action (final cause) from the observed behavior of an agent. In fact, this predictive use of the final cause is commonly encountered in the business-financial news corpus. These epistemological restrictions are summarized in Table 6.3.

As we will see in the following sections, these basic epistemological restrictions on predictions are clearly reflected in the use of modal constructions in the speech acts of prediction. In the course of this chapter, I will come back again to the nature

[15] Here it would be easy to come up with apparent exceptions, but these prima facie counterexamples do not bear closer scrutiny. For instance, someone could infer the future shortage of a raw material from the planned end of the production of a product (e.g. planned end of production of internal combustion engines predicts end of fossil fuels). In reality, this inference is not directly based on an inference from product to material cause, but on guessing the reasoning behind a human *plan*. In turn, the plan is based on a forward looking inference from (lack of) material cause to the (lack of) product. So, the exception is only apparent.

Table 6.3 Epistemological restrictions on causal loci in predictions

Locus	Admissible maxims in predictions
From efficient cause	From cause to effect maxims only
From material cause	From material to product maxims only
From instrumental cause	From instrument to action maxims only
From final cause	No restriction on the directionality of maxims

Table 6.4 Predictive speech acts in the annotated sub-corpus of Italian financial news

			Predictive minimal discourse sequences			
Section of the article	N. section tokens	N. of words by section type.	N. of m.d.s.	N. of words	N. of m.d.s. for 1000 words	Percentage of the text in N. of words
Headline	199	2,355	55	550	23.35 for 1000 w.	23.35%
Highlight	108	1,683	41	483	24.36 for 1000 w.	28.70%
Lead	185	18,366	181	4,636	9.8 for 1000 w.	25.24%
Body text	199	70,577	788	20,727	11.16 for 1000 w.	29.36%
Undecidable			1	19		
Total		92,981	1,066	26,415		28.0%

of predictions in § 6.4.1, to examine more closely the covertly intensional nature the propositional contents denoting future events.

6.2.2.3 Predictions and the genre schema of financial news articles

Let us come back to the headlines in example (7). As first observed in Miecznikowski, Rocci and Zlatkova (2012), headlines containing predictions represent the norm rather than the exception in financial news articles. This study exploited an earlier (and smaller) version of the annotated sub-corpus used in this book to explore the quantitative prominence of predictions in this genre.

It is worth presenting these data again here, updated to the latest version of the annotated 200-text sub-corpus (cf. § 6.1.2.2), which I will later use for the analysis of modal verbs as argumentative indicators in the context of predictions. Table 6.4 summarizes the results of this analysis.

As the table shows, predictive discourse sequences amount to 28% of the words of the sample and are roughly equally distributed in the different sections of an article, including those (headline, highlight, lead paragraph) that play a hierarchically prominent place in the basic genre structure of news articles. In this connection, an additional piece of data worth mentioning is that no less than 113 articles in the sample contain predictions in headline, highlight or lead paragraph.

In fact, it clearly emerges that the articles in the corpus deviate considerably from what would be expected according to the genre schema of the news story proposed by van Dijk (1988) according to which the headline and lead (and/or highlight) are devoted to the summary reporting of the main newsworthy event, while the body of the text retells this main event with more detail, expands it by presenting consequences, verbal reactions of news actors and finally provides some background or historical context to it. According, this model of textual organization *evaluations* and *future expectations* are relegated in the scarcely integrated periphery of the story and are optional.

In financial news the characteristic descending *dispositio* of news articles is maintained: the dominant discourse sequence still appears at the beginning of the text anticipated by title and lead and is then developed and complemented by subordinated discourse sequences. The schema, however, is repurposed to host a completely different kind of hierarchical structure. In this structure the pragmatic predicate *Predict* appears to play a pivotal role. As for the nature of the hierarchy, Miecznikowski, Rocci and Zlatkova (2012) defend the hypothesis that it is essentially an *argumentative* one, where arguments are subordinated to the standpoints they support.

For instance, in example (7), in the short span of the headline and highlight, we find several argumentative moves where the prediction coincides with the standpoint/ conclusion. Expressed in condensed form in the headlines, these arguments are are later expanded in the body of the article.

First, consider how the prediction that the future is bright for BP is supported argumentatively by another prediction, with the argumentative connective being signaled by the colon (:). This argumentation is an instance of the locus from the efficient cause: the high prices commanded by oil will reflect in the company's margins (assuming that costs remain constant) and thus on the earnings (assuming that costs are kept constant). The attribution of the second prediction to the CEO can be framed as a further argumentative move: an expert opinion argument from authority, but also an argument from commitment. This argumentative reading is supported by what we find in the body of the article, where "Lord Browne" is defined "the CEO of the second oil company in the world in terms of capitalization". The body of the text also offers a series further arguments in support of the prediction about oil price, which are also attributed to Browne, such as those found in (9).

(9) 'Ci aspettiamo che il prezzo del petrolio resti alto, - ha affermato ieri l'amministratore delegato Lord John Browne. - I prezzi sono saliti a causa dei problemi nelle forniture dalla Nigeria e a tensioni geopolitiche. Le grandi riserve e l'aumento della capacita di produzione Opec non sono riuscite a frenare l'ascesa del prezzo'. A sostenere il prezzo del petrolio è un'economia mondiale che resta solida, ha spiegato Browne: 'La crescita economica globale sembra robusta. Gli Usa si sono ripresi nel primo trimestre, l'Europa continua a mostrare segnali di un'accelerazione della crescita e l'Asia e l'America Latina stanno crescendo a livelli normali'

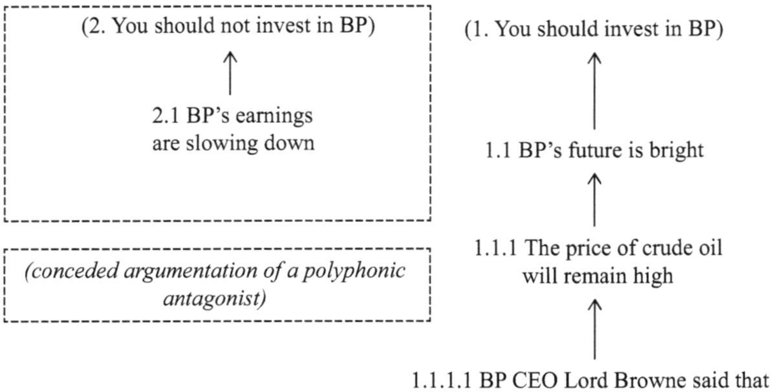

Fig. 6.4 Main argument and polyphonically conceded argument in the headline of example (7)

> We expect the oil price remains high, - he said yesterday CEO Lord John Browne. - Prices rose because of problems in supplies from Nigeria and geopolitical tensions. The great reserves and the increase in OPEC production capacity have failed to curb the price rise '. The oil price is buoyed by a world economy which remains solid, Browne explained: 'The global economic growth seems robust. The US have recovered in the first quarter, Europe continues to show signs of accelerating growth and Asia and Latin America are growing at normal levels.

As observed in Zlatkova (2012), the constant intertwining of unattributed and attributed standpoints and arguments together with the use of attribution itself as an argument *ex auctoritate* is another typical feature of these texts.

If we move back to the headline and highlight in (7) we can see that further argumentative moves can be unpacked by looking at the semantics of the connective *ma* 'but', which appears two times. According to the classic analysis (Ducrot 1980), a connective such as p *but* q, presupposes that p is an argument for a standpoint non-r, and q is an argument for a standpoint r, indicating that q is stronger than p. As it is often the case, r in our headline is implicit. If we put our text within the interaction field of finance, however, the recovery of r becomes rather obvious: it is to identify with the evaluative statement 'BP is a good investment', which is roughly equivalent to the recommendation 'You should buy BP stocks'.

In fact, this proposition directly addresses the investment decision and represents the main practical standpoint of the whole article. The slowing earnings are a potential reason not to buy BP, but a rosy perspective for the medium term (is presented as) a stronger reason to buy. In other words, use of *ma* ('but') marks the first line of argument as a concessive polyphonic representation of the point of view of an antagonist and the second line as the one endorsed by the writer. While the use of the concessive 'but' presupposes that the second argument is stronger, there is no actual refutation of the first line. This double argumentation structure is depicted in Fig. 6.4, above.

The concessive pattern outlined above repeats in the highlight with an interesting twist: BP earnings being down 4% is conceded as a reason against investing in BP,

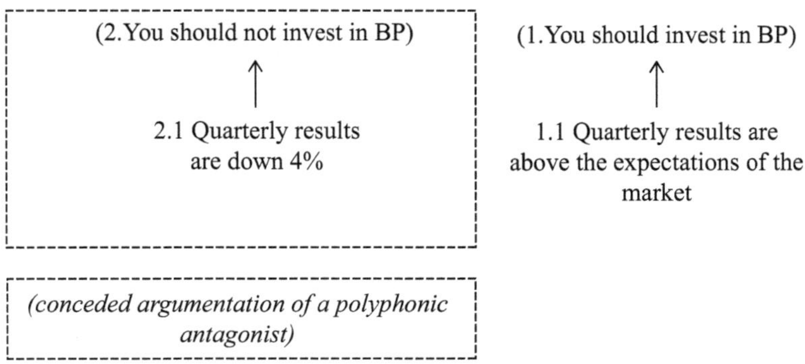

Fig. 6.5 Main argument and polyphonically conceded argument in the highlight of example (16)

but the fact that market expectations have been beaten is seen as a *stronger* reason for investing. Why it should be so?

The point of this argument may easily escape to those who do not partake of the financial interaction field, but it's deeply ingrained into its dynamic. Company earnings are the single most important predictor of returns to equities in the medium and long term and are "an important explanatory factor" also in the short term (Degeorge et al. 1999: 1). Company executives have a strong incentives to beat three kinds of threshold when reporting earnings to external audiences: the first is reporting profits instead of losses, while the other two have to do with "performance relative to widely reported firm specific values" (Degeorge et al. 1999: 3): "performance relative to the prior comparable period and relative to analysts' earnings projections" (*Ibid.*).

In our example, BP's results in the first quarter fail the first benchmark but succeed with respect to the second, which is considered more important. Market expectations concerning earnings, which are concretely embodied by the texts produced by financial analysts and by the company itself, eventually become embedded into the price of the stock. In a sense, if I had bought the stock when these expectations had become public, it means that *I have already paid* for the expected performance (appropriately discounted). So, an investment turns out to be successful when it exceeds the public expectations of the market. Figure 6.5 graphically summarizes the argumentation structure of the highlight.

From the unpacking of these two 'buts' we get an important insight, which amends the working hypothesis of financial news texts as dominated by prediction. What really dominates the discourse-pragmatic these texts is an (almost always implicit) connective predicate of practical evaluation or recommendation, directly connected with the investment decision, while the prediction is the decisive argument in support of the evaluation. In turn, the prediction is supported by various kinds of direct and indirect evidence, including arguments from authority.

Miecznikowski et al. (2012) illustrate through several corpus examples how the argumentation structure template in Fig. 6.6 can be taken as the schematic, generic argumentation structure of financial news articles.

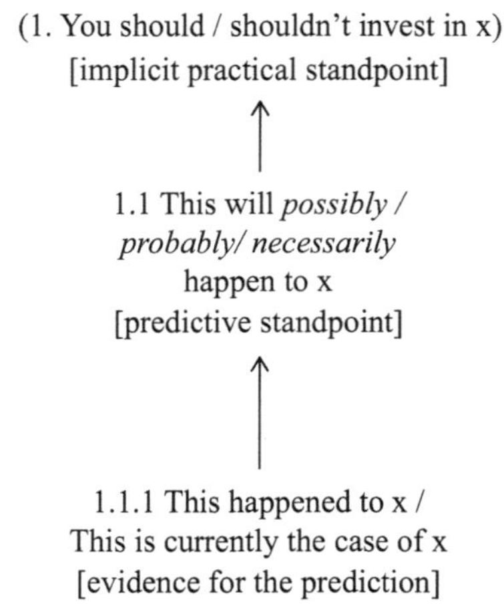

Fig. 6.6 Generic argumentation structure of financial news articles

Such an argumentative connection between prediction and practical evaluation is entirely expectable if we consider the nature of an investment decision: the evaluation of an investment by definition depends on its future returns. There is no reason to invest in a stock other than its future price and its future dividends, which, in turn, will depend on the future earnings of the company.

As observed above, the predictions appearing in the headline and highlight in (7) are resumed and developed in the body text of the article. The passage in (17), which makes up the fourth paragraph of the article, further illustrates the argumentative patterns typical of the genre:

(10) Bp ha fatto sapere ieri di essere pronta ad archiviare i problemi del 2005 e di aspettarsi un anno positivo. La nuova impennata oltre i 75 dollari a barile del prezzo del petrolio avrà un impatto positivo nei mesi a venire e il mercato si attende che i risultati annuali 2006 superino il livello record di 19,3 miliardi di dollari registrato lo scorso anno. A conferma di questo ottimismo, Bp nel primo trimestre 2006 ha speso 4 miliardi di dollari per un buy-back di 349 milioni di azioni.

BP declared yesterday they are ready to file the problems of 2005 and they expect a good year. The new surge above 75 dollars a barrel in oil prices will have a positive impact in the coming months and the market expects that the annual results of 2006 will exceed the record level of $ 19.3 billion recorded last year. To confirm this optimism, BP in the first quarter of 2006 has spent $ 4 billion for a buy-back of 349 million shares

The graph in Fig. 6.7, below, provides a first sketch of a reconstruction of the argument presented in this passage:

The implicit practical standpoint that it is worth investing in BP is supported by (1.1) the expectation that 2006 will be a good year for the oil company. For the reader, the first reason (1.1.1) for adopting such an expectation is that it emanates from BP itself, and that the company declared its commitment to it in a disclosure.

This is an ethotic argument, of the kind AMT situates in the *complex loci* class. It does not stem, however, from the authority of the expert, nor from that of the witness.

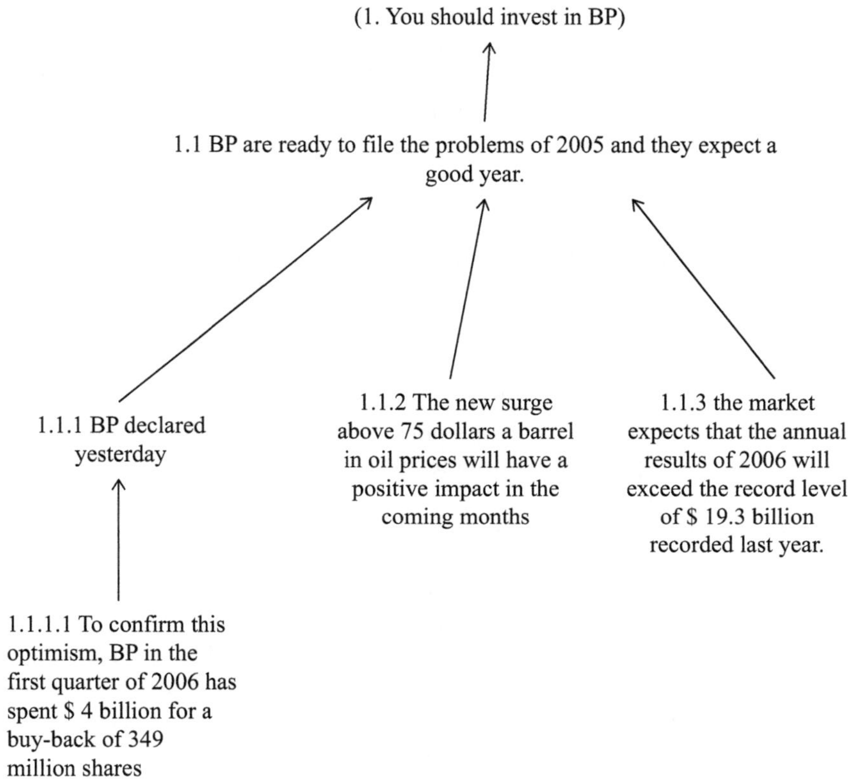

Fig. 6.7 Argumentation structure of example (10)

On the one hand, the argument is based on a credible *commitment*, more akin to the locus *from promising and warning*. I have commented above on the incentives that managers have to beat benchmarks and expectations. Conversely, managers want at all costs to avoid falling short of them. Listed companies that fall short of the earnings expectations they created in the markets will be severely punished when they have to report the actual numbers.

On the other hand, company insiders are privy to the workings of the firm and are in a *position to know* a number of factors influencing expectations that are simply inaccessible to outsiders. In this sense this is an argument from authority. In both cases, we are here in the area of the so-called complex loci.

The *sincerity* of BP in predicting a bright future for themselves is bolstered by argument (1.1.1.1). BP has put their money where their word is: a stock repurchase makes sense only if the managers believe the company is undervalued. This argument is based on a symptomatic use of the *locus from the formal cause*: the action reveals what's in the mind of the actor. This argument supports the predictive standpoint only indirectly, by supporting the ethos of BP.

In contrast, (1.1.2) is an independent causal argument (*locus from the efficient cause*) based on what I will call "economic causality" later in this chapter. With (1.1.3) we go back to indirect argumentation. Market expectations are a peculiarly strong combination of indirect loci, where the argument from expert opinion is fused with the *ad populum*.

6.2.3 Modality and Prediction in Financial News

Contrary to the received self-portrait of journalists as eschewing uncertainty in the pursuit of newsworthiness – as in the saying "If it's only worth a might, it probably isn't a story" (quoted in Coyle 2001) – bare possibility and conditional possibility modals are frequent in financial news stories. Modalization is rife in financial news and the predictions appearing in this journalistic genre frequently take the form of modalized utterances.

Remaining within piece about BP's results discussed above, for instance, we find in (9) the evidential verb *sembra* ('seem'), which Musi (2014, 2015) analyzes as an evidentially constrained relative epistemic necessity modal. In (9) *sembra* introduces the conclusion of an inference based on the locus *from whole and parts*. This conclusion, in turn, provides part of the evidence for the predictive standpoint that the oil price will remain high. This standpoint is introduced by the epistemic attitude predicate *ci aspettiamo* ('we expect'). This part of the argument in (9) is represented by the argumentation structure in Fig. 6.8.

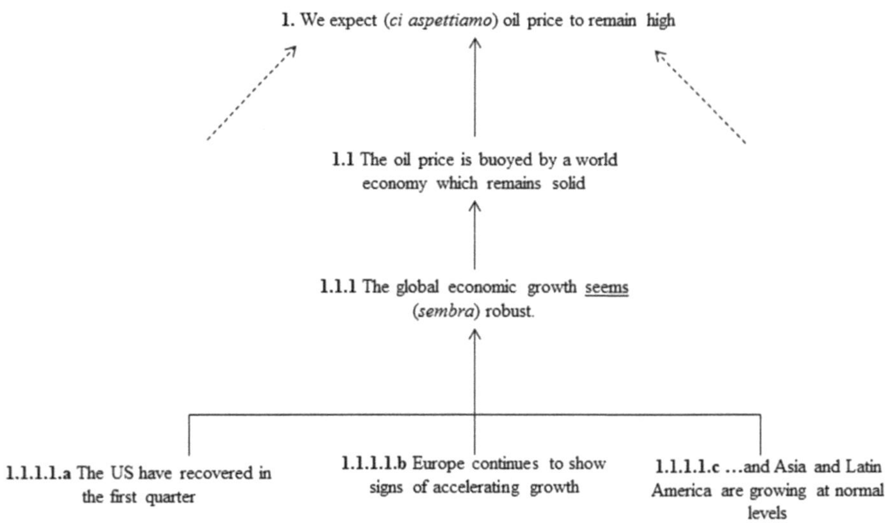

Fig. 6.8 Partial argumentation structure of example (9), showing the inference from parts to whole signaled by evidential verb *sembra*

In the same article, two very different uses of possibility modals appear in example (11). Here BP's CEO forecasts higher price of fuel as a consequence of higher oil price.

(11) Meno rosee le prospettive per i consumatori: secondo Browne il prezzo della benzina **non potrà che** salire data l'impennata del greggio. In Gran Bretagna il prezzo della benzina '**potrebbe** salire oltre una sterlina al litro', ha detto Browne, che ha pero tenuto a sottolineare che BP passa ai clienti una frazione degli aumenti dei costi reali. (*Il Sole 24 Ore*, April 14 2006).

'Less rosy outlook for consumers: according to Browne gasoline prices **cannot but** rise in view of the surge of crude oil price. In Great Britain the price of petrol 'could rise to more than one pound per liter,' said Browne, who was keen to stress however that BP passes onto its customers a fraction of the actual cost increases.'

As we will see later in this chapter the construction with the negated future tense possibility modal (*non potrà che* 'cannot but/ can only') expresses root alethic necessity, while *potrebbe* ('could'/'might'), a conditional mood possibility modal verb, expresses an epistemic modality somewhat stronger than mere possibility. Both modals, in different ways, recover the higher oil price as a premise thanks to the anaphoric functioning of their conversational backgrounds.

Before turning to quantitative data on the prominence of certain modals in predictive discourse sequences in the corpus it is useful to pause on one striking but not untypical example, which is quite revealing of the tone of these texts. The corpus example in (12) is an extract from an article dealing with the possible economic and financial consequences of the indecisive outcome of the 2006 Italian political election:

(12) ***Forse*** qualche hedge fund che si muove sui dati macroeconomici ***potrebbe*** decidere di mettersi "corto" (vendere, anche allo scoperto) sui titoli del debito italiano. '***Forse*** – aggiunge Ragazzi – perché le finanze italiane sono peggiorate e il prossimo Governo ***potrebbe*** trovarsi un buco peggiore di quello che e stato prospettato'. L'***ipotesi*** che i nostri titoli di Stato ***possano*** essere messi sotto pressione non e infatti esclusa da Mattia Nocera, a.d. di Belgrave Capital. Ma sulla Borsa non ***dovrebbe*** succedere nulla di strano: né per l'***incertezza*** legata a una maggioranza risicata, né a causa di un Governo di centro sinistra.

'***Semmai*** la maggior apertura di Prodi all'Europa ***potrebbe*** in qualche modo favorire il processo di aggregazione tra le banche e anche difendere gli interessi italiani nel Continente', sottolinea Nocera.

Maybe some hedge fund that moves on the basis of macroeconomic data might decide to "go short" (to practice shortselling) on Italian debt securities. "Maybe – Ragazzi adds – since the finances have worsened and the next government might find a hole worse that what has been foreseen." The hypothesis that our government bonds might be put under pressure is not, in fact, excluded by Mattia Nocera, CEO of Belgrave Capital. But, as far as the Stock Exchange is concerned, nothing strange should happen: nor because of the uncertainty due to its very narrow majority, nor because of a Center-Left government.

"If anything, the greater openness of Prodi towards Europe could, in some way, favour the process of consolidation among the banks and also defend Italian interests in the Continent" Nocera stresses.

The passage is notable not only because of the abundance of modal verbs (*potrebbe, possano, dovrebbe*), adverbs (*forse* 'maybe', two occurrences) and connectives (*semmai* 'if anything'), but also for the presence of nouns such as *incertezza* ('uncertainty') and *ipotesi* ('hypothesis'), which denounce, so to say, the embedded modality of the subject matter this passage is about. Epistemic concepts such as *uncertainty* and modal notions in general (*risk, opportunity, expectation*) become discourse topics commented upon.

A recent corpus study by Rocci and Musi (2014) on the Italian adjective *possibile* 'possible' confirms the importance of modalities embedded in noun phrases in this genre. In its attributive construction this modal adjective frequently forms NPs with highly generic abstract nouns (e.g. *opzione* 'option', *formula* 'formula', *alternativa* 'alternative', *sviluppo* 'development')to denote a possible state of affairs, or a set of possible states of affairs without actually manifesting the truth conditions of a prejacent proposition:

(13) Il futuro di Lehman Brothers è sempre più incerto e **gli scenari possibili** tutti aperti. (*Milano Finanza*, 12 September 2008, section mercati Globali)
'The future of Lehman Brothers is increasingly uncertain and all the possible scenarios open'

Rocci and Musi (2014: 15) observe that in example (13) the NP denotes a set of propositional objects, the 'scenarios', but the actual truth conditions of the denoted prejacent propositions (what each scenario is about) are in no way accessible to the reader. Basically, we are told that all the possible scenarios are still "open" – where 'open' means still "possible" with an arguably narrower conversational background – but we are not told what these *broadly-possible-scenarios-that-are-still-possible-also-in-a narrower-yet-unspecified-sense* are all about. The reader is left to wonder and fantasize.

The reflexive nature of finance has much to do with this discourse phenomenon: the beliefs of the market, including those that are apparently not anchored to hard information ("market sentiment") *do shape* the financial realities in the markets. So, the question of what the market believes may appear at times just as important as the question of how things are "in reality". The English example reproduced below from the *Wall Street Journal* is particularly telling in this respect:

(14) Such concerns have helped limit the dollar's losses against the euro, despite the Fed's recent moves – which once might have produced a dramatic fall in the dollar. The euro is "still essentially where we were at the start of December," *says* Simon Derrick, the London-based chief currency strategist for the Bank of New York Mellon. "That I find absolutely remarkable."
Mr. Derrick believes investors may be focused more on the risks to growth than on those posed by inflation. They "may well believe that the ECB is not being reactive enough and the euro is too highly valued," he says. ("Rate-policy shift could sap Euro", *WSJ Europe* February 1, 2008).

(14.a) BELIEVE (Derrick, MAY (FOCUS (Investors, RISK ('Economy does not grow'))))

(14.b) BELIEVE (Derrick, MAY (BELIEVE (Investors, 'The Euro is too highly valued')))

Table 6.5 Modals featuring among the highest ranking keywords of predictive speech acts in the annotated sub-corpus of Italian financial news

Rank	Wordform	Gloss	Absolute frequencies (target vs. reference)	Keyness
4	*potrebbero*	CAN cond. pres. 3rd plur.	31 vs. 2	42.54
8	*dovrebbe*	MUST cond. pres. 3rd sing.	51 vs. 5	27.99
9	*potrebbe*	CAN cond. pres. 3rd sing.	84 vs. 9	25.61
19	*potrà*	CAN ind. fut. 3rd sing.	23 vs. 4	15.78
37	*dovrà*	MUST ind. fut. 3rd sing.	11 vs. 3	7.04

The triple or quadruple embedding of modalities and epistemic attitudes displayed in (21.a,b) is "absolutely remarkable", especially if one considers that the also the deeply embedded predicate *to be valued* in (21.b) can be analyzed in terms of a modal structure involving the '*willingness* to pay a certain sum for something'.

The importance of modality in the predictive discourse sequences emerges clearly from the data of the Italian financial news corpus used for this book, in particular for the two core modal verbs *dovere* ('must', 'have to') and *potere* ('can', 'may'). The extraction of the statistically most prominent keywords of the text segments manually annotated as predictions (with respect to the rest of the annotated 200 texts sample) provides a highly suggestive picture.

I have calculated the *keyness* of words occurring in the segments annotated as 'predictions' in the 200-text sub-corpus using the remaining text of the same 200-text sub-corpus as the *reference corpus* for the calculation. The keywords of predictions are those words whose relative frequency in the 'predictions' text is significantly higher than their relative frequency in the reference corpus. The UAM Corpus Tool software. calculates the keyness of a term as the relative frequency of the term in the target corpus (predictive discourse sequences) divided by the relative frequency of the term in the reference corpus (the remaining text of the 200 newspaper articles of the annotated corpus). Relative frequency is the count of the term in the subcorpus divided by the number of terms in that subcorpus[16].

As shown in Table 6.5, there are 5 inflected wordforms of the modal verbs *potere* and *dovere* among the 50 highest ranking keywords, 3 of them among the 10 highest ranking, 4 among the top 20. The *conditional* and the *future* forms of the modals are highly prominent in the predictions that are found in our corpus.

[16]A series of filters and adjustments are then applied to the results of this formula to ensure the relevance of the extracted keywords. In particular, the keyword list generated only includes (a) words that appear more than 8 times both in the target and in the reference corpus, (b) words that appear in more than one text in the target. Additionally, the keyness of words that occur less than 20 times is reduced proportionally to how much less than 20 the count is (e.g. a word occurring 10 times will have its keyness halved).

Table 6.6 Modal adverb probabilmente 'probably' and modal adjective probabile 'probable' in predictive speech acts in the annotated sub-corpus of Italian financial news

Rank	Wordform	Gloss	Absolute frequencies (target vs. reference)	Keyness
33	*probabile*	PROBABLE sing.	7 vs. 1	7.68
35	*probabilmente*	PROBABLY	13 vs. 4	7.58

These data are particularly valuable in providing a sense of perspective on what one should pay attention to if one seeks to understand what kinds of modal expressions really do appear in a given argumentative genre. In this respect, it is useful to compare the modal verbs with what has been, since Toulmin, the prototypical example of modal qualifier.

It is clear that the straightforwardly transparent expressions of probability in Table 6.6, do not have the same incidence in predictive discourse sequences than the modal verbs in Table 6.5, whose quantificational force is ostensibly either possibility or necessity. This is a striking fact, if we pause a moment to consider it.

Economic forecasting and other forms of prediction in finance can achieve, at best, a degree of certainty in the area of probability. At the same time, economic forecasts are constructed from the outset in the language of mathematical probability. Yet, it would seems that *surprisingly* the predictions we encounter in the financial press would be rather cast in terms of *necessity* and *possibility*.

How can possibility and necessity notions function in such a context? Surely, obtaining forecasts with the modal force of *necessary* conclusions would certainly be relevant to inform decisions. Yet, this *rarely* or *never* achieved. As for *possibility*, one could raise the problem of the meaningfulness of qualifying standpoints as mere epistemic possibilities in view of the arguments and the possible alternative of conceiving a separate speech act of *raising a possibility* (cf. Chap. 3, § 3.3.2, 3.3.3).

It seems more reasonable to hypothesize that, in the end, the modals in Table 6.6 *do not* express notions such as the necessity of a conclusion in view of the known fact, nor the mere compatibility of a hypothesis with the known facts. This brings us back to the need of considering the basic quantificational force of a modal as just one of the components of its meaning and of examining the whole contextual functioning of the relational semantics of a modal marker, which has been one recurring motif of this book.

The striking prominence of the conditional forms of the possibility and necessity modal verbs deserves special attention. Three main modal uses of the Italian conditional mood are usually considered to be the following (cf. Miecznikowski and Bazzanella 2007):

(i) a *hypothetical* use, typically in the consequent of a subjunctive conditional construction, where the antecedent is evaluated as weakly possible or counterfactual;
(ii) a *reportative evidential* use indicating that the prejacent has been reported by a named or unnamed source;
(iii) an *attenuative* use where the conditional appears to function as an illocutionary force modifier downgrading the illocutionary force of the speech act.

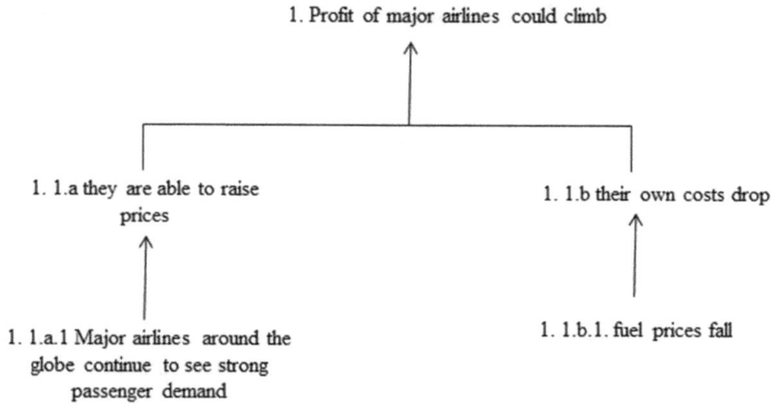

Fig. 6.9 Argumentation structure of example (15.b)

In principle, all three uses of the conditional could be relevant in financial news predictions. The relevance of (iii) is, in principle, fairly obvious as attenuation would make predictions more tentative. But, as it will be shown in the following sections through the detailed analysis of the DOVREBBE$_E$ construction, it is the uses in (i) and (ii) that turn out to be more relevant for understanding predictions.

In financial news predictions and their arguments are often expressed in conditional terms, relativized to plausible or merely possible scenarios[17]. Furthermore, journalists largely attribute predictive standpoints and their supporting argumentation to expert sources (e.g. financial analysts) and sometimes elaborate on these attributions by further indirect argumentation on the source's credibility. Consider, as a partial illustration of the features discussed above, the following English examples, taken from an article appearing in the *Wall Journal Europe* (WSJE) on September 14, 2006[18]:

(15.a) Firm's fortunes **may** rise as commodity prices fall (Headline)

(15.b) Major airlines around the globe continue to see strong passenger demand, so profit **could** climb **if** they are able to raise prices while their own costs drop as fuel prices fall. (From the body text)

The headline in (15.a) consists of a modalized conclusion supported by an argument based on a form of *locus from efficient cause*, while the passage in (15.b) presents a more developed form of the same argument where the modalized conclusion introduced by *could* holds only within a conditional frame (*if they are able to raise prices*). The argument in (15.b) is presented below in Fig. 6.9, with the antecedent of the conditional appearing as a premise, as if we were confronted with an epistemic conditional of the kind discussed in Chap. 4 (§ 4.2.8.4).

[17] See Mc Closkey (1990) on conditional predictions in the discourse of economics.

[18] The whole article contains 12 modal expressions (modal auxiliaries and modal adverbs.

The argument in (15.b) presents an interesting case of isomorphism between the structure of the argument and the conversational background of the modal appearing in the standpoint. The argument is clearly based on causal loci and the causal chain seems to be reflected in the "grounds" of the modal *could*, which appears to have a causal, circumstantial, interpretation, rather than an epistemic one: we have to do with the compatibility of an increase in profit with certain economic circumstance. Such a circumstantial modal is normally an alethic one, and premises 1.1.b, 1.1.b.1 and 1.1.a.1 can be added without any problems to this factual background.

The status of the antecedent 1.1 as a premise, however seems less straightforward than the prototypical examples of epistemic conditional discussed in Chap. 4. The continuing "stong passenger demand" (1.1.a.1) does lend some credibility to the airlines' ability to raise price, but it not a strong enough argument to consider the conditional antecedent in 1.1 simply as a premise that is added to a realistic, conversational background without changing its nature. In fact, the past/ irrealis morphology of *could* seems to signal that the conversational background of the modal contains a *lato sensu* counterfactual element.

In Chap. 3, § 3.4.4, I discussed this example in the context of Toulminian rebuttals. I argued that, just like Toulminian rebuttals correspond to *exceptive conditions* that supposedly are not the case except in extraordinary circumstances, the conditions expressed by antecedents like *if they are able to raise prices* are conversely characterized by a certain degree of expectation, so that they are supposed to hold in normal, unexceptional, circumstances. Toulminian rebuttals and normal conditions are logical duals and can be converted one in the other via external and internal negations (cf. ***Unless*** *they are* ***not*** *able to raise prices*). As shown in Chap. 3, any such condition can be easily turned in its corresponding rebuttal. In the following sections, thanks to the detailed analysis of the $\text{DOVREBBE}_{\text{E}}$ construction, I will be able to provide an integrated picture of these non-factual conditional antecedents considering their relation to the relative-modality semantics of the modal predicate and their function in argumentation.

The prominence of conditional form modals in the Italian financial news sample could also, in principle, be connected with the reportative evidential function (ii) of the Italian conditional. The explicit attribution of standpoints and arguments to expert sources – financial analysts, rating agencies, etc , to corporate insiders, as well as the direct or indirect evocation of rumors, of investor sentiment, and "analyst consensus" creates another kind of shifted discourse domain, which interacts with conditional structures and epistemic modals. This genre-specific interplay of semantic factors is briefly illustrated by another English example in (16):

(16) A reduction of that percentage to 30% **would likely** lead Standard & Poor's to raise the company's corporate credit rating to "stable" from "negative," **according to** primary credit analyst Mary Ellen Olson. (WSJE, February 13, 2007)

The relative prominence of the future forms of the modals alongside the conditional ones, might seem unsurprising and relatively uninteresting given that we deal with predictions. In fact, these data become more surprising and interesting than they seem at first blush, once one realizes that the future forms of *dovere* and *potere* cannot take epistemic conversational backgrounds (cf. Rocci 2005a: 245–255 for a detailed linguistic discussion). All these occurrences of future tense modals in predictions are based on alethic, deontic or anankastic conversational backgrounds.

In the following sections I will show that these non-epistemic future tense occurrences – and, more generally, non-epistemic occurrences of the modals with a future prejacent – on the one hand are closely tied to the underlying modal semantics of the propositional content of acts of prediction, while, on the other hand, they manifest, in an unconditional version, the conversational backgrounds on which the conditional DOVREBBE_E construction is based.

6.3 $DEVE_E$ vs. $DOVREBBE_E$: Two Kinds of Apparently Epistemic Readings for the Necessity Modal Verb

Having set the scene, let us now turn to examining the two modal constructions which occupy the center stage in the case study, that is the apparently epistemic-evidential readings that the Italian modal *dovere* can acquire (i) in the *present*, *imperfect* and *remote past*[19] tenses of the *indicative* mood (DEVE_E construction), and (ii) in the *present* of the *conditional* mood (DOVREBBE_E construction).

As illustrated by (17.a) and (17.b), they can both function as indicators of argumentative connective predicates where the prejacent is presented by the speaker as a conclusion *inferable from a set of premises* that the addressee is invited to supply, and which are at least in part recoverable from the preceding or following co-text. It becomes natural to think that these propositions are anaphorically or cataphorically *included in* or *added to* the conversational background B of the modal:

---

[19] The possibility of this reading in the remote past tense is often omitted in the semantic literature on modality in Italian as well as in the grammars, but it is perfectly natural and appears rather frequently in novels as a point of view marking device. This use is exemplified in the following discourse sequence:

Giovanni ***dovette*** prendere una scorciatoia. Perché arrivò là prima di tutti quel giorno.
Giovanni ***must***-ind-rem-pst-3rd-sing take a shortcut. Because he arrived there before everybody else that day'
Arguably, Giovanni took a shortcut. Because he arrived there before everybody else that day'.

In these remote past uses of *dovere* the inference is never situated in the past and remains anchored to the origo of the utterance (speaker and speech time). In other words, the past tense morpheme reads like a raised constituent taking scope on the prejacent alone rather than on the modal. As shown in Chap. 5, the origo (the starting reference point for deixis) can indeed shift to the past (or even to the future) when the epistemic modal is embedded under indirect reported speech or under an attitude verb. Under appropriate contextual conditions in narrative discourse the same shift can be caused by free indirect discourse (cf. Rocci 2005b: 239–259; Hacquard 2010). Interestingly, these shifted interpretations with embedding of the modal or with free indirect discourse are only possible with the imperfect form of *dovere* (*doveva*). Thus the remote past cases remain unembeddable and always anchored to the *hic et nunc* of the utterance.

(17.a) Il monte Hiei, tradotto da Antonietta Pastore (e purtroppo il suo impegnativo, accurato, lavoro **deve** aver subito l'intervento di qualche malaccorto redattore che ha fatto scempio di nomi e concetti letterari e del buddismo esoterico) è anch'esso un romanzo autobiografico. (*Il Sole 24 Ore*, 23/4/2006)

'Mount Hiei, translated by Antonietta Pastore (and unfortunately her hard, accurate work **must** have suffered the intervention of some clueless editor, who wrought havoc of names and concepts both literary and of esoteric Buddhism) also is an autobiographical novel.'

B ∪ {'Names and concepts of Buddhism etc. where messed up in the text'} □ 'The hard work of A. Pastore suffered the intervention of some clueless editor'

(17.b) [...] Uno dei barometri più incerti sul futuro dell'espansione, a detta della stessa Fed, e proprio quello del settore immobiliare. L'aumento nelle vendite di nuove abitazioni in marzo rappresenta un parziale recupero dopo le flessioni subite nei due mesi precedenti, pari al 10,9% in febbraio e al 5,5% in gennaio. Ma il settore **dovrebbe** affievolire progressivamente la propria spinta sotto la pressione dei rincari dei mutui. Il Beige Book della Fed ha notato ieri un 'raffreddamento' nel mercato residenziale. (*Il Sole-24 Ore*, 27/4/2006)

'[...] One of the most uncertain barometers of future expansion, according to the Fed itself, is precisely the real estate sector. The increase in sales of new homes in March represents a partial recovery after the slump suffered in the past two months: 10.9% in February and 5.5% in January. But the sector **should** gradually weaken its impulse under the pressure of rises in mortgages. The Fed's *Beige Book* noted yesterday a 'cooling down' in the real estate market.'

B ∪{'Mortgages prices rise'} □ 'The real-estate sector gradually weaken(future)'

In (17.a) the premise can be reconstructed from the relative clause ("who wrought havoc of names and concepts both literary and of esoteric Buddhism"), (17.b) a first premise is manifested *within the clause* by a complement featuring two connected nominalized eventualities[20] (*pressione* and *rincari*). This micro-argument[21] is closely linked with the larger discourse context. The conclusion introduced by DOVREBBE_E contrasts, thanks to the connective *ma* 'but' with the concessive

---

[20] This Chap. uses the term *eventuality* to indicate the genus of the states, processes or events (cf. Bach 1986) denoted by finite and non-finite clauses and by nominalizations (e.g. abstract nouns). Eventualities are located in time and have specific temporal properties according to their type (e.g. bounded events vs. unbounded processes). It has been frequently observed that the type of eventuality denoted by the prejacent of a modal verb, jointly with its temporal location, plays a role in restricting the interpretation of the modal (cf., for Italian, Bertinetto 1979; Pietrandrea 2005). In the present Chap. the question of the precise nature of these constraints on interpretation will remain in the background. Note, also, that this Chap. never uses the term *eventuality* in the unrelated sense of "possibility", as in, for instance, when one speaks of the *eventuality* (= possibility) reading of the French subjunctive mood.

[21] The possibility for the modals to signal sentence internal, and even clause-internal (Hobbs 2010) argumentative relations had already been observed in Chap. 5, § 5.2.1. It seems to occur, in particular, with DOVREBBE_E in contexts where nominalized eventualities appear as causal arguments for the prejacent proposition. A similar pattern is described for the conditional possibility modal *potrebbe* in Miecznikowski (2011).This kind of micro-argumentation relying on nominalized eventualities for the manifestation of its premises is, however, not the only possibility: inter sentential argumentative relations can be easily signaled by *dovere*.

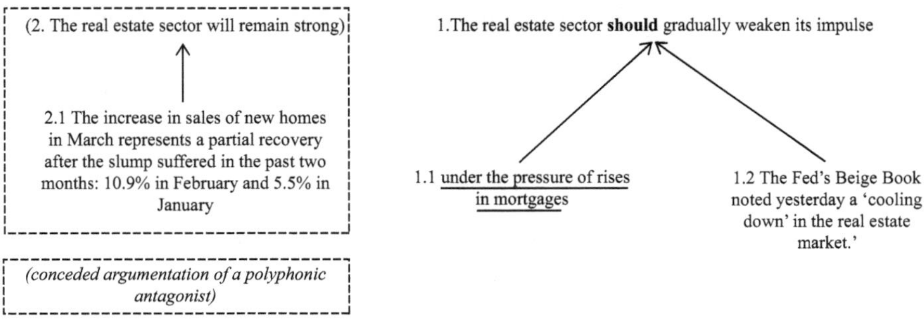

Fig. 6.10 Argumentation structure of example (17.b)

counter-argument expressed in the preceding sentence (the temporary increase in house sales), and is further reinforced by the additional premise in the following sentence (the Fed says there are already signals of cooling down) (Fig. 6.10).

At the level of the loci invoked to support the inference, (17.a) is based on the locus *from the efficient cause*, with a maxim from the effect to the cause. Argument (1.1) in example (17.b) is based on the same locus *from the efficient cause*, but, this time the inference moves from the cause to the effect. As for argument (1.2), it is based on the complex locus *from authority*.

The relative modality analyses appended to each example are not entirely satisfying. If we assume that the conversational background B is epistemic – the known facts – and that *dovere* is a necessity modal, the attitude expressed would be one of certainty (p being a necessary consequence of what is known), while both DEVE$_E$ in (17.a) and DOVREBBE$_E$ in (17.b) express inferences that are somewhat less than certain.

Furthermore, as we have seen in § 6.1.3.1, DEVE$_E$ and DOVREBBE$_E$ are in some respects very different, and most of the time cannot be substituted in a discourse passage without damaging discourse coherence, or at least radically altering the interpretation of the passage. In (17.a) substituting *deve* with *dovrebbe* would make the incidental remark totally incomprehensible unless we are willing to think that the conclusion that the editor was clueless has nothing to do with observing typos in the book, but is instead (strangely) derived by more general beliefs (e.g. *Normally, editors are clueless*).

Tasmowski and Dendale (1994: 51) examining the epistemic uses of the conditional form of French *devoir* (DEVRAIT$_E$) observe that this easily accepts to be modified by adverbs such as *normalement* ('normally') and *en principe* ('in principle'). They observe that these adverbs correspond to a condensed version of a conditional antecedent of the form 'If everything goes as expected / as arranged'. This idea of *normal conditions* plays an important part also in the analysis of DOVREBBE$_E$ proposed here.

In (17.b) the substitution of *dovrebbe* with *deve* would make the weakening of the real-estate sector an inevitable consequence of rising mortgage prices, which, incidentally, would make the sentence inconsistent with the uncertainty and the contrary signals concessively reported in the previous sentence. Interestingly, the contextually inappropriate meaning of *inevitability* that indicative *deve* would take in the manipulated passage is clearly different from the probability meaning usually associated with DEVE_{E}.

6.3.1 Constraints on Future Reference, Causality and Argumentative Connective Predicates

6.3.1.1 DEVE_{E} and FUTURE_{E}: Non-predictive Inferential Evidentials

The divergent distribution observed in the corpus and the lack of mutual substitutability manifested in (17.a) and (17.b) are partly explained by a well-known temporal-aspectual constraint on DEVE_{E}: its prejacent cannot refer to a future state of affairs (Pietrandrea 2005; Rocci 2005a)[22]. More precisely, its prejacent can be a *present state* (18.a), an *habitual* (18.b), or any kind of *past eventuality* (18.c).

With present indicative *dovere*, non-stative eventualities shift the reference of the prejacent to the future and simultaneously block the possibility of an epistemic-evidential reading, as in (25.d), leaving open only *deontic*, *anankastic* or *alethic* readings.

(18.a) Giovanni **deve** *conoscere* una scorciatoia. Perché è arrivato là prima di tutti. [stative, epistemic]
'Giovanni **must** *know* a shortcut. Because he arrived there before everybody else'.

(18.b) Giovanni **deve** *prendere* una scorciatoia. Perché, di solito, arriva là prima di tutti. [habitual, epistemic]
'Giovanni **must** *take* a shortcut. Because he usually arrives there before everybody else'.

(18.c) Giovanni **dovette** prendere una scorciatoia. Perché arrivò là prima di tutti quel giorno. [past prejacent, epistemic]
Arguably, Giovanni took a shortcut. Because he arrived there before everybody else that day'.

(18.d) Giovanni non conosce la strada. ***Deve** *perdersi*. [non-stative, future prejacent, non-epistemic]
'Giovanni does not know the way. *He **must** *get lost.*'

It is impossible to interpret the second sentence in (18.d) as 'I infer that he will get lost'. The only possible interpretations are a *deontic* one, which destroys discourse coherence, and, very marginally, an *alethic* interpretation where the modal indicates an inevitable consequence of factual circumstances. This alethic interpretation

---

[22] For reasons of space here I do not discuss the ultimately temporal or aspectual nature of the constraint or whether it can be reduced to more general principles of interaction between modality and tense-aspect semantics. See Pietrandrea (2005: 133-186). In every case, it is difficult not to think that this constraint must have something to do with the epistemic and ontological asymmetry existing between the future and the past (cf. § 6.4.1 below).

would be made more easily accessible by accompanying the modal with harmonic adverbials like *necessariamente* 'necessarily' as it happens in (19):

(19) [...] i sindacati, che continuano in questa loro miope politica dello sciopero. Evento che **deve necessariamente** destabilizzare il complessivo aspetto logistico del trasporto locale, [...](*Italia Oggi*, 27/4/2006)

'[Also] the trade unions [are to blame], which continue in their short-sighted policy of the strike. An event that is necessarily going to destabilize the overall logistics of local transport,[...]'

A similar inferential-evidential reading is also available for the Italian future tense morphology both in the simple future (20) and in the so-called "anterior future" (21).

(20) Il 5 per mille all'Ulivo non piace. **Sarà** perché l'ha inventato il ministro dell'economia, Giulio Tremonti, giudicato comunque troppo creativo. Oppure perché in realtà non c'è nessuno che possa valutare i requisiti delle associazioni che accedono direttamente alla ricca torta, convincendo i singoli contribuenti.(*Italia Oggi*, 5/4/2006)

The Olive Tree [party] does not like the idea of "5 per thousand". **Maybe** it's because it was invented by the Minister of Economy, Giulio Tremonti, always judged too creative. Or because, in reality, there is no one who can assess the credentials of the associations that have direct access to this rich cake by convincing individual taxpayers.

(21) Quelle immagini televisive davanti agli occhi di tutti hanno scosso di commozione il mondo intero. Sono tentato di pensare che un sussulto o anche solo un pur minimo brivido di pacificazione **avrà sfiorato** il sentimento d'amore che ciascun uomo - lo accetti o no - porta con se. (*Il Sole 24 Ore* 2/4/2006)

Those TV images before the eyes of all have shaken the whole world with emotion. I'm tempted to think that a jolt or even a slightest shiver of peace **will/ must have grazed** the sentiment of love that each human being - you accept it or not - carries within. (Il Sole 24 Ore 2/4/2006)

This well-known (cf. Bertinetto 1979, 1986; Bozzone-Costa 1991; Rocci 2000, 2005b; Squartini 2008)-evidential reading of the Italian future morpheme– henceforth FUTURE$_E$– is associated with inferences that the speaker draws with different degrees of certainty, as shown by the contrast between the English translations of (20) and (21). At the high end of the confidence spectrum FUTURE$_E$ appears to be interchangeable with DEVE$_E$ and translatable with *must*, at least in certain contexts. FUTURE$_E$ is subject to the same temporal-aspectual constraints of DEVE$_E$ and is thus available only when the prejacent does not refer to a future state of affairs. When the future tense actually refers to future eventualities, the FUTURE$_E$ reading is no longer possible and the tense ceases to function as marker of inferential evidentiality, losing its evidential constraints.

While a discussion of diachronic issues falls clearly outside the scope of the present chapter, it might nevertheless be relevant here to mention that in Italian, as well as in other Romance languages, both the future tense and the conditional mood are etymologically modal constructions of necessity. Both derive from a Latin modal construction, formed with the auxiliary *habere* ('to have') and an infinitive – the type *amare habeo* 'I have got to love' – which could be used to express various shades of necessity, including deontic obligation and future-oriented alethic inevitability. While the Romance future corresponds to the form with the auxiliary in the present tense (*amare habeo*), Romance conditionals derive either from the imperfect (*amare habebam*) – as in French, Spanish, Catalan and Portuguese, or, in the case of Italian, from the perfect form (*amare habui*).

The lack of constraint on inferential evidentiality in the Italian future tense when it refers to future eventualities can be demonstrated by showing that, contrary to its properly evidential counterpart, it accepts embedding under a reportative as in (22):

(22) In borsa, <u>a quanto si apprende</u>, **finirà** il 30–40% del capitale e la quota **sarà formata** in parte da azioni di nuova emissione e in parte da titoli messi in vendita dagli attuali azionisti.(*MF*, 4/4/2006)
In the stock exchange, <u>according to what transpires</u>, there **will** eventually end up 30–40% of the capital and this share **will** be composed in part by newly issued stocks and in part by securities offered for sale by the current shareholders.

In fact, when referring to future eventualities the Italian future tense accepts also to be embedded under clauses such as *Nessuno pensa che* (Nobody thinks that), as in (23).

(23) Nessuno pensa che Luigi verrà
Nobody thinks that will come

which suggests that it also lacks a proper m-performative epistemic component. When referring to present states or to past eventualities the future tense rejects both kinds of embedding as shown by (24) and (25). For a more detailed discussion see Rocci (2005a: 279–283)[23].

(24) *Nessuno pensa che Giovanni sarà andato via
(Nobody thinks that Giovanni must/will have left)
(25) ??Luigi ha detto che Giovanni sarà andato via
(Luigi said that Giovanni must/will have left).

The above observations are not to imply that in Italian future tensed sentences referring to future eventualities never appear as the conclusion of an inferential path. Of course, they do at times. The point I am making is that when the future tense refers to future eventualities it does not impose a constraint on interpretation such that we are obliged to take the sentence as the result of an inference. And the linguistic encoding of evidentiality consists precisely in the conventional association of linguistic forms with constraints on interpretation of this type. More broadly, concerning the epistemicity of the Italian future morpheme, I can subscribe what Sweetser (1990: 55) says of English *will*: “The one mistake I can clearly identify in some past analyses is the idea that future will is always epistemic, and concerns future truth value”.

[23] These observations on the future reference uses of future tense should suggest a reconsideration of the use of Italian data in theoretical discussions of the relationship between tense and modality. For instance, while I am sympathetic with Ludlow’s (1999) thesis that the Italian future tense morpheme is, not only etymologically but also semantically, a modal, his statement that in Italian “when the future is used, it is most likely being used to express possibility or uncertainty” (Ludlow 1999: 159) is not entirely warranted.

6.3.1.2 DOVREBBE$_E$: Prediction and Causality

Thus neither FUTURE$_E$ nor DEVE$_E$ can express predictive inferences, a constraint that makes their rarity in economic-financial news something to be expected. DOVREBBE$_E$, is not subject to this constraint and not only *can*, but most of the times *does* have future prejacents, as in (26). Prediction is the typical use of DOVREBBE$_E$.

(26) La costanza con cui i fondi investono nel settore finisce per restringere notevolmente la disponibilità. Un fenomeno che si **dovrebbe** avvertire soprattutto sul mercato dell'argento quando sarà avviato l'Etf del gruppo Barclays. (*Il Sole-24 Ore*, 20/4/2006).

'The constancy with which funds invest in the sector ends up significantly restricting availability. A phenomenon that **should** be felt especially on the silver market when the ETF [=Exchange-Traded Fund] of the Barclays group will be launched.'

As shown in Figs. 6.11 and 6.12, in (26) we have, again, an argument supporting a predictive conclusion invoking a locus *from the efficient cause*, in the direction from cause to effect.

Free to occur in predictions, DOVREBBE$_E$ is subject to a different constraint, directly affecting the possibility of establishing argumentative discourse relations, which concerns precisely the direction of inference in the causal loci:

(27.a) Giovanni ha lavorato molto. **Dovrebbe** essere stanco.
'John worked a lot. He should be tired'

(27.b) Giovanni è stanco. ***Dovrebbe** aver lavorato molto.
'John is tired. He should have worked a lot'

(27.c) E' tutto rosso in faccia. *Dovrebbe essere fuori di sé.
'He's all red on his face. He should be out of his mind'.

As shown by examples (27.b) and (27.c), DOVREBBE$_E$ cannot express argumentative discourse relations where the conclusion denotes an eventuality which, at the content level, is taken to be the cause of the eventuality denoted by the textually

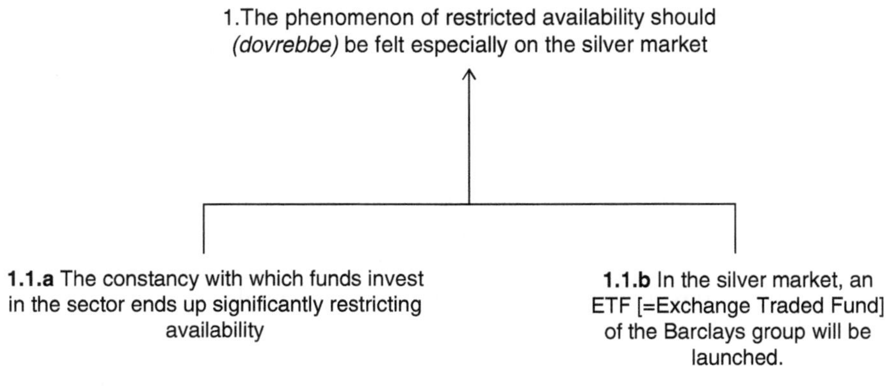

Fig. 6.11 Argumentation structure of example (26)

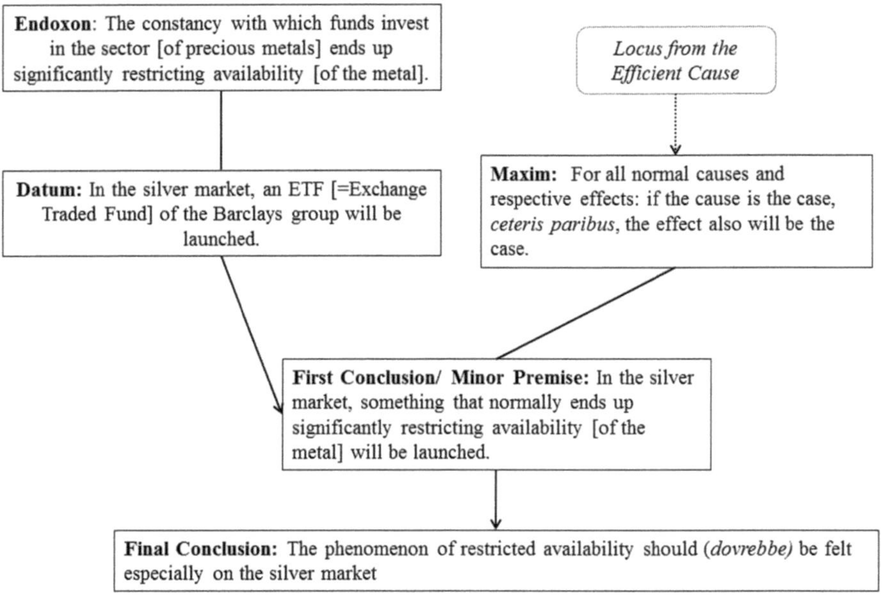

Fig. 6.12 Inferential configuration (Y-structure) of example (26)

available premise. These occurrences correspond, for instance, to arguments based on the *locus of the efficient cause* that invoke a maxim enabling to infer from the effect to the cause. In the AMT framework, a deductive, necessary version of this maxim could be formulated as follows:

(Def. 2) Locus from the efficient cause. Maxim from the effect to the cause (deductive): For all *necessary* causes and respective effects: necessarily if the effect is the case the cause also is or has been the case.

In contrast, as shown by (26) and (27.a), DOVREBBE$_E$ is perfectly acceptable when an argument from the locus of the efficient cause invokes a maxim *from the cause to the effect*. Already in Chap. 5, § 5.3.1, Fig 5.5, the deductive version of this maxim was formulated as follows:

(Def. 3) Locus from the efficient cause. Maxim from the cause to the effect (deductive): For all *sufficient* causes and respective effects: necessarily if the cause is the case the effect also will be the case.

Note that the deductive versions of the maxims require that the cause is a necessary cause in the inference from effect to cause, and that the cause is a sufficient cause in the inference from cause to effect. Non-monotonic, defeasible, versions of the two maxims would loosen this requirement, as it happens with the analysis of example (26) in Fig. 6.12.

In the corpus it is very easy to find examples, like (17.b) and (26) above, where the co-textual premises that saturate the B of DOVREBBE$_E$ are linked with the pre-jacent by what we might call an *economic causality*. They are typical examples of discourse focusing on the economy – which would deserve in-depth semantic consideration by itself – where events in the markets are not seen as the result of human actions, but as quasi-natural events causally related in virtue of "economic laws", which operate blindly. The locus from efficient cause seems to operate identically for purely physical and for economic causality.

This kind of naturalistic economic discourse, effacing human agency and reasoning and downplaying the institutional nature of economic facts, clearly corresponds to the ontology and rhetoric of classical economics, as wittily pointed out by J.R. Searle (2005:1):

> When I was an undergraduate in Oxford, we were taught economics almost as though it were a natural science. The subject matter of economics might be different from physics, but only in the way that the subject matter of chemistry or biology is different from physics. [...] At no point was it ever suggested that the reality described by economic theory was dependent on human beliefs and other attitudes in a way that was totally unlike the reality described by physics or chemistry.

In contrast with DOVREBBE$_E$, DEVE$_E$ is not subject to any constraint on the direction of causality. Arguments *from the effect to the cause* like (28) seem particularly frequent, but arguments *from the cause to the effect* like (29) are also found in the corpus.

(28) Ritirata strategica prima di un nuovo attacco. Il management del Nasdaq ***deve*** aver trovato spunto nell'*Arte della Guerra* di Sun Tzu per la strategia di conquista del London Stock Exchange (Lse). A fine marzo la società, che gestisce la Borsa americana dei titoli ad alta tecnologia, ha ritirato l'offerta da 950 pence per azione sull'Lse senza tante spiegazioni. Ieri la notizia che il Nasdaq controlla una quota pari al 14,99% della società-mercato britannica. (*Il Sole 24 Ore*, 12/4/2006).

'A strategic retreat before a new attack. The management of the Nasdaq **must** have found inspiration in the *Art of War* by Sun Tzu for their strategy for conquering the London Stock Exchange (LSE). In late March, the company, which operates the American Stock Exchange of high-tech securities, has withdrawn its bid of 950 pence per share on LSA without much of an explanation. Yesterday the news that the Nasdaq owns a share of 14.99% in the UK market-company.'

(29) E a giudicare dalle plusvalenze e dai rendimenti ottenuti in questi ultimi tre anni, sembra che le scelte effettuate siano state azzeccate. Dunque, grande **deve** essere la soddisfazione tra i soci. (*Il Sole 24 Ore*, 8/4/2006.).

'And judging from the gains and efficiencies achieved in the last three years, it seems that the choices made were justified. So there **must** be great satisfaction among the shareholders.'

It is also possible to have DEVE$_E$ with inferences that involve no causality at all, such as those from a distinctive feature to a species or class. According to the taxonomy of loci presented in Chap. 2, § 2.2.3.2 these inferences belong to the area of definitional loci, more specifically, to the locus *from proper and accident*:

(30) Bella la prima, ma non è una vipera, giusto? La vipera ha una testa triangolare, questo ***dev***'essere un serpentello data la forma tondeggiante del musetto.

'Nice the first one, but it isn't a viper. The viper has a triangular head, this must be a grass snake because of the rounded shape of the nose.' (Forum post recovered through *Google*. The poster is commenting a photograph.)

In order to develop a semantic hypothesis capable of accounting for the divergent argumentative constraints of the two construction I will now turn briefly to proposals that have been put forth to illuminate similar phenomena in English and French.

6.3.2 Hypotheses on French DEVRAIT$_E$ and English SHOULD$_E$

A similar constraint on causality was singled out by Rivière (1981) in a parallel analysis of French DOIT$_E$ vs. DEVRAIT$_E$ and English MUST$_E$ vs. SHOULD$_E$. Rivière connects the constraint with a difference in modal *force* between *must* and *should*. For him inferences from cause to consequence are weaker and favour the weaker modal *should*, while those from consequence to cause are too strong to be compatible with *should*. The explanation seems unconvincing, at least for Italian *dovere*.

I see two main arguments against Rivière's analysis. One is theoretical and the other is observational. From a theoretical viewpoint, it is difficult to imagine why inferences *from cause to effect* should be, in general, weaker than inferences *from the effect to the cause*. Both loci can have deductive as well as defeasible maxims, as shown in the previous section. In fact, there are reasons to think that, ceteris paribus, the reverse should be the case. In fact, it seems that in real life it is not easy to find *necessary causes* of observable states of affairs and thus maxims from the effect to the cause are typically presumptive and thus non-monotonic. The applicability to real-life situations of the clean, deductive, maxim *from an effect to the necessary cause of such effect* (Def. 2) seems limited. A fictional example of application of such a maxim could be the following quote from *Star Wars (Episode IV)*:

(31) Obi-Wan Kenobi: "And these blast points, too accurate for Sandpeople. Only Imperial Stormtroopers are so precise".

In every case, from finding genuine examples of deductive *from effect to cause* to concluding that these inferences are, generally speaking, stronger than those from the cause to the effect there is quite a long way.

From an observational point of view, in the occurrences of DEVE$_E$ in the corpus we find no sign that inferences from the effect to the cause, such as (28), are perceivably stronger than those where causality goes in the other direction, as in (29). Actually, DEVE$_E$ in (28) seems to express a particularly weak and highly subjective inference.

Salkie (1996) and Dendale (1999) also present arguments against Rivière's analysis of English SHOULD$_E$ and French DEVRAIT. Salkie (1996) uses data from parallel corpora to review and criticize Rivière's (1981) contrastive analysis. Salkie devises a number of counterexamples to Rivière's hypothesis that the differences between MUST$_E$ and SHOULD$_E$ have to do with the strength of the inferences conveyed. The most straightforward test he provides is based on examples like the following:

(32.a) If you know Prof. Fen, you must/*should live in Oxford
(32.b) If you know Prof. Fen, you might live in Oxford.

In the examples of inference *from effect to cause* where *must* cannot be substituted with *should,* as in (33.a), the even weaker modal *might* behaves just fine as shown by (33.b). Obviously, the meaning changes but the modal *might* can occur in that context and the presence of an inference from the effect to the cause remains clearly felt. One important conclusion of Salkie's work is that "the assumption that there is one 'scale' of epistemic modality" (Salkie 1996: 391) is unfounded. Such a conclusion appears in broad agreement with the general attitude adopted in the present study of avoiding reduction of the argumentatively relevant aspects of modal meanings to scales of strength or degrees of commitment and to reduce the latter to the basic quantificational force of the modal expression involved.

Discussing French DOIT$_E$ vs. DEVRAIT$_E$, Dendale (1999) prefers to formulate a different, non-causal. constraint on the kind of premises admissible for the conditional modal: DEVRAIT$_E$ selects "*in absentia* premises", that is "premises that the speaker does not gather at the moment of utterance, but he has previously collected, thus creating a temporal gap between them and the conclusion" (Dendale 1999: 21).

Looking at the examples discussed by Dendale, it seems, in fact, that the notion of *temporal gap* is ambiguous between an ontological interpretation (= 'the premises *as facts* precede temporally the conclusion') and an epistemological interpretation (= 'the premises *as beliefs* were acquired some time before the drawing of the inference'). The first interpretation is covered by the causal constraint. The second reading alludes to a feature of DOIT$_E$ and DEVE$_E$ that is rather difficult to pinpoint precisely: their tendency to refer back to textual premises consisting of *specific observable facts*, often facts observed by the *speaker* close to the *time* and *place* coordinates of the utterance. Such a tendency might correspond to what students of evidentiality call "inference based on results" (cf. Fig. 6.4, § 6.2.2.2).

As observed in Rocci (2005a: 308–310), DEVE$_E$ differs from the Italian epistemic construction based on the future tense morphology (FUTURE$_E$) in that DEVE$_E$ is incompatible with inferences lacking a piece of specific evidence and based exclusively on what is *generally to be expected* (see also Squartini 2008 on "generic" vs. "circumstantial" inferences):

(33)	A: Dov'è lo zucchero?	B: Sarà nel solito posto.
		B: ? Dev'essere nel solito posto.
		B: Dovrebbe essere nel solito posto.
	A: 'Where's the sugar?'	B: 'It will/ ?must/ should be in the usual place'

As we can see in (33), with respect to the contrast between specific evidence and general expectation DOVREBBE$_E$ sides with the epistemic future[24].

The different behavior of DEVE$_E$ and DOVREBBE$_E$ as markers of argumentative discourse relations of *Conclusion* can be accounted for in terms of different

[24] I note, in passing, that with respect to causality and to several other relevant criteria, the epistemic future is much closer to DEVE$_E$ than to DOVREBBE$_E$.

underlying meaning construction processes through which these two "epistemic-inferential" readings are obtained.

The analysis proposed here is inspired, in part, by Kronning's (2001) hypothesis on the difference between the "epistemic" interpretations of French *devoir* in the indicative and in the conditional. For Kronning, epistemic *devrait* is not directly related to the genuinely epistemic readings of *devoir* in the indicative but represents a *hypothetical* version of a future-oriented non-epistemic reading of the modal. For Kronning, the "source" of *devrait* is an *alethic* reading – similar to the one we observed in (26).

Below I will adopt, in a revised form, the central idea of Kronning's analysis, hypothesizing that DOVREBBE_E is a conditional version of the future-oriented non-epistemic readings of *dovere*. I will recast Kronning's hypothesis in terms of the relative modality framework outlined in Chaps. 4 and 5. The result provides an insightful account of the constraints discussed above and of other features of the functioning of DOVREBBE_E in argumentative discourse.

The first step in building this analysis will be to look more closely at the future-oriented non-epistemic readings of the Italian modals in predictions examining how they make accessible to the addressee intended argumentative discourse relations.

6.4 The Argumentative Relevance of Non-epistemic Modal Verbs in Predictions

6.4.1 Futurity as Modality and the Propositional Content of Predictive Standpoints

The interactions between reference to future eventualities and modality have been an object of interest for philosophy since the Antiquity, with a major focus on the logical-ontological issue of (in-)determinism and of the truth-conditions of statements about the future inaugurated by Aristotle (*De Interpretatione*, 9)[25], with the famous discussion of the sea-battle example:

(34.a) There **will be** a sea battle tomorrow.
(34.b) There **won't be** a battle tomorrow.

Aristotle wonders whether we can say that a statement like (31.a) is either true or false *now*. For, if we establish that it is true we establish *ipso facto* that it is

[25] As it sometimes the case with Aristotelian texts, in *De Interpretatione*, 9 it is not easy to say what exactly Aristotle wants to say – as Hintikka (1973: 145) puts it: "what is the discussion all about?". Historically established interpretations have become entrenched, but they do not necessarily hold up on closer examination, as argued by Hintikka (1973) for this passage. Here I do not enter in the detailed exegesis of the Aristotelian passage, as my goal is far more modest. I introduce the problem of future truth as the backdrop of for a view of all statements about the future as modal.

necessary, inevitable given the present state of the world. If, on the contrary, we suppose it is false, we establish ipso facto that it is *impossible*, already ruled out by the present state of the world. For Aristotle, there will be a moment in time at which the truth or falsity of this statement will be settled: necessarily: (34.a) will turn out either as true or as false, *but* neither option is necessary as for now, things could turn out to be different and the future is still indeterminate (cf. Kaufmann et al. 2006: 91–92). According to the traditional interpretation of the Aristotelian passage (cf. Hintikka 1973: 149), the philosopher is worried that "the unrestricted applicability of *tertium non datur* to statements about future events seems to commit us to holding that all future events are predetermined, and thus lead us to determinism". There are three main ways of responding to the problem of future truth (cf. Bonomi 1980: 357):

(i) renouncing bivalence and adopt a trivalent semantics which accounts for future indeterminacy and avoids determinism;
(ii) keeping a two-valued logic and adopting a "Ockhamist notion of truth" according to which future statements are indeed either true or false "even though it may be impossible even in principle to know the truth values ahead of time" (cf. Kaufmann et al. 2006: 92);
(iii) adopting a "Peircean notion of truth" (*Ibid.*) according to which until the facts settle the matter both (34.a) and (34.b) are false. This option, however, does not destroy the principle of bivalence because (34.a) and (34.b) are analyzed as modal statements of necessity vs. impossibility and thus *contrary* rather than *contradictory*.

As reminded by Kaufmann et al. (2006: 92), two modal necessity statements, in view of some conversational background B, such as (35. a, b) can be both false without wrecking the Principle of Non-Contradiction. They are false in case the truth of p *in the future* is unsettled.

(35.a) $\Box_B$ p
(35.b) $\Box_B \neg$p

This modal view of the future accounts for our deep intuition that the future is not completely settled without abandoning bivalence.

According to the commonsense human ontology underlying the modal view of future statements, there is an asymmetry between the past and the future. This asymmetry has both an ontological and an epistemic facet. Ontologically, the future is open to the realization of different possibilities while the past is sealed. We can thus have a unique timeline up to the present moment t_0 which then branches into several alternative future timelines. We can represent alternative timelines as a set w_1-w_n of branching possible worlds which are identical up to t_0 and diverge afterwards (cf. Werner 2005: 249) (Fig. 6.13).

Since the past is ontologically settled (*historical necessity*), possibilities in the past can only be either *epistemic possibilities*, stemming from our incomplete knowledge of what actually happened, or *counterfactual possibilities*, stemming from our deliberate decision to abstract from certain facts of the past. There are no

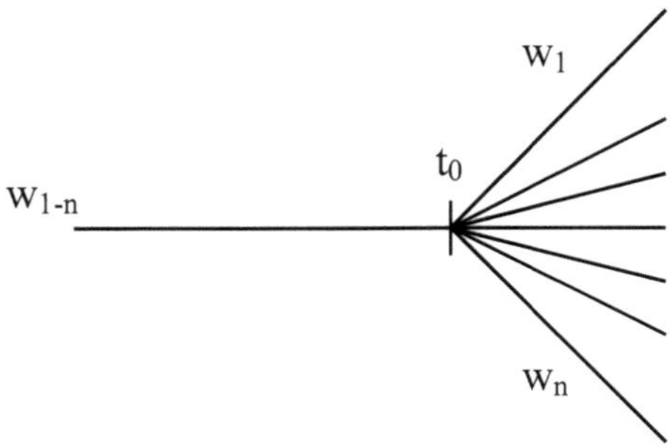

Fig. 6.13 A "broomstick" representation of a set of branching possible worlds (cf. Werner 2005: 250)

genuine, "real possibilities" in the past. Future events, on the other hand, are not only epistemically *uncertain*, but also genuinely *unsettled* from an ontological viewpoint (cf. Kaufmann et al. 2006), so besides epistemic and counterfactual ones, also real possibilities are open.

Real possibilities correspond to Peirce's notion of *substantial* possibilities, that is possibilities in view supposedly complete information "of the present in the present, including among the objects known all existing laws as well as special facts" (Peirce 1998 [1931]:44). While with respect to the past everything that is substantially possible is also substantially necessary, the future involves alternative substantial possibilities. Substantial/ real alternatives always correspond to epistemic alternatives (*lack of foreknowledge*), while epistemic alternatives do not always correspond to real/ substantial alternatives, as they may simply depend on the incompleteness of the information available to the subject.

Should we hypothesize that linguistic structures such as the Romance future tense morphemes and English future modal *will* always denote real/substantial necessity? I do not think that this kind of conversational background would give us the right semantics in most of their uses, where they seem to correspond to a weaker degree of modality. In Rocci (2000, 2005a: 261–341) I systematically defended the hypothesis that the Italian future tense, much like the modal verbs is a context dependent modal predicate, which can pick-up different conversational backgrounds. Only in rare instances of philosophical discourse the B is made to correspond to Peircean substantial modality. Additionally, in Rocci (2005a: 305–314) I argued that the necessity operator of future tense operates on a complex conversational background, which includes a set of assumptions on *normal conditions* of the kind that is typically encountered also in indicative conditionals. While these assumptions are not known to be true, neither they are known to be false. Strictly counterfactual assumptions are excluded from the conversational background of the future.

When the prejacent refers to future eventualities the basic conversational background is typically either a restricted kind of *alethic* modality or one of several kinds of lato sensu *deontic* modality: including schedules and plans, orders and laws, desires, promises and threats, etc. Example (36), below, correspond to the selection of a deontic conversational background of plans to which individual or corporate agents are to some extent committed. This case is particularly frequent in predictions in the corpus.

(36) Più in dettaglio l'operazione, denominata Progetto Stadium, che gli advisor hanno presentato il 3 aprile scorso al consiglio di amministrazione di Agria Holding (la capogruppo che controlla il gruppo Arena-Roncadin), offrirebbe 1.500 nuove azioni Roncadin ogni mille euro di valore nominale del bond. Dei 135 milioni, 55 **saranno convertiti** in azioni, il resto rimborsato cash in due tranche: la prima, pari a 25 milioni, a meta giugno del 2009, la seconda a giugno 2011.

More in detail the operation, called Project Stadium, which the advisors have submitted on April 3 to the Board of Directors of Agria Holding (the parent company that controls the group Roncadin - Arena), would provide 1,500 new Roncadin shares per thousand Euros of nominal value the bond. Of the 135 million, 55 **will be converted** into shares, the remainder refunded cash in two tranches: the first, amounting to 25 million, in mid-June of 2009, the second in June 2011.

When future prejacents are ruled out, the FUTURE_{E} construction (see 6.3.1.1, above) is invoked and a special kind of epistemic-evidential conversational background is selected.

The first characterization of the speech act of prediction as a connective predicate, which was given in § 6.2.2.1, did not pay too much attention to the presuppositions of the connective predicate on the propositional content. In fact, viewing all statements about the future as modal has an interesting consequence regarding the *propositional content* of speech acts of prediction. If we go back to Freeman's (2000, 2005: 93–112) classification of statements introduced in Chap. 2, § 2.3.3 we can observe that, being intensional, the propositional contents of predictions cannot be *descriptive* in the precise sense that Freeman (2005) gives to this term. In fact, they can only be interpretations.

It was noted in Chap. 2 that the different types of statement affect the ways in which we go about determining whether statements have an "epistemic presumption" (Freeman 2005: 21–37) in their favor. Freeman's semantic classification of statement types also serves as a heuristics to investigate what he calls "basic belief generation mechanisms": descriptions are naturally linked to sensory *perceptions*, to the *memory* of our sensory perceptions, and, in certain cases to *introspection*. Descriptions can be also associated with the "interpersonal belief generating mechanism" in the form of simple (non-expert) testimony. We can observe that there is a certain match between these "belief generation mechanisms" and the types of evidence of the linguistic tradition (cf. § 6.2.2.2).

While in his 2005 book Freeman is interested in finding out what are the primary presumptions from which we take premises that are not in need of argumentative support, the main lines his statement classification can be brought back to their rhetorical roots and be used to examine how different semantic types of standpoint constrain the types of loci that can be invoked to license the inference and, indirectly, the types of premises that are congruent with these loci.

What does Freeman say of statements about future eventualities? Surprisingly, in Freeman (2005: 124–125) examples of prediction are given as one of the sub-types of description. He provides the following two examples of descriptive future statements:

(37.a) The program *will* come on in a half-hour
(37.b) The sun *will* rise tomorrow

Descriptions can be either reports – like (38.a) or (extensional) generalizations – like (38.b).

(38.a) A bus is passing my office window
(38.b) All subjects in the experiment displayed cold symptoms

And, according to Freeman, reports can be about particular events "either present, past, or future". And this is where I stop following Freeman.

Clearly, if one accepts the modal solution to the problem of future truth that has been briefly evoked in this section, (37.a, b) cannot be descriptions because they are modalized and thus intensional statements. Specifying the conditions for their truth *does* require reference to possible worlds: all the worlds in the broomstick in the limit case where the future tense is interpreted as Peircean substantial necessity, a different set of worlds where other restricted alethic or deontic saturations of the conversational background are accessible.

But we do not need to assume acceptance of a modal theory of futurity to criticize the view that (37.a, b) are descriptions. I believe that including instances future statements among descriptive statements is a mistake on the part of Freeman, chiefly because they do not fit the evidential role that Freeman assigns to descriptions. Certainly, one can reject a modal theory of the future and treat sentences like (37.a, b) as purely extensional with the proviso that *it may be impossible even in principle to know their truth values ahead of time*. But if we take this route the connection between semantic types and belief generation mechanisms, which forms the backbone of Freeman's theory would be lost. Future statements like (37.a,b) would make a subclass of extensional statements for which the belief generation mechanisms normally associated with descriptions *do not work.*

I have already introduced this topic in § 6.2.2.2 by considering which types of evidence – among those recognized as basic by studies on grammaticalized evidentiality – are available in predictions. Referring this time to Freeman's belief generating mechanism – , which only partially overlap evidentiality types – perception, clearly, is not available, and neither are memory and introspection. Simple testimony is not possible, because perception, introspection and memory of things to come are unavailable to others just as they are to me.

This is readily recognized by Freeman (2005: 126), who concludes that future descriptions of particular events *do not have basic belief generating mechanisms*: they are based on "induction as a form of inference" and thus cannot be but non-basic secondary beliefs. Here is how Freeman describes the inference licensing the belief that (37.a) is the case:

> Presumably I already believe that 7:00 pm is the announced time of the program. Presumably, I also believe, say through perceiving my watch, that the time is now approximately 6.30 pm. Putting these two pieces of information together clearly generates by inference my belief that the program will come on in a half hour.(Freeman 2005: 126)

Interestingly, if we look at the two premises of this inference we find the two component of a broadly Kratzerian deontic conversational background: the *normative component* B_V (the commitment generated by the program announcement) and the *circumstantial component* B_C ('it's 6.30 pm')[26]. In all worlds consistent with the commitment to start at 7:00 pm *and* with the fact than now it's 6.30 pm the program starts in a half-hour. Example (37.a) is, in the end, pretty similar to (36), being based on a deontic conversational background of commitments.

If predictive future statements are intensional and cannot be descriptions, what are they? I believe that the answer is pretty straightforward and is basically already given in Freeman (2005): they are *contingent, non-evaluative, intensional* statements, that is *interpretations*. Even if they do not immediately fit Freeman's examples of interpretation they perfectly fit its definition.

According to Freeman (2005: 108), interpretations "either presuppose nomic statements" or "assert such statements". As expounded already in Chap. 2, § 2.3.3, and more technically in Chap. 5, § 5.3.1 nomic statements are modal "subjunctive" conditionals stating a nomic generalization, holding for a certain set of worlds, where certain conditions conditions hold. In terms of Relative Modality, their antecedent is added to a conversational background of statements that are true in all the worlds of the relevant set of worlds. Refining what Freeman (2005) says of the relationship between interpretations and nomic regularities, interpretive statements can be classified as follows (Fig. 6.14):

Both statements of causality and singular modal alethic (and anankastic) statements presuppose the underlying nomic regularity. In causal statements the antecedent of the presupposed nomic regularity is evoked by the subordinate causal clause. In the case of modals, the antecedent of the presupposed nomic regularity may appear as an anaphoric antecedent or be completely tacit.

Whether they have a syntactic or anaphoric antecedent or not, modals presuppose, just like causal statements, their conversational background. Thus, non-epistemic modal statements in general, and, specifically predictions, are to be considered interpretive. Predictions are thus inherently modal. As we will see in the following sections, the modals occurring in predictions function as a diverse and intermittent manifestation of this inherent modality.

---

[26] See the discussion in Chap. 5, § 5.3.2 and, in particular, the analysis of the deontic conversational background in terms of a set of value judgments (B_V) and a set of circumstances (B_C) offered by (Def. 5.a, b).

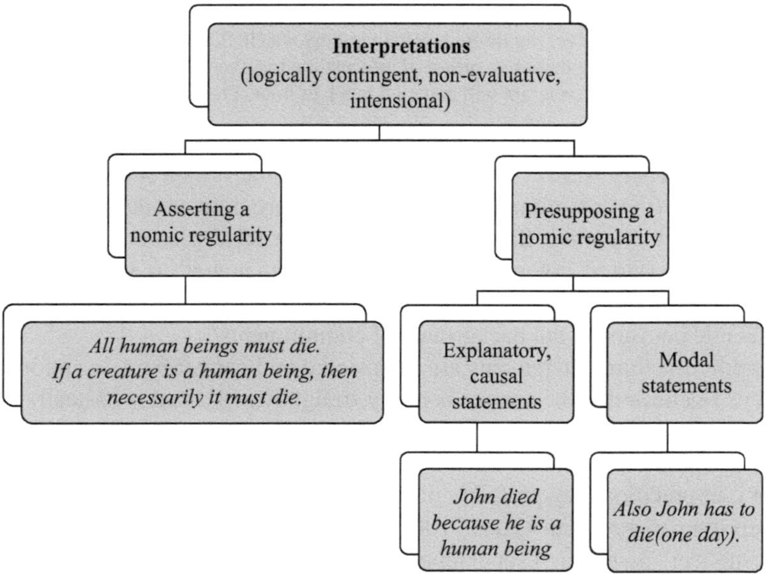

Fig. 6.14 Refining the classification of interpretive statements

6.4.2 *Evidential Implications of Future-Oriented Alethic and Deontic Modals*

All of the occurrences of the Italian modal verbs *potere* and *dovere* pertaining to the indicative mood that we encountered in predictive speech acts in the corpus – most of them in the future tense, some in the present tense – express root modalities and can be classified as deontic, alethic circumstantial or anankastic.

Methodologically it will be important here to confine the discussion to the indicative mood, leaving aside the epistemic reading of conditional mood *potrebbe* and *dovrebbe*. In the subsequent analysis of the DOVREBBE$_E$ construction as a conditional version of the future-oriented root readings of *dovere* it will be clear how the semantics of the conditional mood adds a further layer of epistemicity to the root modals encountered in predictions[27].

As I have observed above, the epistemic inferential reading of indicative *dovere* totally blocked with future prejacents, so it is expected that in predictions we only encounter non-epistemic uses of this modal. As for *potere*, epistemic readings of the possibility modal are subject to a different set of constraints than their necessity counterpart. I summarize them briefly here, referring the reader to Rocci (2005a, b) for a fuller treatment.

[27] Partly similar considerations could be made for the semantics of the conditional of the possibility modal (*potrebbe*). This analysis, however, exceeds the limits of the present study. The issue has been addressed in Miecznikowski (2011).

6.4.2.1 A Note on the Epistemic Interpretations of Italian *potere* in the Indicative Mood

In the literature epistemic readings of *potere* in the indicative parallel to DEVE_{E} are widely believed to exist, since the earliest semantic analyses of Italian modals (cf. Parisi et al. 1975; Bertinetto 1979). The constructed examples adduced by these early studies can be illustrated by the following parallel series, taken from Bertinetto (1979):

(39) a. Giovanni deve essere uscito
'John must have left.'
b. Giovanni può essere uscito.
'John may have left.'
c. Devono essere le cinque, visto che si sta facendo buio.
'It must be five p.m., given that it's darkening.'
d. Possono essere le cinque, visto che si sta facendo buio.
'It may be five p.m., given that it's darkening.'
e. Quel tale deve chiamarsi Giovanni.
'That fellow must be named John [= his name must be John]'.
f. Quel tale può chiamarsi Giovanni.
'That fellow may be named John'(= his name may be John).

These series typically consist of utterances where the present tense indicative modal embeds a proposition relating an event with *past reference*, or a *stative eventuality*. It has already been observed in §6.3.1 that the prejacent of DEVE_{E} can be a *present state*, an *habitual*, or any *past eventuality* regardless of its aspectual class.

However, as observed in Rocci (2005b: 232), while this semantic environment makes it possible for *dovere* to take up the characteristic epistemic and inferential evidential reading that I have labeled as the DEVE_{E} construction, with *potere*, the epistemic interpretations of the examples in (39), while intelligible, are much more awkward, even unnatural. Other alleged markers of epistemic possibility such as the adverb *forse* ('perhaps'/ 'maybe') in this type of utterances seem much more natural:

(40) a. Forse Giovanni è uscito.
'Perhaps John has left'
b. Forse sono le cinque, visto che si sta facendo buio.
'Perhaps it's five p.m., given that the sky is darkening
c. Forse quel tale si chiama Giovanni.
'Maybe that fellow is named John'

Interestingly, the impersonal constructions of *potere* with the infinitives of *essere* 'to be' and *darsi* 'to be the case' as a matrix clause embedding a complement clause (usually in the subjunctive mood) are perfectly natural and appear similar to the examples with *forse*.

(41) a. Può essere / darsi che quel tale si chiami Giovanni
Lit. 'It may be (the case) that that fellow is named John'
b. Può darsi che siano le cinque, visto che si sta facendo buio.
Lit. 'It may be (the case) that it's five p.m., given that the sky is darkening'

In Rocci (2005b), to give more substance to what I had called "strong but opaque intuitions" about the possibility of a truly epistemic reading of *potere,* I turned to examining a corpus of Italian conversations (LIP, De Mauro et al. 1993). It turned out that nearly all the epistemic interpretations of *può* (3rd person sing., present indicative of *potere*) in the corpus came from the impersonal constructions exemplified in (41) and its variants in short replies and reduced sentences (cf. Rocci 2005b: 233). The residual occurrences, featuring an epistemic reading of personal *può* could, in fact, be re-analyzed as circumstantial alethic modalities and capable of acquiring an epistemic meaning by pragmatic implicature. Example (42), from the LIP corpus, used in that investigation, illustrates this dynamic:

(42) Tirana è piccola vi ho detto **può** avere un trecentomila abitanti non più (LIP : FD17, 123–124)
'Tirana is small, as I told you, it **can** have some 300, 000 inhabitants, not more'

The most accessible saturation of B in (42) is the restricted alethic *in view of the size (surface) of the city* rather than the epistemic *in view of all I know/ believe.* If we look carefully at the semantic structure of (42) we find that it refers to a quantity with an upper bound: in view of its size Tirana *cannot have more* than 300,000 inhabitants. Thus, (42) should rather be treated on a par with Toulmin's famous *cannot* sentences, which has clear consequences for inferring an indirect epistemic evaluation. Once an impossibility has been established on the basis of a certain set of facts it remains there monotonically as the B is expanded to include other known facts: all alethic impossibilities semantically entail epistemic impossibilities. Here, once the upper bound of a population of 300,000 has been established on the basis of size alone, it stays there monotonically if we look at the prejacent from an epistemic point of view.

According to the working hypothesis formulated in Chap. 5, these modals can work to establish argumentative discourse relations of *Conclusion* indirectly, by referring to the ontological relations on which the locus of the argument draws. According to this hypothesis, truly epistemic readings of the modals are different, in that they relate directly to premises and conclusions *qua* elements in a mental act of inference (cf. Chap. 5, § 5.3.4).

For French, Tasmoswki and Dendale (1994) hypothesize for *pouvoir* a PEUT_{E} construction functioning as an inferential evidential establishing inferential relationships between utterances in discourse much like a weaker version of DOIT_{E} providing a small number of constructed examples in the personal construction. For Italian, however, the corpus data do not seem to support the hypothesis of a personal PUÒ_{E} construction functioning as an inferential evidential.

In Rocci (2005b) I tried to better understand the difference between the personal and impersonal constructions of *potere* by looking at their functioning as inferential evidentials. Impersonal constructions of *potere* do appear in discourse contexts where they behave much like $DEVE_E$.in signaling that the prejacent is the result of an act of inference and estabishing an inferential, potentially argumentative, discourse relation with anaphorically recovered propositions in the discourse context. In examples such as (43), if we substitute the personal construction (b) to the impersonal one (a), the possibility of understanding an argumentative discourse relation between the two utterances is compromised:

(43) a. Luigi è in ritardo. Può darsi che abbia perso il treno.
'Louis is late. It may well be the case that he missed his train'
b. Luigi è in ritardo. ??Può aver perso il treno.
'Louis is late. ??He can have missed the train' (tentative translation)[28]

Why it should be so? In Rocci (2005b: 235–236), on the basis of corpus examples, I observed that when the personal version of *potere* is used instead of the impersonal construction the resulting modality is weaker: a "mere possibility" reading that cannot support the establishment of an inferential discourse relation. I also observed how personal *potere*, unlike English MAY_E, is subject to negation and dialogical contradiction and thus clearly corresponds to a propositional modality (cf. Chap. 3, § 3.3.3, Chap. 5, § 5.3.4.2). On the other hand, the impersonal construction cannot fall under the scope of negation or dialogical contradiction, similarly to $DEVE_E$.

The so-called non-propositional epistemic modals suggest that only the *prejacent proposition* is *made relevant* as content in a given discourse move. As suggested in Rocci (2005b: 244), the modal in these cases the modal functions as (a partial manifestation of) an higher level connective predicate which situates the proposition on the backdrop of the inferential processes of the speaker and, if it is the case, relates it with other utterances in an argument. In short, the epistemic impersonal *potere* and $DEVE_E$ are both partial manifestations of an argumentative connective predicate of *Conclusion* (cf. Chap. 5, § 5.3.4.3) presenting the prejacent as a standpoint/ conclusion inferable from contextually recoverable arguments/ premises.

In Rocci (2005b), I tied the strengthening of the modality expressed by the epistemic impersonal *potere* precisely to its extra propositional status of an indicator of the argumentative connective. I did so through Joan Hooper's (1975) notion of an *assertive* predicate (cf. Chap. 3 § 3.3.3). When we use an assertive predicate, like *guess* in (44),

[28] English translation of this example is a problem, for the simple reason that the English modals are not the Italian modals. If one translates *può* with *may* the problem in the Italian version completely disappears. If we use *can* for the translation we might be recreating something similar to the problem in the Italian original, but it's hard to exclude the presence of unrelated issues brought in by the English modal. The possibility or impossibility of epistemic *can* and its possible functioning as a marker of inferential relations could be the object of an independent investigation, and a complex one at that, because it would require considering recent diachronic change in the use of this modal, as well as dialectal variation between British and American English (cf. Coates 1995).

(44) I *guess* you're in decent company there.

We basically do two things at once: (i) we assert the prejacent proposition, (ii) we comment, as Urmson put it, on the reliability of the evidence that we have for the statement:

(44.i) *I assert*: You are in decent company there.
(44.ii) *I signal that* this is a guess

If a non-propositional modal such as the impersonal construction of *potere* is an assertive predicate, then its prejacent counts as asserted and the modal is a comment on the quality of the evidence for the assertion. Being asserted – I argued in Rocci (2005b: 244) – is incompatible with mere possibility, even if we have to do with a weak assertion. This is particularly true for argumentative connective predicates presenting the propositional content of the utterance as a standpoint. Such connective predicates presuppose that the utterer can provide (or has already provided) arguments *in favor* of the standpoint. Consider, then, example (45):

(45) Luigi è in ritardo. *Può darsi* che abbia perso il treno.
'Louis is late. It may/can be (the case) that he missed his train'
(i) *I advance the standpoint:* Louis missed his train
(ii) *I signal that this standpoint is (*merely) compatible with the evidence* B
(iii) B = Louis is late (via discourse anaphora)

In my earlier paper I assumed that the necessity of avoiding the incongruity between (i) and (ii) was sufficient to exclude the mere compatibility reading and to obtain pragmatically a reading of the *può essere / può darsi* impersonal constructions as indicating a stronger epistemic attitude. Now, I think this is only part of the story and that the argumentative functioning of epistemic possibility modals needs to be accounted for also at a semantic level. I will briefly come back to this issue in the concluding section of this book.

There are 52 occurrences of *può* (*potere* 3rd sing. pres. ind.) in the 200 text subcorpus of business-financial news, 21 of which in a prediction context. Examining these occurrences we find that only one of them is unmistakably epistemic. Unsurprisingly, this single epistemic occurrence is an instance of the impersonal complementizer construction '*può darsi che* + subj.'. It does occur in a prediction context. The impersonal complementizer construction is insensitive of the temporal and aspectual features of the prejacent:

(46) *Può darsi* che il quadro normativo possa favorire un settore o un altro, ma nessun governo può influenzare le scelte del mercato in un quadro normativo europeo e in uno scenario di globalizzazione e apertura dei mercati (*MF*, April 14, 2006)
'It may be that the regulatory framework can favor one sector or another, but no government can influence market decisions in a European regulatory framework and in a scenario of globalization and open markets'

As for the personal construction of *potere*, we can observe that, with future prejacents, the distinction between root and epistemic readings becomes apparently more blurred, especially in the area of alethic circumstantial uses of the modal. Coates

(1983, 1995), who observed a similar phenomenon for English *can,* referred to these uses as *mergers*. Consider the following example:

(47) **Può** scoppiare un temporale.
'A storm **may** / ?**can** break out.'

If (47) is uttered under a darkened cloudy sky it can be taken either to mean that the breaking out of a storm is compatible with what 'we' (or the relevant epistemic community) know about the world or that this state of affairs is compatible with the current atmospheric circumstances and with the nomic regularities of physical (and specifically meteorological) causality.

Thus we can attribute to the modal either an *epistemic* or a *circumstantial* conversational background. Interestingly, when the modal *potere* is inflected in the future tense, as in (48), the impression of overlap or merger between epistemic and circumstantial reading disappears and only the circumstantial reading remains.

(48) **Potrà** scoppiare un temporale.
'**It will be possible** for a storm to break out.'

In (48) the future tense necessity operator clearly takes scope over the possibility modal and the latter selects a shifted conversational background of future atmospheric circumstances, rather than of present knowledge. Similar effects can be achieved if we add temporal adverbials to (47): the adverbials, as shown in (49), take scope over the modal in a manner quite unlike what will be expected with epistemic modality:

(49) **Può ancora** scoppiare un temporale.
'A storm **can still** break out.'

These remarks prompt us to hypothesize that, at least in the indicative, also for personal *potere* no genuinely epistemic readings are possible with future prejacents. In fact, in the corpus of predictions we have not found any likely candidate for the epistemic status, neither in the present nor in the future tense. What we find, both with *potere* and *dovere* is a variety of *circumstantial* and *deontic* conversational backgrounds. What is interesting, however, is to look at how these non-epistemic conversational backgrounds relate to the kinds of evidence on which predictions are based.

6.4.3 Economic Causality: Necessity and Impossibility

Purely alethic circumstantial interpretations of the modal *potere* in the present and future tenses are encountered in the corpus. These readings select a conversational background composed entirely of facts of the world belonging to a certain loosely and contextually defined kind (and of the associated nomic regularities relevant for the domain). Most of these alethic circumstantial readings clearly involve causal relations, but they are not concerned with physical causality. Rather, they signal the kind of *economic causality* briefly examined in § 6.3.1.2. They are typical examples of discourse focusing on the economy where events in the markets are not seen as the result of human actions, but as quasi-natural events causally related in virtue of "economic laws", which are imagined to operate blindly.

The negative form of *potere* is used to express economic *impossibility*. The combination of this same possibility modal with *only* type quantifiers is, remarkably, used to express economic *necessity*, as shown in (50) and (51).

(50) Meno rosee le prospettive per i consumatori: secondo Browne il prezzo della benzina **non potrà che** salire data l'impennata del greggio. In Gran Bretagna il prezzo della benzina 'potrebbe salire oltre una sterlina al litro', ha detto Browne, che ha pero tenuto a sottolineare che BP passa ai clienti una frazione degli aumenti dei costi reali. (*Il Sole 24 Ore*, April 14 2006).

'Less rosy outlook for consumers: according to Browne gasoline prices **cannot but** rise in view of the surge of crude oil price. In Great Britain the price of petrol 'could rise to more than one pound per liter,' said Browne, who was keen to stress however that BP passes onto its customers a fraction of the actual cost increases.'

(51) 'Solo se ci fossero segnali di possibili ribassi del costo del denaro in America si potrebbero giustificare rendimenti decennali così contenuti - osserva Holger Schmieding, co-responsabile settore economico europeo di Bank of America. Ma siccome non ci sono questi segnali, anzi la tendenza è quella opposta, i rendimenti decennali **non possono fare altro che** aumentare'. Il ragionamento è semplice: chi investe su una scadenza decennale vuole solitamente interessi più elevati rispetto a chi acquista obbligazioni a breve scadenza perché il rischio"temporale" è maggiore' (*Il Sole 24 Ore*, April 14 2006).

'Only if there were signs of possible declines in the cost of money in America one could justify so limited ten-year yields - observes Holger Schmieding, co-head of European economic sector Bank of America. As we do not have these signs, and, in fact, the trend is the opposite, the ten-year yields **can only** increase. 'The reasoning is simple: investors on a ten-year maturity typically want higher interest rates than those who buy short term bonds because the temporal risk is greater.'

The *only* quantifiers in (50) and (51) assert that there is no proposition in a contextually relevant contrast set C other than the prejacent that is compatible with the conversational background. These quantifiers also presuppose that the prejacent is indeed compatible with B (cf. Horn 1996 on the semantics of *only*). *Only,* by asserting the impossibility of the alternatives to the prejacent in C, transforms the modality expressed by *potere* in an indirectly expressed alethic circumstantial *necessity* ($\neg\diamond\neg$ p), under the assumption that the alternatives in C are exhaustive.

It is worth pausing on these two examples to observe how the functioning of these two expressions of alethic circumstantial necessity relates to the argumentative manifestation of the evidence on which the predictions are based. As we know from Chaps. 4 and 5, the key lies in observing their behaviour as context-dependent expressions that seek to saturate the B variable in the discourse context. In the examples (50) and (51) the saturation is provided, in part, by the causal clauses that precede the modal ('Given the surge in crude oil price', 'Since these signals are not there and, actually, the trend is the opposite') which, in combination with the other circumstances of discourse and the relevant nomic regularities, yield the causal necessity.

This causal chain is at the same time an argumentative one and constitutes the evidence on which the prediction is based[29]. Contrary to what genuine epistemic inferential modal expressions do, these alethic circumstantial modals do not refer directly to the speaker's knowledge, beliefs or reasoning processes but to the underlying real-world relations –efficient cause in this case– that license the inference.

The extended text of example (51) is noteworthy also for another reason: it contains, in the last sentence, a shift between what we have called economic causality and its main, underlying, more basic form of connection: *economic rationality*.

The latter concerns the goal-oriented behaviour and practical reasoning of human agents (investors in this case) and arguably involves *deontic-practical* rather than *alethic*, modalities (cf. Portner 2009: 185–6). From an argumentative point of view this would imply a shift from the locus of *efficient cause* to the locus of the *final cause*. However, under the assumptions of classic, non-behavioral, economics, these rational connections can be equated with blind causal connections because the agent's goals are considered as given (utility maximization) and the rationality of the agent is presumed[30].

6.4.4 Quantificational Readings of the Possibility Modal

Let us now consider example (52) below. It selects an alethic circumstantial conversational background of economic facts including *oil price* and *rising interest rates* and relates them to *consumer spending* in a temporally generic statement:

(52) Gli altri rischi che gravano sull'evoluzione del commercio internazionale sono di natura macroeconomica: prezzo del petrolio e rialzo dei tassi d'interesse *possono* influire sui consumi. (Il Sole-24 Ore, April 12, 2006).
"The other risks that loom on the evolution of international trade are macroeconomic in nature: oil price and rising interest rates can influence consumer spending".

This reading is close to the so-called "sporadic" (Kleiber 1983) or "quantificational" (Portner 2009) reading of possibility modals, canonically exemplified by *Lions can be dangerous* (see Chap. 4, § 4.3.3 and Chap. 5, § 5.2.4.2). It does not directly express a prediction concerning a singular event ('oil price and rising interest rates

[29] More precisely, we have here an instance of the argumentation scheme from the cause to the effect. This argumentation scheme exploits the basic entailments of the commonsense ontology of causation to infer a conclusion. If we admit that 'p *causes* q' entails that, for all the worlds where certain physical laws hold, wherever p is the case, q is also the case, we can use this entailment to infer q from p. For an in-depth discussion of contemporary theories of argument schemes and a theoretical proposal making explicit the connection between argument schemes and commonsense ontologies see Rigotti and Greco Morasso (2010).

[30] Additionally, the incompleteness of the information available to the agents is ignored as the markets are presumed to be informationally efficient.

will possibly influence consumer spending' or 'consumer spending will possibly decrease'). Instead, it offers a generalization (possibly inductively based on previous comparable occurrences) that supports the implicit prediction with an implicit weak epistemic evaluation[31].

6.4.5 *Economic Circumstances and Agent's Goals*

In the corpus the necessity modal *dovere* is never found with a strictly alethic circumstantial conversational background corresponding to what we have called economic necessity. This appears to be consistent with the use of the combination of *potere* with quantifiers of the type of *only* to represent this notion that we have seen in (50) and (51). Certain purely alethic circumstantial readings of *dovere* are possible, especially when dealing with physical causality and non-human events, as when uttering (53) under a cloudy sky[32]:

(53) **Deve** piovere.
'It's going to rain.'

In contexts such as (49) and (50), however, purely alethic circumstantial interpretations of indicative *dovere* appear more difficult to access than the deontic ones. We have, however, a number of occurrences of future-tense *dovere* that come *close* to expressing alethic circumstantial necessity, but are subtly different, such as for example (54).

(54) Giovedì si è passata per la prima volta in quattro anni la soglia del 5% per i tassi a dieci anni. La conseguenza più immediata di questo aumento ricade sui tassi per i mutui immobiliari. Chi aveva contratto mutui a tassi variabili - e sono stati in molti - si trova alla scadenza del primo periodo e **dovrà** rinegoziare tassi di 200 o 300 punti superiori a quelli di un paio di anni fa. Questo significa che una famiglia media con un mutuo di 400mila dollari potrebbe trovarsi a dover pagare anche fino a mille dollari in più al mese.
'On Tuesday the threshold of 5% for fixed ten-year interest rates was passed for the first time in four years. The most immediate consequence of this increase will be on mortgage rates. Those who had subscribed variable rate mortgages – and there are many – are at the end of the first period and **will have to** renegotiate rates of 200 or 300 points higher as compared with those of a couple of years ago. This means that an average family with a loan of 400 thousand U.S. dollars could have to pay even up to a thousand dollars more a month.'

Two factual propositions are anaphorically recovered by the preceding discourse context and are added to the conversational background: 'Ten-year fixed interest rates have climbed over the 5% threshold' and 'Those who had subscribed variable

---

[31] Compare with the inferential chain: *Lions* ***can*** *be dangerous* & *Lambert is a lion*, therefore *Lambert* ***may*** *be dangerous.*

[32] See Chap. 5, § 5.2.4.1 for a discussion of similar alethic readings of French *devoir*.

rate mortgages are at the end of the first period'. These are social, economic facts. These facts alone, however, are not sufficient to make necessary the renegotiation of the mortgage with rates of 200 or 300 points higher than those of a couple of years before. Other institutional facts are possible as outcome.

In order to make the proposition necessary, we need to add a nonfactual proposition: a goal like 'if they want to keep their houses', 'in order to keep their houses'. What we have here is a combination of an alethic circumstantial conversational background and a teleological conversational background typical of deontic-practical modality. Deontic conversational backgrounds containing ideals, goals, values, or laws[33] always need to combine with actual circumstances in order to make singular propositions necessary. Alternatively, if we assume a speaker who is completely detached from the goal that the mortgage subscriber keep their houses – and talk about the economy is often not far from such a detached authorial voice – the goal becomes a mere *telos* in the degraded sense associated to anankastic modality, and the occurrence of *dovere* is to be considered anankastic rather than deontic-practical. What makes examples such as (54) special, however, with respect to other examples of deontic-practical or anankastic modality, is the fact that it is taken for granted that the agents *will* rationally act to fulfil their goal. This additional premise makes the modal a predictive one, equivalent to the alethic modal.

Example (55), below is of the same type:

(55) Un aumento dei tassi a lunga inoltre potrebbe scoraggiare gli investimenti, l'unico vero supporto alla crescita rimasto nel contesto macroeconomico americano. Il mercato immobiliare che negli ultimi anni ha offerto performance stellari, si e già leggermente indebolito. Ma il cuore del problema è il pericolo di un indebolimento della domanda interna. Questo capiterà nel momento in cui il consumatore dovrà già destinare una parte crescente del suo reddito disponibile alla copertura degli aumenti del prezzo del greggio. Le stime per il costo della benzina nel corso dell'estate sono di circa il 20% superiori rispetto all'estate dell'anno scorso. L'auspicio è che l'aumento dell'occupazione compenserà le diminuzioni dei consumi.

'A raise in long term interest rates, moreover, might discourage investments, the only true support to growth left in the American macro-economic context. The real estate market, which offered stellar performances during the last few years, is already slightly weakened. But the heart of the matter is the danger of a weakening of internal demand. This will happen when consumers ***will have to*** devote an increasing share of their available income to cover raising [crude] oil prices. The estimates for the cost of gasoline are about 20% higher compared to the summer of last year. The hope is that the increase in the rate of employment will compensate the decrease in consumption'.

6.4.6 *Deontic Readings of* potere

Deontic readings of *potere*, such as the future-tense reading in (56), can also give rise to predictions:

[33] Not *all* deontic conversational backgrounds need to combine with circumstantial ones: directive and commissive ones already originating from singular commands or singular promises do not.

(56) Europa (e mondo) attenti, a Citigroup sono state tolte le redini. La principale banca americana e stata perdonata dalla Federal Reserve e, per crescere, **potrà** tornare a fagocitare prede sui mercati globali: dopo un anno di forzata moratoria sulle grandi acquisizioni, imposta davanti alla scoperta di scandali e inadeguati controlli interni, la Fed ha concluso che il colosso dei servizi finanziari ha "compiuto significativi progressi" nella governance e nella gestione del rischio, sufficienti a togliere i freni a piani di conquista. Anche se ha mantenuto un avvertimento: 'Esamineremo attentamente ogni proposta di espansione di Citigroup'. (*Il Sole 24 Ore*, April 5, 2006)

Europe (and world) beware, Citigroup has been unleashed. The major U.S. bank was pardoned by the Federal Reserve and, in order to grow, **will** again **be able/allowed** to engulf prey on global markets: after a year of enforced moratorium on large acquisitions, set on the wake of the discovery of scandals and inadequate internal controls, the Fed concluded that the financial services giant has "made significant progress" in the governance and risk management, sufficient to remove the brakes to plans of conquest. However, the Fed maintain a warning: 'We will carefully review any proposal for expansion of Citigroup'.

The deontic conversational background of *potrà* in (56) originates from the directive power of the Fed, which has eased a prohibition issued against Citigroup. This is equivalent to permission. In order to understand how this modality works in the prediction, we have to pause for a moment on the nature of the prejacent which concerns 'Citigroup taking over other companies'. Corporate takeovers are institutional actions, rather than natural actions, and norms have causal power over them. This means that the deontic prohibition of the Fed amounts to a special kind of alethic impossibility: it is not that a takeover without the Fed's consent would be *illegal*, it would be *void* or *null*. Thus, the deontic modal ends up counting as a circumstantial alethic one. The circumstantial possibility is then combined with an additional premise concerning the intention ("in order to grow", "plans of conquest", etc.) to derive an indirect prediction. The predictive inference in this case is based on a maxim of the locus *from the final cause* rather than on the kind of impersonal economic causality found in the previous examples. The inferential configuration underlying this prediction can be reconstructed as in Fig. 6.15.

From the reconstruction in Figure 15, it can be observed that the deontically modalized proposition functions as a minor premise – a datum – supporting the (implicit) act of prediction of the same prejacent.

6.4.7 *Deontic Readings of* dovere*: Schedules and Plans*

The necessity modal *dovere* is routinely used in predictions in the corpus with different deontic conversational backgrounds. One type of deontic conversational background found in predictions consists of formal or informal commitments made by institutions, and particularly companies, in the form of *plans* or *schedules*. In (57) below we find a basic example of this recurrent pattern:

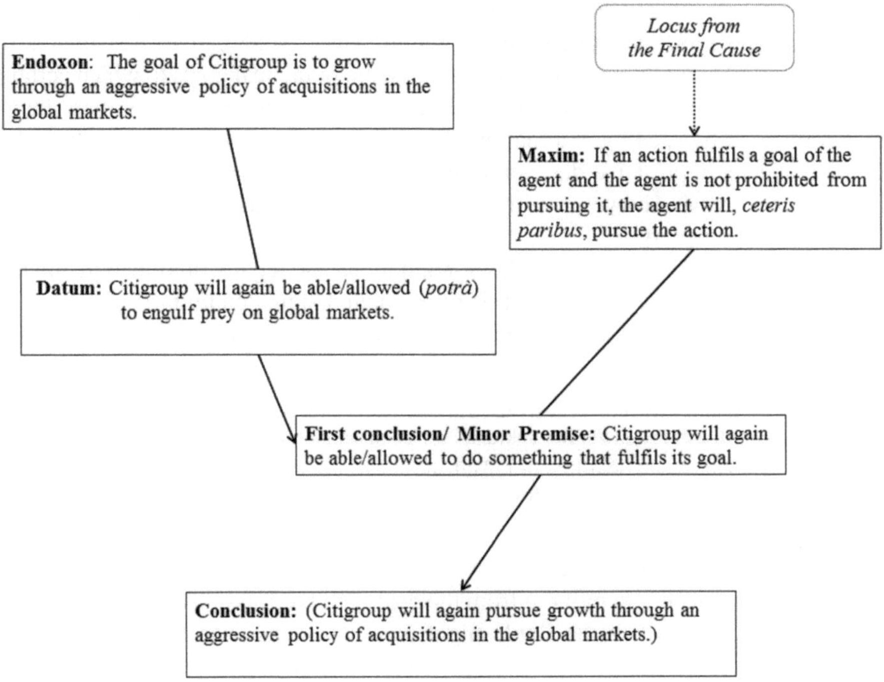

Fig. 6.15 Inferential configuration (Y-structure) of the prediction implied by deontic *potrà* in example (56)

(57) Il prossimo 20 aprile gli azionisti ***dovranno*** votare sul fatto che sussistano ancora o meno i requisiti di onorabilità richiesti al manager per presiedere una banca.
'On April 20 the shareholders **are to** vote on whether the manager still satisfies the requirements of honorability necessary to chair a bank.'

Other researchers working on Romance modal verbs with journalistic corpora have remarked on these uses and provided divergent interpretations.

Kronning (1996, 2001) insists on the predictive nature of similar uses of French *devoir* and for this reason he considers them a subtype of alethic modality, a "restricted" alethic modality. According to Kronning, these uses are based on narrative scenarios to which the future is expected to conform: the prejacent p is true in all the worlds consistent with the *scenario*.

Squartini (2004) examines similar examples in French and Italian, coming to a diametrically different conclusion. For Squartini, examples like (56) represent evidence of a *reportive evidential* reading of *dovere/devoir*.

In fact, the conversational background B invoked by *dovere* in examples such as (57) is neither *alethic* (propositions that are facts in the world) nor primarily *repor-*

tative[34] (meta-represented propositions asserted by another speaker), but rather *deontic* (propositions denoting norms, commitments or goals). Plans formulated by companies and other organizations denote strategic or tactical goals to which certain members are committed. At the same time, plans take the form of written documents that can be used by external observers as sources of information. Thus, the deontic and the reportative do overlap in the journalist's perspective. The deontic commitment remains however primary with respect to its use as evidence in a prediction by a third party.

Regulative rules, in the form of laws, regulations and legally binding agreements (e.g. contracts) can also form the basis of prediction signalled by indicative *dovere* on the basis of the assumption that the subject can and will probably abide the regulative rules:

(58) Gli statunitensi posseggono il 15% di Lukoil, precisa il gestore, e in base agli accordi con i russi ***devono*** salire al 20% acquistando titoli sul mercato.
'The Americans hold 15% of Lukoil, the money manager elaborates, and according to the agreement with the Russians, they *are to* climb up to 20% buying stocks on the market.'

Neither kind of deontic background warrants an attitude of absolute certainty: in (58) the Americans might pull out from the deal with the Russians by paying a penalty, in (57) the board of the Bank might just cancel the shareholders' meeting. Interestingly, while the addressee may well assess the degree of certainty of the prediction on the basis of the nature of the deontic background involved, the writer, by using a deontic modality, does not commit himself directly to a given degree of epistemic certainty towards the prediction. Should the implicated prediction turn out to be false, the speaker could always claim that the deontic modality was correct.

[34] In terms of the typology of conversational backgrounds that I have adopted in Chap. 5, a reportative conversational background can be considered a sub-type of the epistemic conversational background, corresponding to the beliefs to which the source is committed (the beliefs of the source's social persona, independently of their sincerity). The speaker may or may not associate herself with this belief set. Faller (2011) argues that certain reportatives fall outside the scope of epistemic modality and should be based on a separate *informational* conversational background because they can be used also when the speaker does not believe the prejacent. This is consistent with her adoption of a strict definition of epistemic modality in terms of knowledge. In fact, the broader definition of epistemicity in terms of a set of *beliefs associated with an information source* which was adopted in Chap. 5, can be more advantageous in dealing with issues of evidentiality, subjectivity vs. intersubjectivity, polyphony and point of view that arise with epistemic modals.

6.5 Semantic Analysis and Argumentative Functioning of $DEVE_E$ and $DOVREBBE_E$

6.5.1 Refining the Hypothesis in the Light of the Non-epistemic Readings of the Modals

The business newspaper corpus can provide interesting data relevant for testing and refining the hypothesis that $DOVREBBE_E$ is not directly related with $DEVE_E$ and represents a conditional, hypothetical, version of a non-epistemic reading of the modal *dovere*. More precisely, according to Kronning's hypothesis, $DOVREBBE_E$ should be a hypothetical version of an *alethic* modal.

A number of examples of $DOVREBBE_E$ of the type of (59) lend credence to the hypothesis that the modal has indeed a conversational background consisting of particular facts, both physical and economic facts, and general statements of economic laws presupposing an underlying notion of economic causality.

(59) THAILANDIA. L'economia di questo paese è stata colpita duramente dagli effetti dello tsunami abbattutosi sulle sue coste nel dicembre del 2004 ma una ripresa ***dovrebbe*** registrarsi già a partire da quest'anno grazie al flusso di nuovi investimenti e al recupero dei consumi privati. Il piano di investimenti in infrastrutture metterà ***tuttavia*** alla prova la solidità delle istituzioni finanziarie, mentre la bilancia commerciale risente del rialzo del prezzo del petrolio e di un calo dei flussi turistici. (*MF* 7/4/2006)

'THAILAND. The economy of this country has been hit hard by the effects of the tsunami that hit its shores in December 2004 but a recovery **should** be observed as early as this year thanks to the flow of new investment and the recovery of private consumption. The plan for infrastructure investments, **however**, will test the soundness of financial institutions, while the trade balance is affected by the rising oil prices and a decline in tourism.'

In (59) the 'flow of new investment' and the 'recovery of private consumption' appear as good candidate propositions for an alethic conversational background causally supporting the prediction of economic recovery.

As shown by the immediately following counterargument introduced by *however*, these facts are insufficient to necessitate the conclusion: shaky financial institutions under stress, raising oil prices and declining tourism may still affect the outcome (Fig. 6.16).

The possibility of being the target of a counterargument appears to be another feature of the conclusions introduced by $DOVREBBE_E$ that sets them apart from those marked by $DEVE_E$, where such a discourse relation would be impossible, as shown by the altered version of example (29) presented below:

(29.b) E a giudicare dalle plusvalenze e dai rendimenti ottenuti in questi ultimi tre anni, sembra che le scelte effettuate siano state azzeccate. Dunque, grande **deve** essere la soddisfazione tra i soci. ?? Essi potrebbero, **però**, essere invece preoccupati per il livello di indebitamento dell'impresa. (alteration of example 36 in this chapter)

'And judging from the gains and efficiencies achieved in the last three years, it seems that the choices made were justified. So there **must** be great satisfaction among the shareholders. ?? They could be, **however**, rather worried by the debt situation of the company'.

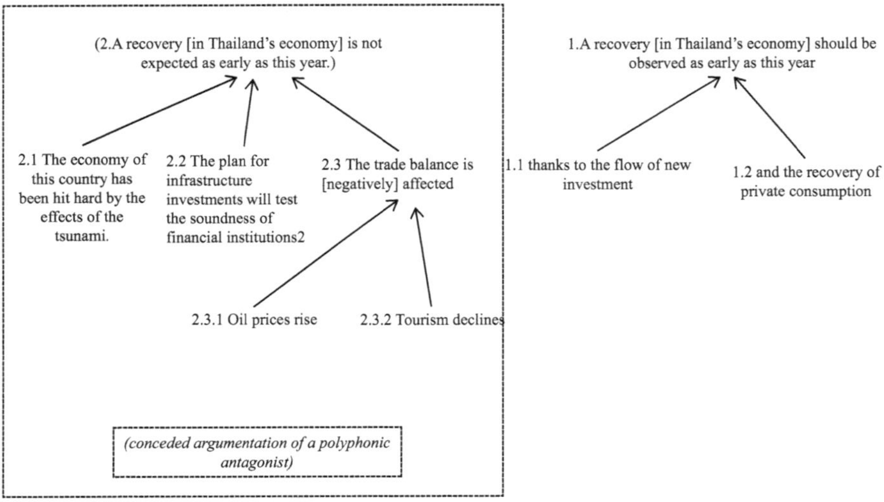

Fig. 6.16 Argumentation Structure of Example (59)

The only possibility of saving (29.b) is assuming that the shareholders can be *satisfied* and *worried* at the same time. Example (29) is interesting also because, being an argument from the cause to the effect and having a prejacent of the modal consisting of a present stative eventuality ('there is great satisfaction among the shareholders'), provides one of the relatively rare contexts where DEVE_{E} and $\text{DOVREBBE}_{\text{E}}$ are mutually substitutable. Interestingly, if we substitute $\text{DOVREBBE}_{\text{E}}$ for DEVE_{E} in the modified example in (36.b) the counterargument becomes acceptable, as shown in (29.c):

(29.c) E giudicare dalle plusvalenze e dai rendimenti ottenuti in questi ultimi tre anni, sembra che le scelte effettuate siano state azzeccate. Dunque, grande **dovrebbe** essere la soddisfazione tra i soci. Essi potrebbero, **però**, essere invece preoccupati per il livello di indebitamento dell'impresa. (alteration of example 36 in this chapter)

'And judging from the gains and efficiencies achieved in the last three years, it seems that the choices made were justified. So there **should** be great satisfaction among the shareholders. They could be, **however**, rather worried by the debt situation of the company'.

It has been observed (cf. § 6.2.3) that forecasts issued by financial institutions, and, in general, predictions in economics are conditional in nature, valid as long as certain background assumptions hold, a feature which is reflected in the business news. Therefore, the unconditional examples of alethic *dovere* that represent the "source" are expected to be rare. As shown in § 6.4.5, however, it is still possible find examples like (54) where *dovere*, especially in the future tense of the indicative, comes *close* to an unconditional prediction based on economic circumstances:

(54) Giovedì si è passata per la prima volta in quattro anni la soglia del 5% per i tassi a dieci anni. La conseguenza più immediata di questo aumento ricade sui tassi per i mutui immobiliari. Chi aveva contratto mutui a tassi variabili - e sono stati in molti - si trova alla scadenza del primo periodo e ***dovrà*** rinegoziare tassi di 200 o 300 punti superiori a quelli di un paio di anni fa. Questo significa che una famiglia media con un mutuo di 400mila dollari potrebbe trovarsi a dover pagare anche fino a mille dollari in più al mese. (*Il Sole 24 Ore,* 15/4/2006, reproduction of example 54)

'On Tuesday the threshold of 5% for ten years interest rates was passed for the first time in four years. The most immediate consequence of this increase will fall on mortgage rates. Those who had subscribed adjustable-rate mortgages – and there are many – are at the end of the first period and **will have** to renegotiate rates of 200 or 300 points higher than a couple of years ago. This means that an average family with a loan of 400 thousand U.S. dollars could have to pay even up to a thousand dollars more a month.'

{'Ten years interest rates have climbed over the 5% threshold', 'Those who had subscribed adjustable-rate mortgages are at the end of the first period', ...}$_{B\text{-}A}$ □ 'They renegotiate rates of 200 or 300 points higher than a couple of years ago'

Not all examples of DOVREBBE$_E$ found in the corpus correspond to a background of economic circumstances. A number of examples similar to (60) are also found:

(60) Stando a quanto emerso ieri nella riunione del cda Bnl, Bnp *sarebbe* orientata a lanciare la prossima settimana l'Opa, che ***dovrebbe*** concludersi tra il 15 e il 20 maggio. Secondo indiscrezioni la banca di Parigi *avrebbe predisposto* tutto per annunciare già stasera l'ok della Consob e i dettagli dell'operazione, con le date di inizio e di conclusione. (*Il Sole 24 Ore*, 13/4/2006)

'According to what transpired from yesterday's meeting of the board of BNL, BNP will reportedly be inclined to launch next week the takeover bid, which should be completed between the 15th and 20th of May. According to rumors, the Paris bank prepared everything to announce this evening the OK of Consob and the details of the transaction, with the dates of commencement and conclusion.'

These examples invoke a plan or some kind of schedule – hence a very specific kind of deontic conversational background – rather than any form of causality. They may appear in close proximity with *reportive evidential* uses of the Italian conditional mood (the forms *sarebbe* and *avrebbe* in the example) so that one may wonder whether the presence of the conditional in them might have relationship with that evidential value of the conditional. At the same time, we have seen in 4.6, above, that these uses are closely matched by examples with *dovere* in the indicative, especially with the indicative future tense as in (57), reproduced here, which can also be used to refer to planned/scheduled events:

(57) Il prossimo 20 aprile gli azionisti ***dovranno*** votare sul fatto che sussistano ancora o meno i requisiti di onorabilità richiesti al manager per presiedere una banca.

'On April 20 the shareholders are to vote on whether the manager still satisfies the requirements of honorability necessary to chair a bank'

In conclusion, the basic conversational background of DOVREBBE$_E$ can be either circumstantial *alethic* or *deontic* and the meaning of its conditional morpheme can be *evidential* or *hypothetical.* The following sections will give a more precise formulation to this analysis and explore some of its further consequences for the analysis of argumentative discourse relations.

6.5.2 Dovrebbe *in the Corpus*

We first present the semantic analysis we propose for the modal and then move to examining how this analysis accounts for the evidential and argumentative properties of this construction in the predictions found in the corpus. In order to do so we have conducted a systematic analysis of the conversational backgrounds of all the occurrences of *dovere* in the conditional mood in the 200 texts of the annotated sub-corpus.

As shown in Table 6.7, the conditional of *dovere* occurs in the 3rd person only, mostly in the singular, and nearly always has a predictive function which can be traced back to the $\text{DOVREBBE}_{\text{E}}$ construction. The remaining four occurrences are purely deontic, comparable to English deontic *ought* (cf. von Fintel and Iatridou 2008), and will not be discussed here.

6.5.3 *Semantic Analysis*

Here I finally propose a more explicit semantic analysis of DEVE_{E} and $\text{DOVREBBE}_{\text{E}}$. The analysis proposed here is based on a context-dependent semantics inspired by Relative Modality but differs from basically monosemic treatments of modal verbs by postulating that, while all readings of the modal *dovere do* share the same basic context-dependent relational core, certain broad classes of interpretations are also encoded in the language, conventionally associated with the presence of certain formal or semantic constraints. As observed in § 6.1.3.3, this approach can be qualified as *procedural* and *constructional.*

The modal constructions DEVE_{E} and $\text{DOVREBBE}_{\text{E}}$ are *procedural* linguistic units because they function as a virtual instruction to the addressee to examine the common ground to fill in the variable B along *preferential paths that constrain the saturation*. These paths are conventionally associated with certain formal and/or semantic restrictions in the construction.

A general procedure for the contextual saturation of modal conversational *backgrounds* was proposed in Chap. 5, § 5.2.4.4. Drawing on Hacquard's (2010) idea of *event-relativity*, the procedure selects the composition of the conversational background B in relation to an event or state of affairs *r* functioning as deictic or anaphoric reference point. The traditional distinction between *epistemic* and *root* modals is thus reformulated in terms of the selection of the reference eventuality *r*:

Table 6.7 The conditional of dovere in the annotated 200 texts sub-corpus

Wordform	Morphological gloss	Tokens	In prediction
Dovrebbe	cond. pres. 3rd sing.	56	53
Dovrebbero	cond. pres. 3rd plur.	11	10
Total		67	63

epistemics are relative to a speech or mental event *s*, while root modals relate to an event *e* corresponding to the prejacent VP-event.

6.5.3.1 Semantics of DEVE$_E$

Formally, the construction DEVE$_E$ is specified at the level of verbal morphology and is further specified by the temporal-aspectual semantic restriction on its infinitival complement. In this sense it represents a marked path of saturation of the modal. I propose that this construction is associated with the following semantics:

(Def. 4) DEVE$_E$ (*s*, *Spk*, B, *p*):

Presuppositions:

a) *s* is a speech event
b) Spk is the speaker of *s*
c) The VP-event of the prejacent *p* is an *e* such that $e \leq s$
d) $B = B_E \cup B_{Dox}$
e) B_E is a set of propositions corresponding to factual evidence available to *Spk* at the moment of *s*
f) B_{Dox} is a set of propositions corresponding to assumptions of *Spk* at the moment of *s*
g) $p \notin B$

Semantic entailment: B ❑ *p*

The presuppositional precondition (c) describes the incompatibility of DEVE$_E$ with prejacents denoting future eventualities. The analysis states thathe conversational background of DEVE$_E$ is saturated deictically with respect to the speaker and the utterance event (*s*). The B combines two distinct sets of propositions that the addressee is prompted to recover: the first (B_E) – where *E* stands for "epistemic" but also for "evidential" – contains factual evidence available to the speaker, the second (B_{Dox}) – Dox stands for doxastic – corresponds to a set of assumptions of the speaker at the moment of utterance, which the speaker might entertain with a variable degree of certainty. This enables to capture the fact that DEVE$_E$ is highly subjective referring to processes taking place in the mind of the speaker, yet anchored to the presence of specific external evidence, and that it is compatible with different forms of deductive and abductive reasoning of variable strength according to the kind of beliefs and assumptions making up B_{Dox}[35]. Finally, the presupposition '$p \notin B$' means that the prejacent proposition *p* is not already part of the evidence, nor of the

[35] In deductive reasoning B_{Dox} will typically contain more or less probable or plausible propositions that are invoked as major premises in the deduction (cf. also Desclès and Guentchéva 2001: 115–116). In abductive reasoning, B_{Dox} will crucially also contain the assumption that besides p there are no other explanatory hypotheses worth considering . See the discussion of abduction in footnote 21, above.

assumptions of the speaker, that make up B, even if it is necessarily entailed by them[36].

This is essential in accounting for the inferential-evidential nature of $DEVE_E$ and of its incompatibility both with direct perceptual evidence, and with information previously acquired through testimony. An evidential marker, $DEVE_E$ becomes argumentative when the inferential path of the speaker is made, to some extent, available to the hearer and the hearer is implicitly invited to follow it himself (See Chap. 5, § 5.3.4).

6.5.3.2 Semantics of Root *dovere*

The distinguishing feature of the *root* meanings of *dovere*, according to my reconstruction of Haquard's proposal, is that the saturation of the B takes as reference point not the speech event but the eventuality denoted by the prejacent proposition. The choice of the propositions selected as B depends, in part, on the ontology of the eventuality at issue.

For natural events that do not involve human agents only factual circumstances causally necessitating the event tend to be relevant, so that a causal alethic interpretation is easily obtained. When the eventuality is a human action, the full ontology of action comes into play making relevant the *efficient causes* of the event – that is both the external and the internal circumstances to the agent – and the *final causes*, that is the *goals* of the agent. Interactions, and any human action placed in a social setting provide a further kind of B, giving access to social *commitments* of all kinds. Instead of providing a semantics with full-blown presuppositions characterizing each argument place, I formulate in instructional terms the underspecified linguistic meaning associated with root modals:

(Def. 5)	DOVERE_{root} (B, *p*):	
	Instruction:	Saturate B looking for salient circumstances of the eventuality *e* denoted by *p*.
	Semantic entailment:	B ❑ *p*

I will not discuss here whether root interpretations of *dovere* involve more specific conventionalized constructions for *alethic*, *deontic* and other kinds of modality. I

[36] It is fair to mention here that the notion that a proposition p can be *entailed by* B without *being part of* B requires us to go beyond the ontology of propositions presupposed by possible worlds semantics, and thus beyond the Kratzerian view of the conversational background. If each proposition of B is *just* a set of possible worlds at which it is true, and B is *just* the inter of all these sets, giving the set of possible worlds at which *all* the propositions are true, for any proposition p being a consequence of B and being a set of worlds included in the inter are one and the same thing. In order to distinguish between the two one needs either to consider propositions as mental representations and see B as a set of mental representations (cf. Papafragou 2000), or to model the semantics of the modals dynamically as test operations on the state of information (see Portner 2009: 85-105 for a concise and relatively accessible presentation of the dynamic semantics literature). I will not pursue this matter further in this paper.

will instead immediately move to examine what is added by the conditional mood morphology to the interpretation of DOVREBBE$_E$.

6.5.3.3 Semantics of the Italian Conditional Mood Construction

Here I will use an analysis of the semantics of the Italian conditional mood morpheme in terms of relative modality first proposed and discussed in Rocci (2006). This analysis is in substantial agreement with other recent analyses of the semantics of Romance conditional morphemes, such as Kronning (2002) and Miecznikowski and Bazzanella (2007), which, while not cast in terms of relative modality, propose a relational and context-dependent "basic semantics" to account for the variety of uses of this form. The basic idea of these analyses is that the conditional marks a proposition as a consequence of an antecedent, which is non-factual (or "distant from the modal-deictic centre", as Bazzanella an Miecznikowski 2007 put it). The context-dependent aspect of this undespecified meaning is reprented by is represented by the contextual saturation of the antecedent. Kronning's (2002) proposal explicitly includes an instructional component in the semantics providing constraints on the saturation of the antecedent in syntactic, discursive and situational context in order to derive the different readings of the conditional. This kind of semantics can be presented in terms of relative modality as follows:

(Def. 6) CONDITIONAL (*Spk*, w_0, C, p):

Presuppositions:

a) *Spk* is the speaker of the utterance
b) w_0 is the actual world
c) C is a set of propositions so that they are not facts for *Spk* in w_0
d) p is either the propositional content of the utterance or the performance of the speech act by *Spk*

Semantic entailment: $C \Box p$

The semantics proposed treats the morpheme of the Italian conditional mood as a necessity modal: *p* holds true in all the worlds at which the non-factual propositions making up C are true. Given the status of C, the actual world w_0 may well not be part of the set of worlds over which the modal quantifies.

Three major modal[37] uses of the Italian conditional are generally recognized: *hypothetical*, *reportive* and *attenuative*. In the *hypothetical* use the condition C is identified either with a syntactic non-factual antecedent (an if-clause), as in (61.a), or with a contextual or anaphorically recovered discourse antecedent, as it happens in (61.b) where a non-factual modalized proposition provides the saturation for C (cf. Roberts 1989; Geurts 1999 on "modal subordination"). In the *reportive* use,

[37] I do not discuss here the purely temporal use of the Italian conditional, where the morpheme is used as a mark of "future in the past". In Rocci (2006) I briefly discuss how the temporal meaning of posteriority could be derived from a modal meaning of causal necessity, in parallel with a modal analysis of the Italian future tense presented in Rocci (2000) and Rocci (2005b).

exemplified in (61.c), the propositions in C are identified with those making up the discourse of someone other than the speaker. Finally, in the *attenuative* use, exemplified by (61.d), the set C is identified either with (some of) the proper illocutionary preparatory conditions of a speech act, or with broader social preconditions for the performance of the speech act (cf. Miecznikowski and Bazzanella 2007).

(61.a) Lo [= un fallimento] ***sarebbe se*** in lui non ci fosse quella tale passione, quel gusto e quel piacere di vivere, nonostante tutto (*Il Sole 24 Ore*, 30/4/2006).

'It ***would be*** a failure ***if*** he did not have that king of passion in him, that gusto, that pleasure for living despite everything'

(61.b) [...] tutti i partner dello studio legale ***potrebbero*** investire nel fondo, ma ***sarebbero vincolati*** per dieci anni. (*MF*, 14/4/2006)

'[...] all the partners of the legal practice ***could*** invest in the fund, but they ***would be bound*** for ten years [to the investment]

(61.c) Alcune finanziarie ***avrebbero fatto*** sottoscrivere contratti a tassi di interesse annui prossimi al 30 per cento. (*Il Sole 24 Ore*, 26/4/2006)

'Certain financial firms ***reportedly had*** [pensioners] subscribe contracts with interest rates close to 30 percent'

(61.d) ***Vorrei*** rispondere all'articolo intitolato [...] apparso sul Sole-24 Ore del 20 marzo scorso (*Il Sole 24 Ore*, 13/4/2006)

'I ***would like*** to reply to the article entitled [...] which appeared in the Sole-24 Ore of March 20'

In all its uses, the conditional mood, while marking morphologically the consequent of an intensional conditional, presupposes an *epistemically evaluated antecedent*. The range of epistemic evaluations compatible with the conditional *includes counterfactuality proper* and excludes the speaker's *full commitment* to the factuality of the antecedent, allowing for a variety of degrees of certainty in between.

6.5.3.4 Semantics of DOVREBBE$_E$

DOVREBBE$_E$ is a non-compositional construction which partially inherits the basic semantics of the conditional mood and the semantics of the root interpretations of *dovere*, adding further constrains on the contextual saturation of one of its arguments. We can represent it as follows:

(Def. 7) DOVREBBE$_E$ (*Spk*, *e*, w_0, C, B, *p*):

Presuppositions:

(a) *Spk* is the speaker of the utterance

(b) *e* is the VP-event of the prejacent *p*

(c) w_0 is the actual world

(d) C is a set of propositions that are not facts for *Spk* in w_0 and are weak presumptions in the relevant epistemic community.

(e) B is a set of propositions relative to the circumstances *c* of *e*, which either are true in w_0 (*alethic*) or correspond to commitments in w_0 (*deontic*).

(f) $e \geq c$.

Semantic entailment: C∪B ❑ *p*

The construction involves a double conversational background consisting of the conjunction – set theoretic union[38]– of an alethic or deontic *modal base* (B) and a *conditional restriction* (C).

The presuppositional condition (d) that DOVREBBE$_E$ imposes on the saturation of C is important for the explanation of the evidential and argumentative functioning of the construction.

First, C is composed of propositions that are *not known to be facts in the world.* They are non-factual, or counterfactual in a weak sense: either they may turn out to be false or, on occasion, are already known to be false (strict counterfactuality).

Second, C corresponds to propositions for which there exist "weak presumptions". The following section will examine the composition of B and C in all the occurrences of DOVREBBE$_E$ in the annotated sub-corpus of 200 texts. It will turn out that the conditional restriction C always belongs to one of a very restricted inventory of weak presumptions, which includes conditions like: 'the trend continues', 'the situation remains stable', 'expectations are confirmed', 'hypotheses/ theories are correct', 'plans are realized', 'what insiders say is true', 'what experts predict is correct'. In the absence of prominent sets of conditions recoverable from the co-text or context, C is identified with a set of propositions which are *normal conditions*[39].

[38] Explaining how people manage to constantly adapt their idea of "what is normally the case" is certainly a very complex matter, but perhaps this complexity is wholly pragmatic and should not be saddled onto the semantics of modal expressions. This pragmatic solution contrasts with the semantic one proposed by Kratzer (2012 [1981]) with *ordering semantics* (see Chap. 4, § 4.3.4 for a brief discussion of the basics of ordering sources and Chap. 5, § 5.3.2.3 for a more in-depth application to practical reasoning).

[39] As mentioned above, the importance of presumptions of normality had already been pointed out in relation to the functioning of DEVRAIT$_E$ by Tasmowski and Dendale (1994: 50-51). Conversational backgrounds corresponding to normal conditions feature prominently also in Kratzer's (1981) classic paper. Kratzer calls these conversational backgrounds *stereotypical.* Kratzer (1981: 47) characterizes as follows the kind of worlds picked by a stereotypical background: "Worlds in which the normal course of events is realized are a complete bore, there are no adventures or surprises. The concept of a normal course of events is analogous to the concept of 'frame' which plays an important role in psychology and artificial intelligence". In Kratzer's analysis stereotypical backgrounds intervene as secondary conversational backgrounds in the *ordering semantics*. They are *ordering sources* restricting the world quantification operated on the worlds of the basic conversational background. In Kratzer's analysis a *stereotypical epistemic necessity* modal would not quantify over all possible worlds compatible with what is known, but only on the subset of worlds that are closest to being stereotypical. That is on those at which the greatest number of stereotypical propositions (that is, more concretely, frames, scripts, stereotypes, etc.) are the case. As observed above in footnote 29, the present analysis of DOVREBBE$_E$ assigns a greater role to pragmatics – and to local processes of discourse coherence – tries not to put into the semantics of the modal a mechanism as complex and as cognitively implausible (Papafragou 2000) as the ordering of possible worlds. Instead of using Kratzer's set of "boring worlds", at which all stereotypes are true, as a yardstick to define a set of "closest to boring yet realistic" worlds on which DOVREBBE$_E$ quantifies, the present analysis simply has a locally salient set of conditions C providing the restriction. This set only contains those stereotypes (those Aristotelian *eikota* and *endoxa*) that are recoverable in the discourse context. This means, however, that the expression *recoverable in the discourse context* should be given a not too vague meaning. As mentioned in the conclusion, a tighter integration of the semantic analysis of the modals with explicit accounts of discourse structure and processing is needed.

As I have done throughout the book, here I assume that no inconsistencies arise from the set-theoretic union of B and C. It is assumed that the pragmatic inference mechanisms that determine the saturation of B and C (to the extent to which it is made precise) are powerful enough to rule out that inconsistencies of this type arise. Suppose, for instance, that Marco has been hit by a tiny fragment of an old Russian satellite falling on Earth and survives. The doctor informs his relatives that (62)

(62) Marco **dovrebbe** ristabilirsi in 30 giorni
'Marco should recover in 30 days'.

Arguably, the doctor takes B to be alethic and to consist of general facts about human physiology and of the specific circumstances of Marco's injury. It seems also sensible to take C as corresponding to "what is normally the case". Theoretically, a contradiction can arise here. One could say that B is inconsistent with C "what is normally the case" because *it is not normally the case* that people are hit by a tiny fragment of a Russian satellite falling on Earth. In practice, however, the set of normal circumstances that the speaker invokes will be already a contextually adapted one, or, better, one contextually created by invoking the relevant propositions.

In many instances both the premises in B and the conditions in C of DOVREBBE$_E$ can be recovered within the sentence or in the immediate co-text, as shown in (63) and (64):

(63) I risultati ottenuti dalla presidenza Da Silva sono fondamentalmente dovuti a una politica economica ortodossa, dal punto di vista macroeconomico (riforma delle pensioni e la stabilizzazione finanziaria), coniugata con politiche sociali che, secondo le intenzioni, ***dovrebbero*** diminuire il divario tra ricchi e poveri, con i progetti 'Fame zero' e 'Agenda famiglia' (*Milano Finanza*, 4/8/2006)

'The results obtained by Da Silva's presidency are basically due to an orthodox economic policy from a macroeconomic perspective (pension reform and financial stabilization), combined with social policies, which, according to the intentions, ***should*** decrease the gap between rich and poor, with the projects 'Zero Hunger' and 'Family Agenda'.'

{'social policies', 'projects 'Zero Hunger', 'Family Agenda' ...}$_B$ ∪ {'social policies realize their intended effect'}$_C$ $\Box$ 'The gap between the rich and the poor is reduced'

(64) Il dato relativo alla vendita di nuove case negli Usa a febbraio ha fatto registrare un vero e proprio crollo (-10,5%, a 1,08 milioni di unita), il calo più forte da nove anni. Aumenta anche il numero degli alloggi invenduti, un fatto che - se confermato in futuro - ***dovrebbe*** riflettersi in una riduzione dei prezzi degli immobili, con effetti di raffreddamento sulla crescita dell'inflazione. Questa statistica ha sostanzialmente ribaltato quella relativa alle case esistenti, che aveva messo in mostra una crescita del 5,2% a febbraio. (*Il Sole 24 Ore* 3/4/2006)

'The sales figures of new homes in the U.S. in February showed a real slump (−10.5% to 1.08 million units), the strongest decline in nine years. The number of unsold housing increases, a fact that – if confirmed in the future – **should** be reflected in reduced house prices, with cooling effect on the growth of inflation. This statistic has essentially reversed that relating to existing homes, which had exposed a growth of 5.2% in February.'

{'February sales figures of new homes in the US showed a slump.', 'The number of unsold housing increases', …}$_B$ ∪ {'The Real estate sales slump continues'}$_C$ $\Box$ 'Reduced house prices'

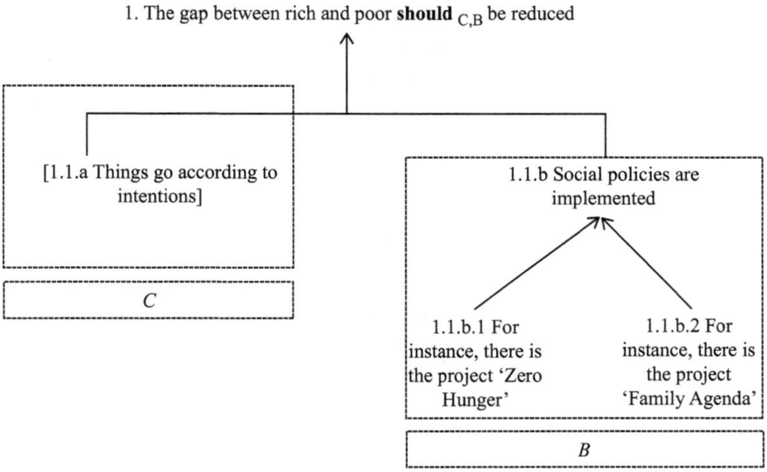

Fig. 6.17 MICRO-ARGUMENTATION structure of example (63)

In (63) the saturation of C is provided by the phrase *secondo le intenzioni* 'according to intention', which restricts the consequence relation between social policies and the reduction of the gap between the rich and the poor to the worlds where the outcomes of the policies correspond to their intention. The saturation of the conversational background B is provided by the syntactic subject of the modal *dovrebbero*: *politiche sociali* 'social policies'. The nature of B is further specified by the phrase *con i progetti 'Fame zero' e 'Agenda famiglia'.*

It is remarkable observe how in (63) a condensed argumentation structure is realized at the level of clause-internal syntax (cf. Fig. 6.17).

This same clause internal micro-argumentation, which I had highlighted earlier in connection with example (17.b), has been systematically observed by Miecznikowski (2011) in her study of the argumentative functions the conditional possibility modal *potrebbe*. Both modal verb constructions seem to play an important role in the manifestation of what Hobbs (2010) calls clause-internal discourse relations (cf. Chap. 5, § 5.2.1).

In (64), B is again alethic and is identified with the current economic circumstances and with the "causal" laws of the economy. The relevant economic circumstances are recovered anaphorically from the preceding sentence and from the main clause of the sentence in which DOVREBBE$_E$ appears as part of an appositive clause. Here again syntax plays a role along with anaphora in guiding the saturation of B, as the subject position in the modal construction is occupied by the anaphoric nominal *un fatto* 'a fact', referring back to the increase in unsold housing[40]. As for

[40] It should be noted that the presence of the causative *riflettersi in* as the lower verb in the modal construction arguably plays an important role in (62). Miecznikowski (2011), in her analysis of the conditional possibility modal *potrebbe*, observed its frequent occurrence with event nominals as surface subjects and causatives as lower verbs. Space reasons impose that I save a discussion of this phenomenon and, more generally, of issues of syntax-semantics mapping for another occasion.

the conditional restriction C, it is straightforwardly identified with the incidental verbless if-clause *se confermato in futuro* ('if confirmed in the future').

It is important to appreciate how the semantics proposed for $DOVREBBE_E$ in (Def. 7), and, in particular, the restriction on the saturation of C, captures the role of this construction as an argumentative indicator and, at the same time, offers the means to contrast it with related Italian modal constructions, including (i) the Italian subjunctive conditional construction, (ii) the predictive uses of the future tense referring to future eventualities, (iii) the inferential evidential $DEVE_E$ and (iv) $FUTURE_E$ constructions.

(i) **$DOVREBBE_E$ vs. subjunctive conditional construction.** Consider the following constructed example of subjunctive conditional construction, based on the corpus example in (64):

(65) Se l'aumento del numero degli alloggi invenduti, **fosse confermato** [subj. imp passive 3rd sing], ciò si **rifletterebbe** [reflect cond. pres. 3rd sing.] in una riduzione dei prezzi degli immobili.
'If the increase in the number of unsold housing, **were confirmed**, this **would be reflected** in a reduction in property prices'

It is important to observe here how the conditions C in $DOVREBBE_E$ do not accept the full range of epistemic evaluations available to C in the semantics of the basic *conditional mood construction* (Def. 6). The Italian *subjunctive conditional construction* exemplified in (65) inherits[41] the properties of the basic conditional mood construction and identifies the antecedent C with a syntactic antecedent expressed in the imperfect of the subjunctive mood (see Fig. 6.18).

In the Italian *subjunctive conditional construction*, which covers the areas traditionally called of the "unreal" and the "potential" (cf. Harris 1986), C is either strictly counterfactual (known to be false) or a mere possibility, often a remote or marginal one. Now, while C in $DOVREBBE_E$ is compatible strictly counterfactual conditions, it is however not compatible with mere, let alone remote, possibilities.

If we were to add a finite verb to the verbless antecedent (*se confermato in futuro*) in (64) it would rather be an indicative future form (*se verrà confermato in futuro*) used for the indicative conditional, rather than the imperfect subjunctive invoked by the subjunctive conditional construction (*se venisse confermato in futuro*).

(ii) **$DOVREBBE_E$ vs. predictive future tense.** The above affinity between $DOVREBBE_E$ and the indicative conditionals featuring a predictive future tense as well as other similarities with deictic uses of predictive future prompt the question on what separates the *predictive future* from future-oriented $DOVREBBE_E$.

[41] In construction grammar *asymmetric inheritance* is used to represent the semantic motivation. Inheritance links are posited "between constructions which are related both semantically and syntactically [...] to capture the fact that two constructions may be in some ways the same and in other ways distinct" (Goldberg 1995: 72).

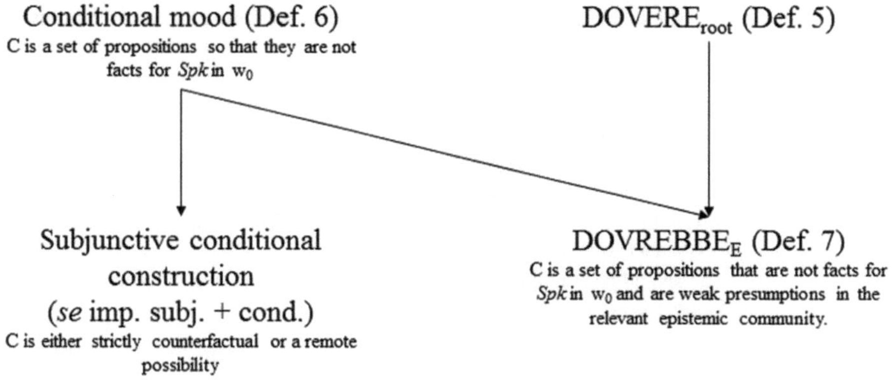

Fig. 6.18 Inheritance links between modal constructions

Following (Rocci 2000, 2005a), I maintain that the basic semantics of the future tense contains a conditional restriction similar to C of DOVREBBE$_E$. This condition C of the future is identified by default with *normal conditions*. Yet, it sharply differs from that of DOVREBBE$_E$. because it cannot be strictly counterfactual .

(iii) **DOVREBBE$_E$ vs. FUTURE$_E$**. *When* DOVREBBE$_E$ refers to present or past eventualities a *parallel contrast is reproduced with* FUTURE$_E$. *Consider the following constructed examples:*

(66)

(a) **Sarà** a casa. Di solito, nel fine settimana non esce.
'She will be at home. Usually, she doesn't go out during the week-end'

(b) **Dovrebbe** essere a casa. Di solito, nel fine settimana non esce.
'She should be at home. Usually, she doesn't go out during the week-end'

Both (66.a) and (66.b) are quite natural as inferences drawn from premises that function as weak presumptions because they describe what is normally, usually the case.

There is, however, a sharp semantic difference between the two constructions, as only (66.b) allows discourse continuations such as (67) where a rebutting defeater introduced by *ma* ('but'), which forces us to recognize a posteriori the argumentative relation signaled by the modal as a concessive move. The same holds for the kind of revisions of the discourse context introduced by a connective such as *in realtà* ('in reality', 'in fact') in (68) (cf. Mayer 1990; Rocci 2009).

(67)

(a) **Sarà** a casa. Di solito, nel fine settimana non esce. *Ma la sua macchina oggi non è nel parcheggio.
'She will be at home. Usually, she doesn't go out during the week-end. *But today her car is not in the parking lot'

(b) **Dovrebbe** essere a casa. Di solito, nel fine settimana non esce. Ma la sua macchina oggi non è nel parcheggio.

'She should be at home. Usually, she doesn't go out during the week-end. But today her car is not in the parking lot'

(68)

(a) **Sarà** a casa. Di solito, nel fine settimana non esce. ***In realtà**, oggi è andata a trovare sua sorella..

'She will be at home. Usually, she doesn't go out during the week-end. *In reality, today she went to visit her sister'

(b) **Dovrebbe** essere a casa. Di solito, nel fine settimana non esce. **In realtà**, oggi è andata a trovare sua sorella..

'She should be at home. Usually, she doesn't go out during the week-end. In reality, today she went to visit her sister'

While the use of FUTURE_E *can be based entirely on assumptions about what is normally to be expected in a situation, it is not possible to cancel the epistemic commitment associated to its use by pointing out that the current situation happen to be one of those that are not normal.*

The point is that FUTURE_E*, just like* DEVE_E, functions as an m-performative (cf. Nuyts 2001, Faller 2002) epistemic marker pointing deictically to the beliefs of the speaker at the moment of utterance. So, while the inference signaled FUTURE_E by may be based on nothing more than default assumptions of normality – what the linguistic research tradition on evidentiality calls "conjectural inference", these assumptions are taken up by the speaker as their "best bet" at the moment of utterance.

As a consequence, FUTURE_E does not allow what Mayer (1990) calls "perspectivization" of the evidential, that is the detachment of the provisional inference drawn from a certain type of evidence from the beliefs of the speaker and from the picture of the world that the speaker's discourse is conveying. To put it in terms of the theory of utterance polyphony developed by Scandinavian linguists (ScaPoLine, cf. Nølke et al. 2004), FUTURE_E does not allow the dissociation of the point of view of the speaker at the moment of utterance from the point of view represented by the expectations of normality in C.

Interestingly, the fact that C is in every case kept distinct from the beliefs of the speaker at the moment of utterance, and is instead associated with *an external epistemic subject* is obviously consistent with the merger of deontically based DOVREBBE_E with reportive evidentiality that we have observed.

The separation between C and the beliefs of the arguer at the moment of utterance, can contribute to explain certain polyphonic configurations of DOVREBBE_E , which typically evokes the presence of a virtual antagonist in the discourse. If we look back at examples such as (17.b) and (59), examined above, we can see that DOVREBBE_E can occur as an indecisive counter-argument to a previous argument (17.b), or be followed followed by a counter-argument, as in (59).

(iv) **DOVREBBE_E vs. DEVE_E** . The above remarks allow us also to better understand *a fortiori* the opposition of DOVREBBE_E with DEVE_E . M-performative by default, unless clearly embedded by a cognitive attitude or speech predicate tying it to another thinking subject, DEVE_E lacks completely the reference to expectations of normality, and, in contrast, requires the emergence of a clearly identifiable *datum* – a notion that includes the idea of the "inference from results" of the scholars of evidentiality, but is broader than that.

Let us compare the conclusions introduced by DOVREBBE$_E$ with those introduced by DEVE$_E$. For, DEVE$_E$, which is immediately tied to the speaker's conclusions – subjective as they may be – the function of indecisive counter-argument (17.b) and especially the role of provisional conclusion followed by a counter argument would have been extremely awkward, as observed above, using example (29.b).

Conclusions introduced with DEVE$_E$ may depend on very uncertain assumptions of the speaker and may, in fact, correspond to an abductive inference, but they are not presented as provisional or defeasible by stronger arguments.

6.5.4 *A Corpus Study of the Conversational Backgrounds of* DOVREBBE$_E$ *in Financial News*

In order to put to the test of real discourse data the semantic analysis proposed for DOVREBBE$_E$ in the previous section and, at the same time, in order to exploit it to gain a better understanding of this construction as an argumentative indicator, I will examine the composition of B and C in all the occurrences of DOVREBBE$_E$ in the annotated sub-corpus of 200 Italian financial news texts.

6.5.4.1 Alethic Conversational Backgrounds: Economic Causality

I start from the basic conversational background B, for which I have hypothesized either an alethic or a deontic saturation. Consider example (69) below:

(69) Il dato relativo alla vendita di nuove case negli Usa a febbraio ha fatto registrare un vero e proprio crollo (-10,5%, a 1,08 milioni di unita), il calo più forte da nove anni. Aumenta anche il numero degli alloggi invenduti, un fatto che - se confermato in futuro - ***dovrebbe*** riflettersi in una riduzione dei prezzi degli immobili, con effetti di raffreddamento sulla crescita dell'inflazione. Questa statistica ha sostanzialmente ribaltato quella relativa alle case esistenti, che aveva messo in mostra una crescita del 5,2% a febbraio. (*Il Sole 24 Ore* 3/4/2006)

'The sale figures for new homes in the U.S. in February showed a real slump (10.5%, to 1.08 million units), the strongest decline in nine years. The number of unsold houses is increasing, a fact that, if confirmed in the future, should be reflected in reduced house prices, with a cooling effect on the growth of inflation. This statistic is essentially in reverse to that relating to existing homes which showed a growth of 5.2% in February.'

In (69) DOVREBBE$_E$ signals a consequence based on economic causality, and can therefore be interpreted as based on an alethic circumstantial conversational background similar to the causal uses of *potere* and, to a lesser extent, to the quasi-causal

Table 6.8 Distribution of alethic conversational backgrounds in the 200 texts sub-corpus

Saturation of B	N. of occurrences
Economic Causality	32
Complex Causality	6
Calculation	2
Human Psychology and Reasoning	1
Total:	41

uses of *dovere* in the indicative examined in the previous sections. Here, however, the causal necessity is conditional on the continuation of the slump (which would be, by the way, normal and expected).

The majority of the occurrences in the 200 texts sample (41/63) have an alethic B, in most cases clearly identifiable with what we have called economic causality, as demonstrated in Table 6.8.

A small number of predictions (6) are based on complex causal processes which involve both economic facts (with economic nomic regularities) and other kinds of propositions (including their relevant nomic regularities).

These include *physical causation*, the (psychological and organizational) *weight of habits*, and the *economic rationality* of agents. Backgrounds combining economic events with the *policy* of central banks also belong here.

While *economic rationality* and *policies* are, in principle, a teleological and a deontic concept they are not really treated as such. Economic rationality we have already discussed. As for central bank policies, they are regarded more as nearly stable elements of the functioning of the system than as deontic commitments of agents.

In two cases, the prejacent of $\text{DOVREBBE}_{\text{E}}$ is the result of a calculation (e.g. short-term interest rates), performed according to a conventionally accepted method, from a more basic value (e.g. primary interest rates). We should note here that the role of calculation is not simply epistemic (as it would be in the natural sciences) but properly causal: banks do not *discover* short-term interest rates when they calculate them, they *establish* them. The interest rates come into being as institutional facts according to the calculations.

Finally, there is a sole example of psychological causality, which is, in fact, used meta-argumentatively (*non dovrebbe sorprendere* 'it should not come as a surprise').

6.5.4.2 Deontic Conversational Backgrounds

Not all examples of $\text{DOVREBBE}_{\text{E}}$ found in the corpus correspond to a background of economic circumstances. The other 22 of the 63 occurrences of predictive $\text{DOVREBBE}_{\text{E}}$ in the 200-text sample are analogous to (70):

(70) Stando a quanto emerso ieri nella riunione del cda Bnl, Bnp *sarebbe* orientata a lanciare la prossima settimana l'Opa, che ***dovrebbe*** concludersi tra il 15 e il 20 maggio. Secondo indiscrezioni la banca di Parigi *avrebbe predisposto* tutto per annunciare già stasera l'ok della Consob e i dettagli dell'operazione, con le date di inizio e di conclusione. (*Il Sole 24 Ore*, 13/4/2006)

'According to what transpired from yesterday's meeting of the board of BNL, BNP would be inclined to launch the takeover bid next week, to be completed between the 15th and 20th of May. According to rumors, the Paris bank prepared everything to announce this evening the OK of Consob and the details of the transaction, with the dates of commencement and conclusion.'

They invariably refer to the *plans* of corporations and other institutions that play a role in the markets. In just one case these plans make reference to the precise calendar scheduling of events. In most occurrences with DOVREBBE$_E$, plans are inside corporate information reported off-record by anonymous insiders (*indiscrezioni*). This contrasts with the more public plans and scheduling that we find with deontic indicative *dovere* (as well as with the simple future tense).

Not surprisingly, in (70), we find instances of the reportative conditional mood (*sarebbe orientata, avrebbe predisposto*) in clauses immediately preceding and following the one with DOVREBBE$_E$.

The relationship between *dovrebbe* and the reportative use of the Italian conditional mood will be better understood if we look more closely at the nature of the conditional restrictions C, discussed in the following section.

6.5.4.3 The Nature of Conditional Restrictions

In most of the occurrences of DOVREBBE$_E$ in the corpus it is relatively easy to identify propositions in the immediate discourse context that make the conditional restriction C explicit.

This is in itself a striking confirmation that the context-dependency of this modal is resolved anaphorically in discourse or even exploiting the micro-context of sentential or clausal syntax. In turn, the possibility for the analysts of singling out the likely saturations of these modal constructions makes them *particularly significant* indicators of connective predicates of *Conclusion*. Finally, these relations, accurately described thanks to the information provided by the modal construction, give a trustworthy basis for understanding (a) the dialectical commitments of the arguers and (b) the argumentative *loci* invoked to license the inferences.

These recoverable "antecedents" DOVREBBE$_E$ can take the form of

(i) if-clauses introduced by the conditional conjunction *se* 'if' with the indicative mood or non-finite verb, as in (64) (*se confermato in futuro* 'if confirmed in the future');
(ii) other kinds of subordinate clauses; as well as

Table 6.9 Conditional restrictions for the alethic B in the 200 texts sub-corpus

Conditional restriction of alethic B	Tokens
Continuation of a trend, stability of a situation	5
Hypothesis (Including the hypothesis of the stability of a situation)	4
Confirmation of an expectation	4
Reportative (Expert predictions, Insider predictions)	14
Rebuttal (Unless improbable event occurs, risk materializes)	5
Modal subordination (*dovrebbe...dovrebbe*)	2
Generic reference to a margin of uncertainty	7
Total:	41

(iii) independent sentences introducing an explicit or implicit modality to which $\text{DOVREBBE}_{\text{E}}$ points anaphorically, realizing a form of modal subordination[42].

Given what has been observed in § 6.5.3.4, it should come as no surprise that these antecedents do not include if-clauses with '*se* + imperfect subjunctive mood', which signal the Italian *subjunctive conditional* construction.

For similar reasons, we do not find[43], in the case of modal subordination, discourse antecedents epistemically modalized with a mere possibility modal (in contrast with the admissibility of modal subordination sequences '*might...would*' in English and '*potrebbe* ...conditional mood' in Italian).

If we look at the occurrences in the 200-text sub-corpus, we find that C always belongs to one of the semantic types from a very restricted inventory. When B is alethic, we find the distribution of conditional restrictions represented in Table 6.9.

As the table shows, in a certain number of occurrences C is simply identified with the *continuation of a trend* or the *stability of a situation*, where the latter are taken as defaults in economic causality, just as they are in naïve physics[44].

Another group involves the introduction of an epistemic layer of hypothesizing with antecedent structures of the type *If the hypothesis X is confirmed.* Since the *hypothesis* is nominalized, it is usually not clear who the subject of the hypothesizing is. The content of the hypothesis is again always a default. Among these defaults we find again, for instance, the (hypothesis of) the stability of the situation.

[42] On modal subordination see Roberts (1989). Modal subordination is briefly discussed in Chap. 4, § 4.2.9.

[43] For instance, if we were to add a finite verb to the verbless protasis in (18) it would rather be an indicative future form (*se verrà confermato in futuro*) than an imperfect subjunctive one (*se venisse confermato in futuro*).

[44] I have already mentioned how in our texts economic causality is, in certain respects, treated as if it were physical causality. The pervasive and sometime subtle use of physical metaphors in economic and financial discourse addressing experts and laypeople is an important issue (cf. Richardt 2003) that overlaps significantly with the expression of alethic causal notions in predictions through the use of *modal* as well as of *aspectual* lexicon and morphology.

Table 6.10 Conditional restrictions for the deontic B in the sub-corpus

Conditional restriction of deontic B	Tokens
Realization of the plan (including approval by a deliberative body)	13
Reportative (Insiders' disclosures, rumors)	5
Rebuttal (Unless plan changes due circumstance X)	1
Hypothesis (scenario)	1
Generic reference to a margin of uncertainty	1
Unclear, indeterminate	1
Total:	22

Confirmation of an expectation is another C that contains an explicit epistemic evaluation whose subject is not spelled out.

The single most common type of C is the reportative. In the case of an alethic B, it nearly always is an *expert prediction* – in one case it's an insider one. So, in the latter uses of DOVREBBE$_E$, the prejacent p is the case in all the worlds where a set of factual economic circumstances with the relevant economic nomic regularities is the case (B) *and* the experts are right (C). The experts' being right is another default, or weak presumption.

Defaults can be formulated negatively in the form of exceptive conditions (*Unless improbable, event X happens*), which are – as shown in Chap. 3, § 3.4.4 – one and the same thing with Toulminian rebuttals. Not surprisingly, one of these exceptions is the *disappearance of a trend.* Finally, the weakly presumed antecedent can be derived anaphorically via modal subordination as the truth of the prejacent of a preceding *dovrebbe.*

When B is deontic, the saturation of the conditional restriction C is as in Table 6.10.

With a deontic B, the most frequent non-factual default is represented by plan realization (*If everything goes according to plan*) which also includes the official sanctioning of the plan by a deliberative body.

The plans that we find with DOVREBBE$_E$ are not the content of an official corporate disclosure or publicly available schedules. Rather, they correspond to inside information reported from unnamed sources and to rumour. In some cases this origin is made partially explicit by phrases such as *according to rumours* or *if rumors are confirmed.* Thus, in these occurrences, the prejacent follows from a set of rumoured plans (B) and the non-factual assumption that the rumours are true (C) – the truthfulness of what is said being a (Gricean) *normal* condition of communication.

Furthermore, plans can involve alternative scenarios so that the prejacent follows from the set of plans (B) and the non-factual assumption that one of the alternative scenarios addressed in the plans is indeed the case (C). Finally, exceptional circumstances might dictate that the original plans are abandoned or revised. This gives rise to a conditional restriction in rebuttal form, similar to those found with alethic B.

6.5.5 *The Endoxical Nature of the Normal Conditions C in DOVREBBE$_E$*

The exploration of conditional restrictions of DOVREBBE$_E$ can provide a deeper insight into the evidential and argumentative functioning of this modal construction in predictions.

Both B and C point to premises of different kinds that are to be recovered anaphorically in discourse context, extracted from subordinate clauses of different types and even from sub-clausal elements such as abstract nominals (cf. Hobbs 2010).

Thanks to the constraints on the saturation of B and C, DOVREBBE$_E$ ends up pointing to a limited set of combination of argument schemes, or argumentative loci supporting the prediction. Either you have a causal locus (most often the *locus from efficient cause*), possibly combined with an argument from authority based on *expert opinion* (*locus from authority*), or you have an argument based on commitments (*locus from promising and warning*), that are further grounded in the authority of the insiders' being a position to know.

Two distinct argumentative roles of reports emerge: the *opinion of the experts* and the insiders' *position to know* about the plans. In both cases the recourse to evidence from reports is combined with, rather than alternative to, the recourse to inference.

It is important to stress that the *weak presumptions* in C, corresponding to defaults, normal conditions, *do function as premises* in the reasoning supporting the prediction. Contrary to the conversational background of other epistemic modals, these presumptions are not directly associated with the beliefs of the speaker at the moment of utterance, but rather with a generic epistemic community, a socialized *doxa* from which the speakers can distance themselves more or less markedly.

It could be tempting here to evoke two notions from Aristotelian rhetoric and dialectic introduced in Chap. 2, § 2.2.3.1, in the characterization of rhetorical and dialectical premises: the notion of *eikós* and the closely related notion of *endoxon*. In her monograph on the enthymeme, Francesca Piazza (2000) provides an in-depth look at *eikós* as a kind of enthymematic premise. The Greek word *eikós* is the neuter participle of the verb *eoika*, meaning *to seem* or *to be like*. Piazza (2000: 117) observes that the impersonal form *eoike* ('it seems') is used in Platonic dialogues to mark dialogical acceptance of a proposition, but also, in other contexts what is right and normal. Thus, *eikós*, to play on a parallelism between Greek and English, refers both to the *likely* and to the *seemly*.*Eikota* are contingent generalizing propositions that are to be expected because they happen most of the time. In the *Rhetoric* Aristotle defines them as follows:

> A probability (*eikós*) is a thing that happens for the most part—not, however, as some definitions would suggest, anything whatever that so happens, but only if it belongs to the class of what can turn out otherwise, and bears the same relation to that in respect of which it is probable (*eikós*) as the universal bears to the particular. (*Rhetoric*. 1357a 34-b 1, transl. by W. Rhys Roberts, Oxford Revised Translation in Aristotle 2014)

Eikóta are closely related to *endoxa*: in the Prior Analytics (70a, 4–5), Aristotle defines the *eikós* as a proposition rooted in the common opinion (*prótasis éndoxos*), corresponding to what is known to normally happen and is therefore more credible. In this connection, Piazza (2000: 120) observes that the notion of *eikós* has both a subjective layer – which coincides with its being an *endoxon*, rooted in the beliefs of the audience – and an objective layer as the shared beliefs of the audience are, in turn, based on what we would nowadays call *observed statistical regularities*. Piazza, however, warns against a wholesale identification of *eikós* with the probable in the statistical sense, as Aristotle has no interest in quantifying objective or subjective probabilities.

The *eikós* with its combination of *normal recurrence* and *endoxical status* seems well apt to capture the nature of the C restrictor in the semantics of $DOVREBBE_E$. The fact that *eikós* can also refer to what is normal or expected in a moral or social sense and thus *ought to be the case* is also fitting, because it captures the slight moral overtone that characterizes this modal construction.

The notion of *endoxon*, on the other hand, is useful to capture the kind of epistemic subject to which the weak presumptions in C are associated. According to the definition that Aristotle gives in the *Topics*,

> Those opinions are reputable (*endoxa*) which are accepted by everyone or by the majority or by the wise—i.e. by all, or by the majority, or by the most notable and reputable of them." (*Topics* I 100b 21–23, transl. by W.A. Pickard-Cambridge, Revised Oxford Translation in Aristotle 2014).

Thus, *endoxa* are eminently socialized beliefs. In contrast with the set of personal conjectures, assumptions and rules of thumb, that the utterers of $DEVE_E$ invoke on the spur of the moment to complement insufficient data in search of an explanation, the beliefs invoked in $DOVREBBE_E$ to make up for a gappy natural or social ontology and issue predictions are *the point of view of a community*.

The financial news examined in this study, while clearly argumentative, are written for the most part in what Martin and White (2005: 185) call a "reporter voice", which imposes strict limits on the expression of subjectivity. The limitations on the expression of subjectivity in the "reporter voice" regime are somewhat less strict for Italian journalistic writing than for its British counterpart (see Pounds 2010). In the writing of an Italian business-financial news reporter the use of evidential strategies such as $DEVE_E$ and $FUTURE_E$, while admissible, remains somewhat exceptional, a temporary concession to the subjectivity of conjectures, informed guesses on the intentions and attempts at reading the mind of the news actors. In contrast, the business-financial reporters in the corpus can roll on with their chains of $DOVREBBE_E$ without perturbing the "reporter voice" in the slightest.

In Chap. 5, § 5.2.4.4, I have introduced a general procedure for the saturation of modal conversational backgrounds. According to this procedure, epistemic conversational backgrounds are reconstructed with reference to a speech or thought event *s*, which, by default, is identified with the speaker's mind and the *hic et nunc* of the speech event; that is with the m-performative case. This is, more or less, how $DEVE_E$ works. Only clear syntactic embeddings and highly marked stylistic devices can shift *s* to the minds and words of other subjects. In contrast, while there is little doubt that the C set in the $DOVREBBE_E$ construction is a meta-representational epistemic conversational background, it is clear that in this particular construction the default for the deictic recovery of the *s* is set to the minds and words of others.

It is tempting, at this point, to adopt the categories of *polyphony theories* (cf. Chap. 3, § 3.2.4) to capture the kind of epistemic subjects behind the use of $DOVREBBE_E$ as a qualifier in an argument. Scandinavian polyphony theorist in the French tradition (cf. Nølke et al. 2004: 39) would call ON – from the French impersonal third person pronoun *on* – the collective epistemic subject distinct from the speaker at the moment of utterance. They distinguish two variants ON, or rather two poles to which actual utterances may be more or less close. On the one side there is the viewpoint of a concrete collective of individuals in the discourse world, while on the other side there is indeterminate ON of the *doxa*, of the *idées reçues*. It appears that the conditions in the C set of $DOVREBBE_E$ can be of either type.

6.5.6 The Inferential Function of C in $DOVREBBE_E$

When I first applied the semantic analysis of $DOVREBBE_E$ to the argumentative reconstruction of example (63), reproduced here, I chose to represent both B and C as premises in the argument (cf. § 6.5.3.4, Fig. 6.17).

(63) I risultati ottenuti dalla presidenza Da Silva sono fondamentalmente dovuti a una politica economica ortodossa, dal punto di vista macroeconomico (riforma delle pensioni e la stabilizzazione finanziaria), coniugata con politiche sociali che, secondo le intenzioni, ***dovrebbero*** diminuire il divario tra ricchi e poveri, con i progetti 'Fame zero' e 'Agenda famiglia' (*Milano Finanza*, 4/8/2006)

'The results obtained by Da Silva's presidency are basically due to an orthodox economic policy from a macroeconomic perspective (pension reform and financial stabilization), combined with social policies, which, according to the intentions, ***should*** decrease the gap between rich and poor, with the projects 'Zero Hunger' and 'Family Agenda'.'

{'social policies', 'projects 'Zero Hunger', 'Family Agenda' ...}$_B$ ∪ {'social policies realize their intended effect'}$_C$ □ 'The gap between the rich and the poor is reduced'

This choice can be now reassessed and better justified in the light of the corpus study of the composition of C and in view of its characterization as *eikós*.

The kind of conditional structures appearing in the context of $DOVREBBE_E$ do not correspond to the idea of epistemic conditional discussed in Chap. 4, § 4.2.8.4. These antecedents are not presented as factual and thus they are not simply part of the basic alethic or epistemic conversational background. They are non-factual, or counterfactual in a weak sense and adding them to the conversational background substantially alters its nature.

Yet, there is reason to treat them as a special kind of premise. Their role in the inference reminds of the role of presumptive premises in Nicholas Rescher's (2006a, b) account of default reasoning. According to Rescher (2006b: 174), "the inference processes of default logic can be assimilated to a deductive pattern" with a structure that "does clearly obtain as valid in traditional logic":

In all ordinary (normal, standard, commonplace) cases, whenever P, then Q
P obtains in the case presently at hand
<The present case is an ordinary (normal, standard, etc.) one>

∴ Q obtains in the present case.
(Rescher 2006b: 174–175)

Rescher observes that the premise in angle brackets "usually tacit and thereby enthymematic" plays a pivotal role in the argument and concludes:

> Default reasoning [...] rests on arguments which would be valid if all their premisses – explicit and tacit alike – were authentic truths, which they are not since at least one of the critical premisess of the argument is no more than a mere presumption.

We should imagine that, as long as the arguer sides with the doxa of C, the premise in angle brackets is tacitly added. When the arguer, on the basis of new data, detaches his/her viewpoint from that of the arguer, the premise is not understood to be there anymore.

The general procedure of default reasoning is captured as follows by Rescher (2006b: 175):

> To treat what is generally (normally, standardly, usually, etc.) the case as if it were the case always and everywhere, and therefore applicable to the present instance.

And this is exactly what happens with the *eikota* making up the restriction C of the conversational background of DOVREBBE$_E$ when the arguer is aligned with the *eikota*. A tentative translation of this kind of presumptive inference in terms of the Y-structure of the AMT model is offered below in Fig. 6.19.

We can observe that, by adding the premise C to the endoxon, one can keep a strong, monotonic, necessary, which requires a sufficient cause (Def. 3). The plausible, presumptive, nature of the inference is wholly derived by the non-factuality of premise C. This approach fits well the idea of DOVREBBE$_E$ as having an underlying alethic *necessity* modal. The approach also fits Rescher's stance on presumptive reasoning, which is often considered reductionist and "deductivist" because of the philosopher's unwillingness to touch the strength of the consequence relation itself (cf. Woods 2010).

The treatment of defeasible inference has yet to be approached systematically within the AMT theory, and I will reserve it for another investigation, in keeping with the promise I made in the *Introduction* (§ 6.2) of *not* addressing presumptive and defeasible reasoning.

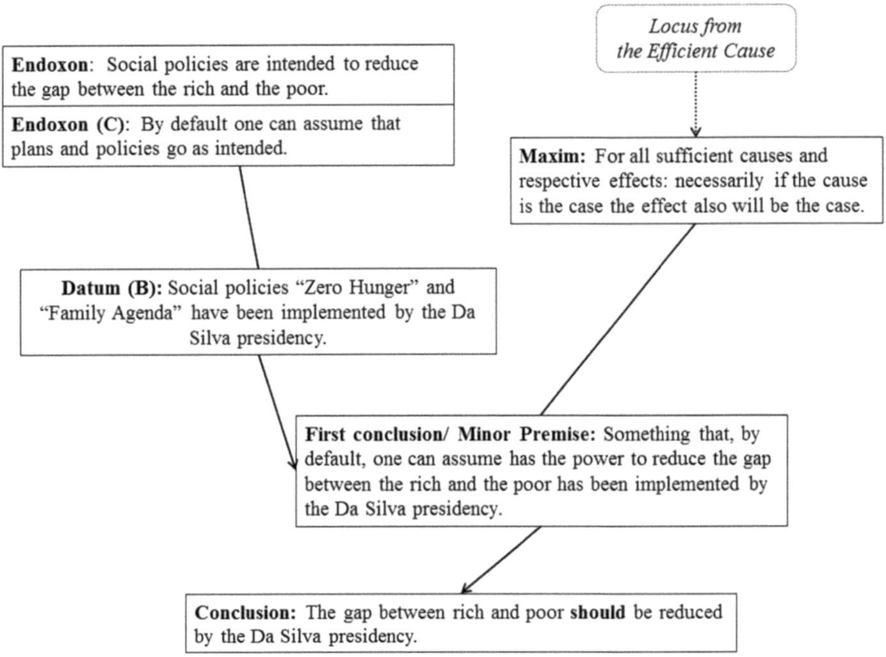

Fig. 6.19 Inferential configuration (Y-structure) of example (63)

References

Aristotle (2014) The complete works of Aristotle: The revised Oxford translation, one-volume digital edition. Jonathan Barnes (ed). Princeton University Press, Princeton (originally published in 1984 as 2 volumes)

Bach E (1986) The algebra of events. Linguist Philos 9(1):5–16

Bakhtin M (1986) The problem of speech genres. In: Speech genres and other late essays. Austin, University of Texas Press, pp 60–102

Barone-Adesi G (2002) The role of inside information. Financial disclosure and value creation. In: Russ-Mohl S (ed) Business journalism, corporate communications, and newsroom managemen. Università della Svizzera italiana, Lugano, pp 63–68

Béguelin M-J (2002) Clause, période, ou autre? La phrase graphique et la question des niveaux d'analyse. In: Charolles M, Le Goffic P, Morel M-A (eds) Y a-t-il une syntaxe au-delà de la phrase? Actes du colloque de Paris 3-Sorbonne Nouvelle, 21-22 septembre 2000, Verbum, XXIV, pp 85–107

Bermejo-Luque L (2011) Giving reasons. A linguistic-pragmatic approach to argumentation theory. Springer, Dordrecht

Bertinetto PM (1979) Alcune ipotesi sul nostro futuro (con osservazioni su potere e dovere). Rivista di grammatica generativa 4(1-2):77–138

Bertinetto PM (1986) Tempo, aspetto e azione nel verbo italiano: Il sistema dell'indicativo. Accademia della Crusca, Firenze

Bloor T, Pindi M (1990) Schematic structure in economics forecasts. In: Dudley-Evans T, Henderson W (eds) The Language of Economics: The analysis of economics discourse. Modern English Publications and The British Council, London, pp 55–66

Bonomi A (1980) Determinismo e semantiche per logiche temporali. In: Bernini S (ed) Atti del convegno nazionale di Logica (1-5 ottobre 1979). Bibliopolis, Naples, pp 351–379
Boogaart R (2009) Semantics and pragmatics in construction grammar: the case of modal verbs. In: Alexandrer G, Bergs D (eds) Contexts and constructions. John Benjamins, Amsterdam/ Philadelphia, pp 213–241
Bozzone-Costa R (1991) L'espressione della modalità non fattuale nel parlato colloquiale (con particolare riferimento agli usi del futuro). Quaderni del dipartimento di linguistica e letterature comparate 7:25–73
Coates J (1983) The semantics of the modal auxiliaries. Croom Helm, London
Coates J (1995) The expression of root and epistemic possibility in English. In: Bybee JL, Fleischman S (eds) Modality in grammar and discourse. John Benjamins, Amsterdam/ Philadelphia, pp 55–66
Cooper G, Ebeling M (2007) Epistemology, structure and urgency: the sociology of financial and scientific journalists. Sociol Res Online 12(3). doi:10.5153/sro.1558
Coyle D (2001) Making sense of published economic forecasts. In: Clements MP, Hendry DF (eds) Understanding economic forecasts. The MIT Press, Cambridge, pp 54–67
Degeorge F, Patel J, Zeckhauser R (1999) Earnings management to exceed thresholds. J Bus 72(1):1–33
De Mauro T, Mancini F, Vedovelli M, Voghera M (1993) Lessico di frequenza dell'italiano parlato. Etaslibri, Milano
Dendale P (1994) Devoir épistémique, marqueur modal ou évidentiel? Langue française 102(1):24–40
Dendale P (1999) Devoir au conditionnel: valeur évidentio-modale et origine du conditionnel. In: Vogeleer S et al (eds) La modalité sous tous ses aspects. Rodopi, Amsterdam, pp 7–28
Dendale P (2000) 'Devoir' épistémique à l'indicatif et au conditionnel: inférence ou prédiction? In: Englebert A (ed) Actes du 22e Congrès international de linguistique et de philologie romanes (Bruxelles, 23–29 juillet 1998). Niemeyer, Tübingen, pp 159–170
Desclés JP, Guentcheva Z (2001) La notion d'abduction et le verbe devoir "épistémique". Cahiers Chronos 8:103–122
Donohue JP (2006) How to support a one-handed economist: The role of modalisation in economic forecasting. English Spec Purp 25(2):200–216
Ducrot O (1980) Le mots du discours. Le éditions de minuit, Paris
Faller MT (2002) Semantics and pragmatics of evidentials in Cuzco Quechua. Ph.D dissertation. Stanford University, Stanford
Faller MT (2011) A possible worlds semantics for Cuzco Quechua evidentials. In: Li N, Lutz D (eds) Proceedings of SALT 20. CLC Publications, Ithaca, pp 660–683
Freeman JB (1991) Dialectics and the macrostructure of arguments. A theory of argument structure. Foris Publications, Berlin
Freeman JB (2000) What type of statements are there? Argumentation 14(2):135–157
Freeman JB (2005) Acceptable premises: An epistemic approach to an informal logic problem. Cambridge University Press, Cambridge
Geurts B (1999) Presuppositions and pronouns. Elsevier, Amsterdam
Goldberg AE (1995) Constructions: a construction grammar approach to argument structure. University of Chicago Press, Chicago
Groefsema M (1995) Can, may, must and should: a relevance theoretic account. J Linguist 31:53–79
Hacquard V (2010) On the event relativity of modal auxiliaries. Nat Lang Semant 18(1):79–114
Halliday MAK (1994) An introduction to functional grammar. Arnold, London
Harris MB (1986) The historical development of SI-clauses in Romance. In: Traugott EC, ter Meulen A, Reilly JS, Ferguson CA (eds) On conditionals. Cambridge University Press, Cambridge, pp 265–284
Hintikka J (1973) Time and necessity. Oxford University Press, Oxford

Hobbs JR (2010) Clause-internal coherence. In: Kuhnlein P, Benz A, Sidner CL (eds) Constraints in discourse 2. John Benjamins, Amsterdam/Philadelphia, pp 15–34
Horn LR (1996) Exclusive company: only and the dynamics of vertical inference. J Semant 13(1):1–40
Karttunen L (1972) Possible and must. In: Kimball J (ed) Syntax and semantics, vol 1. Academic, New York, pp 1–20
Kaufmann S, Condoravdi C, Harizanov V (2006) Formal approaches to modality. In: Frawley W (ed) The expression of modality. Mouton de Gruyter, Berlin/New York, pp 71–105
Kay P (2004) Pragmatic aspects of grammatical constructions. In: Horn L, Ward G (eds) The handbook of pragmatics. Blackwell, Oxford, pp 675–700
Kay P, Michaelis LA (2008) Constructional meaning and compositionality. In: Maienborn C, von Heusinger K, Portner P (eds) Semantics: an international handbook of natural language meaning. Mouton de Gruyter, Berlin/New York, pp 2271–2297
Kleiber G (1983) L'emploi sporadique de pouvoir. In: David J, Kleiber G (eds) La notion sémantico-logique de modalité. Centre d'Analyse Syntaxique de l'Université de Metz, Metz, pp 183–203
Kratzer A (1981) The notional category of modality. In: Eikmeyer H-J, Rieser H (eds) Words, worlds, and contexts: new approaches in word semantics. Mouton de Gruyter, Berlin/New York, pp 38–75 (Republished in Kratzer 2012: 27–69)
Kratzer A (2012) Modals and conditionals. Oxford University Press, Oxford
Kronning H (1996) Modalité, cognition et polysémie: Sémantique du verbe modal devoir. Uppsala University Press, Uppsala
Kronning H (2001) Nécessité et hypothèse: « devoir » non déontique au conditionnel. In: Dendale P, Tasmowski L (eds) Le conditionnel en français. Centre d'études linguistiques des textes et des discours de l'Université de Metz, Metz, pp 251–276
Kronning H (2002) Le conditionnel « journalistique » : médiation et modalisation épistémiques. Romansk Forum 16:561–575
Ludlow P (1999) Semantics, tense and time. An essay in the metaphysics of natural language. The MIT Press, Cambridge
Martin JR, White PRR (2005) The language of evaluation. Palgrave Macmillan, London
Mayer R (1990) Abstraction, context, and perspectivization. Evidentials in discourse semantics. Theor Linguist 16(2-3):101–164
Mc Closkey D (1990) If you're so smart: the narrative of economic expertise. University of Chicago Press, Chicago
McLaren-Hankin Y (2008) We expect to report on significant progress in our product pipeline in the coming year: hedging forward-looking statements in corporate press releases. Discourse Stud 10(5):635–654
Merlini L (1983) Gli atti del discorso economico: la previsione. Status illocutorio e modelli linguistici del testo inglese. Parma, Zara
Miecznikowski J (2011) Construction types and argumentative functions of possibility modals: evidence from Italian. In: Eemeren FH, Garssen B, Godden D, Mitchell G (eds) Proceedings of the 7th conference of the International Society for the Study of Argumentation. Sic Sat, Amsterdam, pp 1284–1297
Miecznikowski J, Bazzanella C (2007) The attenuating conditional: context, appropriateness and interaction. In: Fetzer A (ed) Context and appropriateness: micro meets macro. John Benjamins, Amsterdam/Philadelphia, pp 203–233
Miecznikowski J, Rocci A, Zlatkova G (2012) L'argumentation dans la presse économique et financière italienne. In: Gautier L (ed) Les discours de la bourse et de la finance. Frank und Timme, Berlin, pp 65–83
Musi E (2014) Evidential modals at the semantic- argumentative interface: appearance verbs as indicators of defeasible argumentation. Informal Logic 34(4):417–442
Musi E (2015) Dalle apparenze alle inferenze: i verbi sembrare ed apparire come indicatori argomentativi. Ph.D. dissertation, Università della Svizzera italiana

Nølke H, Fløttum K, Norén C (2004) ScaPoLine. La théorie scandinave de la polyphonie linguistique. Paris, Kimé
Nuyts J (2001) Epistemic modality, language, and conceptualization: a cognitive-pragmatic perspective. John Benjamins, Amsterdam/Philadelphia
O'Donnell M (2008) The UAM CorpusTool: software for corpus annotation and ex-ploration. In: Bretones Callejas CM et al. (eds) Applied linguistics now: understanding language and mind. Universidad de Almería, Almería, pp 1433–1447
Palmieri R, Miecznikowski J (2016) Predictions in economic-financial news: author's stance and argumentative loci. J Argument Cont 5(1):48–73. doi:10.1075/jaic.5.1.03pal
Palmieri R, Palmieri C (2012) Text types, activity types and the genre system of financial communication. In: Gautier L (ed) Les discours de la bourse et de la finance. Frank und Timme, Berlin, pp 85–105
Papafragou A (2000) Modality: issues in the semantics-pragmatics interface. Elsevier, Amsterdam
Parisi D, Antinucci F, Crisari M (1975) «Dovere», «potere» e il futuro dei verbi. In: Parisi D (ed) Studi per un modello del linguaggio. Quaderni della Ricerca Scientifica-CNR, Roma, pp 238–270
Peirce CS (1998 [1931]) Collected papers of Charles Sanders Peirce, In: Hartshorne C, Weiss P, Burks AW (eds) Harvard University Press, Cambridge, MA
Piazza F (2000) Il corpo della persuasione. Edizioni Novecento, Palermo
Pietrandrea P (2005) Epistemic modality. John Benjamins, Amsterdam/Philadelphia
Portner P (2005) *What is meaning? Fundamentals of formal semantics*. Blackwell Publishing, Oxford/Malden
Portner P (2009) *Modality*. Oxford University Press, Oxford
Pounds G (2010) Attitude and subjectivity in Italian and British hard-news reporting: the construction of a culture-specific 'reporter' voice. Discourse Stud 12(1):106–137. doi:10.1177/1461445609346777
Rescher N (2006a) *Presumption and the practices of tentative cognition*. Cambridge University Press, Cambridge
Rescher N (2006b [1979]) Default reasoning. In: Studies in the history of logic. Nicholas Rescher collected papers, vol 10. Ontos Verlag, Frankfurt, pp 173–184
Richardt S (2003) Metaphors in expert and common-sense reasoning. In: Zelinsky-Wibbelt C (ed) Text, context, concepts: Mouton de Gruyter, Berlin/New York, pp 243–298
Rigotti E (1993) La sequenza testuale. Definizione e procedimenti di analisi con esemplificazioni in lingue diverse. *L'analisi linguistica e letteraria* 1(1):43–148
Rigotti E, Morasso SG (2010) Comparing the Argumentum Model of Topics to other contemporary approaches to argument schemes: the procedural and material components. *Argumentation* 24(4):489–512
Rivière C (1981) Is "should" a weaker "must"? J Linguist 17:179–195
Roberts C (1989) Modal subordination and pronominal anaphora in discourse. Linguist Philos 6:683–721
Rocci A (2000) L´interprétation épistémique du futur en italien et en fançais: une analyse procédurale. Cahiers de Linguistique Française 22:241–274
Rocci A (2005a) La modalità epistemica tra semantica e argomentazione. I.S.U Università cattolica, Milano
Rocci A (2005b) On the nature of the epistemic readings of the Italian modal verbs: the relationship between propositionality and inferential discourse relations. In: Hollebrandse B, van Hout A (eds) Crosslinguistic views on tense, aspect and modality. Rodopi, Amsterdam/New York, pp 229–246
Rocci A (2006) Le modal italien *dovere* au conditionnel: évidentialité et contraintes sur l'inférence des relations de discours argumentatives. In: de Saussure L (ed) Temps, description, inteprétation. Travaux neuchâtelois de linguistique (TRANEL), vol 45, pp 71–98
Rocci A (2009) Per un'analisi semantica del connettore testuale *in realtà*. In: Moretti B, Pandolfi EM, Casoni M (eds) Linguisti in contatto. Ricerche di linguistica italiana in Svizzera. Atti del

convegno. Bellinzona 16–17 Novembre. Osservatorio Linguistico della Svizzera Italiana, Bellinzona, pp 71–91

Rocci A (2011) The Italian modal dovere in the conditional: future reference, evidentiality and argumentation. In: Mortelmans T, Mortelmans J, De Mulder W (eds) *In the mood for mood.* Rodopi, Amsterdam, pp 61–79

Rocci A (2012) Modality and argumentative discourse relations: a study of the Italian necessity modal *dovere*. J Pragmat 44(15):2129–2149. doi:http://dx.doi.org.proxy.sbu.usi.ch/10.1016/j.pragma.2012.09.007

Rocci A (2013) Modal conversational backgrounds and evidential bases in predictions: The view from the Italian modals. In: Jaszczolt KM, de Saussure L (eds) Time: language, cognition & reality. Oxford University Press, Oxford, pp 128–153

Rocci A, Luciani M (2016) Economic-financial journalists as argumentative intermediaries. J Argument Cont 5(1):88–111

Rocci A, Musi E (2014) From modal semantics to discourse: exploring the functioning of the adjective "possibile" in a corpus of Italian financial news. In: Blochowiak J, Grisot C, Laenzlinger C, Durrlemann-Tame S (eds) Papers dedicated to Jacques Moeschler, Geneva, pp 1–21

Salkie R (1996) Modality in English and French. Lang Sci 18(1-2):381–392

Sbisà M (1989) Linguaggio, ragione, interazione. Per una teoria pragmatica degli atti linguistici. Bologna, Il Mulino

Searle JR (2005) What is an institution? J Inst Econ 1:1–22

Searle J, Vanderveken D (1985) Foundations of illocutionary logic. Cambridge University Press, Cambridge

Squartini M (2004) Disentangling evidentiality and epistemic modality in Romance. Lingua 114(7):873–895

Squartini M (2008) Lexical vs. grammatical evidentiality in French and Italian. Linguistics 46(5):917–947

Stone M (1994) The reference argument of epistemic must. Proceedings of IWCS 1:181–190

Sweetser E (1990) From etymology to pragmatics: metaphorical and cultural aspects of semantic structure. Cambridge University Press, Cambridge/New York

Tasmowski L, Dendale P (1994) Pouvoir, un marqueur d'évidentialité. Langue française 102(1):41–55

Toulmin SE (2003) The uses of argument. Cambridge University Press, Cambridge

van Dijk TA (1988) News as Discourse. Lawrence Erlbaum Associates, Hillsdale NJ

van Eemeren FH (2010) Strategic maneuvering in argumentative discourse. John Benjamins, Amsterdam/Philadelphia

von Fintel K, Gillies AS (2011) Might made right. In: Egan A, Weatherson B (eds) Epistemic modality. Oxford University Press, Oxford, pp 108–130

von Fintel K, Iatridou S (2008) How to say ought in foreign: the composition of weak necessity modals. In: Guéron J, Lecarme J (eds) Time and modality. Springer, Berlin/New York, pp 115–141

Walsh P (2004) Investigating prediction in financial and business news articles. In: Facchinetti R, Palmer F (eds) English modality in perspective: genre analysis and contrastive studies. Peter Lang, Frankfurt am Main, pp 135–155

Walsh P (2006) Playing safe? A closer look at hedging, conditions and attribution in economic forecasting. In: Bhatia VK, Gotti M (eds) Explorations in specialized genres. Peter Lang, Berlin, pp 135–154

Werner T (2005) Temporal interpretation of modal sentences. In: Hollebrandse B, van Hout A, Vet C (eds) Crosslinguistic views on tense, aspect and modality. Rodopi, Amsterdam/New York, pp 247–259

Willett T (1988) A cross-linguistic survey of the grammaticization of evidentiality. Stud Lang 12(2):51–97

Wilson D, Sperber D (2012) Linguistic form and relevance. In: Wilson D, Sperber D (eds) Meaning and relevance. Cambridge University Press, Cambridge, pp 149–168

Woods J (2010) Defeasible reasoning. In: Reed C, Tindale CW (eds) Dialectics, dialogue and argumentation. An examination of Douglas Walton's theories of reasoning and argument. College Publications, London, pp 239–261

Zlatkova G (2012) Reported argumentation in economic-financial news. In: van Eemeren FH, Garssen B (eds) Exploring argumentative contexts. John Benjamins, Amsterdam/Philadelphia, pp 377–391

Chapter 7
Conclusion

7.1 Tacking Stock

The six previous chapters make for a ponderous book. Hopefully, they also make for an informative one.

Chapter 2 contains a short introduction to the normative and critical study of argumentation and a programmatic discussion of the role of semantic considerations in it. It also contain concise but workable exposition of specific theories such as the Argumentum Model of Topics, James Freeman's typology of statements (Freeman 2000, 2005) and Congruity Theory. Chapter 3 reads as an original contribution to the exegesis of Toulmin's *The Uses of Argument*. It examines this modern classic from the scarcely explored perspective of modality unveiling some latent inconsistencies in the fabric of the book and proposing some new interpretations of the elements of the Toulmin model. Chapter 4 contains an accessible presentation of the Relative Modality approach to the semantics of modality on the backdrop of possible worlds semantics. Chapter 5 builds a typology of the major "flavors" of modality based on a combination of linguistic principles of semantic interpretation and logico-ontological considerations, relying deontic modalities with issues of practical reasoning and covering the rarely discussed anankastic interpretations of the modals. Chapter 6 provides an original semantic analysis of a segment of the Italian system of modality proposing solutions that combine Relative Modality with ideas bred in procedural as well as constructional approaches to the semantics of polyfunctional, context-dependent linguistic units. At the same time the chapter offers some insights for the discourse analysis of the rarely described genre of business-financial news in the daily press. Even Chap. 1 smuggles in some informative bits about topics such as the square of opposition and the approaches to non-deductive, non-monotonic reasoning. Last but not least, the hundreds of English and Italian examples discussed in the book, many of which authentic, offer a convenient starting corpus for future researchers interested in exploring its themes with fresher theoretical verve.

A. Rocci, *Modality in Argumentation*, Argumentation Library 29,
DOI 10.1007/978-94-024-1063-1_7

There is reason to hope that different kinds of readers will find at least some parts of this book interesting, helpful or worth taking issue with and will be more lenient towards its many shortcomings. However comforting this thought might be, it does not excuse me from re-examining the findings of its book in the light of its stated goals.

It is then time to review the three main research questions that were mooted in Chap. 1

7.2 The Three Questions

I have introduced this book as a research on the contribution of modality to the essential structure of an argument. That is to what makes it an argument and an argument of a certain type. I suggested to see this operation through the prism of an analogy, suggesting that it is a bit like studying the work of verbal *tenses* as expressions of *time* contributing to make *narrative* discourse possible. The analogy implies a strong hypothesis: tenses relate to the semantic-ontological dimension of time, whose experience is, in turn, a defining component of what a narrative is. Furthermore, tenses are known to contribute to non-temporal aspects of narrative, such as point-of-view or focalization.

Can we say that there is a semantic-ontological dimension to which modal words and constructions refer, which, at the same time, is constitutive of what an argument is? Can we say that, on top of that, individual modal expressions contribute to signal other aspects of argumentation?

Looking back at the findings of the previous chapters, it seems to me that fundamentally the analogy holds water, even if the ontological dimension associated with modality, unlike time, does not have a direct physical import. This is more clearly visible if we review the three main research questions that were mooted in the *Introduction* to provide a direction to the investigation.

7.2.1 (Q1) How Are Modals Similar to Arguments?

The question about the isomorphism between modal meanings and arguments finds an answer that is in some ways quite unlike the one suggested by Toulmin and yet vindicates some of its preoccupations.

Contrary to what Toulmin suggested, modals, *in general*, do not mirror argumentation as a pragmatic and dialectical process, their forces are not pragmatic in nature and do not correspond to pragmatically and dialectically relevant phases of an argumentative discussion.

On the other hand, all flavors of modality are intimately connected with arguments because, like conditional structures and causal ones, they convey those relations of *consequence* and *(in-)compatibility* which underlie any kind of inferential connection and form the epistemic basis of all arguments. In this sense, I believe, that Angelika Kratzer is right on the money when she sees the main virtue of Relative Modality in that it "links the semantics of modals to general principles of rational inquiry that apply whenever we reason from a set of premises" (Kratzer 2012 [1977]: 10).

It is important to note how the mode of presentation of these fundamental relations in modals differs from that of conditional structures. By having the prejacent as their sole syntactic argument, necessity modals differ from necessary conditionals in that they focus on the consequent of the relationship. The antecedent, which is recovered through the saturation of the conversational background, is presupposed. In contrast, conditional sentences primarily assert the conditional as a whole and, at least in the idealizations of logic, do not assert or presuppose their antecedent or consequent. Causal structures typically assert the antecedent and the consequent, presupposing the conditional link.

It is this characteristic mode of presentation of meaning that makes necessity modals, in particular, germane with the presentation of arguments from the point of view of the conclusion. In this scenario, the recovery of the conversational background reveals both the material premises and the underlying consequence relation warranting the inference.

This view of the characteristic mode of presentation of the meaning of conditional, modal and causal structures is highly idealized, though. For one, real conditional constructions do say something about the epistemic status of their antecedents. We have seen in Chap. 4, § 4.2.8.4, for instance, how *epistemic conditionals*, which are interpreted as presupposing the truth of their antecedent, can function as the sole manifestation of a whole argument. In Chap. 6, § 6.5.4.3, we have seen how conditional antecedents evaluated as *normal* can function as weak premises, with a function that is complementary to that of a Toulminian rebuttal.

Furthermore, we have seen that modals regularly combine with conditional and causal structures when particular propositions in the conversational background are manifested by if-clauses or causal clauses (cf. Chap. 5, § 5.2.1).

Bringing back the isomorphism between modality and argumentation to the logical relations of *consequence* and *(in-)compatibility* would have surely disappointed Toulmin as a conservative solution. For one, it brings back the dreaded ghost of logical necessity right at the center of the picture. Additionally, in the book, I have said nothing about relationships other than *consequence* and *(in-)compatibility*, leaving aside any theoretical discussion of probabilistic relations.

We know from Chap. 3, § 3.2.3 that Toulmin's pragmatic alternative falls flat outside of the special case of the m-performative occurrences of epistemic modals: the bulk of alethic, deontic and anankastic relations of *consequence* and *(in-)compatibility* may or may not correspond to the logical relations that the argumentative inference at issue is exploiting. They often remain buried deep in a stratification of premises, as part of the accepted model of the material or moral world. Interestingly, the very strategy of the conversational background can favor this burial. Syntactically unary, the modals may pragmatically recover their presupposed antecedents in discourse in a precise, fine grained way, as we have seen throughout the book in multiple examples of argumentative discourse, but they may as well accommodate, generic, opaque backgrounds of nomic regularities, facts and values, that act as "black boxes" in the common ground and in the structure of an argument.

But when non-epistemic modals *do* correspond to the logical relations that are exploited in argumentative inference, Toulmin's idea of the "field dependence" of modal criteria and arguments is vindicated in an unexpected way: there is a correspondence between the different flavors of non-epistemic modality and the different concrete loci-relations underlying the argumentative inference. These correspondences have been discussed theoretically in Chap. 5, § 5.3.1, 5.3.2, and 5.3.3, and have found a richer exemplification in the analysis of arguments supporting predictions in business-financial news in Chap. 6 especially in § 6.4.3, 6.4.4, 6.4.5, 6.4.6, and 6.4.7 and § 6.5.4. We have seen how notions like an impersonal "economic causality", goals-means practical reasoning, expertise and commitment to plans represent different facets of the folk-ontology and folk-epistemology of the financial domain that are exploited in arguments supporting predictions by business-financial journalists. All these relations are indexed by the modal markers characteristically employed by the genre.

7.2.2 (Q2) Is There a Special Place for Modality in the Structure of Arguments?

We can understand this question, in a strict sense, as asking whether the present investigation found something similar to the *qualifier* slot in the "fundamental layout of arguments" captured by the Toulmin model. The short answer then is *no*. The present study does not propose any updated or revised notion of modal qualifier.

Instead, Chap. 3, § 3.4 points to the obscurity of the Toulminian idea that the qualifier makes "explicit reference to the degree of force which our data confer to our claim in virtue of our warrant"? (Toulmin 2003 [1958]: 93), which gives rise to the two alternative interpretations of the qualifier either as an indication of the *strength of support* that the premises offer to the conclusion, or alternatively, in pragmatic terms, as an indication of the *degree of commitment* of the arguer towards the propositional content of the standpoint.

Both interpretations are conceptually clear and capture information that is relevant, respectively, for *informal logic* and *normative pragmatic* approaches to argumentation. But we can now see that the problem is that neither fits the way in which the semantics of core modal expressions really works in natural language. As a consequence, by stipulating representations featuring "modal qualifiers" that don't mean what real world modals mean, and don't do what real modals do in arguments they encourage a very selective filtering of the information made available by the modal expressions: only epistemic expressions count, only their force of quantification counts, and only as a direct indication of degree on a scale, be a scale of force of commitment or degrees of support offered by the evidence (cf. Chap. 3, § 3.5.5).

Take the *strength of support* view of the modal qualifier. Even if Freeman (1991: 119) does recognize that in ordinary discourse often modal expressions seem to manifest the conclusion relation itself, he insists that this is not their proper job. They are not signs of consequence, but modifiers indicating the strength of the conclusion relations. This means that, semantically they are unary "operator-operators" (Woods 2010: 246) taking a relation of conclusion as their argument and modifying its strength. This fits well with the graphical practice of decorating arrows with modal boxes in the representation of the macrostructure of an argument and may have practical value in recording how arguers view the strength of the inferential steps appearing in their argument, as well as in the computational manipulation of those recorded intuitions.

However, if these unary operator-operators are supposed to reflect, even with a very moderate degree of realism, the actual predicate-argument structures of modal words, they cease to be a stipulation to become a fiction. The increasingly detailed and linguistically precise semantic analyses of Chaps. 4, 5 and 6 should have made clear that, for instance, there are very good reasons of descriptive adequacy for thinking that real-world necessity modals are relational signs of consequence rather than unary operators, and that they impose a variety of presuppositional constraints on their conversational background of premises. As a result, they do not limit their job to qualifying the relation, they make it apparent and may contribute to qualify the epistemic status of their prejacent both via the quantificational force and via the composition and constraints on the conversational background of premises. The analysis of the $\text{DOVREBBE}_{\text{E}}$ construction in Chap. 6 represents the most accomplished concrete example of how these dynamics are captured by a semantic analysis guided by concerns of descriptive adequacy.

There is now a growing body of linguistic research consistently suggesting that the view of the modal qualifiers as unary operators qualifying conclusion relations is a long shot from any plausible representation of the way in which real modal meanings are articulated. In fact, I believe that the kind of linguistic oriented, descriptively fine-grained, work in semantics highlighted in this book has the potential also to inform choices in the kind of representations that are used to capture the "fundamental layout of arguments".

As for the *degree of commitment* view of the modal qualifier, it can be observed that this view is not really at odds with what we know about the semantics of modality. Rather, it is a very selective and unfocused view. The expression of a certain degree of commitment towards the prejacent is a pragmatic effect of the interpretation of an m-performative occurrence of an epistemic modal expression. Chaps. 4, 5 and 6 have shown that is usually not the only one.

In Chap. 6 the comparative discussion of constructions such as $\text{DOVREBBE}_{\text{E}}$, DEVE_{E} and FUTURE_{E} shows that the degree of commitment of the speaker towards the prejacent is not what sets apart the three constructions which are all located in the area of medium-high certainty. Furthermore, this information emerges indirectly as a result of the combination of the quantificational force and of the constraints on the status of premises in the conversational background. In the case of $\text{DOVREBBE}_{\text{E}}$ we have also seen that the event of thought or speech to which the recovery of the conversational background C is deictically anchored is not identified with the mind of the speaker at the moment of utterance, but, polyphonically, with a diffuse epistemic community (cf. Chap. 6, § 6.5.5) with whom the speaker may or may not associate.

In fact, the pragmatic view of modal qualifiers as indicators of degree of commitment fits well with first-person m-performative uses of doxastic predicates such as *I think* or *I believe.* These are *subjective*, in the evidential sense of to Nuyts (2001), referring in a generic way to a global unanalyzed doxastic conversational background. In contrast with the core modals, they are not incomplete meanings, they do not open a slot for the conversational background to be recovered anaphorically in discourse and, as a consequence, do not systematically prompt the establishment of argumentative discourse relations. It is only for constructions such as these that the expression of a degree of certainty towards the prejacent becomes the whole point of employing the epistemic expression.

Leaving aside the different reinterpretations of Toulmin's *qualifier* slot, question (Q2), about the place for modality in the essential structure of arguments can be interpreted more generally as referring to whether modalities should have a distinct slot in the macrostructure of arguments or be part of the propositional microstructure.

Here the answer differs according to the flavor of modality. We have seen in Chap. 5 that non-epistemic modalities contribute to determine the semantic type of propositions, shifting the type of the prejacent proposition, usually to different kinds of interpretations and evaluations. This has clear argumentative consequences as logically determinate statements, interpretations, evaluations and mere extensional descriptions require different kinds of argumentative support (cf. Chap. 2, § 2.3.3). In contrast, we have seen that epistemic modalities do not interact with the "type of statement" leaving the type of the prejacent intact. They are better treated as extra-propositional for the purposes of argumentative reconstruction (cf. Chap. 5, § 5.3.4.2). Even if, in principle, one could see an epistemically modalized proposition as an intensional non-evaluative statement about the cognitive functioning of the epistemic source, i.e. an interpretive statement, such an interpretation is rarely the target either of support or criticism in argumentation. It is normally the prejacent proposition that enters argumentatively relevant relations.

This obviously does not mean that epistemic modals are irrelevant for argumentative reconstruction. Quite the contrary. While the epistemic modal is not part of the propositional material that enters the argumentative relation, it function as a signal of the relation itself and constraints its interpretation both at the pragmatic and at the inferential level.

Truth be told, in the book I did not offer one level of representation of arguments, graphically displayed or otherwise implemented, where modalities could find their place, as it happens with the Toulmin model and with derived schemes. Actually, I used three different kinds of representation for arguments: (a) argumentation structures to capture, with a relatively coarse granularity, the dialectically and inferentially relevant aspects of complex arguments, (b) Congruity Theory's *connective predicates* for a descriptively rich capture of argumentatively relevant moves at an illocutionary level, e.g. the act of drawing a *Conclusion* from premises anaphorically or cataphorically available, (c) AMT's Y-structure to capture the fine grained inferential functioning of an argumentative move. None of them features a special slot for modality.

It is, however, instructive to devote some space to re-examine how modal meanings interact with (b) and (c). As for argumentation structure (a), it can be quickly dealt with together with (c).

7.2.2.1 Modals and Connective Predicates

In Chap. 5, § 5.3.4.3 and 5.3.4.4, we have seen that m-performative occurrences of an epistemic-evidential modal like *must* can become performative in the fullest sense: the *m-performative* act of inference becomes a true *performative* act of argumentation, inviting the addressee to conclude from the reasons made available. In this case *must* becomes a partial manifestation of the connective predicate *Conclude* (Def. 16, 19, in Chap. 5).

The prejacent of the modal is identified with the propositional content of the utterance U_0 in the connective predicate frame, meanwhile the process of contextual and discursive saturation of the conversational background contributes to make the *phoric* arguments of the connective predicate accessible to the hearer (cf. Chap. 2, § 2.3.4.2). Furthermore, the presupposition of the *Conclude* connective predicate that the propositional content of U_0, nor its negation are already part of the common ground of the speaker and the hearer is perfectly matched by Kartunnen's presupposition that the prejacent of epistemic *must* is not "an established fact" (cf. Chap. 5, § 5.3.4.3, Def. 18, but also Chap. 6, § 6.5.3.1, Def. 4 for Italian $DEVE_E$).

Finally, we have seen that any connective predicate of *Conclusion* presupposes that some locus-relation holds between the premises and the conclusion. An inferential evidential like *must* basically adds its evidential constraints to this presupposition, which exclude direct testimony as well as third party testimony and most types of *locus from authority* (cf. Chap. 5, § 5.3.4.3).

In sum, what we can see – rather than the modal occupying one slot, or specifying one aspect, of the argumentative connective predicate *Conclude* – is a quite systematic slot-by-slot mapping between the modal and the connective. As we have seen in Chap. 5, § 5.3.4.3, the modal is in some ways less informative than the generic connective predicate template, while in other departments it adds information to the template. All in all, what we have, at least in the special case of epistemic-evidential necessity modals, is much closer to the view from Toulmin's first essay, where modals are *illocutionary markers* signaling different kinds of illocutionary acts, rather than to the view from the second essay where *probably*, as a device for "guarding" commitment, plays the role of a mere *illocutionary force modifier* (see the rich discussion in Chap. 3, § 3.3.1).

7.2.2.2 Modals and the Y-Structure

Throughout the book I have often mapped the elements of the conversational background of a modal that where syntactically or anaphorically recoverable onto premises in an argumentation structure. These structures include *compound* (coordinative), *subordinative* and, with possibility modals, also *multiple* structures. Something similar happens at the level of the Y-structure with the difference that here the reconstruction is not limited to the elements of the conversational background that leave textual traces.

Needless to say, Kratzer's notion of the modal conversational background as a "set of propositions" or "premise set" is very abstract and removed both from argumentative practices and from the ontology underlying certain modalities. Sets of propositions do not have structure. Naïvely, they bring to the mind the picture of a bag of sentences haphazardly thrown together, even if, in Kratzer's possible worlds semantics they should rather look like bags of bags of worlds (but also this image works up to a point, as the "bags of worlds" intersect one another!). From a cognitive point of view, we do not sift our assumptions through bags of propositions, but we link them with what we know through chains of reasoning.[1]

From an ontological point of view, not all propositions that concur in determining the conversational background of, say, an *alethic circumstantial modal* sit at the same level of the ontology. Some will be mere facts, mere circumstances; as, when I say that *Björk cannot sing*, the unfortunate circumstances are that she has a sore throat. For these circumstances to actually rule out the singing a nomic regularity must be in place in the biological domain such that a certain inflammation can cause nodules in the vocal cords and impede singing. Interestingly, the nomic regularity is itself an alethic modal proposition, but a more abstract one: clearly its conversational background needn't include the circumstances that are necessary to rule out the prejacent. In the Y-structure, the circumstances will likely sit in the *datum* slot, while the relevant nomic regularity will occupy the place of the *endoxon*. And what about the proposition occupying the *maxim* slot? If we go back to the discussion of this example in Chap. 5, § 5.3.1, we find something like this:

> For all sufficient causes and respective effects, *necessarily* if the cause is the case then also the effect will be the case.

Here again we have to do with an alethic modality, but this time it is a very abstract one. We cannot say it is a circumstantial modality nor that it is causal in nature. In fact, it is a conceptual necessity enshrining a semantic axiom that is

---

[1] What degree of cognitive realism is desirable for a semantic theory (and how to determine whether and how much a proposal is cognitively realistic) is a complex and hotly debated issue that will not be solved here. It is perhaps worth reminding how Chap. 4, § 4.3.4.3 and Chap. 5, § 5.3.2.3 drew a line at the level of ordering sources, rejecting that part of Relative Modality because it blurred the connection between modal relations and notions of logical consequence and compatibility of the prejacent with a set of premises by introducing a ranking of worlds which appeared cognitively implausible without clearly achieving the stated aim of modeling how we rank our alternatives in less than ideal situations.

constitutive of the very notion of efficient cause. From an ontological viewpoint, conversational backgrounds are modally stratified, and the Y-structure happens to offer a surprising glimpse into that.

More obviously, from a cognitive viewpoint conversational backgrounds also are stratified, and Y-structures can manifest its articulation.

7.2.3 (Q3) What Does the Use of Modal Expressions in an Ordinary Discourse Tell Us About the Arguments Being Put Forth in the Discourse?

With the third research question and the study of modal expressions as *argumentative indicators* we come to the book's main contribution to argumentation analysis from a practical point of view.

The stated goal was to show that the exclusive focus on force and degrees of commitment presents a *far too restrictive* view of modals as argumentative indicators, and that there is much more that they can tell us about the workings of arguments (cf. Chap. 1, § 1.3). This goal was met both through the general strategies for the analysis of modalities in arguments provided mostly in Chaps. 4 and 5, and through the limited but highly detailed semantic analyses of Italian modal constructions in Chap. 6.

It is worth briefly going through the chapters *highlighting the wealth of analytical considerations* that can have a practical impact on guiding the reconstruction of arguments in view of their critical evaluation.

Already in **Chap. 3,** § **3.4.4**, the treatment of Toulminian rebuttals as exceptive conditions presumed not to hold, by showing their substantial equivalence with positive conditions presumed to hold, introduces strategies of reconstruction that will be important to account for the manifestation of the C conditions of $\text{DOVREBBE}_{\text{E}}$ and their functioning as weak premises in presumptive reasoning.

Chapter 4, § **4.2.4** and the following sections introduce relative modals as context dependent "gappy", meanings, outlining a general view of the strategy of context dependency in language in § 4.2.5. The vagueness and ad hoc-ness of modal conversational backgrounds (§ 4.2.4.2) should be by itself of interest for argumentation scholars, if only for its importance in the basic task of *determining the propositional commitments of the arguers* and for the potential for fallacy inherent in that vagueness. In many ways, the contextual specification of modal conversational backgrounds is comparable to specifying the domain of *quantifiers*. Specifying the exact contribution of context dependent modal meanings to propositional content can impact, in turn, further aspects of the reconstruction of an argumentative discussion, notably including the semantic type of the propositions functioning as standpoints, the *loci*-relations supporting the inference from the premises to the standpoint, and the macrostructure of the argument.

Chapter 4, § 4.2.8.4 discusses *epistemic conditionals* and their argumentative relevance on the backdrop of the interaction between modals and conditionals in Relative Modality. In the classic examples of epistemic conditional the antecedent is presupposed and becomes a full-fledged premise. From the viewpoint of argumentative reconstruction this is interesting because with them we can have *an entire argument* whose only linguistically explicit part is the conditional. Additionally, it is not necessary to have an antecedent fully accepted in the common ground for a conditional to function as the enthymematic expression of an argument. In natural discourse conditional antecedents are always epistemically evaluated and this evaluation guarantees that they can function on occasion as weakly presumed premises, endowed with a variable degree of certainty. This point on the role of conditionals in reconstruction becomes important in Chap. 6 as it allows to treat the condition C of $\text{DOVREBBE}_{\text{E}}$ as a weak premise.

Chapter 4, § 4.3.3 shows that the quantificational force of propositional modals can guide the analyst in deciding whether a complex argumentation structure is multiple or coordinative, along the lines suggested by Snoeck Henkemans (1992) for the nominal quantifiers. Through the discussion of several examples it is shown that, under the appropriate conditions possibility modals behave like the existential quantifier as an indirect indicator of multiple argumentation structure, while necessity modals work as indirect indicators of coordinative (compound) structure.

Chapter 4, § 4.3.5 goes back to the issue of the vagueness of conversational backgrounds to spell out more clearly its consequences for the reconstruction of the confrontation stage, discussing cases of covert divergence of the conversational backgrounds as well as shifts over the course of the discussion.

Chapter 4, § 4.3.6 finally approaches the big target, presenting an initial treatment of the indicative potential of the "phoric" properties of context dependent modals. It is shown how core context-dependent modals – such as the modal verbs in Italian and English – have a behavior in discourse that closely resembles that of *anaphoric connectives*, as defined by Webber et al. (2003), Forbes-Riley et al. (2005). Conversely, some classic examples of anaphoric connectives, such as *otherwise*, are found to have a modal semantics. The "phoric" saturation of the conversational background of context dependent modals allows the discourse interpreter to establish discourse relations, which, in turn, translate into relations of support in an argumentation structure.

Chapter 5, § 5.2.1, building on cases discussed in the previous chapter, provides a typology of different syntactic and discourse devices that can provide a partial linguistic manifestation of the conversational background. Being able to recognize these devices as systematically working jointly with the modal itself is a decisive step for the analyst for extracting the rich indicative potential of *prima facie* vague modal expressions.

Chapter 5, § 5.2.4 and, particularly, 5.2.4.4, while ostensibly one of the most densely linguistic sections of the book, is also one of the section with the greatest potential for developing new accounts of the indicative potential of individual modal expressions. It provides a principled account of the main types of modal "flavor", developing at the same time a general procedure, similar to those used in the interpretation of verbal tenses, for anchoring deictically the search of the conversational background of a modal. Following Hacquard (2010), the two main classes of conversational backgrounds are associated respectively to the *circumstances of* the prejacent's VP-event (root modality) or to the *content of* a speech or thought event identified by default with the speaker's utterance (epistemic modality). This offers a most general blueprint for mapping the interpretation of modal expressions, how their conversational backgrounds are recovered and what they are made of.

Chapter 5, § 5.3.1, 5.3.2, and 5.3.3 draws a detailed map of modal flavors based on this rationale, relying it to the argumentatively relevant semantic typology of statements proposed by James Freeman. These sections deal chiefly with two related aspects of the indicative potential of non-epistemic modal expressions.

On the one hand, the different "flavors" of non-epistemic modality affect the semantic type of the proposition turning it into an *interpretation* or an *evaluation* of a certain kind. In turn, the semantic type constrains the kinds of relevant argumentative support. This is true at the level of *loci* for the more general types, but it should not be forgotten that the precise determination of the conversational background affects the relevance of specific types of evidence. For instance, with reference to the typology of deontic conversational backgrounds in Figure 8, § 3.2, statutes or contracts provide adequate evidence provided that the deontic conversational background invoked by the modal is institutional. Moral obligations do not originate from thetic acts enshrined by documents and require a different kind of support.

On the other hand, exploring the different conversational backgrounds of non-epistemic modals helps us to capture those arguments where the consequence or (in-)compatibility relationship established by the modal between the prejacent and the conversational background is one and the same thing as the *locus*-relation exploited by the inference. This is the case, for instance, for causal intrinsic *loci* such as efficient and instrumental cause with causal alethic root modalities as well as with deontic-practical modality and the *loci* of practical reasoning.

There are aspects of the functioning of the core modals that may lead us to believe that their utility as argumentative indicators in these areas is limited. They are polyfunctional, context dependent and, after all, people can remain vague as regards the saturation of the conversational background. This may well put into question the practical utility of knowing that, say, a deontic-practical interpretation of a modal is constitutively tied to the *locus from the final cause*, or that abstract alethic modals expressing conceptual necessity have a natural tie with definitional *loci*. If the linguistic expressions are not univocal pointers but need to be interpreted in context what is the point for argumentation analysts of paying attention to them?

In fact, I do not think indicators need to be one-to-one, univocal pointers to be relevant for the practice of argumentative reconstruction. This is simply not how argumentative reconstruction works. Van Eemeren and Grootendorst (2004: 117) stress that in the effort of providing a "maximally reasonable reconstruction" the analysis "maintains an open character" and "no absolute certainty can be reached". Striving to provide a pragmatically optimal reconstruction, argumentation analysts seek to put forward interpretive hypotheses that satisfy in the most reasonable way a multitude of contextual and linguistic constraints as ordinary language users do, while applying a special charity in ensuring that their reconstruction of discourse meaning is sufficiently logically determinate to be actually evaluated as an argument. While engaging in this process, the knowledge of the ways in which the saturation of the modal conversational backgrounds can be resolved and of the argumentative consequences of the different resolutions is indeed a precious asset. By providing a fairly detailed reconstruction of a real example from the famous *I have a dream* speech (Chap. 5, § 5.3.2.2) I wanted to highlight this dynamic: if making sense of a modal's RM semantics requires deep engagement with discourse interpretation by the analyst, it is also true that the RM semantics provides a scaffolding for the search of a satisfactory interpretation, which will inevitably be not only an interpretation of the modal, but also of the discourse and argument as a whole.

The indicative potential of m-performative epistemic modals as signs of *Conclusion* is finally addressed in **Chap. 5,** § **5.3.4**. I have insisted above at some length on the close mapping between the semantic structure of the modal and that of the concerned connective predicate. I mention here the further dimension of the evidential constraints that deeply characterize many epistemic expressions in languages without extensive grammaticalization of evidentiality such as English or Italian. Evidential constraints confer to them a new source of indicative power. More removed from the underlying ontological relations, epistemic conversational backgrounds concern propositions as representations in thought and discourse, but, thanks to specific constraints on the status and role of those propositions in the acquisition of knowledge, the *loci* may well re-enter the picture.

Finally, **Chap. 6** provides two kinds of exemplification in context of the principles and strategies for analysis introduced in the previous two chapters. The discussion of Italian non-epistemic modals in predictions as indicators of the evidential basis of the prediction (§ 6.4.2) is a direct continuation of the sections of Chap. 5 on the indicative power of non-epistemic modals. The semantic analyses of the constructions DEVE_{E} and $\text{DOVREBBE}_{\text{E}}$ bring together many strands not only of the previous two chapters but of the whole book to capture different kinds of finely grained distinctions relevant to the reconstruction of arguments, including not only restrictions on the possible *loci* and maxims, but also information on the status of premises with respect to the common ground of the speaker and the hearer.

These two analyses are intended as examples of descriptive studies of the indicative potential of individual modal constructions. I obviously believe them successful examples and illustrative of the potential of the approach defended in this book. Yet, I feel they may come as a letdown for some readers. It's only two constructions based on the inflection of the same lexeme! And the argumentatively relevant information has to be teased out from a web of semantic interactions that are specific of the Italian linguistic system.

In short, I can imagine that some readers will think this is too much work for too little result. We are certainly very far from a systematic coverage of the domain. It is an understandable reaction and I have pondered at length what should be my *prolepsis* to these imagined readers' objection. You find it the following and very last section of this book.

7.3 Developments

The present work is not only a long book. It was also long in the making, spanning the good part of a decade between the seminal ideas and the final version. In the meantime, some parts saw publication as individual articles (Rocci 2008, 2012, 2013) while others circulated only among close colleagues and students in the form of provisional drafts. One interesting consequence of this otherwise undesirable circumstance is that this period saw the development of two research projects[2] partly related to the book, a PhD thesis (Musi 2015) developing some of its ideas in a methodologically innovative (cf. Musi et al. 2016) and consistent manner and a handful of articles (Miecznikowski 2011; Miecznikowski et al. 2013; Miecznikowski and Musi 2015; Musi 2014; Musi and Rocci 2016; Rocci and Zlatkova 2012) mostly insisting on semantic analyses of different Italian and English epistemic-evidential markers (*epistentials*, according to the terminological flavor of the day), which consistently show that the richness of argumentatively relevant constraints found for DEVE_{E} and $\text{DOVREBBE}_{\text{E}}$ are by no means exceptional or accidental, but correspond to a general proclivity of the class. What is more interesting, these linguistic items manifest constraints of occurrence and statistical preferences that map nicely onto the typology of *loci* proposed by the AMT approach.

I resist here the temptation of reporting extensively on any of these studies (some of which, having been reported at conferences and seminars, are still been worked on for journal publication) and I provide only the most general hints at their topics and results.

One very apparent gap in the case study in Chap. 6 is that it focuses on two construction where the underlying modal force is necessity. Can possibility-based modal expressions become argumentative indicators when they are interpreted epistemically? This gap is partially filled in the early study by Miecznikowski (2011) on Italian conditional possibility modal *potrebbe*, based on the same corpus of economic-financial news. It turns out that the conditional possibility modal can assume a strengthened meaning, beyond mere possibility, and become an indicator of weak conjectural standpoints. The corpus based studies of epistemic adverb *forse* ('perhaps', 'maybe') presented in Rocci and Zlatkova (2012) and Miecznikowski et al. (2013) further develop this line of research by providing motivated semantic analyses for three distinct constructions in which the adverb enters: a *conjectural* construction, where it behaves as an inferential evidential with a preferences for inferences *from effect to the cause*, most typically introducing hypotheses on intentions based on observed behavior (almost a weaker version of DEVE_{E}), a *concessive* construction where the possibility of the prejacent works as a conceded argument for the antagonist, followed by a stronger counterargument and a *rhetorical*

[2] As noted in the *Introduction*, these are the Swiss National Science Foundation projects *Modality in Argumentation. A semantic-argumentative study of predictions in Italian economic-financial newspapers* (directed by Andrea Rocci, Grant: 100012-120740/1) and *From perception to inference. Evidential, argumentative and textual aspects of perception predicates in Italian.* (directed by Johanna Miecznikowski, Grant n. 141350).

interrogative construction, where a negative answer is elicited about the prejacent while rhetorically envisaging its epistemic admissibility by a fictitious antagonist. The analyses show how different epistemic constructions can select different epistemic subjects for the deictic anchoring of the *s* of the epistemic conversational background providing glimpses into the mind of real of fictitious antagonists.

Far more systematic is the work carried out by Elena Musi (2015) for her PhD thesis on the Italian evidential verbs *sembrare* ('to seem') and *apparire* ('to appear'). The study, based on a journalistic corpus, combines a modal analysis of the two evidentials in terms of Relative Modality with an extensive corpus annotation of the *loci* wherever the verbs introduce a local standpoint. The results are strikingly neat once the painstaking separation of grammatically and semantically distinct constructions is taken care of. For instance, *sembrare* ('seem') can appear either as a *reportative* or as an *inferential* evidential. The quantitative results of the corpus annotation show that, when manifesting inferential evidentiality, *sembrare* is compatible exclusively with the AMT class of *intrinsic loci* (cf. Chap. 2, § 2.2.3.2). For instance, the verb can occur, and frequently occurs, with the locus *from whole and parts*, but is never found with, say, a locus *from analogy*. As Musi (2015: 5) explains, this feature is consistent with the perceptual origin of this evidential verb: "categorizations and indirect perceptions place premises and conclusions in the same frame (e.g. by inferring causes from effects or category membership from the presence of typical features)" while it appears to be incompatible with extrinsic loci "in which the events denoted by the premises and by the standpoint belong to tow different frames". Among the loci compatible with *sembrare*, the distribution is uneven: almost half the occurrences are tied to variants of the locus *from definition*, followed by *whole and parts* and by the *causal* loci. Furthermore, constraints on the directionality of the maxims are clear: *sembrare* can be used when inferring the belonging of an entity to the definiendum from the presence of (some) defining traits, but not vice versa; to infer the whole of a frame from the presence of its parts, but not vice versa. Musi (2015) associates this type of directionality to the notion of *abductive inference.*

In an ongoing research, whose preliminary results have been concisely published in Musi and Rocci (2016) a similar systematic process of corpus annotation at the level of *loci* is applied to apparently similar epistemic-evidential adverbs in Italian (and English): *evidentemente* vs. *ovviamente* (Eng. *evidently* vs. *obviously*). While the two adverbs are both situated at the certainty extreme of the epistemic scale and both seem to make reference to obviously apparent evidence, their functioning in arguments is strikingly different. In Italian, in particular, *evidentemente* anchors the thought event *s* to the mind of the speaker (the prejacent is evident to the speaker), while *ovviamente* anchors it to the addressee (the prejacent should be obvious to the addressee). Moving to the available *loci* for each construction, the results are again striking: Italian *evidentemente* turns out to be a highly specialized indicator almost exclusively associated with the *causal loci* and strictly tied with the direction *from effect to cause*. In contrast, *ovviamente* does not manifest a clear specialization for a set of loci.

The encouraging results of these studies suggest that there is still much to learn about the indicative potential of epistemic-evidential constructions and confirm the interest of conducting quantitative corpus investigations using the categories of the AMT model of topics.

References

Forbes-Riley K, Webber B, Joshi A (2005) Computing discourse semantics: the predicate-argument semantics of discourse connectives in D-LTAG. J Semant 23:55–106

Freeman JB (1991) Dialectics and the macrostructure of arguments. A theory of argument structure. Foris Publications, Berlin

Freeman JB (2000) What type of statements are there? Argumentation 14(2):135–157

Freeman JB (2005) Acceptable premises: an epistemic approach to an informal logic problem. Cambridge University Press, Cambridge

Hacquard V (2010) On the event relativity of modal auxiliaries. Nat Lang Semant 18(1):79–114

Kratzer A (1977) What "must" and "can" must and can mean. Linguist Philos 1: 337–355 (republished in Kratzer 2012: 4–20)

Kratzer A (2012) Modals and conditionals. Oxford University Press, Oxford

Miecznikowski J (2011) Construction types and argumentative functions of possibility modals: evidence from Italian. In: Eemeren FH, Garssen B, Godden D, Mitchell G (eds) Proceedings of the 7th conference of the international society for the study of argumentation. Sic Sat, Amsterdam, pp 1284–1297

Miecznikowski J, Musi E (2015) Verbs of appearance and argument schemes: Italian *sembrare* as an argumentative indicator. In: Garssen B, van Eemeren FH (eds) Reflections on theoretical issues in argumentation theory. Springer, Amsterdam, pp 259–278

Miecznikowski J, Rocci A, Zlatkova G (2013) Le funzioni inferenziali e polifoniche dell'avverbio epistemico italiano *forse*. In: Pirazzini D, Schiemann A (eds) Dialogizität in der Argumentation. Eine multidisciplinäre Betrachtung. Peter Lang, Frankfurt am Main, pp 201–230

Musi E (2014) Evidential modals at the semantic- argumentative interface: appearance verbs as indicators of defeasible argumentation. Informal Logic 34(4):417–442

Musi E (2015) Dalle apparenze alle inferenze: i verbi sembrare ed apparire come indicatori argomentativi. PhD dissertation, Università della Svizzera italiana

Musi E, Rocci A (2016) Obviously epistentials are argumentative indicators: evidence from an Italian and English corpus of newspaper articles. In: Proceedings of the workshop on the Foundations of the Language of Argument at COMMA, University of Potsdam, 13 September 2016

Musi E, Ghosh D, Muresan S (2016) Towards feasible guidelines for the annotation of argument cchemes. In: Proceedings of the third workshop on argument mining (ArgMining2016), pp 82–93. http://aclweb.org/anthology/W/W16/W16-2810.pdf. Accessed 8 Nov 2016

Nuyts J (2001) Subjectivity as an evidential dimension in epistemic modal expressions. J Pragmat 33(3):383–400

Rocci A (2008) Modality and its conversational backgrounds in the reconstruction of argumentation. Argumentation 22:165–189. doi:10.1007/s10503-007-9065-8

Rocci A (2012) Modality and argumentative discourse relations: a study of the Italian necessity modal *dovere*. J Pragmat 44(15):2129–2149. doi:http://dx.doi.org.proxy.sbu.usi.ch/10.1016/j.pragma.2012.09.007

Rocci A (2013) Modal conversational backgrounds and evidential bases in predictions: The view from the Italian modals. In: Jaszczolt KM, de Saussure L (eds) Time: language, cognition & reality. Oxford University Press, Oxford, pp 128–153

Rocci A, Zlatkova G. 2012. The Italian epistemic modal adverb *forse*; from semantics to argumentation. In: Canatrini S (ed) Dialogue: state of the art. Studies in memory of Sorin Stati. LINCOM Europa, pp 232–53

Snoeck Henkemans F (1992) Analysing complex argumentation. SicSat, Amsterdam

Toulmin SE (2003 [1958]) The uses of argument. Cambridge University Press, Cambridge

van Eemeren FH, Grootendorst R (2004) A systematic theory of argumentation: the Pragma-Dialectical approach. Cambridge University Press, Cambridge

Webber B, Stone M, Joshi A, Knott A (2003) Anaphora and discourse structure. Comput Linguist 29(4):545–587

Woods J (2010) Defeasible reasoning. In: Reed C, Tindale CW (eds) Dialectics, dialogue and argumentation. An examination of Douglas Walton's theories of reasoning and argument. College Publications, London, pp 239–261

Index

A. Rocci, *Modality in Argumentation*, Argumentation Library 29,
DOI 10.1007/978-94-024-1063-1

D

MIX
Papier aus verantwortungsvollen Quellen
Paper from responsible sources
FSC® C105338

If you have any concerns about our products,
you can contact us on
ProductSafety@springernature.com

In case Publisher is established outside the EU,
the EU authorized representative is:
Springer Nature Customer Service Center GmbH
Europaplatz 3, 69115 Heidelberg, Germany

Printed by Libri Plureos GmbH
in Hamburg, Germany